SPACE SCIENCE: NEW RESEARCH

SPACE SCIENCE: NEW RESEARCH

NICK S. MARAVELL
EDITOR

Nova Science Publishers, Inc.
New York

For permission to use material from this book please contact us:
Telephone 631-231-7269; Fax 631-231-8175
Web Site: http://www.novapublishers.com

Library of Congress Cataloging-in-Publication Data

Space science : new research / Nick S. Maravell, editor.
p. cm.
Includes bibliographical references and index.
ISBN 1-60021-005-8
1. Space sciences--Research. I. Maravell, Nick S.
QB500.262.S636 2006
500.5072--dc22 2006000596

Published by Nova Science Publishers, Inc. ✢New York

CONTENTS

PREFACE

Space science, often considered the next frontier for exploration, is a body of scientific knowledge as it relates to space exploration; it is sometimes also called astronautics. Space science draws on the conventional sciences of physics, chemistry, biology, and engineering, as well as requiring specific research of its own. This new book presents the latest research in this dynamic and growing field.

As presented in Chapter 1, although Mercury, Venus, Earth/Moon, and Mars originated from similar material accreted in close regions of the planetary nebula, the worlds that developed after 4.6 billion years of Solar System evolution are not much alike. Mercury and the Moon, for example, are without a protecting atmosphere and were not exposed to liquid water on their surface for at least 4 billion years. The surfaces of Venus and Mars are desiccated and are presently not a suitable habitat for life, but reservoirs of liquid water remain in the atmosphere of Venus and the subsurface of Mars, and microbial organisms may have adapted to survive in these ecological niches. The search for water and life on any of the terrestrial planets is intrinsically connected to their geological history. Missions should be designed to explore any potential past and present habitats. In addition, ices on Mercury and the Moon should be explored for remnant biogenic material from the early evolution of life on Earth (and elsewhere). Especially desirable are sample return missions and missions that include hierarchical architectures with scalable degrees of mission operation autonomy, which will allow optimal reconnaissance of planetary environments, including surface and subsurface environments.

Much research has been done on the influences of solar activity on processes on Earth and in near-Earth space. While there is a general consensus about the reality of such influences, the results are often contradictory even if highly statistically significant, in many cases the physical mechanism of the influence is unclear, and the mechanism of solar activity itself is not yet fully understood. Obviously there are additional factors affecting solar activity and its influences on the Earth which are not accounted for in the present models. Chapter 2 investigates one of these possible factors – the dynamics of the Sun and its significance for the solar-terrestrial influences.

One of the most controversial problems in solar-terrestrial physics is the relation between solar activity and climate. Different authors have reported positive, negative or missing correlations between solar activity and surface air temperature in the 11-year sunspot cycle. The authors demonstrate that the sign of the correlation changes regularly in consecutive secular (Gleissberg) solar cycles, and depends on the long-term changes of North-South solar activity asymmetry which in turn is determined by the movement of the Sun about the barycenter of the Solar system.

This movement of the Sun about the barycenter of the Solar system is also related to the dynamics of both the Sun itself and of the Earth. The asymmetry in the rotation of the Northern and Southern solar hemispheres correlates very well with the variations in the Earth rotation rate, and their dominant common periodicity is the periodicity of the rotation of the Sun about the Solar system barycenter. The correlation of the solar dynamics and the Earth dynamics is mediated by the solar wind carrying momentum and magnetic fields and modulating the electromagnetic core-mantle coupling torques responsible for the variations in the Earth's rotation.

The variable solar differential rotation affects also the way in which the solar drivers interact with the Earth's magnetosphere. The periodicities in the interplanetary magnetic field at the Earth's orbit in any period coincide with the periodicities in the latitudinal gradient of the differential rotation in the more active solar hemisphere, and the azimuthal components of the interplanetary magnetic field are proportional to the solar equatorial rotation rate.

The most geoeffective solar driver are the magnetic clouds – coronal mass ejections with smooth rotation over a wide angle of the magnetic field inside the structure. The portion of coronal mass ejections which are magnetic clouds varies throughout the solar cycle, and is determined by the amount of helicity transferred from the solar interior to the surface, and by the surface differential rotation. They demonstrate the influence of the surface differential rotation on the helicity and geoeffectiveness of magnetic clouds.

As explained in Chapter 3, the plasmasphere is the high altitude extension of the ionosphere. To ensure accuracy and reliability of communications, navigation and other satellites stationed in this region, there is a considerable interest to understand the plasmaspheric environment and its dependence on external parameters.

Dynamical simulations have been developed at the Belgian Institute for Space Aeronomy to simulate the deformation of the plasmasphere during geomagnetic substorms and other variations in the level of geomagnetic activity. These simulations are based on the mechanism of plasma instability and on the E5D empirical electric field model. The results of these simulations are compared with satellite observations of the plasmasphere. Comparisons with observations of the IMAGE spacecraft launched in March 2000 are especially useful to understand the dynamics of the plasmasphere since the satellite provides the first global comprehensive images of the Earth's plasmasphere. The EUV (Extreme UltraViolet) instrument gives intensity maps of the 30,4 nm emissions of Helium ions integrated along the line of sight. It revealed the global behavior of the plasmasphere and its dynamics influenced by the solar wind and interplanetary magnetic field variations. New structures in the plasmasphere such as shoulders, channels, notches and the formation of plumes were discovered. The four Cluster spacecraft launched in 2000 also provide useful data to compare with the simulations.

Chapter 4 the authors represent a survey of the present state and analytical methods of the theory of transversally small-scale standing MHD perturbations in the inner magnetosphere of the Earth, as well as their own views on this matter. They restrict our consideration by two important types of such perturbations: flute and ballooning modes.

In the most general case arbitrary three-dimensional transversally small-scale standing MHD perturbations in ideal plasmas with nested magnetic surfaces are described by a pair of Dewar-Glasser equations. In some papers such equations are derived from the MHD equations by application of differential operators. This casts certain doubts upon the correctness of the resulting spectra of perturbations. The authors choose a different path, using just the condition of transversally small-scaleness and longitudinal elongation of perturbations, which, nevertheless, appeared to be sufficient to derive mentioned equations.

In addition they used ideal MHD approximation, neglected the convection and considered the equilibrium to be static. Obtained equations were applied to a dipolar magnetic configuration, approximately representing the inner magnetosphere of the Earth. This made the equations much simpler and after introducing dimensionless variables they reduced to linear homogeneous ordinary differential equations of second order.

When hydrodynamic pressure is considerable flute and ballooning perturbations are generated. The most unstable of them are the perturbations with a transversal to the magnetic surfaces polarization of the magnetic field.

Obtained equations were supplemented with ionospheric boundary conditions derived in the same approximation. Considered perturbations appeared to be affected only by integral Pedersen conductivity of the ionosphere, which was approximated by a thin spherical layer. Moreover, when Pedersen conductivity of the ionosphere is finite flute modes can appear in the magnetosphere, determining its stability in this case. Using a modified energetic principle they derived the corresponding stability criterion, which sets stronger restrictions on the stability than well known Gold criterion.

The existence of features indicative of shorelines of ancient oceans on Mars has been proposed for several authors. In Chapter 5 the authors revise the topography of possible Martian paleoshorelines, and their consequences for the amount of water infilling ocean basins when water load is considered. They show that a re-evaluation of paleoshorelines is need. For example, the putative Meridiani shoreline could be the same feature as some portions of the Arabia shoreline. Indeed, elevations in the Meridiani shoreline are roughly similar to that of the Arabia shoreline in northeast Arabia, Utopia (not taken into account the Isidis basin), Elysium, and Amazonis regions. This is still far of an equipotential surface, but a paleoshoreline through these regions and the Meridiani shoreline would be better candidate to represent a paleoequipotential surface than the Arabia shoreline *sensu strito*. Moreover, the elevation of the Arabia shoreline in northern Arabia Terra after is intriguingly close to the mean elevation of the Deuteronilus shoreline, and it cannot be discarded a "mixed" Arabia/Deuteronilus shoreline, which would include the Arabia shoreline in northern Arabia Terra, and the Deuteronilus shoreline elsewhere.

If reality of global shorelines is accepted, as increasing evidence suggests, then present-day topographic variations in these features postdate shorelines formation. So, their topographic range should provide information on large-scale vertical movement of the lithosphere, which in turn provides information on the thermal evolution of Mars. They describe the application of thermal isostasy concept to constraint the ancient thermal state of the lithosphere from present-day paleoshoreline topography. For the ~1.1 km total elevation range of the Deuteronilus shoreline, the relative amplitude of heat flow variations (the ratio between maximum and minimum heat flow) is ≤1.6. This value is clearly lower than that presently observed on continental areas on the Earth. If heat flow variations on Mars are currently greatly disappeared, then the obtained heat flow variations upper limits must be mostly related to the paleoshoreline formation time: the present-day elevation range along Deuteronilus shoreline suggests that differences in the thermal state of the lithosphere in regions along this putative paleoshoreline have been relatively small since the feature was formed, and therefore the absence of lithospheric tectonothermal events by the latest ~3 Gyr, at least. If the Deuteronilus shoreline is a combination of portions of several paleoshorelines, then the total elevation range, and the implied heat flow variations, would be lower, and the lithosphere stability higher.

Chapter 6 reviews the information on mineralogy and chemistry of Martian soils from the data of Landers and remote sensing. Compositional information from Mars Pathfinder Landers indicates a mixture of weathered local rocks of mafic composition. If the Martian soil is composed by sediments only, porosity can be 50% at the surface. In other areas of interest a surface porosity of 20% can be present for lavas.

Martian soil analogues, in dry and frozen conditions, are investigated in the present work as far as thermal conductivity and temperature variation, along its depth, are concerned. The thermal conductivity is predicted theoretically with the cubic cell model, which considers the dry soil as a cubic cell, with a solid cubic particle at the center, and gas around it. The thermal conductivity of the porous medium is evaluated from the solution of the one-dimensional Fourier conduction equation with the assumption of isothermal lines. The model needs to know the thermal conductivities of the solid particle and of the materials present, i.e. atmospheric gas and/or frozen ice, and the porosity of the soil analogue. Martian soil analogue is simulated with a kind of olivine tested in laboratory. The soil mineral composition allows to evaluate experimentally the thermal conductivity of the olivine particle, with the help of the theoretical method, which results equal to $K_s = 2.94$ W/m K. The thermal conductivity of dry Martian soil analogue, evaluated with the theoretical model in the porosity range $p = 0 - 1$ and at several temperatures, increases with the temperature and decreases with the porosity. The Martian atmospheric pressure is about 6 mbar and the Martian atmospheric gas is composed of CO_2 (95%), N_2 (3%), Ar (1.5%) and traces of water vapour. Heat capacity of soil analogue is evaluated with the knowledge of its physical properties, the porosity and the specific heats of the materials present.

Thermal diffusivity, calculated as the ratio of thermal conductivity and heat capacity, is a function of porosity and ice mass content of the soil analogue.

The temperature of Mars surface is assumed variable during the day in the range: 148 K – 298 K. Temperature variations in dry and partially frozen soil analogues are predicted during a Martian day. The temperature variation at different depth is attenuated, as compared to the surface variation, and a phase delay is present depending on the soil thermal properties. The temperature variation, as well as the derivative of the temperature variation with the depth, is dependent on the thermal diffusivity of the soil analogue. In conclusion, the temperature measurement along the depth of a Martian soil analogue can be used to verify its physical status, i.e. dry or partially frozen.

Starting with the discovery of the asteroid Ceres, during the first night of January 1801, by Piazzi in Palermo (Sicily), the discipline of asteroid-discovery evolved quickly into large searches of similar bodies in the solar system. At this very moment, more than130.000 new objects have been numbered and the more of them classified into *families.* A sreported in chapter 7, one of these groups of asteroids, the NEO's (Near Earth Objects), subdivided in three substantially different *subfamilies* (Aten-, Amor- and Apollo-objects) have presently become search objects of primary importance; since they may collide with Earth, resulting in disastrous situations on our planet and (posing) serious risks for the preservation of life on its surface.

Effect of electromagnetic radiation on dynamics of arbitrarily shaped cosmic dust particles is investigated. Process of extinction, scattering, absorption and thermal emission is taken into account. Equation of motion is derived and physical meaning of the important terms is emphasized and discussed. The equation of motion is written in a relativistically covariant form. The equation of motion yields, as a special cases, both the results presented by Einstein in his relativistic paper written 100 years ago, and, also the Poynting-Robertson effect, well-known in astrophysics.

Applications of the derived equation of motion are shown for various systems in the Universe. The significance of the general equation of motion is always stressed by presenting results both for spherical and nonspherical dust particles. While Poynting- Robertson effect holds for spherical particles, nonspherical particles may behave in a different way than it would correspond to the Poynting-Robertson effect. The first application in Chapter 8 deals with the observed flux of interstellar dust entering the Solar System. Spherical interstellar dust grains can be captured only in small distances from the Sun and the grains are thermally destructed. The flux of interstellar dust may produce capture and surviving of nonspherical particles. The second application concentrate on motion of dust grains near commensurability resonances with a planet - mean motion resonances (MMR; a particle is in resonance with a planet when the ratio of their mean motions is the ratio of two small integers) - and possible capture of the grains in the resonances. Capture times for spherical particles can be larger than for nonspherical grains. However, nonspherical grains may exhibit an increase of semimajor axis before capturing in the MMR. This is caused by the effect of electromagnetic radiation on nonspherical dust grain and this is the first known case when a cosmic body can be captured in the MMR during the process of an increase of semimajor axis.

Another application, of the general equation of motion for interaction between incident electromagnetic radiation and dust grain, concerns evolution of meteoroids ejected from comets. While observations yield that small dust grains are observed not only in the orbital plane of comet Encke near its aphelion, Poynting-Robertson effect cannot explain such a behaviour. The effect of electromagnetic radiation on nonspherical grains produces also change of orbital inclination. Thus, nonspherical dust grains near cometary aphelion have to be present also in the plane normal to the orbital plane of the comet.

Chapter 9 discusses magnetic reconnection, which is a topological restructuring of magnetic field lines, leading to a conversion of magnetic energy into kinetic energy and a heating of the plasma. It takes place in regions with strong magnetic gradients: the Earth's magnetopause and the plasma sheet in the magnetotail, during solar and stellar flare events, and in laboratory plasmas on the Earth. It is also involved in astrophysical phenomena like the generation of relativistic jets and processes in accretion disks. The Earth's magnetosphere is the only region where signatures of magnetic reconnection can be measured in-situ by satellites. Based on the method developed by Semenov et al., the authors solve the direct problem of magnetic reconnection and calculate the magnetic field and plasma flow disturbances in a compressible plasma caused by transient reconnection. In order to get the reconnection rate from measured data, they invert the problem. Since the inverse problem is ill–posed, it is treated with the method of regularization. Application of this method to Cluster measurements from September 8th, 2002, where a series of Earth-ward propagating 1–minute scalemagnetic field and plasma flow variations were observed outside of the plasma sheet, showed good agreement for the z-component of the reconstructed magnetic field. The reconnection rate is about 1 mV/m and the reconnection region is located at about 24–25 Earth radii in the magnetotail.

Crushing and evaporation are basic physical processes at the entrance of natural space bodies of size 0.01-100 m into atmospheres of planets. The relative role of evaporation is characterized by value of mass loss parameter, which is proportional to the relation of kinetic energy of a body mass unit to effective enthalpy of evaporation. Examples of real large bolides for various values of this parameter are given in Chapter 10. The offered general approach helps to understand a plenty of observant data of a different level of reliability, and also allows to separate preliminary real hypotheses from fantastic.

A qualitative discussion of the background, the present status, and the future prospects of space weather research, monitoring, and services is presented in Chapter 11. The elements of the interconnection between the Sun and the Earth are described, and their relations with space weather phenomena are explained. The effects of space weather from the near Earth space to Earth's surface are discussed, and a short summary of present instruments, space facilities, and space weather services is given. Aspects of future needs and directions are also discussed.

In: Space Science: New Research
Editor: Nick S. Maravell, pp. 1-34
ISBN: 1-60021-005-8

Chapter 1

GEOLOGY OF THE TERRESTRIAL PLANETS WITH IMPLICATIONS TO ASTROBIOLOGY AND MISSION DESIGN

Dirk Schulze-Makuch[1a], James M. Dohm[2,5b], Alberto G. Fairén[3c], Victor R. Baker[2,5d], Wolfgang Fink[4e], Robert G. Strom[5f]

[1] Dept. of Geological Sciences, Washington State University
[2] Dept. of Hydrology and Water Resources, University of Arizona, Tucson, Arizona
[3] Centro de Biología Molecular, CSIC-Universidad Autónoma de Madrid, 28049-Cantoblanco, Madrid, Spain.
[4] Division of Physics, Mathematics and Astronomy, California Institute of Technology, Pasadena, California
[5] Lunar and Planetary Laboratory, University of Arizona, Tucson, Arizona

Abstract

Although Mercury, Venus, Earth/Moon, and Mars originated from similar material accreted in close regions of the planetary nebula, the worlds that developed after 4.6 billion years of Solar System evolution are not much alike. Mercury and the Moon, for example, are without a protecting atmosphere and were not exposed to liquid water on their surface for at least 4 billion years. The surfaces of Venus and Mars are desiccated and are presently not a suitable habitat for life, but reservoirs of liquid water remain in the atmosphere of Venus and the subsurface of Mars, and microbial organisms may have adapted to survive in these ecological niches. The search for water and life on any of the terrestrial planets is intrinsically connected to their geological history. Missions should be designed to explore any potential

[a] E-mail address: dirksm@wsu.edu, Corresponding author: dirksm@wsu.edu/mantid5@aol.com, Tel: 509-335-1180, Fax: 509-335-7816, Webster Hall, Dept. of Geology, Washington State University, Pullman, WA 99164-2812
[b] E-mail address: jmd@hwr.arizona.edu
[c] E-mail address: baker@hwr.arizna.edu
[d] E-mail address: agfairen@cbm.uam.es
[e] E-mail address: wfink@autonomy.caltech.edu
[f] E-mail address: rstrom@pirl.lpl.arizona.edu

past and present habitats. In addition, ices on Mercury and the Moon should be explored for remnant biogenic material from the early evolution of life on Earth (and elsewhere). Especially desirable are sample return missions and missions that include hierarchical architectures with scalable degrees of mission operation autonomy, which will allow optimal reconnaissance of planetary environments, including surface and subsurface environments.

Keywords: Terrestrial Planets, Venus, Mars, Geological Processes, Exobiology, Mercury, Mission Architectures, Mission Autonomy.

Introduction

We are at a threshold point of planetary exploration. Observations from Earth-based telescopes, the Hubble space telescope, and space missions have provided some initial "reconnaissance-based" understanding about the environments of the inner planets of the Solar System. We are now transitioning into non-traditional autonomous (robotic) exploration that has the potential to improve upon our current understanding of the evolutional histories of the terrestrial planets, including geology, paleohydrology, and paleoclimatology, and whether life had an opportunity to gain a foothold. An initiative has emerged to spearhead the means of providing spacecraft with the capability to operate on their own, such that unconstrained science-driven missions loom on the near-term horizon. Past space missions that had well conceived plans and instruction lists on how to take and retrieve pictures of the Earth, as well as other planetary bodies in the Solar System, were not designed to respond to changes in science objectives while en route. Rather, these previous missions relied on ground-based efforts, which included data processing by numerous technicians, data analysis by multiple scientists, and ultimately the planning of new observations collectively by both engineers and scientists, all of which required tremendous energy (e.g., human power, time, and cost). In addition, ground exploration on Mars has been safety and engineering constrained, greatly limiting exploration of the martian surface.

Thus, one desirable objective is to design robotic missions that can home in on regions and/or features of extreme scientific interest in order to better understand environmental conditions (e.g. geology, hydrology, climate, etc.) of the terrestrial planets that neighbor Earth, and to evaluate their potential for harboring life, both as a place where life could have originated independently from Earth and where terrestrial species, from microbes to humans, may find a suitable environment to thrive. Here, we review what has been learned about the geological and environmental conditions on our neighboring planets and the Moon, especially with respect to water availability, evaluate their habitability for life, and provide suggestions for future missions.

Spacecraft Reconnaissance of the Terrestrial Planets

A number of planetary missions have provided extremely valuable information on the terrestrial planets. Space exploration to planetary bodies in the Solar System began in earnest with the quest to explore the Moon and its geologic history, expressed in the landforms and the rock compositions at its surface (Wilhelms, 1987). The pioneering work of Eugene Shoemaker and many others resulted in several successful missions to the Moon that provided very detailed information on the surface structure, composition, and gravity field. The Apollo landings from 1969-72, for example, returned numerous samples from a variety of landing sites. They also made numerous other measurements of the lunar environment and interior. Other important missions included the five Lunar Orbiter missions in the mid 1960's that provided high-resolution images of the entire satellite, and the Clementine mission in the mid 1990's that provided high-resolution, multi-spectral images of the surface. The Clementine mission allowed the determination of compositional variations across the surface, and provided high-resolution gravity data (Nozette et al., 1994).

There have been several important missions to Mars. The first spacecraft to visit Mars was Mariner 4 in 1965. Mariner 4 took 22 pictures of the surface and revealed a cratered "Moon-like" surface with little hope for dynamic activity and life on Mars. Mariner 6 and 7 were identical spacecrafts sent for Mars flybys to collect data about the surface and atmosphere of Mars. The spacecrafts took more than 200 images of the Martian surface. The Mariner 6 and 7 images showed more geological variety, but confirmed the picture of Mars with little dynamic activity. Mariner 9 changed this perception greatly. It was the first orbiter around Mars and returned more than 7,329 images from its global mapping mission. Due to its limited sample size, the flybys missions (Mariner 4, 6 and 7) missed most of the evidence for dynamic activity on Mars such as the valley channel networks relevant to astrobiology. Together, the Mariner missions to Mars provided the first images of the volcanoes of Mars (e.g. Olympus Mons) and the south polar region, determined the atmospheric surface pressure, measured the moment of inertia of Mars, and provided a record of the surface dichotomy of old, crater-rich highlands and the northern low plains.

The Mariner 9 mission and Viking orbiters and landers in the early to mid 1970's provided complete imaging coverage of the planet at moderate to high resolution (generally from meters to hundreds of meters per pixel), which formed the basis for the Mars Geological Mapping (MGM) program (largely spearheaded by Joseph Boyce and David H. Scott). The primary goals of the MGM program were (1) to determine the geologic and paleoclimatic histories at global to local scales, (2) to provide constraints on theoretical models relating to the time-space relations of magmatic-tectonic activity (which includes geophysical modeling of the planet's interior), and (3) to provide foundational information to select candidate landing sites for future science-driven missions. The landers provided information on the atmosphere, the fine-scale surface structure, surface composition, some constraints on the presence or absence of life, and the only ever life detection experiments conducted on another planetary body. The Viking life detection experiments were designed to detect metabolism of putative martian microbes. Although the life detection experiments revealed a positive result, most involved scientists concluded that the positive responses at both lander sites were caused by some chemical agent in the soil and not by microorganisms. The result still remains controversial (Levin and Straat, 1981; Miller et al., 2001). Although the experiments were well-designed, especially given the technology and knowledge at that time, the major lesson learned was not to conduct a life detection experiment before a thorough knowledge of the tested environment is gained. A test for metabolic activity appears especially unsuited in this respect. Ending a long hiatus of missions, Mars Pathfinder landed successfully on Mars on July 4, 1997. Since Pathfinder, both Mars Global Surveyor (MGS) and Mars Odyssey spacecraft have returned significant imagery, topographic, geophysical (gravity and magnetic), spectral (e.g. visible to thermal wavelengths), and elemental information with very high-resolution images of selected targets of high interest. The current Odyssey mission, for example, is providing information on the water content of the surface and shallow subsurface, as well as certain elemental abundances (Boynton et al., 2002). High- to moderate-resolution thermal images are also providing valuable information on the physical properties of the surface (Christensen et al., 2001). In addition to the MGS and Odyssey spacecraft, explorations through Mars Express and the Mars Exploration Rovers (Opportunity and Spirit) are currently returning information on the composition of the surface, which includes the possible presence of ancient and present water. With these missions little doubt remains that water and not carbon dioxide was the fluid that carved the outflow channels observed on Mars (e.g., see Baker, 2001 and Hoffman, 2000, respectively), and that water has periodically been an active agent on Mars. These results make a sample return mission all the more important, in that water is a significant ingredient for life. For additional information on previous missions to Mars the reader is referred to Godwin (2000).

Most of our understanding of Venus comes from orbiters carrying radar instruments and probes that have descended through the atmosphere. Mariner 2 was the first flyby, launched by the United States in 1962, followed by Mariner 5 in 1967 and Mariner 10 in 1974. The former USSR sent flybys and orbiters to Venus, some of which were combined with entry probes. These included multiple Venera missions during the 1960s, 70s, and 80s, and two Vega missions, which were destined to observe Halley's comet in 1984, but also passed by Venus and released descent capsules. Several of these probes were successful in reaching the surface of the planet. The United States sent two Pioneer Venus missions in 1978, which included orbiters and surface probes. One of the orbiters, Pioneer Venus 1, measured the upper atmosphere for several years. The Magellan spacecraft of the United States collected a tremendous archive of radar imagery of the planet during the early 1990s. It provided high-resolution radar images of about 98 percent of the surface, and altimetric and gravity information for most of the planet. For additional information on missions to Venus, the reader is referred to Bougher et al. (1997). The Venus Express Mission launched in 2005 will allow a global investigation of the atmosphere and plasma environment from orbit including sounding of the structure, composition, dynamics, and variability of the global cloud layer.

The only spacecraft to have surveyed Mercury to date was Mariner 10. In 1974 and 1975, the spacecraft had three flybys of the planet. It imaged only about 45 percent of the surface at moderate to low resolution, and provided information on the magnetic field, composition of the exosphere, and the charged particle environment. However, the Messenger mission, launched on 3 August 2004, will image the entire surface on its three flybys before it goes into orbit in March 2011. Once in orbit, it will image the surface in stereo at high resolution through 12 filters, determine the composition of the surface and exosphere, measure the magnetic field, determine the topography and gravity field of the northern hemisphere, and determine the size of the fluid iron core. Also, the BepiColombo mission is in the planning stage, which is the first large-sized Europe-Japan joint mission. This mission consists of two orbiters (Mercury Magnetosphere Orbiter (MMO) and Mercury Planetary Orbiter (MPO)), which will obtain detailed information on Mercury's magnetic field, magnetosphere, interior, and surface. For more detailed results of the Mariner 10 mission and Mercury in general, the reader is referred to Strom and Sprague (2003).

Geology and Environmental Conditions of the Terrestrial Planets

Geologic, hydrologic, geophysical, and climatic records of a planetary body are significant in the evaluation of its potential habitability for life. Here, we provide a detailed discussion of these current records in the following subsections and a summary in Table 1 based on numerous works (e.g. Wilhelms, 1987; Kieffer et al., 1992; Bougher et al., 1997; Dohm et al., 2001a; Strom and Sprague, 2003).

Mercury

Mercury is the innermost planet in the Solar System and has a diameter of 4878 km. It has no known satellites. To date, Mariner 10 is the only spacecraft that has explored the planet. As a consequence, there are still many questions concerning its history and evolution. Mercury has the most elliptical (0.2 eccentricity) and inclined (7 degrees) orbit of any planet in the inner Solar System. As such, its distance from the Sun varies from 0.30 AU at perihelion to 0.46 AU at aphelion. Mercury's rotation period is 58.6 Earth days and its orbital period is 87.9 Earth days, therefore creating a unique 3:2 spin-orbit coupling (it rotates exactly three times around its axis for every two orbits around the Sun). Mercury has the greatest range (day to night) of surface temperatures of any planet or satellite in the Solar System because of its proximity to the Sun and its long solar day, ranging from about 723 K at the equator when it is closest to the Sun to about 90 K at night just before dawn.

Mercury's small size and relatively large mass ($3.301x10^{23}$ kg) means that it has a very large density of 5.44 g/cm^3; only slightly less than Earth (5.52 g/cm^3), and larger than Venus (5.25 g/cm^3). Because of the large internal pressures of Earth, however, its uncompressed density is only 4.4 g/cm^3 compared to the uncompressed density of Mercury (5.3 g/cm^3). This means that Mercury contains a much larger fraction of iron than any other planet or satellite in the Solar System. Thus, the iron core must be about 75 percent of the planet diameter, or some 42 percent of its volume. Its rocky outer region is only about 600 km thick.

Mercury may have always been void of an atmosphere, since numerous fresh impact craters with bright radial aureoles, which provide a relative age estimated to date back as far as 4 Gy (Strom and Neukum, 1988), would otherwise have been modified or erased even under atmospheric pressures as low as those recorded for Mars. Although Mercury does not have a significant atmosphere, it does have an exosphere where atoms rarely collide. The exosphere consists primarily of sodium and potassium, which could be released from surface rocks by their interaction with solar radiation, or by impact vaporization of micrometeoroid material (Wurz and Lammer, 2003).

Table 1. Primary Traits of the Terrestrial Planets

Process or Property	Mercury	Venus	Earth	Moon	Mars
Impact Cratering	*Yes (Large)*	*Yes (Scarce)*	*Yes (Very scarce)*	*Yes (Large)*	*Yes (Duality Large/Scarce)*
Surface Age	*Very old, recording the early history of the Solar System*	*Young, recording only the last ~600-700 m.y.*	*Very young*	*Very old, recording the early history of the Solar System*	*Very diverse, the only planet that may record the whole history of the Solar System*
Geologic State	*Inactive*	Globally active	*Globally very active*	*Inactive*	*Local and temporally active*
Volcanic Features	*Probably flood basalts; intercrater plains and smooth plains*	*Flood basalts; regional plains; lava flow fields (central and fissure-fed activity); shield volcanoes; some viscous extrusions such as lava domes (> 70% are flood basalts)*	*Flood basalts; lava flow fields (central and fissure-fed activity); wide-ranging types of volcanic constructs, which include composite, shield, cinder, and silicic-rich lava domes; hot spots*	*Flood basalts; intercrater plains and smooth plains*	*Flood basalts; intercrater plains and smooth plains; lava flow fields (central and fissure-fed activity); mostly shield volcanoes;* some putative viscous (silicic-rich) extrusions (e.g., highly eroded lava domes)
Intrusive Features	(?)	*Igneous plateaus and basins; largely mantle plume-related deformation, including coronae and other uplifted/blistered surfaces; convection cell deformation*	*Igneous plateaus and basins, and exposed intrusives; plume-related (ranging from mantle plumes to superplumes) deformation, which includes uplifted terrains, ranging from mantle to superplumes; convection cell deformation*	*(?)*	*Igneous plateaus and basins; plume-related (ranging from mantle plumes to superplumes) deformation, which includes the formation of coronae-like features (e.g. Alba Patera and Syria Planum) and localized uplifts (e.g. central Valles Marineris; convection cell deformation*
Magmatis m	*Probably flood basalt volcanism, as well as impact-induced volcanism (?)*	*From oldest to youngest: (1) global resurfacing event, (2) stagnant-lid, episodic (pulse-like) magmatism that results in regional to local uplifts (e.g. Coronae) and downdwellings (e.g. Basins) and regional plains development; magmatic-driven tectonic activity and regional plains development that transitions into centralized volcanism and more local tectonism*	Mostly efficient internal heat release during plate tectonism with major pulses of paleomagmatic activity (e.g. Ontong Java and Decaantraps)	*Flood basalt volcanism; impact-induced volcanism*	*Putative early earth-like conditions that transitions into a stagnant-lid regime with episodic (pulse-like) magmatism such as observed at Tharsis and in the development of the regional plains; magmatic-driven tectonic activity and regional plains development transitions into centralized volcanism and more local tectonism*
Tectonic Structures	*Lobate scarps (thrust faults) and impact-related tectonic structures*	*Structure belts, fracture, fault, rift, and wrinkle ridge systems, broad ridges (possible thrust faults); plateaus and topographic basins; pit crater chains and lava channels marking, in part, subterranean withdrawal of magmas and associated collapse; densities of structures following global resurfacing is an order of magnitude greater than late-stage activity*	*Convergent, divergent, and strike-slip plate boundaries; fracture, fault, rift, and canyon systems; mountain ranges, plateaus, and topographic basins; collapse features of diverse types, indicating the migration of fluids into and away from geologic systems, as well as subterranean voids*	*Local extensional and contractional (e.g. wrinkle ridges), which includes impact-related tectonic structures; lava channels*	*Fracture, fault, rift, canyon, and wrinkle ridge systems, broad ridges, and mountain ranges (ridges and mountain ranges interpreted to result from contractional deformation including thrust faulting), plateaus, and topographic basins; densities of structures in the south-central part of the Tharsis region during the Noachian are an order of magnitude greater than the density observed during the Late Hesperian; collapse features of diverse types, indicating the migration of fluids into and away from geologic systems, as well as subterranean voids*

Table 1. Continued

Process or Property	**MERCURY**	**VENUS**	**EARTH**	**MOON**	**MARS**
Tectonism	*Global; early expansion (tension); late contraction (compression)*	*Possible early global plate tectonics, followed by stagnant-lid activity, which includes localized extension and contraction; regional and local magmatic-driven uplift (mainly extensional, but also contractional); an order of magnitude difference is recorded in the densities of structures of older geologic terrains compared to younger regional plains materials*	*Global to local; plate tectonics; extensional and contractional deformation related to both plate tectonics and magmtic-driven vertical tectonism*	*Local; basin subsidence; mainly tension*	*Possible early global plate tectonics followed by stagnant-lid activity, which includes localized extension and contraction; regional and local magmatic-driven uplift (mainly extensional, but with contraction); When comparing fault and fault-length densities of faults in the Thaumasia region of Mars (southeast part of Tharsis), an order of magnitude difference is recorded from the Noachian to Late Hesperian; also during the Late Hesperia, magmatic-tectonic activity becomes more localized*
Hydrosphere	Probably never in its history	*Past: possible, if the geological evolution included CO_2 recycling;* *Present: no, but some water present in the atmosphere*	Yes, covering 2/3 of the surface and water in the atmosphere and the subsurface	*Probably never in its history*	*Possibly early oceans and ice sheets; episodic inundations in the northern lowlands ranging from oceans to lakes; presently none, but probably subsurface ice and water*
Weathering	*No*	*Very little, if any*	Extensive	No	*Moderate (heightened erosion during climatic perturbations); erosion rates decreased significantly sometime during the Middle/Late Noachian*
Atmosphere	*No significant*	*Thick CO_2 (93 bars)*	*Moderate N_2, O_2 (1 bar)*	*No*	*Thin CO_2 (~8 mbar)*
Polar Deposits	*Yes, ice at the poles (cometary?)*	No	*Yes, ice sheets and glaciers*	*Very little*	*Larger ice bodies than present: polar caps presently comprise H_2O and CO_2*
Magnetosphere	*Weak (could be active?)*	No	*Strong magnetic field*	*Controversial remnant magnetism (possibly related to large impacts)*	*Early Magnetic Field/* No Present Magnetic Field
Size ***Earth Radius = 1*** ***Earth Volume = 1***	*Small* *0.38* *0.06*	*Large* *0.95* *0.85*	*Large* *1.0* *1.0*	*Small* *0.27* *0.02*	*Intermediate* *0.53* *0.15*

In general, the surface of Mercury can be divided into three major terrains: (1) heavily cratered regions, (2) intercrater plains, and (3) smooth plains. The heavily cratered uplands record the period of heavy meteoroid bombardment that ended about 3.8 billion years ago. This event may have been a catastrophic event that peaked about 3.9 billion years ago. Therefore the age of this surface is probably about 3.9 Ga. The two plains units of Mercury have been interpreted to be volcanic (Spudis and Guest, 1988). The older intercrater plains are the most extensive terrain on Mercury. They were emplaced over a range of ages contemporaneous with the period of late heavy bombardment. They are thought to be volcanic plains erupted through a fractured crust during the first 800 million years of Mercury's evolution.

The younger smooth plains are primarily on the interior and exterior of large impact basins. They are similar to the lunar maria, and are therefore interpreted to be lava flows erupted relatively late in the geologic history of Mercury (Strom and Sprague, 2003). They may have an average age of about 3.8 billion years as indicated by crater abundances.

The tectonic framework of Mercury is unique in the Solar System. It consists of a system of thrust faults called lobate scarps. They are more-or-less uniformly distributed over the part of Mercury as viewed by Mariner 10. Presumably they occur on a global scale. This suggests that Mercury was subjected to global compressive stresses. Stratigraphic evidence indicates that the system of thrust faults formed after the emplacement of intercrater plains materials, relatively late in the geologic history of Mercury. The faults were probably caused by a decrease in the size of the planet due to cooling.

Mercury appears to exhibit no evidence of any exogenic activity other than impact cratering and almost surely formed under mostly anhydrous conditions. Its surface appears similar to the Moon with no evidence of any volatile or geologic cycling, but rather surfaces are marked by impact craters, and in places there are ridges that resemble thrust faults on Earth (Murray et al., 1981). Also, similar to the Moon, but contrasting with Venus, Earth, and Mars, there is little to no evidence, as of this writing, for concentrated magmatism at the surface in the form of prominent shield volcanoes and centers of uplift (Table 1).

Radar signals show very high reflectivities and polarization ratios centered on the poles. The reflectivity and ratio values are similar to outer planet icy satellites and the residual polar water-ice cap of Mars. Therefore, Mercury's polar radar features have been interpreted to be water ice. The radar anomalies occur in permanently shadowed portions of polar craters where water ice is stable over the lifetime of Mercury. The deposits are probable water ice derived from comets of water-rich asteroid impacts. However, the impact velocity of comets at Mercury (~50 km/sec) may be so high that the water may be disassociated into H and O and lost to the planet. Therefore, water-rich asteroids are the most likely source because they contain bound water and their impact velocities are lower. A recent study of regolith formation and the burial rate of the polar deposits, combined with the radar characteristics of the polar deposits suggests that the deposits are less than 50 million years old (Crider and Killen, 2005). If this is the case there is a possible source on the side of Mercury not imaged by Mariner 10. Earth-based radar detected a fresh 87-km-diameter impact crater at 58 degrees N latitude, 345 degrees longitude, which is a present-day source of sodium. This feature could be the result of a comet or water-rich asteroid that caused the polar deposits.

Venus

The second planet from the Sun, Venus, is enveloped by clouds, making it impossible to observe its surface from Earth and spacecraft without the use of radar. The atmosphere is extremely thick (93 bars compared to 1 bar for Earth), consisting of 97 percent carbon dioxide (CO_2). The surface temperature on Venus is extremely hot, reaching 735 K. The high surface temperature is hypothesized to be the result of a runaway greenhouse effect, which caused any oceans to evaporate into the atmosphere (Wildt, 1940; Sagan, 1960; Ingersoll, 1969; Pollack et al., 1980; Kasting, 1988; Hunten, 1993). The venusian atmosphere is acidic with sulfuric acid as a significant component of observed cloud particles. In comparison, the atmosphere of Earth also contains sulfuric acid particles in the stratosphere, though only in minor amounts. On Earth, rain carries with it sulfuric acid to react with surface materials. On Venus, the acid evaporates at the base of the cloud cover (about 50 km above the surface), remaining in the atmosphere. The upper parts of the clouds extend 80 km above the surface. Pioneer Venus 1 studied the upper atmosphere and ionosphere in great detail. On Earth, this region is very hot, while not for Venus. On the night side of Venus, the upper atmosphere is extremely cold.

Geologic mapping of the surface of Venus reveals a past dynamic planet, as indicated by a variety of coronae, igneous plateaus, large volcanoes, ridge and fracture belts, tesserae, impact craters, and other volcanic and structural features distributed within a plains setting (Head and Wilson, 1992; Head et al., 1992; Stofan et al., 1992; Basilevsky and Head, 1998; Phillips and Hansen, 1998; Bilotti and Suppe, 1999), affording study of their detailed age relations and evolutionary development. In particular, at least five major stages of geologic activity (each of indefinite duration) are recorded (Tanaka et al, 1997; Basilevsky and Head, 2000; Ivanov and Head, 2001; Dohm et al., 2002a, 2006c). These stages can be characterized by the following dominant materials, from oldest to youngest: (1) tessera material, which may also include rafted and highly deformed crustal materials recording earlier evolutional phases of planetary development; (2) coronae and older plains units; (3) widespread intermediate plains units; (4) younger regional plains units; and (5) flow materials of large shields, local flows, and smooth and rough plains units. Each of the stages following the formation of tessera material resulted in a significant preserved record of areal resurfacing. Tectonically, some parts of coronae and various structure belts may have been active for most of the available geologic record, though activity waned or became more localized over time. In addition, impact cratering and dome and shield-field formation may have occurred over a relatively large span of the exposed geologic record, and structure belts may mark lithospheric zones of weakness that formed through time, perhaps prior to the last definitive global resurfacing event.

The impact cratering record on Venus is unique in the solar system and indicates that the planet has experienced at least one, and probably multiple, episodes of global resurfacing (Strom and Schaber, 1994). There are 967 impact craters and most of them are in pristine condition (84%). Only about 3% have been flooded with lava and 11 % disrupted by fractures. There are also 400 craterless splotches that have resulted from the shock waves of objects that never hit the surface, and none of them have been embayed by lava. The craters are randomly distributed across the surface of the planet. They are also randomly distributed with respect to elevation. The random distribution is also independent of size; small craters are just as randomly distributed as large ones. The embayed craters are concentrated in zones of recent volcanism, and the heavily fractured craters are associated with the major rift systems in the equatorial highlands. These characteristics strongly indicate that Venus has been resurfaced on a global scale, followed by reduced and concentrated endogenic activity, as evidenced by the noticeable decline in the formation of tectonic structures and the stratigraphically young shield volcanoes, respectively. Except for tesserae outcrops, most of the geologic units observed today were formed since the last global resurfacing event, which ended between 500 ma to 1 G.a. ago. Geologic activity is probably still continuing today, but at a much lower rate than during the earlier recorded geologic history.

It is difficult to determine the exact number of global resurfacing events that Venus has undergone. López et al. (1998) theorized eight global resurfacing events based on calculations of the maximum possible amount of CO_2 outgassed in each event, approximately 5.6×10^{19} kg or 1/8 the total CO_2 present in the atmosphere, assuming that the planet lacks efficient mechanisms for CO_2 recycling. The observed eruptions on the surface of Venus (e.g., the Sapho Patera region) indicate that, in such a dense, hot atmosphere, a high abundance of magmatic volatiles had to be present to overcome the atmospheric pressure. Water would likely be a major constituent of these magmatic volatiles. Though inconsistent with the extreme desiccation of rocks near the surface, the reportedly low-viscosity lava flows on Venus could be explained by a water-enriched magma source (Schulze-Makuch and Irwin, 2002).

Venus formed in the same general region of the Solar System as Earth, and thus abundant solar energy and primordial waters on Venus can be assumed. Evidence for a proto-ocean were advanced by (1) Donahue et al (1982) because of the venusian enhanced D/H ratio, (2) Abe and Matsui (1988) based on the evolution of H_2O-CO_2 atmospheres, (3) Matsui and Tajika (1991) because of the amount of ^{40}Ar in the present venusian atmosphere, and (4) Pollack (1991) based on comparative analyses of the similar amounts of CO_2 and N_2 in the near-surface volatile reservoirs of Venus and Earth. The early venusian ocean was likely a hot sea (Kasting, 1989) that evaporated eventually during a run-away moist greenhouse effect. How long the ocean or oceans were present is a controversial matter. Hydrodynamic escape and a strong solar wind could have eroded the early Venusian ocean very quickly (Kasting and Pollack, 1983; Donahue and Hartle, 1992; Chassefiere, 1996, 1997 a & b, Lundin et al., 2004). Others (e.g. Marov and Grinspoon, 1997; Grinspoon, 2003) argue that the ocean or oceans could have been present on Venus for billions of years. Catling (2004) points out that significant scattering of sunlight back to space by marine stratus clouds could have cooled the surface and may have helped to delay a runaway greenhouse over a considerable period of geological time. The atmospheric change is usually thought to have occurred gradually due to a positive feedback loop that amplified the greenhouse effect (Walker et al., 1981).

Earth

Earth is the only planet in the Solar System where liquid water is stable on its surface for extensive periods of time. Earth is an extremely active, water-enriched planet where high erosion rates, magmatic-driven activity, and plate tectonism have modified its surface over relatively short time periods (processes ranging from days to millions of years). Earth presently transports heat through two primary modes: plate tectonism and plumes of diverse dimensions and origins ranging from superplumes (e.g. Maruyama, 1994; Fukao et al., 1994;

Li et al., 2003) to mantle plumes (e.g. White and McKenzie, 1989; Mege and Ernst, 2001) . The diverse geologic processes and histories are recorded in the landforms, stratigraphic sequences, and rocks, sediments, and soils. These histories record endogenic-, exogenic-, climatic/atmospheric-, and biologic-related activities, making the surface environment of Earth extremely heterogeneous, as exemplified with its oceans, continents, mountains, basins, plateaus, lakes, river systems, deserts, aquifers, glaciers and ice sheets, subsurface voids/caverns, seasonal changes, latitudinal changes, temperature difference between poles and equator, etc.

The history of Earth can be summarized into four major evolutional phases based on a collection of geologic, paleotectonic, geomorphic, topographic, geophysical, spectral, and elemental information, from oldest to youngest: (1) shortly after accretion, Earth differentiated into a liquid metallic core, a mantle boundary layer (MBL) of high-pressure silicate mineral phases, upper mantle, magma ocean, thin komatiitic crust, and convecting steam atmosphere; (2) Earth cooled to condense its steam atmosphere resulting in the formation of oceans and the transformation from its mode of mantle convection to plate tectonics. Subduction of water-rich oceanic crust initiated arc volcanism and transferred water, carbonates, and sulfates to the mantle; (3) the core dynamo initiated and the associated magnetosphere was produced; and (4) accretion of thickened, continental crust and subduction of hydrated oceanic crust to the mantle boundary layer and lower mantle of Earth occurred. Earth, similar to both Mars and Venus, has recorded spiked periods in endogenic-driven activity, as recorded in the evolution of Ontong Java (Richardson et al., 2000), a significant component of the Southern Pacific Superplume (Maruyama, 1994), the Decaantraps (Coffin and Eldholm, 1994), and the Karnataka Plateaus of the southern peninsular shield in India (Kailasam, 1979).

As the surface and subsurface of Earth underwent significant changes during its history, so did its atmosphere. The earliest atmosphere of Earth, derived mostly from mantle degassing, must have contained large amounts of nitrogen, carbon dioxide, and methane. The greenhouse effect at that time was much larger than it is today, otherwise mean surface temperatures during the Archaean should have been below the freezing point of seawater due to a 30 % less luminous Sun (Sagan and Chyba, 1997). It is likely that methanogenic bacteria mineralized most of the organic matter in the Archean and that the global activity of these bacteria increased the methane content of the atmosphere substantially (Habicht et al., 2002). Pavlov et al. (2001) suggested that the early atmosphere of Earth contained organic haze and a CH_4/CO_2 ratio of about 1 to explain the extremely low ^{13}C kerogens in late Archaean sediments. Oxygenic photosynthesis by cyanobacteria led then to a dramatic increase of global oxygen content about 2.4 to 2.1 G.a. ago (Klein et al., 1987; Holland, 1999; Farquhar et al., 2000; Canfield et al., 2000; Bekker et al., 2004). From that time on, the oxygen content continued to increase to a current level of about 21 % with carbon dioxide and methane being atmospheric trace compounds only.

The Moon

The Moon is the largest satellite relative to its accompanying planet in the Solar System. It is 1738 km in diameter and has a density of 3.34 g/cm^3. Its orbit is inclined 5° to the plane of the ecliptic, and it rotates around its axis once every 27 days. Therefore, its rotation period is the same as its orbital period, and thus always keeps the same face turned toward Earth. The surface gravity is too low, and the daylight temperatures are too high for the Moon to retain an atmosphere. The maximum and minimum surface temperatures are about 390 K and 104 K.

The Moon is thought to have formed 4.5 billion years ago by a gigantic impact of a Mars-sized body with the Earth (Hartmann et al., 1986). The satellite accreted from the rocky mantle of the Earth and impactor when a fraction of it was ejected into an orbit around the Earth. The impactor's iron core was assimilated into the Earth's core (Hartmann et al., 1986; Cameron and Benz, 1991). This model explains the high spin of the Earth-Moon system, the inclined lunar orbit, its low density compared to Earth, its composition similar to Earth's mantle, and the very low volatile and high refractory elemental abundances. At least, the outer part of the Moon was molten and differentiated into an anorthositic crust and basic to ultrabasic mantle. If the Moon has an iron core, it is very small (no more than about 400 km diameter). The center of mass is offset from the center by about 1.8 km toward the Earth. This means that its anorthositic crust is thinner on the frontside than the farside.

All of the terrestrial planets, including the Moon, experienced a period of heavy bombardment that ended about 3.8 G.a ago. Thus, the Moon has heavily cratered highlands similar to Mercury and Mars. This heavy bombardment was followed by a period of basaltic volcanism that was concentrated in the large frontside impact basins. Because of the thicker crust on the farside, an equipotential surface is closer to the surface on the frontside; for example, magmas that originate at the same level can reach the surface more easily than the farside. That is why the mare basalts are concentrated on the frontside. All internal activity apparently ceased about 2 to 3 G.a ago.

Because of the Moon's origin, lunar rocks have essentially no water, bound or otherwise. Thus, the Moon never had oceans or lakes on its surface, and no significant amount of subsurface water. Any water in the original rocks of the Moon was likely lost in the impact process. Like Mercury, the Moon has polar ice deposits in the permanently shadowed areas of craters in the polar regions (Ingersoll et al., 1992; Feldman et al., 1998). However, the deposits are considerably smaller than Mercury's, with an ice concentration of only about 1.5% weight fraction (Feldman et al., 1998).

Mars

The atmosphere of Mars comprises primarily carbon dioxide with very small amounts of other gases, such as nitrogen and oxygen. Martian air contains very little water vapor (only about 1/1,000 as much water as Earth's atmosphere), but even this small amount can form clouds that occur high in the atmosphere. Local patches of early morning fog have been noted to form in valleys. Even thin layers of water frost cover the ground each winter, as noted at the Viking Lander 2 site. Geologic mapping and geomorphic-based investigations undertaken by the Mariner and Viking missions indicate that Mars is a dynamic, water-enriched planet (e.g., Scott and Tanaka, 1986). Confirmation of Mars being a water-enriched planet came following the return of the elemental information from the Gamma Ray and Neutron Spectrometers onboard the Odyssey spacecraft that show significant amounts of water (Boynton et al., 2002). Since erosional features (presumably) of liquid water are clearly visible on Mars, there has been abundant speculation that the martian atmosphere was much thicker early in the history of the Solar System. Recent results from the rovers Opportunity and Spirit also support the presence of ancient water on the surface of Mars. The early martian atmosphere appeared to have had a much denser CO_2 atmosphere (> 1 bar) after its origin and during the heavy bombardment (Haberle, 1998; Carr, 1999; Philipps et al., 2001), with substantial CO_2 retained and episodically outgassed via the Tharsis magmatic complex into the atmosphere from at least 5 major pulses stemming from the Noachian Period to recent geological times (Dohm et al, 2001a,b). Dreibus and Waenke (1985) show that the volatile content of Mars is higher than Earth based on plotting martian meteorite elemental abundances as a function of volatililty. Dynamical simulations suggest that Mars acquired less total bulk of water than Earth, about 6 – 27 % of Earth's total oceanic water (Lunine et al, 2003). Still, this estimate represents more water content per volume unit in Mars than in Earth, consistent with the position of Mars in space being richer in volatiles.

There is compelling evidence that liquid water flowed on the surface of Mars in the distant past and also fairly recent (Baker et al., 1991, 2000; Scott et al, 1995; Bridges and Grady, 1999; Carr, 1996; Malin and Edgett, 2000a; Cabrol and Grin, 2001; Baker, 2001; Dohm et al., 2001a-c; Ferris et al., 2002; Malin and Edgett, 2003; Fairén et al., 2003). In addition, the D/H ratio of the atmosphere attests that Mars has lost water to space over time (Yung et al., 1988; Donahue, 1995), perhaps as much as 75% of its original reservoir (Jakosky and Phillips, 2001). Much of this loss of water may have occurred relatively rapidly early in martian history due to impact erosion and hydrodynamic escape (Melosh and Vickery, 1989, Chassefiere, 1997a; Lammer et al., 2003), although there is evidence for (periodic) oceans later in Martian history (e.g. Fairén et al., 2003). The Noachian oceans, and possibly also any subsequent liquid water-mass on Mars, were enriched in iron hydroxides and magnesium sulfate salts, as revealed through MER Opportunity ground-based observations and analysis of the rock materials at Meridiani Planum (Squyres et al., 2004; Öner et al., 2004). The CO_2, SO_2 and water generated from the volcanism and flood outbursts would have produced the acidic conditions to generate extensive sulfate salt emplacement on the martian surface. Aqueous thermodynamic calculations, which consider a CO_2-dominated atmosphere and a steady supply of iron and sulfate to seawater (raising concentrations up to 0.8 mM and 13.5 mM, respectively), indicate acidic oceanic waters with a $pH<6.2$; these acidic aqueous conditions would have precluded carbonate formation by oceanic sedimentation as a widespread global phenomena (Fairén et al., 2004c). Even today liquid water can transiently exist in rock pores near the surface when ice is melted by the sun. Brines can even exist longer at temperatures well below the freezing point of water.

Several lines of evidence (including geologic, topographic, geophysical, hydrologic) collectively point towards a dynamic, water-enriched planet, with somewhat Earth-like conditions during the embryonic stages of the development (estimated to be nearly the first 700 million years or approximately equivalent to the Hadean time period) (e.g., Fairén and Dohm, 2004). The geologic, climatic, and paleohydrologic evolutional histories of Mars are portrayed by the overarching GEOMARS theory (Baker et al., 2002), which describes 7 major evolutional phases (from oldest to youngest): (1) shortly after accretion, Mars differentiated into a liquid metallic core, a mantle boundary layer (MBL) of high-pressure silicate mineral phases, upper mantle, magma ocean, thin komatiitic crust, and convecting steam atmosphere; (2) Mars cooled to condense its steam atmosphere and transformed its mode of mantle convection to plate tectonics. Subduction of water-rich oceanic crust initiated arc volcanism and transfered water, carbonates and sulfates to the mantle; (3) the core dynamo initiated and the associated magnetosphere led to conditions conducive to the stability of liquid water on the martian surface; (4) accretion of thickened, continental crust and subduction of hydrated oceanic crust to the mantle boundary layer and lower mantle of Mars; (5) the core dynamo stopped during Noachian heavy bombardment while plate tectonics continued; (6) initiation of the Tharsis superplume shortly following and/or contemporaneous with the termination of plate tectonism, which included the focused subduction of hydrated crust in the Tharsis region; and (7) superplume/stagnant-lid phase of martian planetary evolution with episodic phases of volcanism and water outflows. The 7th and latest stage may have persisted for probably more than 3.7 G.a. with prevailing cold and dry conditions punctuated by episodic endogenetic-driven activity that resulted in flooding, ponding in the northern plains, and short-lived hydrologic cycles (Baker et al., 1991, 2000; Dohm et al., 2001 a,c; Fairén et al., 2003). Orbital parameters such as changes in obliquity are also important factors concerning the surface modification of Mars (e.g. Touma and Wisdom, 1993; Head et al., 2003).

Astrobiological Implications

Assuming that the search for life is directly linked to the search for water, the study of the history of water on the terrestrial planets is appropriate for tracking planetary habitability. Distance from the Sun, planet size, composition, and variations in internal heat energy, which includes efficiency of release to the surface through space and time, are different for each of the terrestrial planets, affecting the planet's suitability for the origin and persistence of life. The likelihood of the presence of liquid water on the surfaces and in the subsurfaces of Mercury, Venus, Earth, Moon, and Mars in one or more episodes of the past, the nature of their climatic and atmospheric changes, and their current different atmospheres (except Mercury and Moon), provide exciting potential for fossilized and/or extant life in more than one world of the inner Solar System.

Mercury

Mercury's conditions are extremely harsh. The planet is closer to the Sun, and thus during its embryonic stage of formation probably received smaller amounts of volatile compounds than the rest of the terrestrial planets. Irwin and Schulze-Makuch (2001) assigned Mercury a Plausibility of Life (POL) rating of IV (low), based on the assessment that there is a reasonable inference of past conditions possibly suitable for the origin of life, prior to the development of conditions so harsh as to make its perseverance at present unlikely but conceivable in isolated habitats. These isolated habitats would be possible beneath discontinuous near-polar water ice deposits, discovered within permanently shadowed impact craters (Slade, 1992; Harmon and Slade, 1992; Ingersoll et al., 1992; Paige et al., 1992; Slade et al., 1992; Harmon et al., 1994; Harmon et al., 2001; Figure 1), where volatiles are stable for long periods of time at temperatures of about 112 K. The area covered by the ice deposits is estimated to be about 3×10^{10} m^2 (Slade et al., 1992). Here, temperature and pressure conditions would allow water to occur in the liquid state, assuming that the overlying ice deposits would prevent the liquid water from evaporation into space. The high temperatures on Mercury and its probable formation in a region of the Solar System where water is mostly unstable suggest that Mercury formed under anhydrous conditions, so the chance for life on Mercury would be expected being very low with only a slight chance for life remaining beneath the ice deposits.

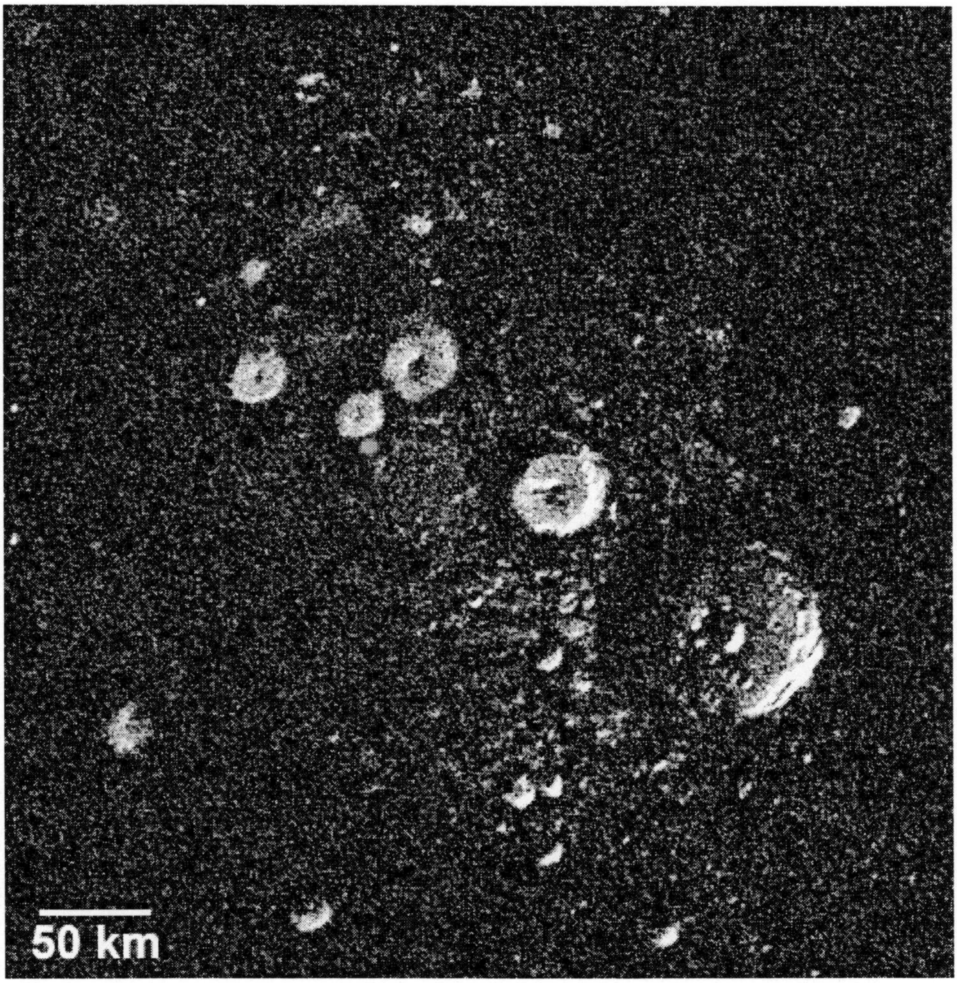

Figure 1A: Radar image of craters in the north pole region of Mercury. The radar signature of ice is denoted in the crater's bright floors. The red spot marks the North Pole position. Image captured in July 25-26, 1999, with the Arecibo radiotelescope.

Though the ice probably originated from relatively recent impact events, it may be possible that the remnant waters could archive early solar system information if the impacts and partial water infilling occurred relatively early in the development of Mercury (e.g., shortly after the formation and hardening of the crust). Considering the latter case, this water ice may hide some clues about the materials that protected and catalyzed the first biological synthesis on Earth, and therefore its analysis must be considered as a first-order objective in an astrobiological survey of the Solar System.

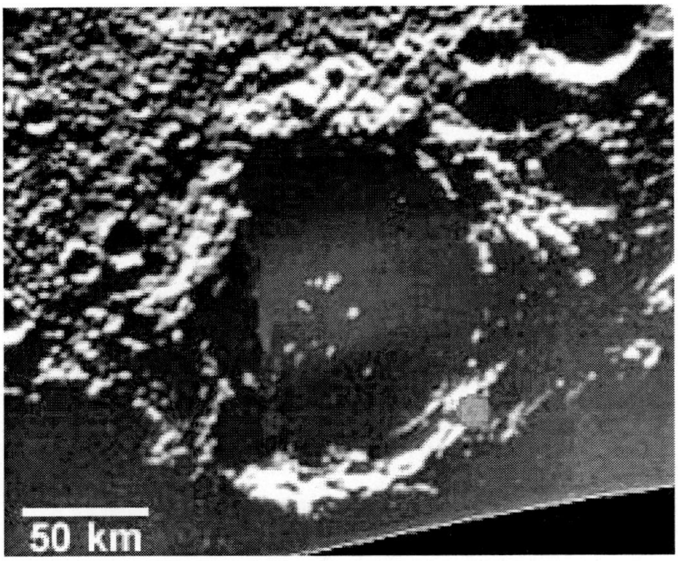

Figure 1B: Crater Chao-Meng-Fu, at the South Pole of Mercury as imaged by Mariner 10. The red spot marks the South Pole position. Arecibo results confirmed that the radar-reflective areas coincide with some craters at both poles of the planet.

Venus

If liquid water prevailed on the surface of Venus during its early history as it appears and also given Venus' dynamic endogenic-driven activity, it seems reasonable that life originated on Venus. Another possibility is that organisms could have been delivered by meteorites from early Earth or Mars, and found a suitable habitat on early Venus. Recent studies suggest that a significant amount of bacteria can survive space travel for millions of years, including lift off from the place of origin and re-entry into the planetary atmosphere of another planet (Davies, 1996; Horneck and Rettberg, 2002). Life could then have retreated into ecological niches progressively through directional selection during the period of greenhouse warming (Schulze-Makuch and Irwin, 2002). Whether an ecological habitat in the atmosphere and associated adaptation can occur is unclear. However, if the run-away moist greenhouse effect occurred gradually rather than catastrophically, adaptation to life in a thick and nutrient-rich atmosphere appears feasible. A number of arguments have been advanced previously in support of the possibility of microbial phototrophic life in the venusian atmosphere. These include the following observations: (1) The clouds of Venus are much larger, more continuous, and stable than the clouds on Earth; (2) the atmosphere is in chemical disequilibrium, with H_2 and O_2, and H_2S and SO_2 coexisting; (3) the lower cloud layer contains non-spherical particles comparable in size to microbes on Earth; (4) conditions in the clouds at 50 km in altitude are relatively benign, with temperatures of 300-350 K, pressure of 1 bar, and a pH of about 0; (5) the super-rotation of the atmosphere enhances the potential for photosynthetic reactions; (6) an unknown absorber of ultraviolet (UV) energy has been detected in the venusian atmosphere; (7) COS is present in the atmosphere, which on Earth is a strong indicator for biological activity; (8) CO is less abundant than expected under venusian atmospheric conditions and could be oxidized as a reactant in plausible metabolic pathways; and (9) while water is scarce on Venus, water vapor concentrations reach several hundred ppm in the lower cloud layer (Schulze-Makuch and Irwin, 2006).

Figure 2: Venus as it appears in reflected UV light. The dark areas on the image are caused by the ultraviolet absorber, which likely consists of various allotropes of sulfur. Putative microbes in the venusian atmosphere may be coated by the sulfur allotropes for UV protection or to convert UV radiation into visible light for photosynthetic reactions (e.g., Schulze-Makuch et al, 2004b). Source: NSSDC, http://nssdc.gsfc.nasa.gov/photo_gallery/photogallery-venus.html.

In addition, Cockell (1999) pointed out the availability of trace compounds in the venusian atmosphere, which are necessary for life. These include carbon, nitrogen, and phosphorus; for example, some unknown P-bearing substance has been detected in the lower cloud base (Andreichikov, 1987). The major problems for any possible microbial life in the venusian atmosphere would be the lack of water, the low pH, and the large amounts of UV radiation that the venusian atmosphere receives. The lack of water could be overcome by microbial organisms if they would have developed a mechanism by which to assimilate water vapor from hydrated sulfur compounds or from the atmosphere, similar to the assimilation of carbon from CO_2 by microbes in the atmosphere of Earth. Although a pH of about 0 in the lower cloud layer may seem extreme, Schleper et al. (1996) isolated microorganisms on Earth that thrive at this pH level. With regard to the high amount of UV radiation in the venusian atmosphere, cycloocta sulfur (S_8) could be used by microbes for protection (Schulze-Makuch et al., 2004b). S_8 has the capability of shielding organic macromolecules such as DNA and proteins at wavelengths most susceptible to UV damage. S_8 is also thermodynamically stable and does not react with sulfuric acid. In their attempt to identify the unknown UV-absorber in the venusian atmosphere (Figure 2), Hapke and Nelson (1975) concluded that S_8 along with some polymorphic sulfur provided the best match. If S_8 is used as a kind of "sunscreen", it would explain why only minute amounts of organic compounds were detected by spectral methods (e.g. Plummer, 1969). As Venus' geologic evolution has likely deleted any possible palaeontological register of ancient life on the surface, the presence of microorganisms thriving in the atmosphere is the only testing bench for the putative biological history of the planet. All these arguments highlight the necessity of considering a survey of the venusian atmosphere as a main objective in the astrobiological exploration of the Solar System

Earth

Earth is the only planet where we are certain that life does exist. Life on Earth is amazingly versatile and appears to thrive in every imaginable ecological niche as long as liquid water and a usable energy source are available, both on the surface, deep in the subsurface, and possibly within the atmosphere. It is found under oxygen-rich conditions near the surface, anaerobic conditions in subsurface soil, high-temperature and high pressure conditions in Earth's crust (Rothschild and Mancinelli, 2001), in the soils of Antarctic dry valleys (Mahaney et al., 2001), in super-cooled cloud water droplets (Sattler et al., 2001), in extremely acidic hot springs (Schleper et al, 1996), in caves (Hose et al., 2000), and many other environments. Since perhaps 4.4 Ga ago, Earth may have recorded the presence of liquid water on its surface (Wilde et al, 2001; Watson and Harrison, 2005), readily available organic compounds, and multiple energy sources. Evidence for life is found in the earliest rock record available (Schopf, 1993, 1999; McNaughton et al., 1999; Rasmussen, 2000; see the opposite point of view to Schopf's works expressed in Brasier et al., 2002), so indicating that life on Earth likely originated at least 3.9 G.a. ago (Mojzsis et al., 1996; van Zuilen et al., 2002). The origin may have occurred in high temperature environments, as indicated by phylogenetic analyses suggesting a hyperthermophillic chemolithoautotrophic organism as the last common or surviving prokaryote ancestor. As environmental conditions on Earth became relatively benign, life spread out rather quickly and colonized the whole planet during the Archaean. Later, with the development of Photosystem II by the ancestors of modern cyanobacteria, the atmosphere of Earth changed to include significant amounts of free oxygen, up to 0.01 of the present atmospheric level at the beginning of the Proterozoic (earlier times have been proposed for atmosphere oxygenation by Ohmoto et al. 1993, and Ohmoto, 1997). This meant a severe environmental crisis for most of the organisms at that time, which were living in an anaerobic environment. However, some of the organisms succeeded to adapt using poisonous free oxygen, which paved the way for complex life on Earth. In this scenario, the origin of the eukaryotic cell is hypothesized as a later development of living systems by a process of symbiogenesis between prokaryotes (Merezhkovsky, 1909; Margulis, 1981; Hedges et al., 2001). In addition, Earth's heterogeneous environment aided in the evolution of complex life (Schulze-Makuch and Irwin, 2004).

Today, the biosphere and the geosphere of the planet are so intrinsically connected that the evolution of life on Earth really has to be understood as a co-evolution of life and its environment. Organisms of Earth are directly responsible for an oxygen-rich atmosphere in chemical disequilibrium due to the constant production of both methane and oxygen. The myriad of organisms plus the presence of a hydrological cycle and an ocean and the geological process of worldwide-scale plate tectonics collectively act as a global thermostat of the planet, actively regulating atmospheric composition and climate (Lovelock and Margulis, 1974; Lovelock, 1989), and so keeping the environment in a temperate-climate suitable for living beings as it is maintained by themselves, via several mechanisms including the emission of methane in the oceans (Jahren, 2002), the control of the weathering rate feedback of carbonates (Volk, 1987), and the release of dimethyl sulfide for cloud condensation affecting planetary albedo (Charlson et al., 1987).

Moon

Due to its mode of origin, the Moon is extremely anhydrous. Given that no water or other volatiles ever formed an atmosphere or were present in the surface or subsurface, the Moon is not a likely candidate for the origin or persistence of life. However, the Moon may hold some evidence for the early evolution of life on Earth, since meteorite impacts on Earth may have delivered biogenic material to the surface of the Moon (Armstrong et al., 2002), and, if by chance this matter was covered by materials such as dust or ice, it would be protected from radiation decay. The ice at the south pole of the Moon (Figure 3) may also serve as a relatively accessible reservoir to analyze cometary ice and to provide water for possible future human habitation of the Moon.

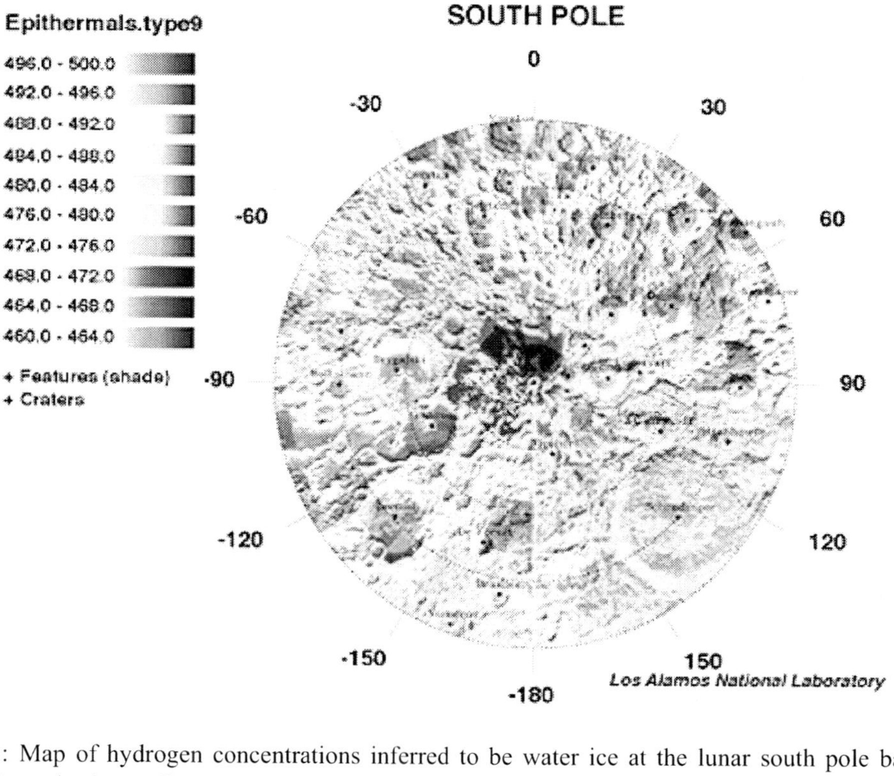

Figure 3: Map of hydrogen concentrations inferred to be water ice at the lunar south pole based on results form the Lunar Prospector neutron spectrometer (purple highest concentrations, green lowest concentrations; from NASA; http://lunar.arc.nasa.gov/printerready/science/newresults/ns.html).

Mars

All theories for the origin of life appear to be compatible with conditions on Mars during its early history (Davis and McKay, 1996). Therefore the origin of life on Mars is clearly plausible (Irwin and Schulze-Makuch, 2001), especially during Early into Middle Noachian, when conditions were quite similar to those on Earth (Baker et al., 2002). For example, during the period of heavy bombardment, impact crater bombardment competed with relatively high erosion rates, as evidenced from sedimentary sequences, which include exhumed impact craters and alluvial and fluvial deposits (Malin and Edgett, 2000b, 2001), observed in ancient geologic provinces of Mars such as Arabia Terra (Dohm et al., 2005). As sedimentary deposits form in topographic basins on Earth, so too it appears for Mars, as the Arabia Terra province displays characteristics that collectively point to a possible ancient impact basin/drainage system at least 1.5 times the size of the Hellas impact basin (Dohm et al., 2004a, 2006b). Another example of a possible ancient large (Europe-sized) basin is identified in the eastern part of the Tharsis magmatic complex, a proposed source of water for the circum-Chryse and northwestern slope valleys outflow channel systems, the northeast and northwest watersheds of Tharsis, respectively (Dohm et al., 2001a,c). Such putative ancient drainage basins, which were later subdued by burial, erosional, and endogenic- and exogenic-related processes (including thermal isostacy related to features such as Tharsis; e.g., Ruiz et al., 2004), may have resulted in long-lived aquifers of regional extent (also see Barlow et al., 2001). As such, these subterranean water worlds could have provided microbial communities with nutrients for long time periods, especially if shielded from later atmospheric conditions through sedimentary deposition (wind and water). In addition, ancient exogenic- and endogenic-related basement structures (e.g., Dohm et al., 2002b), buried impact crater populations (Rodriguez et al., 2005a), and subterranean voids such as caverns (Boston et al.; Rodriguez et al., 2005b), all played a significant role in the ponding and migration of water and other volatiles (e.g., subsurface migration, as well as conduits to reach the surface), and are thus considered prime candidates for life-containing habitats. As such, these arguments highlight the necessity for existing (e.g., MARSIS instrument of the Mars Express Mission) and future subsurface exploration missions to Mars and other volatile-rich planetary bodies..

Water stability during later periods (Hesperian to Amazonian) of martian history raises many questions about life's adaptation and survival during persisting periods of extremely cold and dry conditions, briefly punctuated by moderate-milder epochs (Baker et al., 1991). In fact, the martian substrate could have maintained large amounts of liquid H_2O, as noted above, between dry periods, especially where elevated geotherms have persisted such as in parts of the Tharsis and Elysium regions (Dohm, 2005; Dohm et al., 2005,2006a). If these locations are similar to magmatic/aqueous environments on Earth, then any biologic activity may occur between 0.5 and 1 km depth, as up to 10% of the Earth's biosphere lies 0.5 km beneath the surface (Parkes et al., 2000). These subsurface water bodies may be separated by extensive and thick cryosphere regions. An alternative life style for possible Martian life was suggested by Schulze-Makuch et al (2005b). Martian life may have evolved alternating cycles between dormant forms present close to the surface and active, proliferative forms in protected environments. Periodic liquid water on Mars could provide opportunities for biologic activity at the surface during the short-lived warmer and wetter periods in an otherwise cold and dry Martian environment.

Current life on Mars could exist in the fringe areas of the polar ice caps. Solar energy sources have been continuously available until the present. Analogue environments on Earth are the Antarctic dry valleys (e.g., Mahaney et al., 2001) and cool hydrothermal systems that exist in the Arctic and Antarctic. Cryptoendolithic cyanobacteria live in arid landscapes 0.5 to 5mm below the rock surface for shielding from the hostile environment and are still able to carry out photosynthesis (Friedmann, 1980, Friedmann et al., 1980, Friedmann, 1982; Vestal, 1988, Nienow et al., 1988; Friedmann et al., 1993). In addition, terrestrial anaerobic methanogens utilize H_2 and CO_2 in metabolism and release CH_4 as a by-product (Boston et al., 1992; Chapelle et al., 2002; Sleep et al., 2004). Conversely, multiple terrestrial archaeal groups that live in anoxic cold seep marine sediments oxidize the methane buried beneath the seafloor to CO_2, thus consuming up to 20% of the total methane flux to the atmosphere (Orphan et al., 2002). Similar strategies could account for the large sink of methane in the martian atmosphere. Moreover, methane hydrates are unstable on the surface under current ambient martian conditions, but stable in the subsurface at depths of larger than 15 m (> 140 kPa; Max and Clifford, 2000), and so methane on Mars may reside beneath the cryosphere. The recent discovery of methane in the martian atmosphere (Formisano et al., 2004) may point toward life activity among other alternatives such as volcanic activity or geochemical reactions.

Proposed Mission Designs and Scenarios

Future missions to our neighboring planets should be designed to test the expectations discussed above. NASA (Astrobiology Institute) termed the phrase "Follow the Water" in the search for extraterrestrial life. Water is so pervasive in the inner Solar System and appears to be so intrinsically related to life based on our experience with Earth that its presence seems to be essential for any life on our neighboring terrestrial planets. We suggest the following key points in exploration and mission design.

With regard to the inner planets Mercury, Venus, Earth, and Mars, we are dealing with planetary bodies that exhibit very different types of atmospheres and surface conditions that are challenging in terms of spacecraft landing and exploration. Traditional robotic exploration scenarios favor single lander/rover missions at the expense of mission redundancy, mission safety, and mission science return. Furthermore, single landers/rovers are restricted to small areas of exploration and are not likely to explore potentially hazardous but scientifically interesting terrains, including: (1) canyons (e.g. Valles Marineris on Mars); (2) mountain ranges (e.g. Thaumasia highlands on Mars, Isthar Terra on Venus); (3) sites of suspected magmatic-driven uplift (e.g. central part of Valles Marineris on Mars, and Maxwell Montes on Venus) and associated hydrothermal activity (e.g. Warrego Valles rise of Mars); (4) polar ice caps (e.g. Mars); (5) ice deposits within impact basins (e.g. Mercury and Moon); (6) volcanoes of diverse sizes and shapes (e.g. Venus and Mars); (7) putative ancient accreted terrains and associated volcanism (e.g. Mars); (8) regions indicating potential recent hydrologic activity such as spring-fed seeps (e.g. Mars); and (9) chaotic terrain (e.g. the circum-Chryse outflow channel system on Mars). All of the terrains listed above are particularly crucial for astrobiological-oriented exploration requiring a fundamental change in how we approach future planetary exploration (Fink et al., 2005a-c).

A "tier-scalable" mission paradigm (Fink et al., 2005c) integrates multi-tier (orbit⇔atmosphere⇔surface/subsurface) and multi-agent (orbiter⇔blimps⇔rovers/landers/drillers/sensors) hierarchical mission architectures, not only introducing, for the first time, mission redundancy and safety, but enabling distributed scientific exploration of prime locations on Mars and elsewhere, spanning larger surface areas than previously possible. This will allow for increased science return, and pave the way towards fully autonomous robotic missions.

The multi-layer hierarchical mission architecture overcomes the inherent challenge of geologic planetary surface exploration: airborne units (orbiters in conjunction with balloons/blimps) possess overhead perspectives at different length scales/resolutions that would provide guidance to ground-based agents (e.g., mobile rover units and a web of sensors). For example, one of the airborne units could detect a scientifically interesting feature (e.g., unique rock/rock outcrop or a transient geologic or hydrologic event such as a landslide or water seep, respectively) or a hazard or obstacle hidden from a rover camera by an intervening mountain. This airborne unit would acquire terrain images that would subsequently be processed by automated feature-extraction software packages, such as the "Automated Geologic Field Analyzer (AGFA)" (Fink et al., 2005d). The feature data would be used by science prioritization algorithms to choose potential targets for close examination by the rovers and for determining safe passages to their designated targets. At the respective targets, the rovers would conduct in-situ science experiments and thereby gather data that complement the remote sensing data, obtained by the airborne units. Future rovers should be equipped with a driller to explore the near-subsurface and collect samples to be returned to Earth or the International Space Station for analysis.

In our view, the highest priority for a sample return mission should be a potential hydrothermal site on Mars (e.g. Dohm et al, 2004b; Schulze-Makuch et al., 2004a). The Northwestern Slope Valleys (NSV) region, for example, is a prime candidate site for such future science-driven Mars exploration because it records Noachian to Amazonian Tharsis development in a region that encapsulates (1) a diverse and temporally extensive stratigraphic record, (2) at least three distinct paleohydrologic regimes, (3) gargantuan structurally-controlled flood valleys that generally correspond with gravity and magnetic anomalies, possibly marking ancient magnetized rock materials exposed by fluvial activity, (4) water enrichment, as indicated by Mars Odyssey and impact crater analyses, (5) long-lived magma and ground water/ice interactions that could be favorable for the development and sustenance of life, and (6) potential paleosol development (Dohm et al., 2004b). As such, there is a high probability that this region could yield significant geologic, climatic, and astrobiological information that would revolutionize our understanding of Mars. Interestingly, this region also indicates elevated chlorine (e.g., Boynton et al., 2004) and methane (e.g., Formisano et al., 2004) abundances. Among several possible explanations (Keller et al., 2005), the GRS-based elevated chlorine signature could indicate aqueous processes in the past or present, which includes magma-water interactions (Dohm et al., 2004b, 2005, 2006a; Dohm, 2005). Another region where water may be present at or near the surface is within Valles Marineris due to its geomorphic expression (e.g. Komatsu et al., 2004), which includes dendritic valleys (Magnold et al., 2004), and relatively low topography. At a site difficult to access, such as inside the Valles Marineris, a sample-return mission is feasible (Fink et al., 2006; Schulze-Makuch et al., 2006) using the multi-tier and multi-agent hierarchical mission architecture (Fink et al., 2005a-c) with varying degrees of mission operation automation/autonomy. For example, for a Mars sample return mass of 100 g at an altitude of 0 m, a blimp/balloon would need to be deployed with a radius > 8 m (for details see Schulze-Makuch et al., 2005a). The blimp/balloon would have the ability to descend into valley and canyon systems and collect samples of potential astrobiological relevance.

Evidence has been presented that the atmosphere of a planet may also serve as a primary habitat for microbial life (Dimmick, 1979; Sattler et al., 2001) as discussed before. These findings and the more benign conditions in the venusian atmosphere compared to Earth's atmosphere has led several authors to speculate on the possibility of microbes in the venusian atmosphere. (e.g. Sagan, 1961; Grinspoon, 1997; Schulze-Makuch and Irwin, 2002, 2004, 2006; Schulze-Makuch et al., 2004b). The atmosphere of Earth serves as a transient habitat for both microbes and even macroscopic life, and experiments should be conducted to test whether it also serves as a permanent habitat for microbial life in spite of harsh conditions. Compared to Earth's atmosphere, the atmospheric conditions on Venus are much more benign with respect to temperature, pressure, and particle residence times, and, thus microbial life in the venusian atmosphere is a distinct possibility. Therefore, a sample return mission to the venusian lower cloud layer where the microbes may reside should be considered. Schulze-Makuch et al. (2002) considered various mission options and concluded that a sample return mission involving a Parachute Drop – Balloon Floatation Mission, designed to return astrobiologically relevant material for analysis to the International Space Station, would be the most preferable option. Even if the promising mode 3 particles in the venusian atmosphere were not of biological nature, a sample return mission would tremendously increase our knowledge about the composition and dynamics of the atmosphere. The relative ease of reaching Venus and returning to Earth and the availability of appropriate existing technology make such a mission feasible in short-term. For example, for a Venus sample return mass of 100 g at an altitude of about 50 km, a blimp/balloon would need to be deployed with a radius > 0.4 m (for details see Schulze-Makuch et al., 2005a). The blimp/balloon would have the capability of hovering at an altitude of about 50 km or to descend to lower altitudes and to collect samples of cloud particles with aerogels similar to STARDUST and GENESIS (e.g. Knollenberg and Hunten (1980) report a cloud particle density of 10-100 particles per cm^3 at about 50 km altitude). These cloud particles, once obtained by the blimp/balloon, could be transported into orbit, and from there to the International Space Station or Earth for analysis. As technology for a sample return mission to Venus exists, the mission could be undertaken at present.

Conclusions

Three of the four terrestrial planets of our Solar System (Venus, Earth, and Mars) may have had liquid water bodies on their surface early in their history. Any liquid water on the Venusian surface evaporated, but significant amounts of water remain in the atmosphere. On Earth liquid water is abundant and stable at the surface, as well as in the subsurface and the atmosphere. Due to the thin martian atmosphere today, liquid water in most locations is not stable at the surface of Mars. Large liquid and frozen water reservoirs, however, are likely to exist in the martian subsurface. The search for life on the inner terrestrial planets should follow the presence of liquid water. Clearly, the geological history of the planetary body in question is key to astrobiological considerations. Life is present in the subsurface of Earth, on its surface and, at least transiently, in its atmosphere. It may be present in the venusian atmosphere and the martian subsurface. In addition, ices on Mercury and the Moon should be explored for remnant biogenic material from the early evolution of life on Earth (and elsewhere). Sample return missions and missions that use multi-tier and multi-agent hierarchical architectures with scalable degrees of mission operation autonomy, which will allow optimal reconnaissance of planetary environments, are most suitable for exploring these potential habitats.

References

[1] Abe, Y., Matsui, T., 1988. Evolution of an impact-generated H_2O-CO_2 atmosphere and formation of a hot proto-ocean on Earth. *Journal of the Atmospheric Sciences* 45, 3081-3101.

[2] Andreichikov, B.M., 1987. X-ray radiometric analysis of the cloud aerosol of Venus by the Vega 1 and 2 probes. *Kosmich. Issled.* 25, 721-736 (in Russian)

[3] Armstrong, J., Wells, L., Gonzalez, G., 2002. Rummaging through Earth's attic for remains of ancient life. *Icarus* 160, 183-196.

[4] Baker, V.R., 2001. Water and the martian landscape. *Nature* 412, 228-236.

[5] Baker, V.R., Strom, R.G., Gulick, V.C., Kargel, J.S., Komatsu, G., Kale, V.S., 1991. Ancient oceans and ice sheets and the hydrological cycle on Mars. *Nature* 352,589-594.

[6] Baker, V.R., Strom, R.G., Dohm, J.M., Gulick, V.C., Kargel, J.S., Komatsu, G., Ori, G., Rice J.W., Jr., 2000. Mars' Oceanus Borealis, ancient glaciers, and the Megaoutflow hypothesis (abstract), *Lunar Planet. Sci.* [CD-ROM], XXXI, 1863.

[7] Baker, V.R., Maruyama, S., Dohm, J.M., 2002. A theory of early plate tectonics and subsequent long-term superplume activity on Mars. *Electronic Geosciences* 7, (http://194.94.42.12/licensed_materials/10069/free/conferen/superplu/index.html).

[8] Barlow, N.G., Koroshetz, J., Dohm, J.M., 2001. Variations in the onset diameter for Martian layered ejecta morphologies and their implications for subsurface volatile reservoirs. *Geophys. Res. Lett.* 28, 3095-3098.

[9] Basilevsky, A.T., Head, J.W., 1998. The geologic history of Venus: A stratigraphic view. *J. Geophys. Res.* 103, 8531-8544.

[10] Basilevsky, A.T., Head, J.W., 2000. Geologic units on Venus: evidence for their global correlation. *Planetary and Space Science* 48, 75-111.

[11] Basilevsky, A.T., Head, J.W., 2002. Venus: Timing and rates of geologic activity. *Geology* 30, 1015-1018.

[12] Bekker, A., Holland, H.D., Wang, P.-L., Rumble III, D., Stein, H.J., Hannah, J.L., Coetzee, L.L., Beukes, 2004. Dating the rise of atmospheric oxygen. *Nature* 427, 117-120.

[13] Bilotti, F., Suppe, J., 1999. The global distribution of wrinkle ridges on Venus. *Icarus* 139, 137-157.

[14] Boston, P.J., Ivanov, M.V., McKay, C.P., 1992. On the possibility of chemosynthetic ecosystems in subsurface habitats on Mars. *Icarus* 95, 300-308.

[15] Bougher, S. W., Hunten, D. M., Phillips, R. J., (Eds.), and 100 collaborating authors. 1997. *Venus*, Univ. of Ariz. Press, Tucson, pp. 1362.

[16] Boynton, W.V., Feldman, W.C., Squyres, S.W., Prettyman, T.H., Brückner, J., Evans, L.G., Reedy, R.C., Starr, R., Arnold, J.R., Drake, D.M., Englert, P.A.J., Metzger, A.E., Mitrofanov, I., Trombka, J.I., d'Uston, C., Wänke, H., Gasnault, O., Hamara, D.K., Janes, D.M., Marcialis, R.L., Maurice, S., Mikheeva, I., Taylor, G.J., Tokar, R., Shinohara, C., 2002. Distribution of hydrogen in the near surface of Mars: evidence for subsurface ice deposits. *Science* 297, 81-85.

[17] Boynton, W., Janes, D., Kerry, K., Kim, K., Reedy, R., Evans, L., Starr, R., Drake, D., Taylor, J.,

[18] Wänke, H., d'Uston, C., 2004. The distribution of non-volatile elements on Mars: Mars Odyssey GRS results. *Lunar Planet. Sci. Conf.* [CD-ROM], XXXV, abstract 1950.

[19] Brasier, M.D., Green, R.O., Jephcoat, A.P., Kleppe, A.K., Van Kranendonk, M.J., Lindsay, J.F., Steele, A., Grassineau, N.V., 2002. Questioning the evidence for Earth's oldest fossils. *Nature* 416, 76-81.

[20] Bridges, J.C., Grady, M.M., 1999. A halite-siderite-anhydrite-chlorapatite assemblage in Nakhla: mineralogical evidence for evaporites on Mars. *Meteoritics and Planetary Sci.* 34, 407-415.

[21] Cabrol, N.A., Grin, E.A., 2001. The evolution of lacustrine environments on Mars: is Mars only hydrologically dormant? *Icarus* 149, 291-328.

[22] Cameron, A.G.W., Benz, W., 1991. The origin of the Moon and the single impact hypothesis IV. *Icarus* 92, 204-216.

[23] Canfield, D.E., Habicht, K.S., Thamdrup, B., 2000. The Archaean sulfur cycle and the early history of atmospheric oxygen. *Science* 288, 658-661.

[24] Carr, M.H. 1996. Water on Mars. Oxford University Press, Oxford, UK.

[25] Carr, M.H., 1999. Retention of an atmosphere on early Mars. *J. Geophys. Res.* 104, 21,897 – 21,909.

[26] Catling, D.C. 2004. Personal communication, Assistant Professor, University of Washington.

[27] Chapelle, F.H., O'Neill, K., Bradley, P.M., Methe, B.A., Ciufo, S.A., Knobel, L.L., Lovley, D.R., 2002. A hydrogen-based subsurface community dominated by methanogens. *Nature* 415, 312-315.

[28] Charlson, R.J., Lovelock, J., Andreae, M.O., Warren, S.G., 1987. Oceanic phytoplankton, atmospheric sulphur, cloud albedo and climate. *Nature* 326, 655-661

[29] Chassefiere, E., 1996. Hydrodynamic escape of hydrogen from a hot water-rich atmosphere: the case of Venus. *J. Geophys. Res.* 101, 26,039-26,056.

[30] Chassefiere, E., 1997a. Hydrodynamic escape of oxygen from primitive atmospheres: applications to the cases of Venus and Mars. *Icarus* 124, 537-552.

[31] Chassefiere, E., 1997b. Loss of water on the young Venus: the effect of a strong primitive solar wind. *Icarus* 126, 229-232.

[32] Christensen, P.R., Morris, R.V., Lane, M.D., Banfield, J.L., Malin, M.C., 2001. Global mapping of martian hematite mineral deposits: remnants of water-driven processes on early Mars. *J. Geophys. Res.* 106, 23,873-23,885.

[33] Cockell, C.S., 1999. Life on Venus. Planetary Space and Science 47, 1487-1501.

[34] Coffin, M.A., Eldholm, O., 1994. Large igneous provinces: crustal structure, dimensions, and external consequences. *Review Geophysics* 32, 11-36.

[35] Crider, D, Killen, R.M. 2005. Burial rate of mercury's polar volatile deposits, *Geophy. Res. Letters* 32, L12201.

[36] Davies, P.C.W., 1996. *The transfer of viable microorganisms between planets.* In Brock, G.R., Goode, J.A. (Eds.) *Evolution of Hydrothermal Ecosystems on Earth (and Mars?), Ciba Foundation Symposium* 202, John Wiley, Chichester, pp. 304-317.

[37] Davis, W.L., McKay, C.P., 1996. Origins of life: A comparison of theories and applications to Mars. *Origins Life and Evol. Biosph* 26, 61-73.

[38] Dimmick, R.L., Wolochow, H., Chatigny, M.A., 1979. Evidence for more than one division of bacteria within airborne particles. *Applied Environmental Microbiology* 38, 642-643.

[39] Dohm, J.M., 2005. The Tharsis and Elysium corridor: a marker for an internally active Mars? *Conference Proceedings of the Goldschmidt Conference*, 20-25 May 2005, Moscow, Idaho, in press.

[40] Dohm, J.M., Ferris, J.C., Baker, V.R., Anderson, R.C., Hare, T.M., Strom, R.G., Barlow, N.G., Tanaka, K.L., Klemaszewski, J.E., Scott, D.H., 2001a. Ancient drainage basin of the Tharsis region, Mars: Potential source for outflow channel systems and putative oceans or paleolakes, *J. Geophys. Res.* 106, 32,943-32,958.

[41] Dohm, J.M., Tanaka, K.L., Hare, T.M. 2001b. Geologic map of the Thaumasia region of Mars. *USGS Misc. Inv. Ser. Map* I-2650, scale 1:5,000,000.

[42] Dohm, J.M., Anderson, R.C., Baker, V.R., Ferris, J.C., Rudd, L.P., Hare, T.M., Rice Jr., J.W., Casavant, R.R., Strom, R.G., Zimbelman, J.R., Scott, D.H., 2001c. Latent outflow activity for western Tharsis, Mars: significant flood record exposed. *J. Geophys. Res.* 106, 12,301-12,314.

[43] Dohm, J.M., Tanaka, K.L., 2002a. Geologic mapping of the Metis Regio Quadrangle (V-6), Venus. in *Abstracts of the Annual Planetary Geologic Mappers Meeting, June 20-21, 2002, Tempe, Arizona, USGS Open-File Report 02-412.*

[44] Dohm, J.M., Maruyama, S., Baker, V.R., Anderson, R.C., Ferris, J.C., Hare, T.M., 2002b. Plate tectonism on early Mars: Diverse geological and geophysical evidence. *Lunar Planet. Sci. Conf.*, XXXIII, #1639 (abstract) [CD-ROM].

[45] Dohm, J.M., Barlow, N.G., Williams, J.P., Baker, V.R., Anderson, R.C., Boynton W.V., Fairén, A.G., Hare, T.M., 2004a. Ancient giant basin/aquifer system in the Arabia region, *Mars. Lunar Planet. Sci. Conf.*, XXXV, #1209 (abstract) [CD-ROM].

[46] Dohm, J.M., Ferris, J.C., Barlow, N.G., Baker, V.R., Mahaney, W.C., Anderson, R.C., Hare, T.M., 2004b. The Northwestern Slope Valleys (NSVs) region, Mars: A prime candidate site for the future exploration of Mars. *Planetary and Space Science* 52, 189-198.

[47] Dohm, J.M., Kerry, K., Keller, J.M., Baker, V.R., Maruyama, S., Anderson, R.C., Ferris, J.C., Hare, T.M., 2005. Mars geological province designations for the interpretation of GRS data. *Lunar Planet. Sci. Conf.*, XXXVI, #1567 (abstract) [CD-ROM].

[48] Dohm, J.M., Anderson, R.C., Baker, V.R., Barlow, N.G., Boynton, W.V., Davies, A.G., Fairén, A.G., Ferris, J.C., Glamoclija, M., Keller, J.M., Kerry, K., Marinangeli, L., Miyamoto, H., Ori, G.G., Rodríguez, J.A.P., Schulze-Makuch, D., Strom, R.G., Taylor, G.J., de Pablo Hdez, M.A., Karunatillake, S., and the GRS Team, (2006a) Tharsis/Elysium corridor: a marker for an internally active Mars? to be submitted in a special GRS edition of the *Journal of Geophysical Research.*

[49] Dohm, J.M., Barlow, N.G., Williams, J.P., Ferris, J.C., Miyamoto, H., Baker, V.R., Strom, R.G., Taylor, G.J., Fairén, A.G., Boynton, W.V., Keller, J.M., Kerry, K., Janes,

B., Rodríguez, J.A.P., Hare, T.M., Anderson, R.C., (2006b) Possible ancient giant impact basin/aquifer system in the Arabia Terra province, Mars. to be submitted in a special GRS edition of the *Journal of Geophysical Research*.

[50] Dohm, J.M., Tanaka, K.L., Skinner, J.A., (2006c) Geologic map of the Metis Mons Quadrangle (V-6), Venus. *USGS Misc. Inv. Ser. Map*, scale 1:5,000,000, in review.

[51] Donahue, T.M., 1995. Evolution of water reservoirs on Mars from D/H ratios in the atmosphere and crust. *Nature* 374, 432-434.

[52] Donahue, T.M., Hartle, E., 1992. Solar cycle variations in H+ and D+ densities in the Venus ionosphere: implications for escape. *Geophysical Research Letters* 12, 2449-2452.

[53] Donahue, T.M., Hoffman, J.H., Hodges, R.R., Watson, A.J., 1982. Venus was wet: a measurement of the ratio of Deuterium to Hydrogen. *Science* 216, 630-633

[54] Dreibus, G., Wänke, H., 1985. Mars, a volatile-rich planet. *Meteoritics* 20, 367-381.

[55] Fairén, A.G., Dohm J.M., 2004. Age and origin of the lowlands of Mars, *Icarus* 168, 277-284.

[56] Fairén, A.G., Dohm, J.M., Baker, V.R., de Pablo, M.A., Ruiz, J., Ferris, J., Anderson, R.C., 2003. Episodic flood inundations of the northern plains of Mars. *Icarus* 165, 53-67.

[57] Fairén, A.G., Dohm, J.M., Öner, T., Ruiz, J., Rodríguez, A.P., Schulze-Makuch, D., Ormö, J., McKay, C.P., Baker V.R., Amils, R., 2004a. Updating the evidence of oceans on early Mars. 2nd Early Mars Conference, Jackson, Wyoming, 11-15 October 2004.

[58] Fairén, A.G., Dohm, J.M., Rodríguez, A.P., Schulze-Makuch, D., Baker V.R., Amils, R., 2004b. A model for the hydrogeological evolution and related prime candidate sites for the astrobiological exploration of Mars. *4th European Workshop on Exo/Astrobiology*, London, UK, November 2004.

[59] Fairén, A.G., Fernández-Remolar, D., Dohm, J. M., Baker, V. R., Amils, R. , 2004c. Inhibition of carbonate synthesis in acidic oceans on early Mars. *Nature* 431, 423-426.

[60] Farquhar, J., Bao, M., Thiemens, M., 2000. Atmospheric influence of Earth's earliest sulfur cycle, *Science* 289, 756-758.

[61] Feldman, W.C., Maurice, S., Binder, A.B., Barraclough, B.L., Elphic, R.C., Lawrence, D.J. , 1998. Fluxes of fast and epithermal neutrons from Lunar Prospector: Evidence for water ice at the lunar poles, *Science* 281, 1496-1500.

[62] Ferris, J.C., Dohm, J.M., Baker, V.R., Maddock, T., 2002. Dark slope streaks on Mars: are aqueous processes involved? *Geophys. Res. Lett.* 29, 10.1029/2002GL014936.

[63] Fink, W., Dohm, J.M., Tarbell, M.A., Hare, T.M., and Baker, V.R., 2005a. Next-Generation Robotic Planetary Surface/Subsurface Reconnaissance Missions: A Paradigm Shift [abstract 1977]. In 36th *Lunar and Planetary Science Conference Abstracts* [CD-ROM], Lunar and Planetary Institute, Houston.

[64] Fink, W., Dohm, J.M., Tarbell, M.A., Hare, T.M., and Baker, V.R., 2005b. Next-generation robotic planetary reconnaissance missions: a paradigm shift. In *Abstracts of the 15th Annual V.M. Goldschmidt Conference, Moscow, Idaho. Geochimica et Cosmochimica Acta*, Volume 69, Number 10S, A533.

[65] Fink, W., Dohm, J.M., Tarbell, M.A., Hare, T.M., Baker, V.R., 2005c. Next-Generation Robotic Planetary Reconnaissance Missions: A Paradigm Shift. *Planetary and Space Science* 53, 1419-1426.

[66] Fink, W., Datta, A., and Baker, V.R., 2005d. AGFA: Automated Geologic Field Analyzer. In *Abstracts of the 15th Annual V.M. Goldschmidt Conference, Moscow, Idaho. Geochimica et Cosmochimica Acta*, Volume 69, Number 10S, A535.

[67] Fink, W., Dohm, J.M., Tarbell, M.A., Hare, T.M., Baker, V.R., Schulze-Makuch, D., Furfaro, R., Fairén, A.G., Ferré, T.P.A., Miyamoto, H., Komatsu, G., Mahaney, W.C., 2006. Multi-tier Multi-agent Autonomous Robotic Planetary Surface/Subsurface Reconnaissance For Life [abstract 1433]. In 37th *Lunar and Planetary Science Conference Abstracts* [CD-ROM], Lunar and Planetary Institute, Houston.

[68] Friedmann, E.I., 1980. Endolithic microbial life in hot and cold deserts. *Origins of Life* 10, 223-235.

[69] Friedmann, E.I., 1982. Endolithic microorganisms in the Antarctic cold desert. *Science* 215, 1045-1053.

[70] Friedmann, E.I., Kappen, L., Meyer, M.A., Nienow, J.A., 1993. Long-term productivity in the cryptoendolithic microbial community of the Ross Desert, *Antarctica. Microbial Ecology* 25, 5169.

[71] Friedmann, E.I., LaRock, P.A., Brunson, J.O., 1980. Adenosine triphosphate (ATP), chlorophyll, and organic nitrogen in endolithic microbial communities and adjacent soils in the dry valleys of southern Victoria land. *Antarctic Journal of the US* 15, 164166.

[72] Fukao, Y., Maruyama, S., Obayashi, M., Inoue, H., 1994. Geologic implication of the whole mantle P-wave tomography, *J. Geol. Soc. of Japan*, 100, 4-23.

[73] Godwin, R. 2000. Mars: The NASA Mission Reports, Vol. 1. *Apogee Books Space Series*.

[74] Grinspoon, D.H. 1997 *Venus Revealed: a New Look Below the Clouds of Our Mysterious Twin Planet*. Perseus Publishing, Cambridge, Massachusetts.

[75] Grinspoon, D.H. 2003. *Personal communication*, Research Scientist at Southwest Research Institute, Boulder, Colorado.

[76] Haberle, R.M., 1998. Early Mars climate models. *Journal of Geophysical Research* 103, E12, 28,467-28,479.

[77] Habicht, K.S., Gade, M., Thamdrup, B., Berg, P., Canfield, D.E., 2002. Calibration of sulfate levels in the Archean ocean. *Science* 298, 2372-2374.

[78] Hapke, B., Nelson, R., 1975. Evidence for an elemental sulfur component of the clouds from Venus spectrophotometry. *Journal of Atmospheric Sciences* 32, 1212-1218.

[79] Harmon, J.K., Slade, M.A., 1992. Radar mapping of Mercury: full-disk images and polar anomalies. *Science* 258, 640-643.

[80] Harmon, J.K., Slade, M.A., Velez, R.A., Crespo, A., Dryer, M.J., Johnson, J.M., 1994. Radar mapping of Mercury's polar anomalies. *Nature* 369, 213-215.

[81] Harmon, J.K., Perillat, P.J., Slade, M.A., 2001. High-resolution radar Imaging of Mercury's north pole. *Icarus* 149, 1-15.

[82] Hartmann, W.K., Neukum, G., 2001. Cratering chronology and the evolution of Mars. *Space Sci. Rev.* 96, 165-194.

[83] Hartmann, W.K., Phillips, R.J., Taylor, G.J., 1986 (Eds). *Origin of the Moon*, Lunar and Planetary Institute.

[84] Head, J.W., Wilson, L., 1992. Magma reservoirs and neutral buoyancy zones on Venus: implications for the formation and the evolution of volcanic landforms. *J. Geophys. Res.* 97, 3877-3903.

[85] Head, J.W., Crumpler, L.S., Aubele, J.C., Guest, J.E., Saunders, R.S., 1992. Venus volcanism: classification of volcanic features and structures, association, and global distribution from Magellan data. *J. Geophys. Res.* 97, 13,153-13,197.

[86] Head, J.W., Mustard, J.F., Kreslavsky, M.A., Milliken, R.E., Marchant, D.R., 2003. Recent ice ages on Mars. *Nature* 426, 797-802.

[87] Hedges, S.B., Chen, H., Kumar, S., Wang, D.Y.C., Thompson, A., Watanabe, H., 2001. A genomic timescale for the origin of eukaryotes. *BMC Evol. Biol.* 1, 4-13.

[88] Hoffman, N., 2000. White Mars: A new model for Mars' surface and atmosphere based on CO_2. *Icarus*, 146, 326-343.

[89] Holland, H.D., 1999. When did the Earth's atmosphere become oxic? a reply. *Geochem. News* 100, 20-23.

[90] Horneck, G., Rettberg, P., 2002. A thin meteorite layer protects bacterial spores in space. Astrobiology Science Conference, NASA Ames Research Center, April 7-11, 2002, *Abstract collection*, p. 23

[91] Hose, L.D., Palmer, A.N., Palmer, M.V., Northup, D.E., Boston, P.J., DuChene, H.R., 2000. Microbiology and geochemistry in a hydrogen-sulphide-rich karst environment. *Chemical Geology* 169, 399-423.

[92] Hunten, D. M., 1993. Atmospheric evolution of the terrestrial planets. *Science* 259, 915-920.

[93] Ingersoll, A.P., 1969. The runaway greenhouse: a history of water on Venus. *J. Atmos. Sci.* 26, 1191-1198.

[94] Ingersoll, A.P., Svitek, T., Murray, B.C., 1992. Stability of polar frosts in spherical bowl-shaped craters on the Moon, Mercury, and Mars. *Icarus* 100, 40-47.

[95] Irwin, L.N., Schulze-Makuch, D., 2001. Assessing the plausibility of life on other worlds. *Astrobiology* 1, 143-160.

[96] Ivanov, M.A., Head, J.W., 2001. Geology of Venus: mapping of a global traverse at 30°N latitude. *J. Geophys. Res.* 106, 17,515-17,566.

[97] Jahren, A.H., 2002. The biogeochemical consequences of the mid-Cretaceous superplume. *Journal of Geodynamics* 34, 177-191.

[98] Jakosky, B.M., Phillips, R.J., 2001. Mars' volatile and climate history. *Nature* 412, 237-244.

[99] Kasting, J. F., 1988. Runaway and moist greenhouse atmospheres and the evolution of Earth and Venus. *Icarus* 74, 472-494.

[100] Kasting, J.F., 1989. How Venus lost its oceans. *Oceanus* (Woods Hole) 32, 54-57.

[101] Kasting, J.F., Pollack, J.B., 1983. Loss of water from Venus I. hydrodynamic escape of hydrogen. *Icarus* 53, 479-508.

[102] Keilasam, L.N., 1979. Plateau uplift in peninsular India. *Tectonophysics* 61, 243-269.

[103] Keller, J.M., Boynton, W.V., Evans, L.G., Starr, R., Reedy, R.C., Taylor, J., Squyres, S.W., Baker, V.R., Dohm, J.M., Hamara, D., Janes, D., Kerry, K. and the GRS Team, 2005. Global distribution of chlorine at the Martian near surface measured by Mars GRS, to be submitted in a special GRS edition of the *Journal of Geophysical Research.*

[104] Kieffer, H.H., Jakosky, B.M., Snyder, C.W., Mathews, M.S., (Eds.), and 114 collaborating authors. 1992. *Mars*, Univ. of Ariz. Press, Tucson, pp. 1498.

[105] Klein, C., Beukes, N.J., Schopf, J.W., 1987. Filamentous microfossils from the Early Proterozoic Transvaal Supergroup: their morphology, significance, and paleoenvironmental setting. *Precam. Res.* 36, 81-94.

[106] Knollenberg R.G., Hunten D.M., 1980. The Microphysics of the Clouds of Venus: Results of the Pioneer Venus Particle Size Spectrometer Experiment. *Journal of Geophysical Research* 85, 8038-8058.
[107] Komatsu, G., Dohm, J.M., Hare, T.M., 2004. Hydrogeologic processes of large-scale tectono-magmatic complexes in Mongolia-Southern Siberia and Mars, *Geology* 32, 325-328.
[108] Lammer, H., Lichtenegger, H.I.M., Kolb, C., Ribas, I., Guinan, E.F., Bauer, S.J., 2003. Loss of water from Mars: implications for the oxidation of the soil. *Icarus* 165, 9-25.
[109] Levin, G.V. and Straat, P.A., 1981. A Search for a Nonbiological Explanation of the Viking Labeled Release Life Detection Experiment. *Icarus* 45, 494-516.
[110] Li, Z.X., Li, X.H., Kinny, P.D., Wang, J., Zhang, S., Zhou, H., 2003. Geochronology of Neoproterozoic syn-rift magmatism in the Yangtze Craton, South China and correlations with other continents: evidence for a mantle superplume that broke up Rodinia. *Precambrian Res.* 122, 85-109.
[111] López, I., Oyarzun, R., Márquez, Á., Doblas-Teyes, F., Laurrieta, A., 1998. Progressive build up of CO_2 in the atmosphere of Venus through multiple volcanic resurfacing events. *Earth, Moon, Planets* 81, 187-192.
[112] Lovelock, J.E. 1989. *The Ages of Gaia*. Oxford Univ. Press.
[113] Lovelock, J.E., Margulis, L., 1974. Atmospheric homeostasis by and for the biosphere: the Gaia hypothesis. *Tellus* 26, 2-9.
[114] Lundin, R., 2001. Erosion by the solar wind. *Science* 291, 1909.
[115] Lundin, R., Barabash, S., Andersson, H., Holmstroem, M., Grigoriev, A., Yamauchi, M., Sauvaud, J.-A., Fedorov, A., Budnik, E., Thocaven, J.-J., Winningham, D., Frahm, R., Scherrer, J., Sharber, J., Asamura, K., Hayakawa, H., Coates, A., Linder, D.R., Curtis, C., Hsieh, K.C., Sandel, B.R., Grande, M., Carter, M., Reading, D.H., Koskinen, H., Kallio, E., Riihela, P., Schmidt, W., Saeles, T., Kozyra, J., Krupp, N., Woch, J., Luhmann, J., McKenna-Lawler, S., Cerulli-Irelli, R., Orsini, S., Maggi, M., Muea, A., Milillo, A., Roelof, E., Williams, D., Livi, S., Brandt, P., Wurz, P., Bochsler, P., 2004. Solar wind-induced atmospheric erosion at Mars: first results from ASPERA-3 on Mars Express. *Science* 305, 1933-1936.
[116] Lunine, J.I., Chambers, J., Morbidelli, A., Leshin, L.A., 2003. The origin of water on Mars. *Icarus* 165, 1-8.
[117] Magnold, N., Quantin, C., Ansan, V., Delacourt, C., Allemand, P., 2004. Evidence for precipitation on Mars from dendritic valleys in the Valles Marineris area. *Science* 305, 1931-1934.
[118] Mahaney, W.C., Dohm, J.M., Baker, V.R., Newsom, H.E., Malloch, D., Hancock, R.G.V., Campbell, Iain, Sheppard, D., Milner, M.W., 2001. Morphogenesis of Antarctic paleosols: Martian analog, *Icarus* 154, 113-130.
[119] Malin, M.C., Edgett, K.S., 2000a. Evidence for recent groundwater seepage and surface runoff on Mars. *Science* 288, 2330-2335.
[120] Malin, M.C., Edgett, K.S., 2000b. Sedimentary rocks of early Mars, *Science* 290, 1927-1937.
[121] Malin, M.C., Edgett, K.S., 2001. Mars Global Surveyor Mars Orbiter Camera: Interplanetary cruise through primary mission, *J. Geophys. Res.* 106(E10), 23,429-23,570.

[122] Malin, M.C., Edgett, K.S., 2003. Evidence for persistent flow and aqueous sedimentation on early Mars. *Science* 302, 1931-1934.

[123] Margulis, L. 1981. Symbiosis in Cell Evolution. Freeman and Co., San Francisco.

[124] Marov, M.Y., Grinspoon, D.H. 1998. *The Planet Venus*. Yale University Press, New Haven, USA.

[125] Maruyama, S., 1994. Plume tectonics, J. *Geol. Soc. of Japan,* 100, 24-49.

[126] Matsui, T., Tajika, E., 1991. Early environmental evolution of Venus. *Proceedings of the Lunar and Planetary Science Conference*, vol.22, 863-864.

[127] Max, M.D., Clifford, S.M., 2000. The state, potential distribution, and biological implications of methane in the Martian crust. *J. Geophys. Res.* 105, 4165-4171.

[128] McNaughton, N.J., Rasmussen, B., Fletcher I.R., 1999. SHRIMP Uranium-Lead dating of diagenetic xenotime in siliciclastic sedimentary rocks. *Science* 285, 78-80.

[129] Mege, D., Ernst, R.E., 2001. Contractional effects of mantle plumes on Earth, Mars, and Venus, In Mantle plumes: their identification through time, edited by R.E. Ernst and K.L. Buchan, *Geological Society of America Special Paper* 352, 103-140.

[130] Melosh, H.J., Vickery, A.M., 1989. Impact erosion of the primordial atmosphere of Mars. *Nature* 338, 487-489.

[131] Merezhkovsky, K.S., 1909. Theory of two plasms as the basis of symbiogenesis: a new study on the origin of organisms. *Publications Dept.*, Kazan Imperial University, Kazan.

[132] Miller, J.D., Straat, P.A., Levin, G.V., 2001. Periodic analysis of the Viking Lander Labeled Release Experiment. Instruments, Methods, and Missions for Astrobiology, *SPIE Proceedings*, 4495, 96-107.

[133] Mojzsis, S.J., Arrhenius, G., McKeegan, K.D., Harrison, T.M., Nutman, A.P., Friend, C.R.L., 1996. Evidence for life on Earth before 3800 Myr. *Nature* 384, 55-59.

[134] Murray, B., Makin, M.C., Greeley, R. 1981. *Earth-like Planets; Surfaces of Mercury, Venus, Earth, Moon, Mars*. W.H. Freeman Company, San Francisco, CA.

[135] Nienow, J.A., McKay, C.P., Friedmann, E.I., 1988. The cryptoendolithic microbial environment in the Ross desert of Antarctica: light in the photosynthetically active region. *Microbial Ecology* 16, 271-289.

[136] Nozette, S., Rustan, P., Pleasance, L.P., Horan, D.M., Regeon, P., Shoemaker, E.M., Spudis, P.D., Acton, C.H., Baker, D.N., Blamont, J.E., Buratti, B.J., Corson, M.P., Davies, M.E., Duxbury, T.C., Eliason, E.M., Jakosky, B.M., Kordas, J.F., Lewis, I.T., Lichtenberg, C.L., Lucey, P.G., Malaret, E., Massie, M.A., Resnick, J.H., Rollins, C.J., Park, H.S., McEwen, A.S., Priest, R.E., Pieters, C.M., Reisse, R.A., Robinson, M.S., Simpson, R.A., Smith, D.E., Sorenson, T.C., Vorder-Bruegge, R.W., Zuber, M.T., 1994. The Clementine mission to the Moon; scientific overview. Science 266, 1835-1839.

[137] Ohmoto, H., 1997. When did the Earth's atmosphere become oxic? *Geochemical News* 93, 13.

[138] Ohmoto, H., Kakegawa, T., Lowe, D.R., 1993. 3.4-billion-year-old biogenic pyrites from Barberton, South Africa: sulfur isotope evidence. *Science* 262, 555.

[139] Öner, A. T., Ruiz, J., Fairén, A. G., Tejero, R., Dohm J.M., 2004. The volume of possible ancient oceanic basins in the northern plains of mars. *Lunar Planet. Sci. Conf.* XXXV, #1319.

[140] Orphan, V.J., House, C.H., Hinrich, K., McKeegan, K., DeLong, E.F., 2002. Multiple archaeal groups mediate methane oxidation in anoxic cold seep sediments. *Proc. Natl. Acad. Sci.* 99, 7663-7668.

[141] Paige, D.A., Wood, S.E., Vasavada, A.R., 1992. The thermal stability of water ice at the poles of Mercury. *Science* 258, 643-646.

[142] Parkes, R.J., Cragg, B.A., Wellsbury, P., 2000. Recent studies on bacterial populations and processes in subseafloor sediments: a review. *Hydrogeol. Journal* 8, 11-28.

[143] Pavlov, A.A., Kasting, J.F., Eigenbrode, J.L., Freeman, K.H., 2001. Organic haze in Earth's early atmosphere, source of low ^{13}C late Archean kerogens? *Geology* 29, 1003-1006.

[144] Phillips, R.J., Hansen, V.L., 1998. Geological evolution of Venus: rises, plains, plumes, and plateaus. *Science* 279, 1492-1497.

[145] Phillips, R.J., Zuber, M.T., Solomon, S.C., Golombek, M.P., Jakosky, B.M., Banerdt, W.B., Smith, D.E., Williams, R.M.E., Hynek, B.M., Aharonson, O., Hauck, S.A., 2001. Ancient geodynamics and global-scale hydrology on Mars. *Science* 291, 2587-2591.

[146] Plummer, W.T., 1969. Venus clouds: test for hydrocarbons. *Science* 163, 1191-1192.

[147] Pollack, J.B., 1991. Present and past climates of the terrestrial planets. *Icarus* 91, 173-198.

[148] Pollack, J.B., Toon, O.B., Whitten, R.C., Boese, R., Ragent, B., Tomasko, M.G., Esposito, L.V., Travis, L.D., Wiedman, D., 1980. Distribution and source of the UV absorption in Venus' atmosphere. *J. Geophys. Res.* 85, 8141-8150.

[149] Rasmussen, B., 2000. Filamentous microfossils in a 3,235-million-year-old volcanogenic massive sulphide deposit. *Nature* 405, 676-679.

[150] Richardson, W.P., Okal, E.A., Van der Lee, S., 2000. Rayleigh-wave tomography of the Ontong-Java plateau. *Physics of the Earth and Planetary Interiors* 118, 29-51.

[151] Rodriguez, J.A.P., Saski, S., Dohm, J.M., Tanaka, K.L., Strom, R., Kargel, J., Kuzmin, R.O., Miyamoto, H., Spray, J.G., Fairén, A.G., Komatsu, G., Kurita, K., Baker, V.R., 2005a. Control of impact crater fracture systems on subsurface hydrology, ground subsidence, and collapse, Mars. *J. Geophys. Res.* 110, 10.1029/2004JE002365.

[152] Rodriguez, J.A.P., Saski, S., Kuzmin, R.O., Dohm, J.M., Tanaka, K.L., Miyamoto, H., Kurita, K., Komatsu, G., Fairén, A.G., Ferris, J.C., 2005b. Outlfow chnnel sources, reactivation, and chaos formation, Xanthe Terra, Mars. *Icarus* 175, 36-57.

[153] Rothschild, L.J., Mancinelli, R.L., 2001. Life in extreme environments. *Nature* 409, 1092-1101.

[154] Ruiz, J., Fairen, A.G., Dohm, J.M., Tejero, R. 2004. Thermal isostacy and deformation of possible paleoshorelines on Mars. *Planetary and Space Sci*, 52, 1297-1301.

[155] Sagan, C.. 1960. *The Radiation Balance on Venus*. JPL, Pasadena, CA.

[156] Sagan, C., 1961. The planet Venus. *Science* 133, 849-858.

[157] Sagan, C., Chyba, C., 1997. The early faint Sun paradox: organic shielding of ultraviolet-labile greenhouse gases. *Science* 276, 1217-1221.

[158] Sattler, B., Puxbaum, H., Psenner, R., 2001. Bacterial growth in supercooled cloud droplets. *Geophys. Res. Lett.* 28, 239-242.

[159] Schopf, J.W., 1993. Microfossils of the early Archaen apex chert: new evidence of the antiquity of life. *Science* 280, 640-646.

[160] Schopf, J.W. 1999. *Cradle of Life: The Discovery of Earth's Earliest Fossils*. Princeton University Press, Princeton, NJ.

[161] Schleper, C., Peuhler, G., Holz, I., Gambacorta, A., Janekovic, D., Santarius, U., Klenk, H.P., Zillig, W., 1996. *Picrophilus gen.* Nov., fam. Nov.: a novel aerobic, heterotrophic, thermoacidophilic genus and family comprising archaea capable of growth around pH 0. *Journal of Bacteriology* 177, 7050-7079.

[162] Schulze-Makuch, D., Irwin, L.N., 2002. Reassessing the possibility of life on Venus: proposal for an astrobiology mission. *Astrobiology* 2, 197-202.

[163] Schulze-Makuch, D., Irwin, L.N., 2004. *Life in the Universe: Expectations and Constraints*, Springer-Verlag, Berlin, Germany.

[164] Schulze-Makuch, D., Irwin, L.N., 2006. The prospect of alien life in exotic forms on other worlds. *Naturwissenschaften*, in press.

[165] Schulze-Makuch, D., Irwin, L.N., Irwin, T., 2002. Astrobiological relevance and feasibility of a sample collection mission to the atmosphere of Venus. *ESA-Sp.* 518, 247-252.

[166] Schulze-Makuch, D., Fairén, A.G., Dohm, J.M., Baker, V.R., Boynton, W., 2004a. Strategy for detecting hydrothermal activity on Mars. *Abstract*, Early Mars Conference, 11-15 Oct. 2004, Jacksonhole, Wyoming.

[167] Schulze-Makuch, D., Grinspoon, D.H., Abbas, O., Irwin, L.N., Bullock, M., 2004b. A sulfur-based UV adaptation strategy for putative phototrophic life in the Venusian atmosphere. *Astrobiology* 4, 11-18.

[168] Schulze-Makuch, D., Dohm, J.M., Fairén, A.G., Baker, V.R., Fink, W., Strom, R.G., 2005a. Venus, Mars, and the ices on Mercury and the Moon: astrobiological implications and proposed mission designs. *Astrobiology* 5(6), 778-795.

[169] Schulze-Makuch, D. Irwin, L.N., Lipps, J.H., LeMone, D., Dohm, J.M., Fairén, A.G., 2005b. Scenarios for the evolution of life on Mars. *JGR – Planets* 110, E12S23.

[170] Schulze-Makuch, D., Dohm, J.M., Fairén, A.G., Baker, V.R., Fink, W., Strom, R.G., 2006. Sample Return Missions to Mars, Venus, and the Ices on Mercury and the Moon [abstract 1324]. In 37th *Lunar and Planetary Science Conference Abstracts* [CD-ROM], Lunar and Planetary Institute, Houston.

[171] Scott, D.H., Tanaka K.L. 1986. Geologic map of the western equatorial region of Mars, *USGS Misc. Inv. Ser. Map* I-1802-A (1:15,000,000).

[172] Scott, D.H., Dohm, J.M., Rice, J.W. 1995. Map of Mars showing channels and possible paleolake basins. U.S. Geol. Surv. Misc. Invest. Ser. MAP I-2461.

[173] Slade, M.A., 1992. Mercury radar imaging: Evidence for polar ice. *Science* 258, 635-640.

[174] Slade, M.A., Butler B.J., Muhleman, D.O., 1992. Mercury radar imaging: evidence for polar ice. *Science* 258, 635-640.

[175] Sleep, N.H., Meiborn, A., Fridrikson, Th., Coleman, R.G., Bird, D.K., 2004. H_2-rich fluids from serpentinization: geochemical and biotic implications. *PNAS* 101, 12818-12823.

[176] Spudis, PD, Guest, J.E. 1988. Stratigraphy and geologic history of Mercury. In Vilas, F., Chapman, C.R., Matthews, M.S. (Eds.) Mercury, Univ. of Arizona Press, *Tucson*, pp. 118- 164.

[177] Squyres et al., 2004. In situ evidence for an ancient aqueous environment at Meridiani Planum, Mars, *Science* 306, 1709-1722.

[178] Stofan, E.R., Scharpton, V.L., Schubert, G., Baer, G., Bindschadler,, D.L., Janes, D.M., Squyres, S.W., 1992. Global distribution and characteristics of coronae and related

features on Venus. Implications for origin and relation to mantle processes. *J. Geophys. Res.* 97, 13,347-13,378.

[179] Strom, R.G., Neukum, G., 1988. The cratering record on Mercury and the origin of impacting objects (1988) In Vilas, F., Chapman, C.R., Matthews, M.S. (Eds.) Mercury, Univ. of Arizona Press, *Tucson*, pp. 336-373.

[180] Strom, R.G., Schaber, G.G., Dawson, D.D., 1994. The global resurfacing of Venus. *J. Geophys. Res.* 99, 10,899-10,926.

[181] Strom R. G., Sprague, A. L. 2003. Exploring Mercury, Springer-Praxis,Chichester, UK.

[182] Tanaka, K. L., 1986. The stratigraphy of Mars. *J. Geophys. Res.* 91, 139–158.

[183] Tanaka, K.L., Senske, D.A., Price, M., Kirk, R.L., 1997. *Physiography, geomorphic/geologic mapping, and stratigraphy of Venus.* In Bougher, S.W., Hunten, D.M., Phillips, R.J. (Eds.) Venus, University of Arizona Press, Tucson, Arizona, pp 667-694.

[184] Touma, J., and Wisdom, J., 1993. The chaotic obliquity of Mars. *Science* 259, 1294-1296.

[185] Van Zuilen, M.A., Lepland, A., Arrhenius, G., 2002. Reassessing the evidence for the earliest traces of life. *Nature* 418, 627-630.

[186] Vestal, J.R., 1988. Carbon metabolism of the cryptoendolithic microbiota from the Antarctic desert. *Applied and Environmental Microbiology* 54, 960-965.

[187] Volk, T., 1987. Feedbacks between weathering and atmospheric CO_2 over the last 100 million years. Amer. *J. Sci.* 287, 763-779.

[188] Wagner, R.J., Wolf, U., Ivanov, B.A., Neukum, G., 2001. Application of an updated impact cratering chronology model to mercury's time-stratigraphic system. *Lunar Planet. Sci. Conf.*, XXXII, abstract #8049.

[189] Walker, J.C.G., Hays, P.B., and Kasting, J.F., 1981. A negative feedback mechanism for the long-term stabilization of Earth's surface temperature, *J. Geophys. Res.* 86 (C10), 9776-9782.

[190] Watson, E.B, Harrison, T.M., 2005. Zircon thermometer reveals minimum melting conditions on earliest Earth, *Science* 308, 841-844.

[191] White, R.S., McKenzie, D.P., 1989. Magmatism at rift zones: the generation of volcanic continental margins and flood basalts. *J. Geophys.Res.* 94, 7685-7729.

[192] Wilde, S.A., Valley, J.W., Peck, W.H., Graham, C.M., 2001. Evidence from detrital zircons for the existence of continental crust and oceans on the Earth 4.4 Gyr ago. *Nature* 409, 175-178.

[193] Wildt, R., 1940. Note on the surface temperature of Venus. *Astrophys. J.* 91, 266-268.

[194] Wilhelms, D.E., 1987. The geologic history of the Moon. *USGS Prof. Paper* 1348.

[195] Wurz, P., Lammer, H., 2003. Monte-Carlo simulation of Mercury's exosphere. *Icarus* 164, 1-13.

[196] Yung, Y.L., 1988. HDO in the martian atmosphere: implications for the abundance of crustal water. *Icarus* 76, 146-159.

In: Space Science: New Research
Editor: Nick S. Maravell, pp. 35-81

ISBN: 1-60021-005-8

Chapter 2

SOLAR DYNAMICS AND SOLAR-TERRESTRIAL INFLUENCES

Katya Georgieva*

Solar-Terrestrial Influences Laboratory at the Bulgarian Academy of Sciences,
Bl.3 Academy G. Bonchev str., 1113 Sofia, Bulgaria

Abstract

Much research has been done on the influences of solar activity on processes on Earth and in near-Earth space. While there is a general consensus about the reality of such influences, the results are often contradictory even if highly statistically significant, in many cases the physical mechanism of the influence is unclear, and the mechanism of solar activity itself is not yet fully understood. Obviously there are additional factors affecting solar activity and its influences on the Earth which are not accounted for in the present models. In this paper we are investigating one of these possible factors – the dynamics of the Sun and its significance for the solar-terrestrial influences.

One of the most controversial problems in solar-terrestrial physics is the relation between solar activity and climate. Different authors have reported positive, negative or missing correlations between solar activity and surface air temperature in the 11-year sunspot cycle. We demonstrate that the sign of the correlation changes regularly in consecutive secular (Gleissberg) solar cycles, and depends on the long-term changes of North-South solar activity asymmetry which in turn is determined by the movement of the Sun about the barycenter of the Solar system.

This movement of the Sun about the barycenter of the Solar system is also related to the dynamics of both the Sun itself and of the Earth. The asymmetry in the rotation of the Northern and Southern solar hemispheres correlates very well with the variations in the Earth rotation rate, and their dominant common periodicity is the periodicity of the rotation of the Sun about the Solar system barycenter. The correlation of the solar dynamics and the Earth dynamics is mediated by the solar wind carrying momentum and magnetic fields and modulating the electromagnetic core-mantle coupling torques responsible for the variations in the Earth's rotation.

The variable solar differential rotation affects also the way in which the solar drivers interact with the Earth's magnetosphere. The periodicities in the interplanetary magnetic field at the Earth's orbit in any period coincide with the periodicities in the latitudinal gradient of

---

* E-mail address: kgeorg@bas.bg

the differential rotation in the more active solar hemisphere, and the azimuthal components of the interplanetary magnetic field are proportional to the solar equatorial rotation rate.

The most geoeffective solar driver are the magnetic clouds – coronal mass ejections with smooth rotation over a wide angle of the magnetic field inside the structure. The portion of coronal mass ejections which are magnetic clouds varies throughout the solar cycle, and is determined by the amount of helicity transferred from the solar interior to the surface, and by the surface differential rotation. We demonstrate the influence of the surface differential rotation on the helicity and geoeffectiveness of magnetic clouds.

Introduction

Sun, the star closest to us, like many other stars is a variable star. Apart from satisfying the purely scientific curiosity about what a variable star looks like from a close distance, we have much more practical reasons to study solar variability – actually, we live in the corona of this star, our Sun, so everything that happens on and around the Earth is if not determined, then at least influenced by solar activity. A long way has been gone toward accepting the reality of this influence, even in areas where earlier it was violently rejected, like climate, earthquakes, or human health, however the ways in which the variations in solar activity affect the Earth, as well as the mechanism of solar activity itself are not quite clear yet. Probably there are factors which present theories don't account for. Here we are studying one such possible factor – the solar system dynamics and its influence on the solar activity and on the solar influences on the Earth. The study is far from complete, and many of the facts are still waiting for an explanation. We hope that by presenting them we will provoke further investigations in this interesting and controversial area.

A Brief Historical Overview

Sunspots and Solar Rotation

By the term "solar activity" usually any type of variation in the appearance or energy output of the Sun is understood. Elements of solar activity are sunspots, solar flares, coronal mass ejections, coronal holes, total and spectral solar irradiance, etc. The most prominent evidence of solar activity and with the longest data record, though not geoeffective themselves but related to geoeffective active regions, are sunspots. Very big sunspots can be seen with naked eye, and old chronicles testify that they have been indeed observed even in ancient times. There is evidence that the Greeks knew of them at least by the 4th century BC, and the earliest records of sunspots observed by Chinese astronomers are from 28 BC. However, systematic observations of sunspots began only early in the 17th century after the telescope was invented.

Like many facts about solar activity, the discovery of sunspots is still a matter of controversy. Galileo Galilei and Christoph Scheiner, a German Jesuit astronomer, never stopped arguing about the priority in the discovery of sunspots, but neither or them was aware of the earlier observations of Harriot and Fabricius. The English scholar Thomas Harriot was probably the first to observe telescopically sunspots as evidenced by a drawing in his notebook dated December 8, 1610, and the first published observations (June 1611) entitled "De Maculis in Sole Observatis, et Apparente earum cum Sole Conversione Narratio"

("Narration on Spots Observed on the Sun and their Apparent Rotation with the Sun") were by Johannes Fabricius who had been observing systematically the spots for a few months and, as stated in the title of his paper, had noted also their movement across the solar disc. This can be considered the first observational evidence of the solar rotation. Christopher Scheiner ("Rosa Ursine sive solis", book 4, part 2, 1630) was the first who measured the equatorial rotation rate of the Sun and noticed that the rotation at higher latitudes is slower, so Scheiner can be considered the discoverer of solar differential rotation, and Galileo was the one who found that the solar equator is inclined to the plane of the ecliptic (Abetti, 1965).

St.John (1918) was maybe the first one who established that the solar rotation is not constant. He summarized the published solar rotation rates, and concluded that the differences in series measured in different years can hardly be attributed to personal equation or to local disturbances on the Sun, and are probably due to time variations in the rate of rotation. Hubrecht (1915) first found that the two solar hemispheres rotate differently or, as he remarked, "with regard to the equator symmetry is suspiciously absent". Though he called these differences suspicious, Hubrecht nevertheless found them consistent with Emden's (1902) solar theory.

Sunspot Cycle

The sunspot cycle was discovered by Heinrich Schwabe, a German pharmacist and amateur astronomer. Schwabe was trying to discover a new planet inside the orbit of Mercury which was tentatively called Vulcan. Because of the close proximity to the Sun, it would have been very difficult to observe Vulcan, and Schwabe believed one possibility to detect the planet might be to see it as a dark spot when passing in front of the Sun. For 17 years, from 1826 to 1843, on every clear day, Schwabe would scan the sun and record its spots trying to detect Vulcan among them. He didn't find the planet but noticed the regular variation in the number of sunspots and published his findings in a short article entitled "Solar Observations during 1843". It is worth noting that Schwabe determined very accurately the 11-year sunspot cycle from observations covering only 17 years. This paper at first attracted little attention, but Rudolf Wolf who was at that time the director of Bern observatory, was impressed so he began regular observations of sunspots and studied historical records to make a reconstruction of the number of sunspots as far back in the past as possible. He was the first to note the possible existence in the sunspot record of a longer modulation period of about 55 years – again in quite a short data record.

In 1848 Wolf devised the famous relative sunspot number, also known as the "Zürich sunspot number" which is being used ever since:

$$R=k\ (10g+s),$$

where R is the ("Zürich" or "International") sunspot number; g is the number of sunspot groups on the solar disk; s is the total number of individual spots in all the groups; and k is a variable scaling factor (usually <1) that accounts for observing conditions and the type of telescope (binoculars, space telescopes, etc.).

On the basis of reconstructed sunspot numbers obtained from scattered reports dating back to 1610, when telescopic observations of spots were first made, Wolf refined the average duration of the sunspot cycle to 11.1 years.

But what are the sunspots? Galileo supposed they were big clouds floating in the solar atmosphere. Some, like Scheiner and Schwabe, assumed they could be planets passing in front of the Sun. Only in 1908 George Ellery Hale, an American astronomer, found that sunspots are associated with strong magnetic fields, and four years later offered his "Sun Spot Hypothesis". The next fragment of this short research note demonstrates how little our present understanding differs from the insight of this great scientist: "As the result of an eruption, or some other cause tending to produce rapid convection, a gaseous column moves upward from within the Sun toward the surface of the photosphere. Vortex motion is initiated by the difference in rotational velocity of adjoining zones or by irregularities of structure and is maintained by convection. The circulation in the vortex is vertically upward and then outward, as in terrestrial tornado. Expansion produces cooling at the center of the vortex, and a comparatively dark cloud (the umbra) results." Hale was also the one to find that sunspots tend to occur in pairs, and that on one hemisphere the leading (with respect to the direction of the solar rotation) spots in all pairs have the same polarity, and the trailing spots in the pairs have the other polarity, while on the other hemisphere the polarities are oppositely oriented. The discovery that in subsequent 11-yr sunspot cycles the polarities in the two hemispheres reverse led to the idea of the "23-year magnetic sunspot period" (Hale, 1925).

Sunspots and Geomagnetic Storms

A decennial period in the daily variation of magnetic declination had been reported by Lamont from Munich in 1851, but he did not relate it to the sunspot cycle. Such a connection was first made in 1852 by the Irish geophysist, astronomer and explorer Sir Edward Sabine, who noted that minima in the average rate and size of magnetic disturbances at the widely separated Hobarton and Toronto observatories in 1843 corresponded to a minimum in sunspot numbers, while maxima in 1848 corresponded to a maximum in the decennial sunspot curve, and announced that the 11 year sunspot cycle was "absolutely identical" with the geomagnetic cycle. Since Newton's explanation of the effect of the Sun's gravity on Earth, this was the first new phenomenon of the Sun interacting with the Earth, and is considered the birth of the solar-terrestrial physics. Within months of Sabine's discovery it was reported independently by Gautier and Wolf in Switerland, using a larger dataset.

In September 1859 Carrington saw by chance a bright outburst of light in a group of large sunspots, lasting about five minutes. Carrington's observation was confirmed by Hodgson, an English amateur observing nearby. This was followed 17 hours later by a very powerful magnetic storm. Despite the temporal association of the flare and the magnetic variations, Carrington declined to draw a physical connection between the phenomena, cautiously commenting that "one swallow does not make a summer." A serious problem at that time was the lack of suitable instruments for flare observations, so in the whole 19th century only about 10 independent major flares were documented (Cliver, 1994).

Sunspots and Climate

After finding a relation between sunspot number and geomagnetic disturbances, Wolf fought a similar periodicity in various meteorological phenomena, but without conclusive results. The first one to declare a connection was Sir William Herschel, who noticed that during five prolonged periods of few sunspots the price of wheat at the London stock market was high. He hypothesized that times of many sunspots "may lead us to expect a copious emission of heat and therefore mild seasons," and periods of few spots would signal "spare emission of heat" and "severe seasons" which would raise the price of wheat. In 1801 Herschel presented his finding to the Royal Society, and was severely ridiculed.

In spite of the reaction of the Royal Society, a growing number of scientists continued looking for relations between solar activity and terrestrial climate. Such relations have been indeed found. But, as Sazonov and Loginov (1969) noticed in their book on solar-tropospheric relations, "a researcher first acquainting with the results of the numerous studies can't help falling into desperation realizing that the correlations change in time and space, apparently without any reason". Really, the results, even when highly statistically significant, are quite contradictory. Different authors (and sometimes even the same authors), for different locations (and sometimes even for the same locations) have found positive, negative or missing correlations between sunspot activity and a given meteorological element (Herman and Goldberg, 1978), say the surface air temperature. For example, Köppen (1914), based on data gathered from all available sources, showed that from 1804 to 1909 the mean annual global temperature was lower in sunspot maximum and higher in sunspot minimum. The same result was found for the above period separately for tropical and subtropical regions (Köppen, 1873). Clayton (1923) reanalyzed Köppen's data and confirmed his conclusions. He calculated 5-year moving averages of the temperature deviations in different latitudinal zones and found that everywhere except in the dry subtropical regions (where the results were ambiguous) the temperatures were maximum in sunspot minimum and minimum in sunspot maximum, with the difference in the temperature between sunspot minimum and sunspot maximum being about 0.4° C. A similar relation was found by Walker (1915). Troup (1962) also studied the tropical zone and found a negative correlation between temperature and sunspot number between 1862 and 1920-1925, but positive afterwards. Sazonov and Loginov (1969) also found different signs of the correlation. Studying the number of latitudinal zones with stable winter and summer positive temperature anomalies, they showed that in the period 1884-1920 this number decreased with increasing geomagnetic activity, while from 1921 to 1960 the number of anomalies increased with increasing geomagnetic activity. (As anomalous winter was regarded a winter with the same sign of the deviation of the monthly mean temperature from the long-term normal in all three winter months – December, January and February, and as anomalous summer – a summer with the same sign of the deviations in June, July and August). The same relation was established for the biggest warmings in each year, irrespective of the latitude. Starr and Oort (1973) found a positive correlation between the temperature of the "average air mass" in the Northern hemisphere and the sunspot number in the period 1958-1963.

These contradicting results can be a serious reason to doubt the reality of solar activity influences on the terrestrial atmosphere. If the sign of the correlation changes randomly, these may be only random coincidences without any physical meaning. On the other hand, if there is some regularity in the change of the correlation, most probably the influence is real but there is some additional factor ruling the way in which solar activity affects climate.

Solar Activity and Temperature in the 11-Year Sunspot Cycle

Historical Compilation

Figure 1.1 is a compilation of all available published results on correlations between solar activity and surface air temperature in the 11-year cycle. Each period for which positive or negative correlation has been reported in some paper between surface air temperature and solar activity is presented by a light or a dark bar, respectively. Two facts are obvious from this figure. First, the cases of positive and negative correlations are fairly well grouped, therefore the sign of the correlation depends only on the period studied and not on the location. Second, the sign of the correlation changes from one century to another, and seems related to the secular (Gleissberg) cycle of solar activity - the solid line in Figure 1.1.

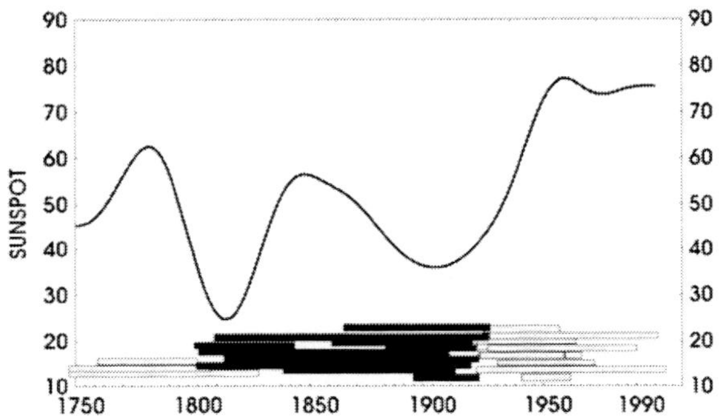

Figure 1. Published results reporting negative (dark shading) and positive (light shading) correlations between solar activity and surface air temperature in the 11-year solar cycle. Solid line - secular solar cycle.

Figure 1 is based on occasional, fragmentary studies. To verify the hypothesis which it implies, we should study global, hemispherical or zonal averages of the temperatures.

Data

Such a study, however, is limited by the data availability. The mercury thermometer was invented in 1714 by Daniel Gabriel Fahrenheit (1686-1736), and the first temperature scale which bears his name - Fahrenheit Scale, was introduced by him in 1724. Some individuals and organizations began creating meteorological measurement networks, mainly for agriculture and weather forecast, but regular measurements of the temperature didn't begin until the end of the 18th century (this is why Herschel had to use the wheat prices as a proxy for the temperature). The first attempts for large-scale monitoring of the temperature were in Europe and North America, and later they spread into other parts of the world. Nowadays mainly instrumental records of ground-based meteorological stations are used for climate studies. There are two reasons for this: first, the instrumental records are direct measurements in fixed points of space and time, and second, during the last two centuries such measurements have been gathered from more than 100,000 points (Peterson and Vose, 1997).

As most instrumental networks have been intended for registering of local weather rather than long-term climatic changes, using of their data is connected to certain difficulties. For example, often the records are not digitized and/or are not easily accessible outside the country in which they have been measured. Second, the stations are not uniformly distributed over the globe. Especially big are the differences in the coverage between land and ocean. An uneven distribution of stations introduces network biases which have significant effects on the estimated temperature trends, particularly on the regional scale. Third, instrumental records often contain errors resulting from their gathering and archiving which reduce their reliability. Finally, the instrumental records are often inhomogeneous because of a change in the location of the station and/or the introduction of a new type of thermometer. Such inhomogeneities introduce non-climatic variations which can distort the long-term trend.

Since the beginning of the last century, efforts have been made to overcome these problems. One of the first and longest running efforts is the World Weather Records (WWR) initiative, which commenced in 1923 and resulted in the regular publication of decadal series of global climate records ever since (Clayton, 1927). Another good example is the National Center for Atmospheric Research's annually published "World Monthly Surface Station Climatology" data set (WMSSC, Spangler and Jenne 1992), which consists of WWR, miscellaneous acquisitions, and the National Climatic Data Center's (NCDC) "Monthly Climatic Data for the World" for more recent records. In the early 1990s, climatologists from NCDC and the Carbon Dioxide Information Analysis Center (CDIAC) undertook a new initiative aimed at creating a data set appropriate for the study of climate change at both global and regional scales. This database, known as the Global Historical Climatology Network, was released in 1992 (Vose et al. 1992) and contained quality-controlled monthly climatic time series from over 6,000 land-based temperature stations worldwide. Its second version which we use here, includes data from already more than 8,000 stations to improve regional-scale analyses, particularly in previously data sparse areas; made are detailed assessments of data quality to increase the confidence in research results; rigorous and objective homogeneity adjustments are made to decrease the effect of non-climatic factors on the time series.

A number of research groups and individual researchers have also compiled their own global, hemispheric and zonal data sets for specific applications. Apart from GHCN, we use three such reconstructions.

The most famous of these is the data set of Jones (1999), which has been used extensively in climate research. The anomalies (deviations from the averages for the period 1961-1990) of the mean monthly surface air temperatures for the Northern and Southern hemispheres and averages for the globe (calculated directly rather than by averaging the two hemispheres) have been calculated for the period 1856-2005 based on a 5° x 5° grid-box basis from over 3,000 land stations and a number on sea surface temperature measurements (Jones and Moberg, 2003). The data are available online at http:/www.cru.uea.ac.uk/cru/data/ temperature.

The second data set that we use is of the Global Mean Surface Air Temperature Anomalies (deviations from the mean for the period 1951-1980) from 1880 to present from NASA Goddard Institute for Space Studies (GISS). This GISS analysis uses data from about

2,000 meteorological stations around the world. The Earth's surface is divided into 80 equal area "boxes", the full dimension of a box side being about 2,500 km. From them, the zonally averaged mean zonal monthly temperatures have been calculated for the zones 0-24°, 24-44°, 44-64°, 64-90°, 24-90° North and South, and 24°S-24°N (Hansen and Lebedeff, 1988).

The third data set contains the deviations from a 1951-1975 averages, spatially averaged monthly temperature records compiled by Vinnikov et al. (1990) from the State Hydrological Institute, Sankt Petersburg, Russia. It covers the period 1881-1990 and gives the zonally averaged values for the zones 0.0°-23.6°, 23.6°-44.4°, 44.4°-64.2° and 64.2°-90.0°, centered at 11.8°, 34.0°, 54.3° и 77.1° respectively, North and South.

Results

All three data sets for global, hemispheric and zonal averages yield similar results which are illustrated in Figure 2. In the upper panel, the mean annual temperature (3 points running mean, detrended) is compared to the International Sunspot Number from the National Geophysical Data Center (available online at *www.ngdc.noaa.gov/stp/SOLAR/ftpsunspotnumber.html*). In the beginning of the period high temperatures are observed in solar minimum and low ones - at high solar activity. About 1920-30 the correlation changes to the opposite. In the lower panel the cross-correlation functions for the two periods are presented.

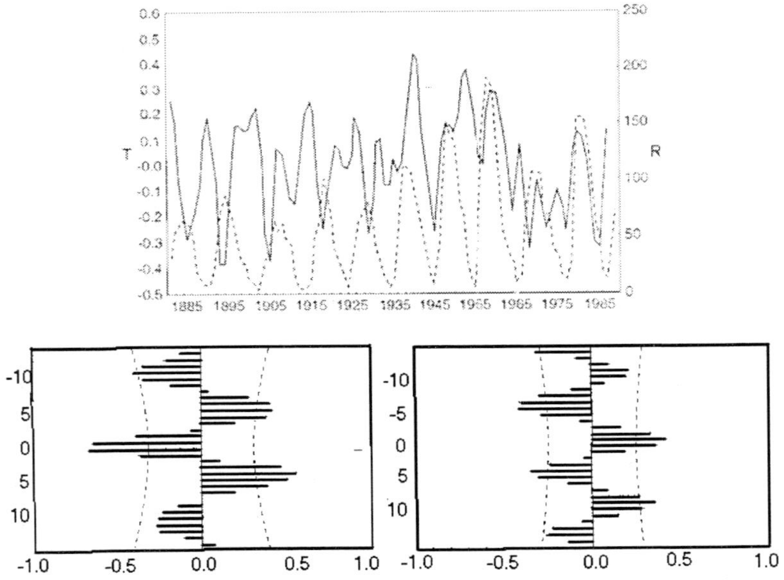

Figure 2. Upper panel: Mean zonal annual temperature from Vinnikov et al. (1990) for the zone 23.6°-44.4° N (solid line) and sunspot numbers (dotted line); 5-point runnung mean, detrended; Lower panel: Cross-correlation functions for the periods before and after 1925.

All global reconstructions are relatively short - the longest data set (Jones et al, 1999) begins in 1856. To study earlier epochs, data from individual meteorological stations with long measurement records have been used. In the Global Historical Climatology Network

Temperature Database of NCDC (Vose et al., 1992), 65 stations were found with almost continuous data records beginning before the middle of the 19th century. The time series for each station was divided into subseries yielding the best correlation with solar activity. In Figure 3 the percentage of stations with statistically significant positive and negative correlation (along the positive and negative parts of the axis, respectively) is shown together with the secular solar activity cycle. In the 18th century, in the vast majority of the available stations the surface air temperature was positively correlated with solar activity in the 11-year solar cycle, this correlation changed to negative in the 19th century and to positive again in the 20th century. (Herschel's data about the wheat price are in agreement with this figure: in the 18th century sunspot maximum was associated with warmer weather, higher wheat yield and lower prices).

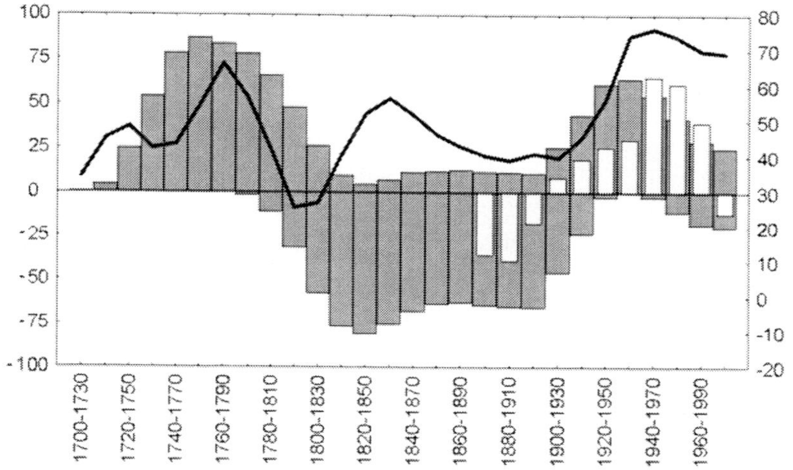

Figure 3. Dark bars - percentage of stations with statistically significant correlations between solar activity and surface air temperature in the 11-year solar cycle - positive or negative, along the positive and negative Y-axis, respectively; Solid line - secular solar cycle; White bars - north-south solar activity asymmetry.

The change in the sign of correlation in consecutive secular solar cycles implies that there is some solar parameter determining the way in which solar activity affects climate, which changes in consecutive secular cycles. One parameter which changes in the end of the 19th and the beginning of 20th century is the north-south asymmetry of solar activity. The white bars in Figure 3 represent the asymmetry of the solar activity, $A=(S_N-S_S)/(S_N+S_S)$ where S_N and S_S stand for the total sunspot area in the northern and southern solar hemispheres, respectively. Negative A seen in the 19th century means more active southern solar hemisphere and is associated with negative correlation between solar activity and surface air temperature in the 11-year cycle, and positive A in the 20th century - more active northern hemisphere and positive correlation.

The different reaction of the atmosphere to solar activity originating from the two solar hemispheres has been already noted, mainly as differences between positive and negative IMF sectors, or between +/- and -/+ passages of the interplanetary sector boundary which is in fact the heliospheric current sheet separating the two solar hemispheres. Smirnov (1967, 1969) found that when the Earth passes from one sector of the IMF into another, the sign of the correlation between the solar wind velocity and the atmospheric parameters changes. Loginov et al. (1975) investigated the relation among IMF sector structure, Ap index of geomagnetic activity, and properties of atmospheric circulation, and showed that meridional circulation in the northern hemisphere increases when the Earth passes through a positive sector and decreases during passage through a negative sector. The opposite relation is observed for zonal circulation. Besprozvannaya et al. (1997) found that stratospheric circulation responses to interplanetary sector boundaries' intersections is dependent on the direction of the transition (+/- or -/+, that is, north to south or south to north). The dependence on the direction of interplanetary sector transition has been also found for geomagnetic activity. Arora and Rangarajan (1981) studied the change in daily Ap in relation to the passage of IMF sector boundaries. The activity associated with a +/- boundary was larger than that associated with a -/+ boundary up to about 1972, then the relation reversed. This was considered to be a consequence of the modulation of the sector structure by the north-south asymmetry in solar activity which changed around this time, but it should be noted that about 1972 solar polarity also reversed, from negative (magnetic field pointing toward the Sun in the north, from the Sun in the south) to positive (from the Sun in the north, toward the Sun in the south).

Secular Cycle of North-South Solar Activity Asymmetry

Waldmeier (1957) based on data up to the 18th solar cycle, found the following relation between the north-south asymmetry of solar activity and the secular solar cycle: solar activity dominates in the northern solar hemisphere during the ascending part of the secular solar cycle, in the southern one during the descending part, and in epochs of secular minima and maxima the asymmetry is small. However, there are at least two episodes when this rule does not hold: the secular maximum in the 20th century when the asymmetry had a maximum rather than a minimum value, and the end of the Maunder minimum when the activity was clearly ascending but all sunspots were concentrated in the southern solar hemisphere. Based on the study of the solar activity effects on climate it was suggested that solar asymmetry changes in consecutive secular cycles, being positive in even secular cycles (if we denote the 20th century secular cycle as "even") and negative in odd ones. It has a maximum positive or negative value coinciding with the maximum of solar activity in the secular cycle, and changes sign around secular solar minimum. This leaves no "anomalies" in the available data (Georgieva and Kirov, 2000). Therefore, we can speak about a "double secular" solar cycle consisting of two secular cycles – one in which more active is the southern solar hemisphere, and a second one in which more active is the northern hemisphere.

The existence of a cycle about twice longer than the Gleissberg cycle has been suggested based on radiocarbon data (see, e.g., Vitinsky, Kopecký, and Kuklin, 1986, and references therein). That this double secular cycle consists of two secular cycles with alternating asymmetry was confirmed by Mursula and Zieger (2001) based on a study on the long-term evolution of the semiannual variation in geomagnetic activity. By geomagnetic activity we mean the variations of the Earth's magnetic field due to the solar-wind interaction with the Earth's magnetosphere. (We will come back to solar wind – magnetosphere interaction in the following sections.) As shown by Dungey (1963), the magnetosphere is "open" and energy can be deposited into it by the solar wind when the interplanetary magnetic field (IMF) is southward (Figure 4b), while when IMF is northward, the magnetosphere is "closed" (Figure 4a).

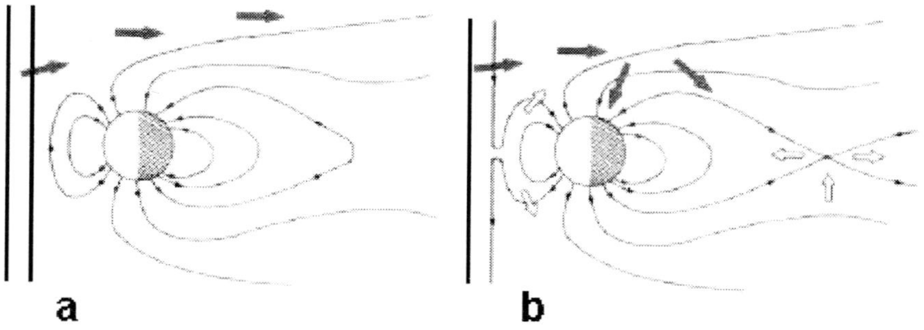

Figure 4. Solar wind interaction with the magnetosphere. (**A**) IMF north, no energy flow into the magnetosphere. (**B**) IMF south allows entry of plasma, momentum, and energy. Energy flow is indicated by solid arrows, magnetospheric convection is indicated by the open arrows. Adapted from Dungey (1963).

Because of the semiannual variation in the effective southward component of the interplanetary field (Russell and McPherron, 1973), the geomagnetic activity is highest around equinoxes and lowest around solstices (Chapman and Bartels, 1940). The solar equator is tilted relative to the ecliptic plane, so that in March the Earth is at highest southern heliolatitudes (that is, mostly exposed to the influence of the southern solar hemisphere), and in September it is at highest northern heliolatitudes. Comparing the geomagnetic activity maxima in March and September, Mursula and Zieger (2001) found that in the 20th century more active was the northern solar hemisphere, while in the 19th century more active was the southern hemisphere. The existence of a secular cycle in the variations in the north-south asymmetry of different solar activity indices was also found by Verma (1996, 2000).

A closer look at Figure 3 reveals that about 1970 the asymmetry became negative and a part of the stations began showing negative correlations between solar activity and surface air temperature in the 11-year cycle. If the above suggestion is correct, this would mean that the next secular cycle has already started. Yoshimura (1979) and Kopecky (1991), using different methods, have both suggested that solar cycle 21 is already on the ascending branch of the next secular cycle. Javaraiah (2003), studying the long-term periodicities in solar differential rotation, suggested that the Gleissberg cycle consists of two "double Hale" cycles. It is remarkable that the end of the previous and the beginning of the current Gleissberg cycles determined by his rotation data also come between sunspot cycles 20 and 21.

Atmospheric Circulation

The problem of solar activity influences on climate is closely related to its influences on atmospheric circulation. Atmospheric circulation is the system of atmospheric motions over the Earth on the scale of the whole globe (general atmospheric circulation), or over a certain region with its specific features (local circulation). These large scale atmospheric motions are caused by the differential heating of the Earth surface. Low latitudes receive more heat because the Sun's rays are more perpendicular, and the thickness of the atmosphere which they have to pass through is less than at high latitudes. Heated air at the equator rises, creating low pressure there, and proceeds south and north toward the poles. If the Earth didn't rotate, this air would reach the poles, descend there, and return to the equator, so there would be two circulation cells, one in each hemisphere (Figure 5).

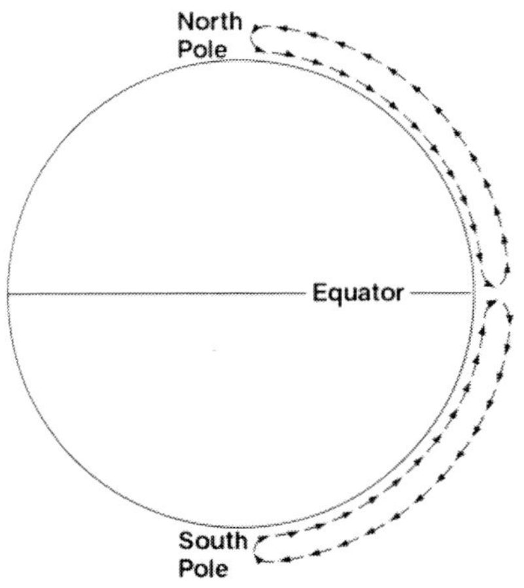

Figure 5. One-cell circulation for a non-rotating Earth.

However, the Earth does rotate, so when the heated air at the equator rises to a maximum vertical altitude of about 14 kilometers (top of the troposphere), and begins flowing horizontally to the North and South Poles, the Coriolis force deflects it, and by about 30° of latitude the air already flows zonally (parallel to the equator) from west to east. This zonal flow is known as the "subtropical jet stream". During its motion from the equator, the air cools, and a part of it sinks back to the surface creating the subtropical high pressure zone. From this zone, the surface air travels in two directions. A portion of the air moves back from the subtropical high pressure zone to the equatorial low pressure zone completing the circulation system known as the Hadley cell. This moving air is also deflected by the Coriolis effect - to the right (that is, to the west) in the northern hemisphere, and to the left (again to the west) in the southern hemisphere. The southwestern winds from the north and the northwestern winds from the south converge at the equator in a zone called "Intertropical Convergence Zone". The net winds are directed from east to west and are known as the "Trade winds" – a name given to them in the old sail-ship times when they facilitated sailing to the west and thus trade.

Another part of the air descending at subtropical latitudes moves to the poles and is also deflected by the Coriolis force, so at about 60° north and south the flow is from west to east forming the "polar jet stream". On the Earth's surface at about 60° north and south, these subtropical westerlies collide with cold air coming from the poles. This collision results in frontal uplift and the creation of the "subpolar lows" - the mid-latitude cyclones. A small portion of the lifted air is sent back to the subtropics after it reaches the top of the troposphere completing the circulation system known as the Ferrel cell. Most of the lifted air is directed to the poles where it creates the polar vortex, moves downward to form the polar high pressure zone, and flows back to midlatitudes in the Polar cell (Figure 6).

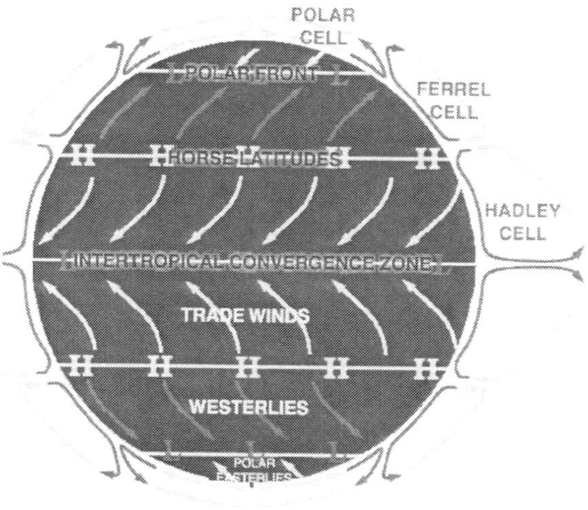

Figure 6. A 3-cell general atmospheric circulation for a rotating Earth (from http://en.wikipedia.org/wiki/Atmospheric_circulation).

The Hadley cell and the Polar cell are "thermally direct" – they exist as a direct consequence of the latitudinal differences in surface temperatures. The Ferrel cell is "thermally indirect" - its existence depends upon the Hadley cell and the Polar cell, and comes about as a result of the eddy circulations (the high and low pressure areas) of the midlatitudes. For this reason it is sometimes known as the "zone of mixing." While the Hadley and Polar cells are truly closed loops, the Ferrel cell is not, the prevailing westerly winds in it depend on the passing weather systems. While upper-level winds are essentially westerly (from west to east), surface winds can vary sharply and abruptly in direction. The predominantly westerly winds there can weaken or can become meridional (in the north-south direction) or even easterly for days. However, what happens in the Ferrel cells is decisive of the general atmospheric circulation and thus of the terrestrial climate.

In the thirties years of the 20th century, Vangengeim (1933, 1952) defined three types of atmospheric circulation in the Atlantic-Eurasian region calculated from the daily atmospheric pressure charts over northern Atlantic-Eurasian region: Westerly (W), Easterly (E), and Meridional (C). Later Girs (1948) applied the same criteria to the circulation in the Pacific-American sector and also found three types of circulation there: one westerly (which he denoted Z=zonal) and two meridional (M1 and M2). The circulation in the southern hemisphere is less studied, mostly because of the lack of systematic observations over big parts of the oceans, but it was found that many of the features observed in the northern hemisphere are present there too. Again there are two basic types of circulation – zonal and meridional, the first one divided into two subtypes, and the second one into five, depending on the location of the crests and troughs of the long atmospheric waves.

W and Z are zonal types of circulation (westerly at midlatitudes) characterized by low amplitude waves in the troposphere moving quickly from west to east. When a meridional circulation type (C or E, M1 or M2) prevails, stationary high amplitude waves are observed and the meridional transfer is intensified. The difference between C and E and between M1 and M2 is in the different location of the waves' crests and troughs.

On hemispheric scales, the atmospheric circulation is from west to east, and its strength is determined by the temperature contrast between the equatorial and the polar regions. For this reason the angular velocity of the atmosphere is higher than that of the solid Earth. The angular velocity of rotation of the atmosphere relative to the solid Earth is called circulation index and is an important characteristic of the zonal circulation.

Girs and Kondratovich (1978) compared the occurrence of the different types of circulation from 1900 to 1980 and suggested that in periods of long-term increase in sunspot activity the occurrence of meridional forms of circulations is increased, while in periods of decreasing sunspot activity more developed are the zonal forms of circulation. However, as in the case of the correlation between solar activity and surface air temperature in the 11-year sunspot cycle, more probably the decisive factor is not the level or the direction of change of the solar activity, but rather the predominantly more active solar hemisphere. Figure 7. presents the long-term variations of the index of the North-Atlantic Oscillation (NAO) and solar activity. NAO index is calculated as the difference in sea-level pressure between two stations in the Atlantic, one at high latitudes (always Iceland) and the other in the subtropics (usually Gibraltar). NAO positive phase, when the difference between surface pressure in the subtropics and in high latitudes is higher than normal, is associated with strong midlatitude westerly winds across the Atlantic onto Europe. In NAO negative phase, the westerly winds are weakened. This index can therefore be used as a proxy for the zonal circulation in the Atlantic-Eurasian sector. Here we use its reconstruction from 1821 to present (Jones et al., 1997). It is clearly seen that in all 19th century when the southern solar hemisphere was more active, the long-term increase in solar activity intensified zonal circulation, while in the 20th century when predominantly more active was the northern solar hemisphere, increasing solar activity was associated with prevailing meridional circulation.

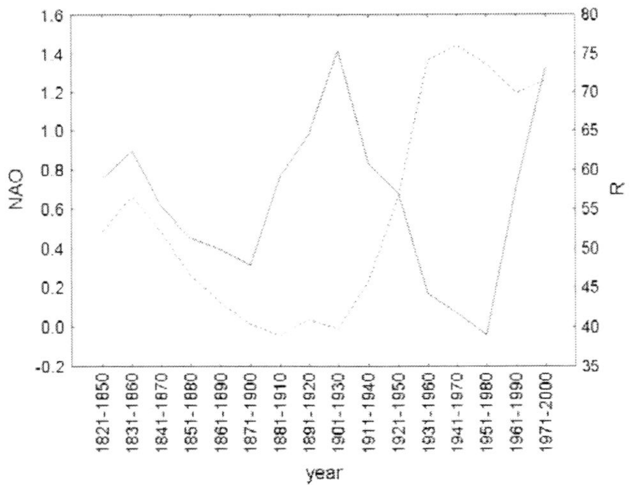

Figure 7. NAO index (solid line) from Jones et al. (1997) and sunspot number (broken line), climatic normals (30-year moving averages with a step of 10 years).

The atmospheric circulation and the rotation of the Earth are strongly related: If the solid Earth and the atmosphere can be considered a closed system, from the conservation of angular momentum it follows that when the zonal winds strengthen, the Earth's rotation rate must decrease, and vice versa. Some authors even propose to use the Earth's rotation rate for which there are longer measurement records as a proxy for the interannual variability in atmospheric circulation (Salstein and Rosen,1986).

Atmospheric Circulation and Earth's Rotation

It has long been known that the Earth doesn't rotate quite uniformly, but its rotation rate varies on different time-scales - from centuries to days. As a measure of the Earth's rotation rate, usually the length of the day (LOD) is used. LOD is of the order of milliseconds and is defined as the difference between the astronomically determined duration of the day and 86,400 SI seconds (24 hours x 60 minutes x 60 seconds). LOD measurements are provided by the International Earth Rotation and Reference Systems Service through their web-site http://hpiers.obspm.fr. The mean annual values of LOD are available since 1623, and daily values since 1962. Details on the computation of the duration of the day are given by Stephenson and Morrison (1984). The accuracy of determination of LOD varies throughout this period, after 1840 the error being less than 0.6 ms (Rivin, 1989) and small enough to relate to the atmosphere (Salstein, 1988). The decadal oscillations obvious in time-series derived from different types of observations agree fairly well to be concluded that they are a real feature of LOD (Gross, 2001).

The relationship of the angular velocity of the Earth ω with LOD is

$$\omega = \Omega^N (1 - LOD/T)$$

where $\Omega^N = 72{,}921{,}151.467064\ 10^{-12}$ radians/s is the nominal rotation rate (corresponding to the mean rotation rate of the epoch 1820) and T is the associated duration of the mean solar day, that is 86,400 s.

On secular time scale the Earth's rotation decelerates, and this deceleration is attributed to tidal friction – the friction of the oceanic tides caused by the Moon with the bottom of the ocean (Munk, 1997). On decadal time scales, the variations in the rotation rate have been shown to be correlated with the long term variation in the Earth's magnetic field (Golovkov, 1977; Djurovic and Stajic, 1985). As the main driving force of the geomagnetic field variations are processes in the core and the mantle of the Earth (Le Mouel and Courtillot, 1982; McFadden and Merill, 1995), this suggests that the core plays an important role in these LOD changes (Hide, 1995).

On subannual time-scale, the seasonal nontidal variations in the Earth's rotation rate were believed to be fully explained by the action of the atmospheric "interhemispheric heat machine", fed by the temperature differences between the summer and the winter hemispheres (Sidorenkov, 1978). In July and in January, when the temperature difference between the two hemispheres is greatest, the Earth rotates fastest. The northern hemisphere is on average warmer than the southern one, so the temperature difference in July is bigger than in January, and the Earth rotates fastest in July. In April and in November, when the temperature differences between the two hemispheres are small, the rotation is slow, the Earth rotates slower, and in spring when the temperatures are almost equal, the rotation rate is smallest. Figure 8 presents the annual cycle of the length of the day.

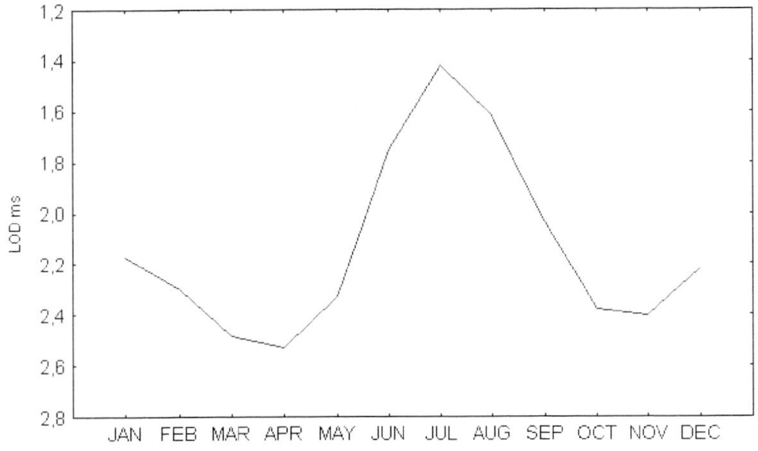

Figure 8. Annual variations of LOD in ms for the period 1962-2000. Note that the LOD scale is reversed to represent rotation rate.

Attempts have been also made to explain the interannual to decadal LOD changes by long-term changes in atmospheric circulation. Okazaki (1977) found that the global zonal wind circulation has anomalous long-term changes that are quite similar to the Earth's rotation velocity changes. Loginov (1992) demonstrated a satisfactory agreement between the changes in the number of days with zonal types of circulation and in the angular velocity of the Earth's rotation. A relation was also found between LOD and large-scale atmospheric phenomena like El Niño (Gross et al., 1996). Monin (1969) calculated that a change in the Earth's rotation rate could lead to a change in the extend of the trade winds and atmospheric waves, and to changes in the equator/pole temperature contrast, hence the zonality of atmospheric circulation. Hunt (1979) modelled the general atmospheric circulation for 5 times faster or slower Earth's rotation, and found differences in the formation of the centers of action of the atmosphere (persistent high or low pressure areas) and in the latitudinal temperature gradient. According to Maximov and Smirnov (1964), changes in the rotation rate cause a deformation in the sea surface resulting in a slope toward the pole or toward the equator. If the Earth rotates faster, the slope is toward the pole, the ocean currents like Gulfstream and Kuroshio strengthen and more warm water is transported to the pole, the equator/pole temperature contrast decreases, and so does the zonal circulation. In the opposite case, the ocean currents and the warm water transport to high latitudes weaken, and negative temperature anomalies are formed there, so the zonal circulation strengthens.

However, all these studies proving the relation between the Earth's rotation rate and atmospheric circulation are confined to the Northern hemisphere and cover only the 20th century, because of data limitations. The intensity of zonal circulation, as mentioned earlier, is determined by the temperature contrast between the equator and the pole. Therefore, as an indirect measure of the strength of the global zonal circulation, the temperature difference between the equatorial and polar regions can be used (Georgieva, 2002). The longest data-set providing zonally-averaged mean annual temperatures is the one of NASA Goddard Institute for Space Studies (Hansen and Lebedeff, 1987), from which the annual mean temperature anomalies with respect to the base period 1951-1980 averaged over the zones 64-90N and 64-90S were taken as the northern and southern polar temperatures, respectively (NP and SP), and over the zone 24N-24S - as the equatorial temperatures (EQ). The 64-90N and 24N-24S temperature anomalies are evaluated since 1866, and the 64-90S one - since 1903. Therefore, our EQ-NP temperature differences were calculated for the period 1866-2000, and EQ-SP - for the period 1903-2000. To validate them, they have been compared to the ones calculated from another long-term data-set, of the St.Peterburg State Hydrological Institute (Vinnikov et al., 1990) giving the northern hemisphere zonal means from 1880 to 1987 and the southern hemisphere ones - from 1957 to 1987. The EQ-NP temperature differences derived from the two data-sets agree within 97.7% with $p<0.01$, and the EQ-SP ones - within 88.9% with $p<0.001$.

The comparison between the long-term changes of EQ-NP - solid line, and LOD - dashed line (31-point running means) is presented in Figure 9.a. In the 20th century, there is a high positive correlation between the decadal variations in LOD and the strength of the zonal atmospheric circulation in the northern hemisphere (which means a negative correlation between the zonal circulation and the Earth's rotation rate). As both variables are not normally distributed, nonparametric statistics was used for the evaluation of the correlation between them (STATISTICA for Windows, StatSoft, Inc.), which yields r=94.7 %. This result is in agreement with earlier findings cited above and with the theoretical considerations that the strengthening of the zonal circulation should result in the deceleration of the Earth's rotation (and hence, in the increase of the length of the day). Surprisingly, however, before 1913 the relation is opposite, and the highest equator/pole temperature contrast is observed during the period with the smallest LOD, respectively the fastest Earth's rotation. The correlation between EQ-NP and LOD for this period is -89.7 %. The confidence levels for both correlations are better than 99.9%.

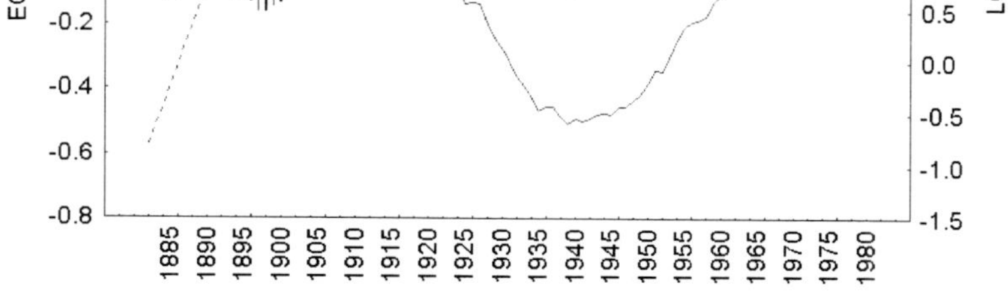

Figure 9.a. Decadal variations of the zonality of the atmospheric circulation in the northern hemisphere presented by the equator-pole temperature contrast (EQ-NP) in degrees Celsius - solid line, and of the length of the day (LOD) in milliseconds - dashed curve. The vertical bars present the North-South solar activity asymmetry.

The data-set for the southern hemisphere is shorter, so the relation between EQ-SP and LOD can only be estimated in the XX century - Figure 9.b.

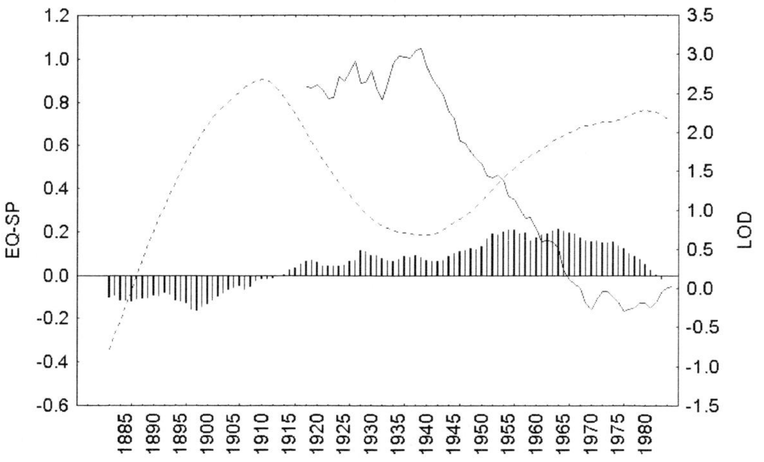

Figure 9.b. The same as Figure 2.5.a for the southern hemisphere equator-pole temperature contrast (EQ-SP).

The irregularities in the equator-pole temperature contrast in the beginning of the period are most probably due to the greater errors in the determination of the mean zonal temperature in the southern high-latitude region at these early times. Nevertheless, clearly seen is the negative correlation between the long-term variations of EQ-SP and LOD in the 20th century. The correlation between the two quantities is r=-87.6 %, with $p<0.01$. This is another surprising result, as the long-term changes in atmospheric circulation in the southern hemisphere in the 20th century are seen to be opposite to the ones in the northern hemisphere, though a close connection between the circulation in the two hemispheres is supposed (Girs and Kondratovich, 1978; Duane et al, 2000), and from the conservation of the angular momentum the variations of the circulation in the two hemispheres should be in phase.

To compare the behavior in the 11-year solar cycle of zonal atmospheric circulation in the two hemispheres as evaluated by the equator-pole temperature contrast, we have used the superposed epochs method (Georgieva, 2000). Figure 10 presents the relation between the strength of the zonal circulation and sunspot activity in the northern and southern hemispheres for odd and even solar cycles.

In even cycles, there is a pronounced minimum in the zonality of the circulation in the northern hemisphere in solar activity maximum, and a maximum in the southern hemisphere 2 years after the solar activity maximum. The picture in odd cycles is quite opposite: a deep minimum in zonal circulation is observed in the southern hemisphere in years of solar activity maximum, and a minimum in the northern hemisphere 2 years after the solar activity maximum. Due to the small number of data samples (5 even and 5 odd 11-year cycles for the southern hemisphere, Zurich numbers from 14 to 23, and 6 even and 7 odd cycles for the northern hemisphere, Zurich numbers from 11 to 23), the statistical significance of this result is not very high, however the differences in the behavior are obvious. This result not only demonstrates the different effect of solar activity on the northern and southern hemispheres, but also confirms that the circulation in the two hemispheres is different. Therefore, the simple considerations for conservation of the angular moment are not applicable to the relation between atmospheric circulation and the Earth's rotation on decadal time-scales as if they were true, the relation would have the same sign in the two hemispheres, and in all periods. Figs. 9 and 10 demonstrate that, as noted by Sidorenkov (1978), on time-scales longer than one year, the Earth/atmosphere system cannot be considered closed. Obviously, there are external factors which have to been taken into account.

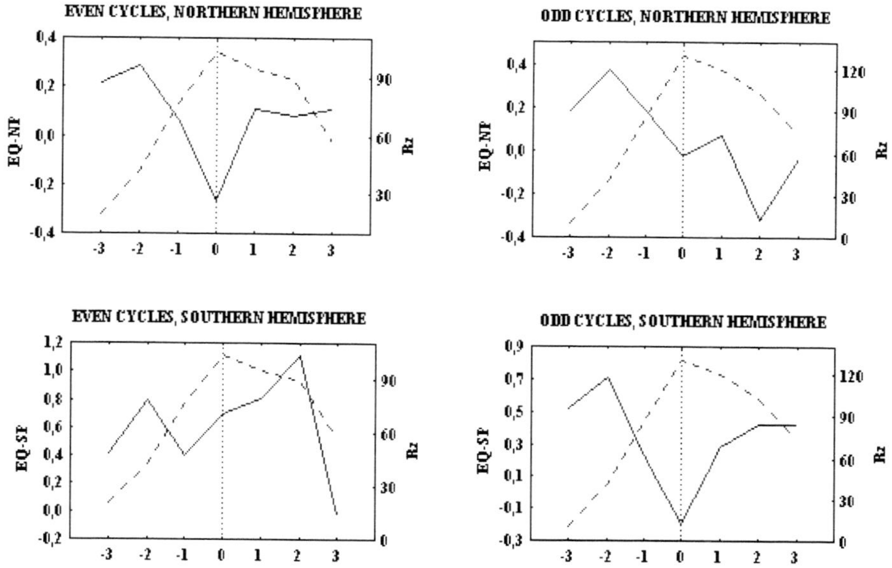

Figure 10. Equator-pole temperature contrast (solid line) and sunspot number Rz (dashed line) in the Northern hemisphere (top panel) and in the Southern hemisphere (bottom panel) for even (left side) and odd (right side) 11-year solar cycles in years of sunspot maximum (year 0), one year before and after sunspot maximum (+1 and -1, respectively), etc.

Solar Rotation and Earth's Rotation

The correlation between the long-term changes in atmospheric circulation and Earth's rotation rate changed sign around 1913. This coincides with the period of a change in the sign of the correlation between solar activity and surface air temperature in the 11-year solar cycle which, at it was shown above, is determined by the long-term changes in north-south solar activity asymmetry. The vertical bars in Figure 9 represent the solar activity asymmetry, Sa=(Sn-Ss)/(Sn+Ss) where Sn and Ss stand for the total sunspot area in the northern and southern solar hemispheres, respectively. More active northern solar hemisphere in the 20th century was associated with positive correlation between the decadal changes of EQ-NP and LOD, and negative correlation between EQ-SP and LOD (and positive correlation between solar activity and surface air temperature in the 11-year solar cycle), and more active southern solar hemisphere in the 19th century - with negative correlation between decadal changes in EQ-NP and LOD (and negative correlation between temperature and solar activity in the 11-year cycle).

Many authors have studied the relation between the Earth's rotation rate and solar activity. Kiselev (1981) showed that the long-term changes in solar activity must be taken into account in order to explain the century-scale retardation in the Earth's rotation rate. This idea was further advanced by Gu (1998) who speculated that the interaction between the solar wind and the magnetosphere of the Earth will lead to additional magnetic pressure in the tail of the magnetosphere which could produce angular deceleration of the Earth's rotation with a magnitude of $\omega = -1.7 \times 10\text{-}22\ s^{-2}$. Djurovic (1983) found common cyclic variations with periods of a few months in LOD, geomagnetic activity and the zonal component of the global atmospheric angular momentum as an argument in favour of the hypothesis that solar activity contributes to the angular velocity of the earth's rotation variations. Kosek (1993) found 27-day oscillations in the Earth rotation rate indicating a possible relationship between solar activity and the short periodic exchange of angular momentum between the atmosphere and the solid Earth. Lin (1998) found a weak linear connection between the Earth's rotation rate and the velocity of the solar wind implying that the Earth's rotation could be influenced by the variation in the velocity of the solar wind with a short time scale of several days. Mörner (2003) suggested that variations in the solar wind affect the Earth's rate of rotation, which in turn affects the oceanic and atmospheric circulation. The relation of the Earth's rotation rate, atmospheric angular momentum and solar activity was studied in detail by Abarca del Rio et al. (2003). Here we will look for a connection between the rotation of the Earth and the Sun.

Solar Rotation

The solar differential rotation can be determined from the full disk velocity data using the standard polynomial expansion (Howard and Harvey 1970),

$$\omega(\varphi) = A + B \sin^2(\varphi) + C \sin^4(\varphi)$$

where $\omega(\varphi)$ is the solar sidereal angular velocity at latitude φ, the coefficient A represents the equatorial rotation rate, and B and C - the latitudinal gradients of the rotation, with B representing mainly low latitudes and C largely higher latitudes. The above equation has no theoretical foundation, but fits very well to the observations.

The determination of the coefficients A, B and C can be made from solar tracers (Javaraiah and Gokhale, 1995), or from Doppler shift measurements (Piddington, 1971).

We have used the parameters of solar differential rotation calculated from the Greenwich data on sunspot groups from 1 January 1879 to 31 December 1976, and the spot group data from Solar Optical Observing Network (SOON) from 1 January 1977 to 10 August 2004, kindly provided by Dr.Javaraiah. Details on the data reduction and determinations of the values of the differential rotation coefficients A and B, are described by Javaraiah (2003).

When determining the solar differential rotation from sunspots data it is sufficient to use only the first two terms of the expansion (Newton & Nunn 1951):

$$\omega(\varphi) = A + B \sin^2(\varphi)$$

We first look for a relation between LOD and the equatorial rotation rate of the Sun. As pointed out by Javaraiah and Gokhale (1997), the yearly data for the solar differential rotation are too noisy, so the 5-year running averages of the solar North and South equatorial rotation rates, An and As, respectively, are used. To comply with An and As data, the 5-year running means of LOD are calculated for the same period. No correlation can be found of the decadal changes in LOD with the decadal changes in the equatorial rotation rate of the northern or the southern solar hemisphere, of the more active or the less active one, nor of the faster or the slower rotating one. The only quantity with which LOD does correlate, is the north-south asymmetry of the solar equatorial rotation Aa=(An-As)/(An+As). Figure 11 presents their detrended 11-point running means. The correlation between LOD and Aa is r=-0.92 with p<0.01. (Note that LOD scale is reversed to represent rotation rate). In Figure 12 the FFT power spectra of Aa (solid line) and LOD (dotted line) are compared. Both show a pronounced periodicity at 47 years, with a squared coherence of 0.83. This periodicity matches the dominant periodicity of 45.5±11.5 years in Aa determined by Javaraiah and Gokhale (1997) who studied in detail the north-south asymmetry in solar equatorial rotation rate. The periodicity of 45.57±0.33 years was found earlier by Bucha et al. (1985) in the solar inertial motion (i.e. the motion of the Sun around the center of mass of the Solar system).

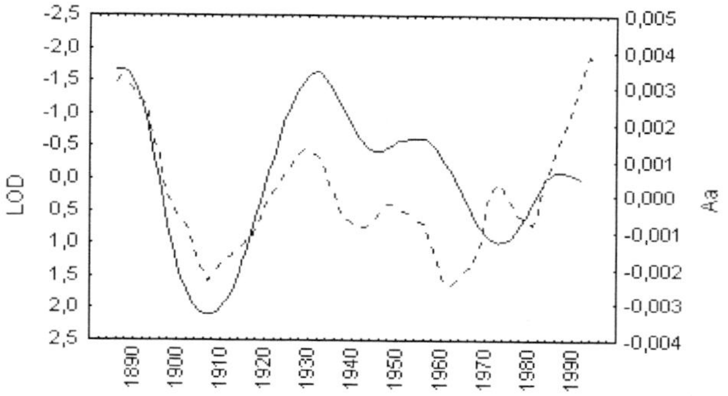

Figure 11. Decadal variations of the length of the day LOD in ms (solid line) and the north-south asymmetry of solar equatorial rotation Aa = (An-As)/(An+As) (dashed line).

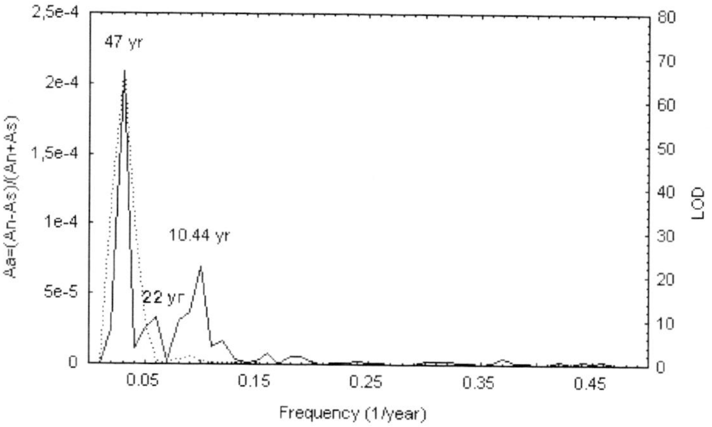

Figure 12. Periodicites in the north-south asymmetry of solar equatorial rotation raye (solid line) and LOD (dashes line).

Javaraiah (1996) suggested that the temporal variations of the solar differential rotation which agree with the known periodicities in the photospheric magnetic field, represent the "even-parity" - i.e. symmetric about the solar equator - torsional oscillations of the Sun (Howard and LaBonte, 1980), while the periodicities in Aa, also found in sunspot asymmetry, represent the "odd-parity" - antisymmetric - torsional solar oscillations, and showed that the amplitudes of the symmetric and antisymmetric torsional oscillations depend upon the contribution of the main planets to the symmetric and asymmetric parts of the angular momentum of the solar system. Charvatova (1997, 2000) calculated the solar inertial motion taking into account the varying positions of the giant outer planets in the Solar system and found that a change between disordered and trefoil-like pattern of the solar motion occurred about 1913 – the period when the correlation between sunspot activity and surface air temperature, between sunspot activity and the number of transformations of the forms of circulation, and between the Earth's rotation rate and the intensity of the zonal circulation all changed sign.

The planetary effects of the movement of Sun about the center of mass of the Solar system and their influence on solar activity and on the eventual solar-terrestrial influences have been studied in many works (Jose, 1965; Wood, K.D.1972; Wood, R.M. 1975; Landscheidt, 1981, 1987; Blizard, 1987; Fairbridge and Sanders, 1997). Javaraiah and Gokhale (1995) showed that the dominant periodicities in Sun's torsional magnetohydrodynamic oscillations match the north-south asymmetries in the parameters of the solar differential rotation, and the "combination periods" of two or more consecutive planets, hence it was suggested that planetary effects can provide the perturbations needed to excite these oscillations. Javaraiah (1996) pointed out the importance of the inclination of the solar equator to the ecliptic. Taking this into account, Juckett (2000) evaluated the momentum exchange between the Solar system dynamics and the Sun's rotational dynamics, and showed that because of the interaction between the orbital and angular momentum of the Sun, solar cycles and prolonged periods of very low solar activity may be caused or modulated by planetary induced spin-orbit interaction. The frequency modulated solar cycle and respectively the 22-year solar magnetic cycle follow from the combination of long-period components of P(t) - the normalized projection of the solar rotational axis onto its orbital radius, and L(t) - the solar orbital angular momentum. The high-frequency components of P(t) due to the orbits of Jupiter and Saturn, have the major contribution to the north-south sunspot asymmetry. Epochs of very low solar activity correspond to the beating generated by the superposition between the dominant 11,86-year cycle of Jupiter, and the frequencies due to the combination between P(t) and L(t).

Solar Wind as a Mediator Between the Solar Rotation and the Earth's Rotation

The close anticorrelation between the long-term variations of LOD and Aa may be due to the direct influence of the solar system dynamics on both the Earth's rotation rate and the North-South asymmetry of solar rotation. A more probable explanation, however, is that the variations in the Earth's rotation rate are induced by the solar wind transferring angular momentum and magnetic field from the Sun. The 45-50 year periodicity found in LOD and in the asymmetry of solar rotation has been also found in the Earth's electromagnetic core-mantle coupling torques (e.g. Greiner-Mai, 1987). Processes in the core and the mantle are considered one of the possible mechanisms for the excitation of LOD fluctuations (Hide, 1995) and the main driving force for the geomagnetic field variations (Le Mouel and Courtillot, 1982; McFadden and Merill, 1995). A close connection has been found between the variations in LOD and in the geomagnetic field (Golovkov, 1977; Djurovic and Stajic, 1985). Jin (1992) studied the cross correlation of the variations in the geomagnetic dipole moment and in the fluctuations of the Earth's rotation rate and found a dominant peak of 51 years in both of them (matching the corresponding peak of 47 years in LOD and Aa demonstrated above, and the 45.5±11.5 years in Aa found by Javaraiah and Gokhale, 1997), with a coherence between the two of above 0.99.

Some of the mechanisms proposed to explain the relation between solar activity and the Earths' rotation assume direct solar activity influences on the core and/or the mantle. Laptukhov (1980) suggested a mechanism which involves heating of the Earth's mantle by an alternating electromagnetic field whose intensity increases with solar activity, as a result the convection rate changes, the gravitational differentiation of the mantle material is amplified, the Earth's moment of inertia decreases, and the angular velocity increases such that the total moment of momentum of the Earth remains constant. Greiner-Mai (1987) found that the dominant periodicity in the electromagnetic core-mantle coupling torques is 22 years, and this periodicity is missing in the mechanical coupling torques. As 22 years is the length of the solar magnetic cycle, this implies that the relation between the rotation of the Sun and the Earth is due to electromagnetic rather than gravitational interactions.

According to other mechanisms, the effects of solar activity on the Earth's rotation rate are mediated by the solar wind–magnetosphere interaction. Gu (1998) calculated that the interaction between the solar wind and the magnetosphere of the Earth will lead to additional magnetic pressure in the magnetic tail which could influence the long-term variations in the Earth rotation with a the magnitude of $\omega = -1.7 \times 10\text{-}22\ s^{-2}$. According to the mechanism of Krymskij (1993), the solar wind interaction with the magnetosphere leads to the generation of a convective electric field and meridional Hall currents are generated in the magnetotail plasma layer which close through the high-latitude E-layer of the ionosphere by the Pedersen current. The plasma collision frequency with neutrals at the altitude of E-layer is high enough for the neutral atmosphere to be dragged by the plasma, so an eastward flow occurs. This rotation is continually transmitted to the Earth, with the moment of momentum being accumulated in the atmosphere during winter and passed to the Earth during spring.

This group of mechanisms also implies a 22-year periodicity, corresponding to the solar magnetic cycle. If a 22-year periodicity already demonstrated in atmospheric circulation (Figure 10) is also found in the solar rotation, in the parameters of the solar wind and interplanetary magnetic field, and in the Earth's rotation rate, this will be a serious indication that the rotation of the Sun and the Earth are not both directly influenced by the Solar system dynamics, but instead the Earth's rotation is influenced by the solar rotation, with the solar wind as a mediator.

Solar Rotation in Positive and Negative Polarity Cycles

After the pioneering work of St.John (1918) who found that the solar rotation is not constant, the time variations in the solar differential rotation have been extensively studied (Balthazar and Wohl, 1980; Arevalo et al., 1982; Balthazar et al., 1986; Hathaway and Wilson, 1990;

Pulkkinen and Tuominen, 1998, among others). Javaraiah (2000) studied the solar differential rotation and its north-south asymmetry for the period 1879-1976 computed from data on sunspot groups compiled from Greenwich Photo-heliographic Results, and found a difference in the rotation in even and odd numbered cycles: in odd numbered solar cycles the Sun's equatorial rotation is significantly larger, the rotation is more differential, and the north-south asymmetry in the equatorial rotation is greater than in even numbered solar cycles. Gigolashvili (2001), on the basis of observations on sunspots and hydrogen filaments during 1950-1990 and 1965-1993, respectively, studied solar differential rotation separately for the northern and southern solar hemispheres, and found that the northern hemisphere rotates faster during the even cycles (Cycles 20 and 22) while the rotation of the southern hemisphere dominates in the odd one (Cycle 21).

To comply with the differences in the annual cycle of geomagnetic activity, we compare the solar rotation determined from the Greenwich data not in even and odd but in positive and negative polarity solar cycles.

In a previous study (Kirov et al., 2002) it was found that in the period 1879-1976 the solar equatorial rotation was faster (at a significance level of 2 sigma) and the latitudinal gradient of the rotation rate was greater (at 4 sigma) in positive polarity cycles than in negative polarity cycles. Because of the high latitudinal gradient, the rotation rate in positive polarity cycles was quickly falling with latitude, and at all latitudes above about 15° the rotation rate was higher in negative polarity cycles than in positive polarity cycles (Figure 13.a). It is known that the solar rotation rate varies on different time-scales, so we calculated separately the rotation in the period 1967-2000 for which we have daily data for solar rotation, solar wind and Earth rotation (Figure 13.b). In this period, in negative polarity cycles Sun rotates faster at all latitudes, and the latitudinal gradient of the rotation rate is bigger. The difference in the equatorial rotation rate is 2 sigma, and in the latitudinal gradient of the rotation rate – 4 sigma.

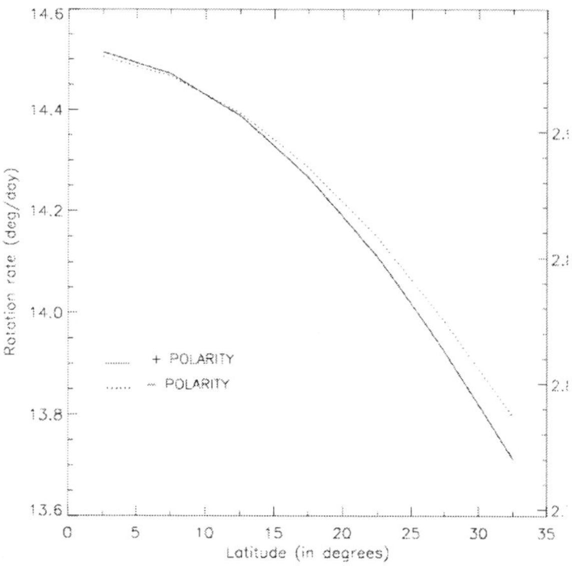

Figure 13.a. Solar differential rotation (average over the two solar hemispheres) in positive (solid line) and negative (broken line) polarity cycles in the period 1879-1976. From Kirov et al. (2002).

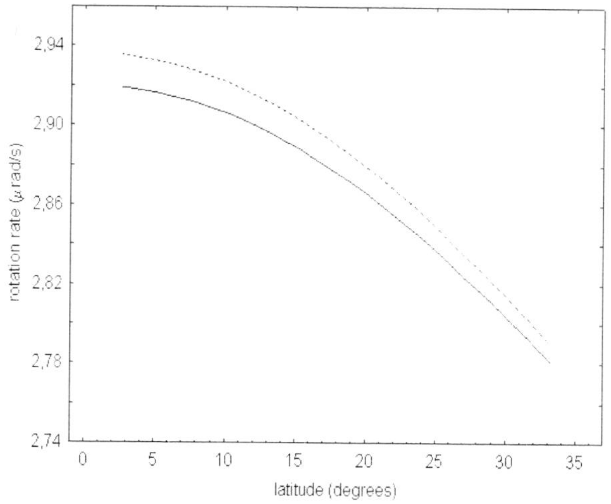

Figure 13.b. Solar rotation (average over the two hemispheres) in positive (solid line) and negative (broken line) polarity cycles in the period 1962-2000.

Earth's Rotation in Positive and Negative Polarity Cycles

As mentioned in the previous section, on time-scales less than a year the Earth-atmosphere system is believed to be closed, so the seasonal variations in the Earth's rotation rate should be fully explained by the redistribution of air masses. However, it was recently shown (Kirov et al., 2002) that the Earth's rotation rate depends on the magnetic polarity of the Sun. Figure 14 presents the seasonal cycle of the Earth's rotation rate in cycles with positive and negative solar magnetic polarity.

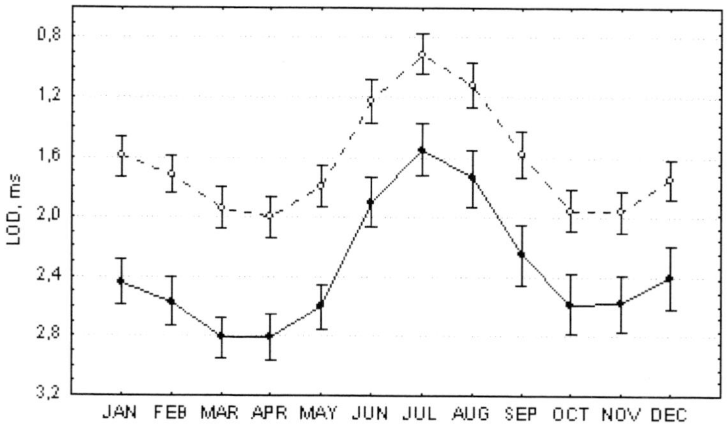

Figure 14. Length of the day (LOD) in cycles with positive (solid line) and negative (broken line) polarity solar magnetic cycles. The vertical bars represent the standard errors. Note that the LOD scale is reversed to represent rotation velocity.

All year long, the Earth rotates faster in negative polarity solar cycles. The biggest differences in the seasonal variations in the rotation rate between positive and negative polarity cycles are in equinox periods. From the theory of the interhemispheric heat machine it follows that in autumn Earth rotates faster than in spring as the temperature differences between the two hemispheres are greater. This is indeed the situation in positive polarity cycles, while in negative polarity cycles the Earth's rotation rate is almost equal in spring and autumn equinoxes. Therefore, in negative polarity solar cycles some external factor both increases the mean annual rotation rate, and additionally speeds up the rotation in spring.

Solar Wind in Positive and Negative Polarity Cycles

Slow solar wind comes from the closed magnetic field regions in the heliospheric current sheet separating the two solar hemispheres with opposite magnetic polarities, from slow coronal mass ejections, and from small low-latitude coronal holes. Fast solar wind mostly originates from the big polar coronal holes on the declining phase of the sunspot cycle and around sunspot minimum, and from solar flares and fast coronal mass ejections around solar maximum. During the declining phase of the sunspot cycle and around sunspot minimum, the polar coronal holes extend toward the equator. In these periods, during spring and fall seasons when the Earth is at highest heliolatitudes, these big polar coronal holes contribute to the solar wind reaching the Earth, and the solar wind speed has a maximum. Zieger and Mursula (1998) found that around positive polarity sunspot minima the solar wind is faster in spring than in fall, and around negative polarity sunspot minima it is faster in fall than in spring. As in spring the Earth is at highest southern heliolatitudes and in fall – at highest northern heliolatitudes, this means that faster solar wind originates from the southern solar hemisphere in positive polarity cycles, and from the northern solar hemisphere in negative polarity cycles. Figure 15 demonstrates the seasonal variation in solar wind speed in positive and negative polarity cycles based on data from OMNI for the period 1964-2000.

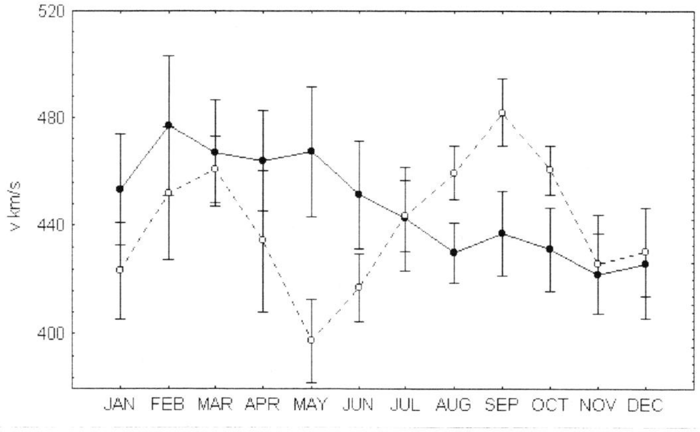

Figure 15. Solar wind velocity (V) in cycles with positive (solid line) and negative (broken line) polarity solar magnetic cycles.

The fact that in autumn the solar wind speed is higher in negative polarity cycles should lead, according to the mechanism of Krymskij (1993), to accumulation of momentum in the atmosphere during the winter, and transmitting it to the solid Earth in spring. This is consistent with the increased Earth's rotation rate in spring in negative polarity cycles (Figure 14).

We could therefore conclude that solar wind is the probable agent transmitting torques and magnetic fields which influence the Earth's magnetosphere and/or processes in the core and mantle leading to variations on different time scales in the Earth's rotation rate.

Solar Wind and the Solar Magnetic Field

The solar wind is the extension of the solar corona, carrying the Sun's magnetic field to the Earth and beyond. Because of the high electrical conductivity, the magnetic field is "frozen" in the wind – when the magnetic pressure is much higher than the plasma pressure (as in the solar corona), the plasma flows along the magnetic field lines, and when the plasma pressure is much higher (as in the solar wind), the plasma carries the magnetic field.

Interplanetary Magnetic Field and Solar Photospheric Field

In Figure 16.a the magnitude B of the interplanetary magnetic field (IMF) at the Earth's orbit is compared to the photospheric magnetic field. Solar wind parameters have been measured by a series of spacecraft since the beginning of space era, mid 1960's, and are compiled in OMNI data base of the National Space Science Data Center (http://nssdc.gsfc.nasa.gov/omniweb/). The photospheric magnetic field has been measured by the Wilcox Solar Observatory (Scherrer et al., 1977) since 1976. The grid of data available online (http://wso.stanford.edu/synoptic.html) consists of 30 equal steps in heliographic latitude - from 75.2^o North to 75.2^o South, and 5 degree steps in heliographic longitude. The northern and southern photospheric magnetic fields MFN and MFS presented in Figure 16.a are integrated over the northern and southern solar hemispheres, respectively.

The IMF magnitude B has a clear 11-year cyclicity while the photospheric magnetic fields in the two hemispheres exhibit the 22-year Hale magnetic cycle. The maximum B values correspond to the polarity reversal periods around sunspot maximum when fields of both polarities are present in each of the hemispheres so their integrated values are close to zero. The minimum of B is in periods when the photospheric magnetic field in the two hemispheres has maximum positive or negative values. Obvious in the figure is the asymmetric evolution of the integrated photospheric magnetic field in the northern and the southern hemispheres in the 23 sunspot cycle. In the polar regions of the northern hemisphere the polarity reversal was completed by February 2000, while over the south pole it didn't occur before October 2000 (Wang et al., 2002). The difference in the hemispherically integrated magnetic field is still greater – it changed polarity in 1997 in the southern hemisphere, and only in 2002 in the northern hemisphere.

The 22-year cycle is characteristic of the components of the interplanetary magnetic field. Bx has a maximum positive value in the maximum of the even sunspot cycle (1990) and maximum negative values in the maxima of odd sunspot cycles (1980 and 2000). An even-odd asymmetry in Bx can also be noted: its magnitude is higher in the maximum of cycle 22 that in the maxima of the adjoining odd cycles 21 and 23 (Figure 16.b).

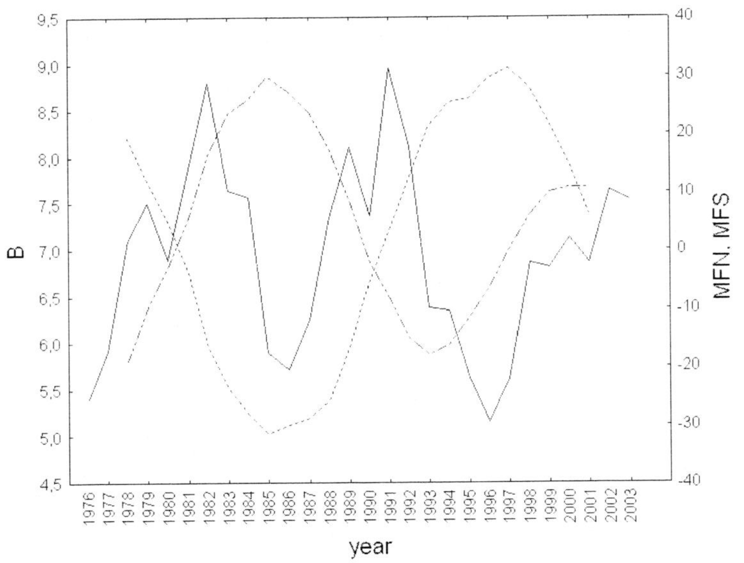

Figure 16.a. IMF at the Earth's orbit (solid line) and the solar photospheric field integrated over the northern (MFN - broken line) and southern (MFS - dash-dotted line) hemispheres.

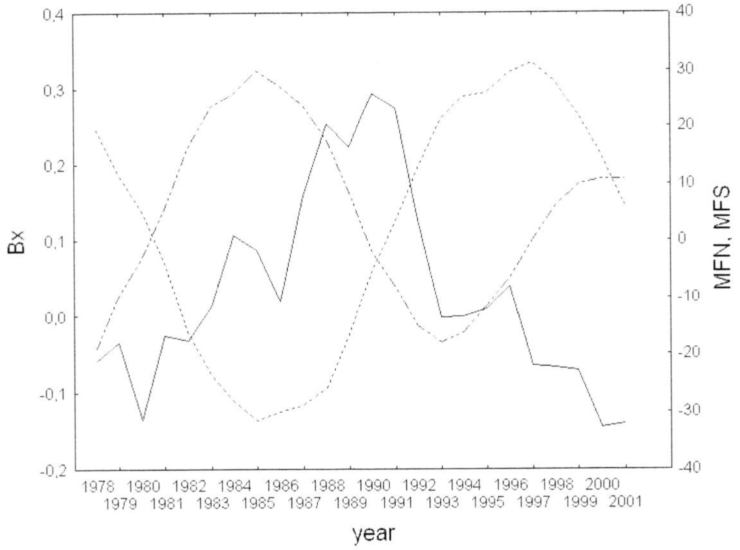

Figure 16.b. IMF B_x component at the Earth's orbit (solid line) and the solar photospheric field integrated over the northern (broken line) and southern (dash-dotted line) hemispheres.

IMF B_y-component also has a 22-year cyclicity but, unlike B_x, it is in phase with the solar photospheric field with a delay of about 2 years (Figure 16.c). B_y has a maximum positive value in sunspot minimum between an even and an odd cycle (i.e. in positive solar magnetic polarity maximum when the magnetic field points away from the Sun in the northern hemisphere and toward the Sun in the southern hemisphere) and maximum negative value in sunspot minimum between odd and even sunspot cycle (negative solar magnetic polarity maximum when the magnetic field points toward the Sun in the northern hemisphere and away from the Sun in the southern hemisphere). We should remind here that in GSM

(Geocentric-Solar-Magnetospheric) reference system in which the parameters of the solar wind and interplanetary magnetic field at the Earth's orbit are measured, the x-axis points from the Earth to the Sun, so B_x is positive when the interplanetary magnetic field is toward the Sun, B_z points to the north, and B_y complements the right-handed system.

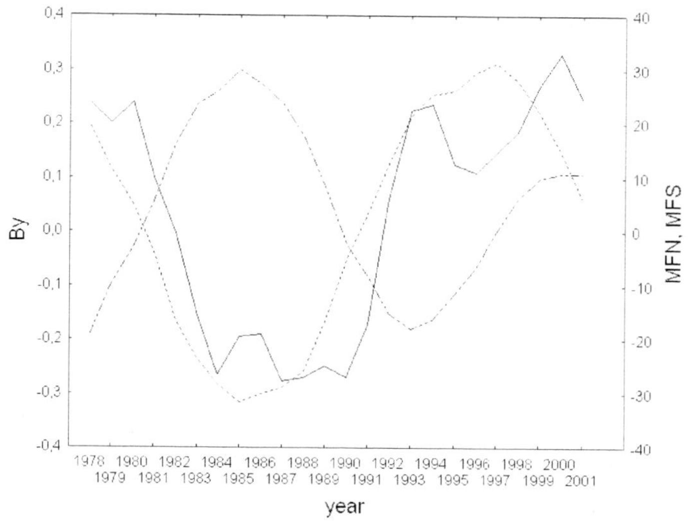

Figure 16.c. IMF B_y component at the Earth's orbit (solid line) and the solar photospheric field integrated over the northern (broken line) and southern (dash-dotted line) hemispheres.

Bz component varies much like B_x (Figure 16.d): it has a high positive value in even sunspot cycle maximum, and its yearly averages are around zero in the adjoining odd sunspot cycles maxima. As pointed out earlier, B_z component is of critical importance for the interaction of the solar wind with the terrestrial magnetosphere (see Figure 4).

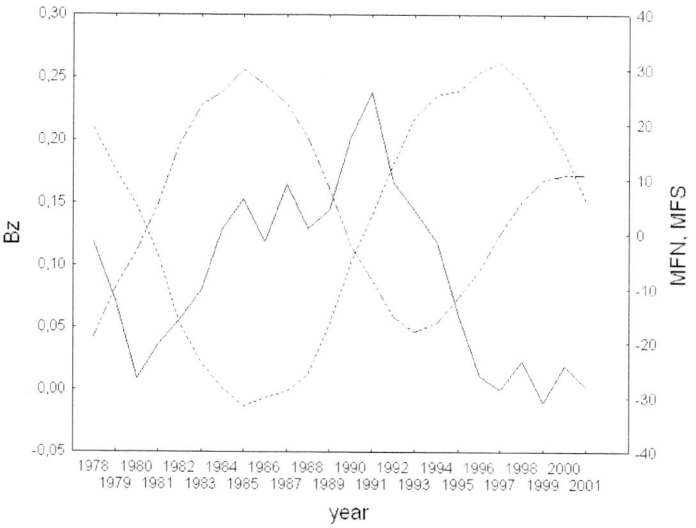

Figure 16.d. IMF Bz component at the Earth's orbit (solid line) and the solar photospheric field integrated over the northern (broken line) and southern (dash-dotted line) hemispheres.

Solar Photospheric Field and the Earth's Rotation

The 22-year Hale solar magnetic cycle is evident also in the LOD variations (Figure 17). LOD is minimum (the Earth rotates fastest) in maximum negative polarity in the sunspot minimum between odd and even sunspot maxima, and LOD is maximum (the Earth rotation rate is lowest) in maximum positive solar polarity (around sunspot minimum between an even and odd sunspot cycle).

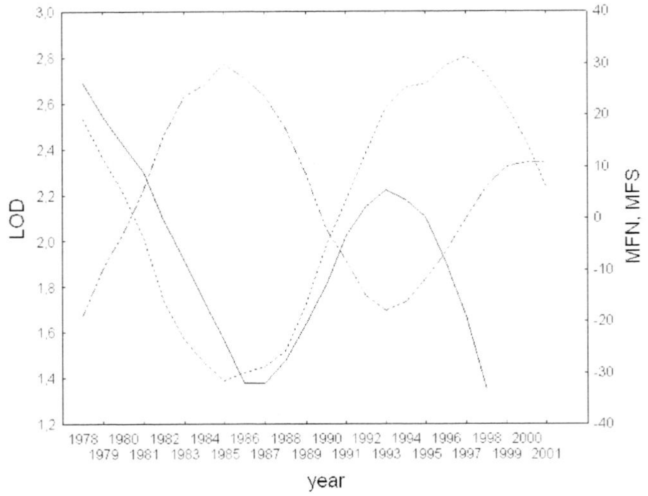

Figure 17. Length of the day (LOD) and the solar photospheric field integrated over the northern (broken line) and southern (dash-dotted line) hemispheres.

The dependence of LOD on the photospheric magnetic field is very similar to the dependence of the B_y component of the interplanetary magnetic field (compare with Fig16.c). Negative B_y corresponds to a shorter day, therefore to faster Earth's rotation, and positive B_y – to a longer day and slower rotation.

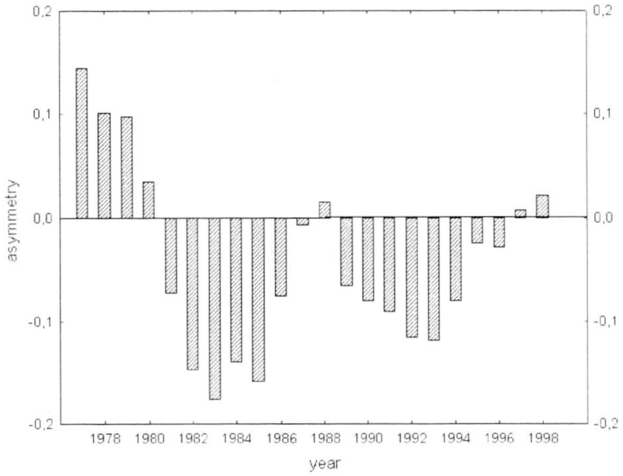

Figure 18. Asymmetry of the number of sunspots in the northern and southern solar hemispheres sefined as $(S_N-S_S)/(S_N+S_S)$, 5-year smoothing, from Temmer et al. (2002).

In the sunspot maximum between cycles 22 and 23, when the polarity reversal of the hemispherically integrated photospheric field in the northern solar hemisphere was delayed with respect to the southern one, LOD was in phase with the variations in the photospheric field in the southern solar hemisphere. As pointed out by Mursula et al. (2003), in the period studied the heliospheric current sheet was displaced toward the north, therefore the Earth was more exposed to solar wind originating from the southern solar hemisphere. Moreover, during this period, the southern solar hemisphere was the more active one. Figure 18 shows the north-south solar activity asymmetry calculated by hemispheric sunspot numbers for the period 1975-2000 (Temmer et al., 2002) provided online by Manuela Temmer at http://www.uni-graz.at/igam5www/daily_ RnRs/monthly_RnRs.html.

Interplanetary Magnetic Field and Solar Equatorial Rotation

As demonstrated in the previous paragraph, the variations in the Earth's rotation rate are related to the variations in the IMF azimuthal component By. On the other hand, By is directly proportional to the solar equatorial rotation rate, so the variations in the solar rotation rate should lead to variations in By and hence in LOD.

The components of the IMF at the Earth's orbit are given by Parker's model (Parker, 1958) where a purely radial solar wind is assumed, emanating with a constant velocity from a uniformly rotating Sun, with the magnetic field frozen in the flow. As this frozen magnetic field is an extension of the solar magnetic field, it is rooted at one end in the Sun, and is twisted into an Archimedean spiral by the combined effects of solar rotation and the outward solar wind flow. The winding angle of the spiral increases with distance from the Sun and at the Earth's orbit is about 45°. In a spherical coordinate system its components are expressed by

$$B_r(r) = B_r(r_0)\left(\frac{r_0}{r}\right)^2$$

$$B_\theta(r,\theta,\phi,t) = B_r(r_0)\left(\frac{r_0}{r}\right)\frac{V_\theta(\theta,\phi',t')}{V_w}$$

$$B_\phi(r,\theta,\varphi,t) = B_r(r_0)\left(\frac{r_0}{r}\right)\frac{-r_0\Omega\sin\theta + V_\phi(\theta,\phi',t')}{V_\omega}$$

where r_0 is the the radius of the "source surface" defined as a sphere around the Sun, inside which no currents flow, and on which and outside it currents restrict the field to being radial, and then stretch toward a spiral (Schatten et al., 1969) - in other words, magnetic field lines from the Sun may loop back to the Sun inside this surface, but lines which have crossed it once proceed into the heliosphere and never come back; Ω is the solar rotation rate, V_ω is the solar wind radial velocity (assumed constant), and V_0 and V_ϕ the polar and azimuthal velocities, respectively, at the source surface $r=r_0$, and $t'=t-(r-r_0)/V_\omega$, where ϕ is the longitude and θ the colatitude. This is a generalization of Parker's formula provided by Giacalone (2001). The nominal Parker's formula is retained by setting $V_0=0$ and $V_\phi=0$.

In Parker's formula the solar rotation rate is assumed to be constant. However, it has been shown to vary on different time-scales (Javaraiah and Gokhale, 1995). The dependence of B_y on Ω is statistically significant, though weak, and with a maximum correlation r=0.056 when B_y is lagged with respect to Ω by 4 days, as expected from the time for the propagation of the solar wind from the Sun to the Earth. In Figure 19 the cross-correlation function is shown between the equatorial solar rotation rate (averaged over the two hemispheres) and the B_y component of the interplanetary magnetic field at the Earth's orbit. The correlation coefficient is not high because in periods of interplanetary disturbances the behavior of B_y doesn't obey Parker's law (Georgieva et al., 2002), but still it is highly statistically significant. The dotted vertical lines in Figure 18 represent the 95% confidence limit.

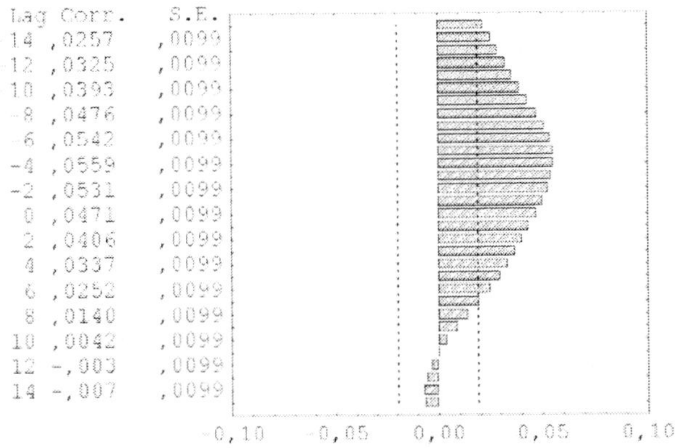

Figure 19. Cross-correlation function of the observed azimuthal IMF component and the equatorial solar rotation rate, with the 0.95 significance level.

Therefore, the variations in the solar rotation should be taken into account when comparing Parker's model to the actually measured IMF components. To take account of the non-constant solar rotation rate, in the above formula we introduce $\Omega=\Omega\ (\theta', t')$ and, following Parker (1958), we keep the zero velocity azimuthal component $V_\phi=0$. Further, we confine our study to the ecliptic plane, so $\sin\theta=1$, and $\Omega\ (\theta', t')=\Omega\ (t')$ is the solar equatorial rotation rate. We now denote the measured in-situ values of the azimuthal and radial components of the solar wind velocity by B_y and B_x, respectively, to distinguish from the model ones B_ϕ and B_ω, and calculate the difference between the observed and the model IMF azimuthal component, $dBy=By-B_\theta$. The yearly averages of dBy (5-year running means) are presented in the middle panel of Figure 20, and in the top panel – the sunspot cycle.

The agreement between Parker's model and observations has been studied by a number of authors. Smith and Bieber (1991), based on data between 1965 and 1987, found that from 0.7 to 16 AU the measured winding angle of the heliospheric magnetic field defined as B_{long} = arctan (B_x/B_y) is always greater than the predicted one. Moreover, they found that the difference between the model prediction and the measured value varies in phase with the sunspot cycle, with a 10° variation between solar minimum and solar maximum, and suggested that this variation may be partly due to the solar cycle variations in solar wind speed. Bruno and Bavassano (1997) confirmed this result for 0.3 – 1 AU and found a definite dependence of the winding angle on the solar wind speed. However, even after taking into account the measured solar wind speed, the winding angle still remained overwound by several degrees beyond the theoretical value.

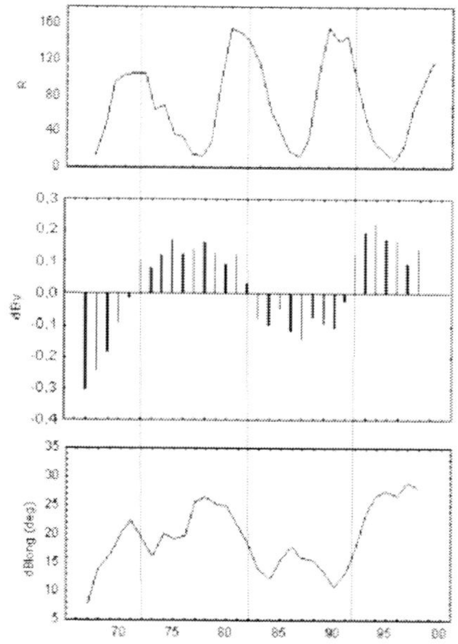

Figure 20. Yearly averages (5-year running means) of the international sunspot number (upper panel), the difference between the observed and calculated IMF azimuthal component (middle panel) and the between the observed and calculated IMF winding angle (bottom panel). The vertical lines denote periods of solar magnetic polarity reversals.

Figure 20 demonstrates that unlike the measured IMF winding angle which has been found by Smith and Bieber (1991) to be always "overwound" with respect to the theory predictions, the measured IMF azimuthal component can be both bigger or smaller than the model one, that is, the difference *dBy* can be both positive and negative. Moreover, the sign of this difference changes shortly after sunspot maximum coinciding with the solar polarity reversals. In positive (negative) polarity cycles, B_y is larger (smaller) than predicted. Therefore, the difference between the measured and calculated IMF azimuthal component is governed by the solar polarity cycle rather than by the sunspot cycle as suggested by Smith and Bieber (1991) and Bruno and Bavassano (1997) for the IMF winding angle.

The conclusion of Smith and Bieber (1991) about the sunspot cycle dependence of the difference between the measured and predicted IMF winding angle has been drawn on the basis of data from 1965 to 1987, and using a constant solar rotation rate. We have recalculated this difference dB_{long} until 2000 taking account of the measured solar rotation rate, and it is presented in the lower panel of Figure 20. It is clearly seen that the difference between the measured and predicted IMF winding angle dB_{long} also demonstrates a 22-year dependence, matching the 22-year solar magnetic polarity cycle: dB_{long} is bigger (more "overwound") in positive polarity cycles, and smaller ('less "overwound") in negative polarity cycles.

The systematic variation of the differences between the measured and calculated IMF azimuthal components means that Parker's formula doesn't properly account for all factors determining them. The neglected or underestimated factor we are looking for should obviously have a 22-year periodicity. To identify it, we have compared the mean values in negative and positive polarity solar cycles of the IMF magnitude (*B*) and solar wind speed (*V*) from OMNI data-base, and the solar rotation parameters: the equatorial rotation rate *a* and the latitudinal rotation rate gradient *b* for the whole sphere, and separately for the Northern and Southern hemispheres, a_n, b_n, a_s and b_s, respectively, from Greenwich data – Table 1.

Table 1. Mean values and standard deviations in positive and negative polarity solar cycles (+ and -, respectively), of IMF magnitude *B*, solar wind velocity *V*, solar equatorial rotation rate and latitudinal gragient for the whole sphere (*a* and *b*) and for the northern and southern hemispheres (a_n, a_s, b_n, and b_s, respectively)

	B	**V**	**a**	**a_n**	**a_s**	**b**	**b_n**	**b_s**
Average in +	5.823	442.33	2.933	2.932	2.934	-0.564	-0.570	-0.557
Sigma in +	1.071	45.359	0.001	0.001	0.002	0.011	0.014	0.015
Average in -	6.096	442.52	2.931	2.928	2.934	-0.499	-0.502	-0.498
Sigma in -	1.136	54.967	0.001	0.002	0.002	0.013	.017	0.019

The following differences are seen between the positive and negative polarity cycles:

In positive polarity solar cycles

- the solar rotation rate is faster
- the latitudinal gradient of the rotation rate is greater
- the IMF azimuthal component is greater than calculated
- the IMF winding angle is more overwound

In negative polarity solar cycles

- the solar rotation rate is slower
- the latitudinal gradient of the rotation rate is smaller
- the IMF azimuthal component is smaller than calculated
- the IMF winding angle is less overwound

The solar rotation rate is included in Parker's formula. What is neglected in it are the solar wind and magnetic field non-radial components. One of the possible reasons for the 22-year periodicity in the difference between Parker's formula and the measured B_y, respectively the IMF winding angle B_{long}, may be the presence of a non-zero azimuthal solar wind component at the source surface, proportional to the solar rotation rate. This probably has some effect, as not only the difference dB_y, but B_y itself show a clear 22-year dependence, however, it cannot explain the negative differences.

Another reason for the 22-year periodicity in the deviations of the IMF azimuthal components from the theoretical predictions can be the presence of a small azimuthal magnetic field component on the source surface, as suggested by Smith and Bieber (1991).

Such azimuthal component is forecasted by the model of Fisk (1996) of the heliospheric magnetic field. According to this model, the differential rotation of the photosphere and the nonradial expansion of the magnetic field and solar wind from the polar coronal holes which are the main source of the open magnetic flux and hence of the IMF at solar minimum, will lead to magnetic field footpoint motions at low latitude on the source surface. These footpoint motions can be in the direction of the solar rotation or opposite to it and, as discussed by Zurbuchen et al. (1998), will lead to systematic overwinding or underwinding of the heliospheric magnetic field, depending on the magnetic polarity. If this hypothesis is true, the result must be a 22-year dependence of the difference between the Parker spiral and the actual heliospheric field, with the biggest deviations occurring in solar minimum. This is exactly what we observe.

Interplanetary Magnetic Field and the Latitudinal Gradient of Solar Rotation

While the solar equatorial rotation rate is important for the IMF azimuthal components, the latitudinal gradient $\boldsymbol{b}$ in the formula of the solar differential rotation $\omega(\varphi) = a + b\ \sin^2(\varphi)$, expressing the differential character of the rotation, is related to the total IMF magnitude. According to the solar dynamo mechanism proposed by (Babcock, 1961), the large-scale solar magnetic field is generated jointly by differential rotation and turbulent convection. As it has been found that the two solar hemispheres rotate in a different way (e.g. Arevalo et al., 1982; Javaraiah, 2003; Gigolashvili et al., 2003, Temmer et al., 2002, 2003), this should result in differences in the magnetic field generated by the differential rotation in the two solar hemispheres, and consequently in differences at the Earth's orbit in the IMF originating from the northern and from the southern hemispheres. Antonucci et al. (1990) analyzed photospheric synoptic charts obtained at the Wilcox Solar Observatory for solar cycles 20 and 21, and pointed out the different behavior of the solar photospheric magnetic field originating from the northern and southern solar hemispheres, suggesting only a weak magnetic coupling between the two. Temmer et al. (2002), based upon full-disc averages of the daily line-of-sight component of the unsigned magnetic field strength measurements from the Kitt Peak National Solar Observatory, confirmed this conclusion for cycles 21, 22 and the current part of cycle 23. In an earlier work (Georgieva et al., 2005) we compared the periodicities in IMF magnitude at the Earth's orbit to the periodicities in the latitudinal rotation gradient in the two solar hemispheres since 1881, and found that the interplanetary magnetic field is related to the differential rotation in the more active solar hemisphere. Figure 21 demonstrates the Fourier spectra of the latitudinal gradients of the differential rotation in the northern and southern solar hemispheres (b_n and b_s, respectively) and of the magnitude of the IMF at the Earth's orbit for three periods – 1881-1912 when the southern solar hemisphere was more active, 1913-1966 when the northern hemisphere was more active, and 1967-2000 when more active was again the southern hemisphere. In the last period for which we have direct satellite measurements of the IMF, we have compared the periodicities in b to the periodicities in IMF magnitude B. In the two earlier periods we have used instead ***aa***-index of geomagnetic activity as a proxy for the IMF magnitude.

In the first period, the dominant periodicity in b_s is 16 years, and the same is the dominant periodicity in ***aa***, while b_n varies with a period of 10.8 years (Figure 21.a). During the second period, from 1913 to 1966, the dominant periodicity in ***aa*** is 10.8 years. A strong peak at this periodicity is seen in $\boldsymbol{b_n}$, and no peak in $\boldsymbol{b_s}$ – Figure 21.b. In the last period the dominant periodicity in the average IMF magnitude ***B*** in this period is 9.3 years coinciding with the dominant periodicity in $\boldsymbol{b_s}$, while the dominant periodicity in $\boldsymbol{b_n}$ is 14 years (Figure 21.c). Therefore, in all periods the dominant periodicity in the IMF matches the dominant periodicity in the differential rotation of the more active solar hemisphere.

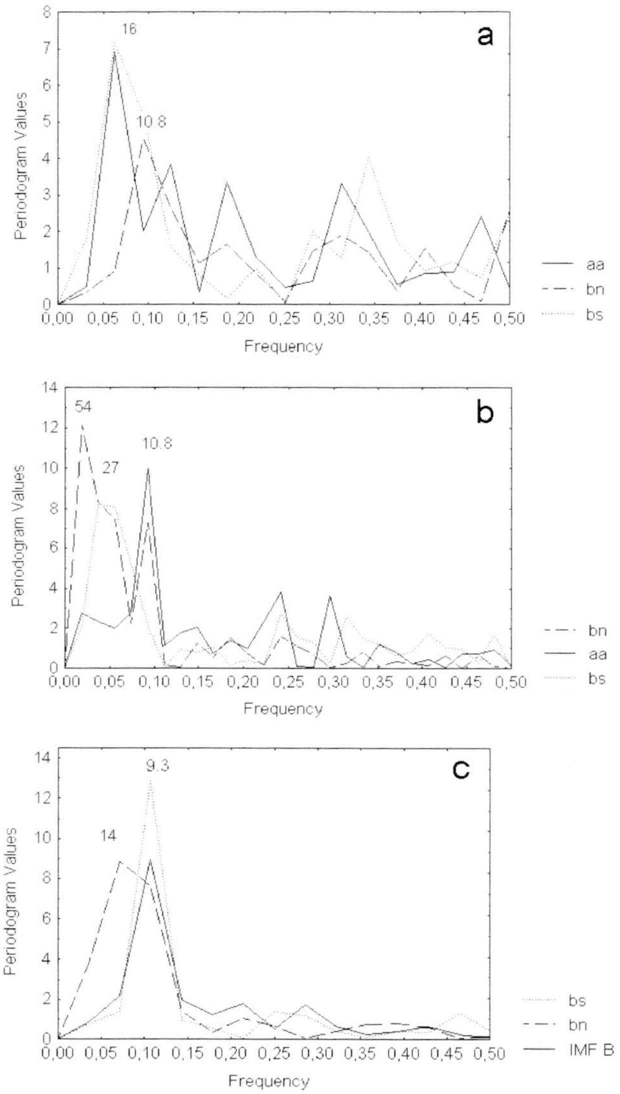

Figure 21. a - Periodicities in the aa-index of geomagnetic activity aa (solid line), and the rotation rate latitudinal gradients in the Northern solar hemisphere b_n (broken line) and in the Southern solar hemisphere b_s (dotted line) in the period 1881-1912;.b - Periodicities in the aa-index of geomagnetic activity aa (solid line), and the rotation rate latitudinal gradients in the Northern solar hemisphere b_n (broken line) and in the Southern solar hemisphere b_s (dotted line) in the period 1913-1966; c - Periodicities in IMF magnitude B (solid line), and the rotation rate latitudinal gradients in the Northern solar hemisphere b_n (broken line) and in the Southern solar hemisphere b_s (dotted line) in the period 1967-2000.

Differential Rotation and Geomagnetic Disturbances

Geoeffectiveness of Different Solar Drivers

While the effects of the solar wind on the Earth's rotation rate are still a subject of controversy, there is no doubt that the solar wind is the main agent driving geomagnetic disturbances (Gonzales et al., 1994). Two main types of structures in the solar wind can cause geomagnetic storms: high speed streams (HSS's) from solar coronal holes, and coronal mass ejections (CME's) from solar flares and disappearing filaments (Feynman, 1982). Coronal holes can live for many solar rotations, and once every solar rotation they come into "geoeffective position" such that the solar wind originating from them can hit the Earth, so they are sources of "recurrent" (regularly reappearing every 27 days) geomagnetic activity. CME's are basically unpredictable bursty phenomena, and are related to sporadic geomagnetic activity. Richardson et al. (2001) studied the sources of geomagnetic storms over nearly three solar cycles (1972-2000) and found that the most intense storms as defined by Kp index at both sunspot minimum and sunspot maximum are almost all generated by CME's; weaker storms are preferentially associated with HSS's from coronal holes at solar minimum, and with CME's at solar maximum, while a small fraction of the weaker storms at both solar minimum and solar maximum are generated by slow solar wind.

These two types of solar drivers are very different in their magnetic field configuration. The coronal holes are regions of open magnetic field lines propagating outwards from the Sun into the interplanetary space, and the high speed solar wind originating from them is characterized by high speed, low density, and high temperature. The coronal mass ejections originate from regions of closed field lines extending far away from the Sun, with unusually low proton temperature or low plasma beta (ratio of plasma pressure to magnetic pressure), composition anomalies, bidirectional suprathermal electron strahls indicating looped magnetic field lines rooted at both ends on the Sun. Fast CME's drive shocks with enhanced plasma speed, density and temperature. An Earth-directed CME is observed as a halo around the Sun, so a "halo CME" has about 50% chance to hit the Earth – the other 50% account for the possibility the CME to be ejected from the back side of the Sun and to be actually moving away from the Earth. A subclass of CME's are magnetic clouds (MC's) distinguished by enhanced magnetic field with smooth rotation inside the structure.

The geomagnetic disturbances caused from the different solar drivers are usually evaluated by the planetary geomagnetic index Kp and the ring current index Dst. Both of them are available online from NASA OMNI database (http://nssdc.gsfc.nasa.gov/omniweb). The same database can be used to identify the three different types of events - HSS's, CMEs and MCs. We define a HSS event as a sharp rise in the solar wind velocity by at least 100 km/s in no more than a day to no less than 500 km/s, accompanied by high plasma temperature and low plasma density. In the period 1997-2001 we have a total of 73 HSS events. In the period 1997-2002 we have 73 MCs defined by high magnetic field magnitude, low proton temperature or low beta, and smooth magnetic field rotation. We have used the list from Georgieva et al. (2005) completed by the events from http://cdaw.gsfc.nasa.gov/geomag_cdaw/. For the CME's we use the list of Cane and Richardson (2003) from which all events identified as MC's have been removed, which leaves us with a total of 128 cases in the period 1997 - 2002.

To compare the average geoeffectiveness of CME's, MC's and HSS's, in Figure 22.a and 22.b a superposed epoch analysis is presented of the daily averaged values of Kp and Dst indices, respectively, on days of events (day 0), one day before and after the events (days -1 and +1), etc.

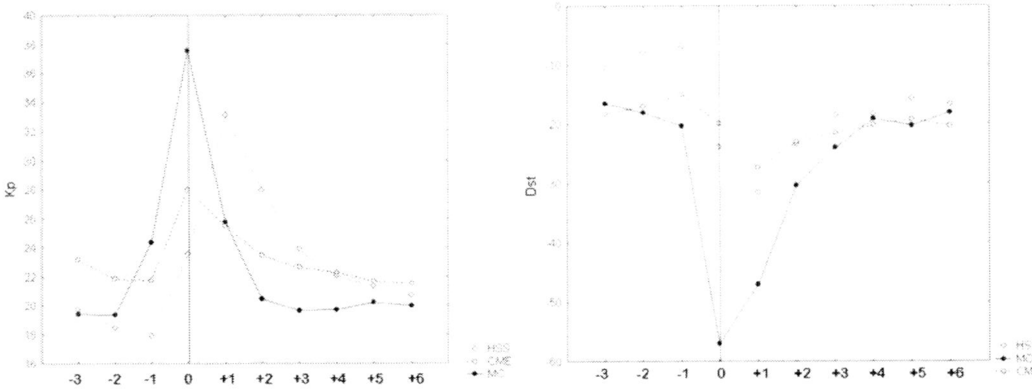

Figure 22. Superposed epoch analysis of average daily (a) Kp index – left panel, and (b) Dst index – right panel, on days with MC's (solid line), MC's (broken line) and HSS's (dotted line) in the period 1997-2002.

Magnetic clouds are the most geoeffective solar driver as expressed by both Kp and Dst magnetic indices. CME's which are not magnetic clouds are as effective as HSS's in terms of Dst index, and less effective than HSS's in terms of Kp index. Geomagnetic storms, as expressed by Kp index, start on the day of the event in the case of MC's and CME's, and one day later in the case of HSS's. This is to be expected as, according to the definition of a HSS which we have adopted, the event is considered to start when the solar wind speed first jumps up; however, the speed usually continues increasing and reaches it maximum a day later (Figure 23). For both HSS's and CME's the maximum phase of the geomagnetic storm as expressed by the Dst index is on the day following the event. This is also understandable as the Dst index is related to the intensity of the ring current, and the ring current needs some time to build up. What needs an explanation in this case is the instantaneous reaction of the Dst index to the encounter of MC's. The start of a MC event is considered the beginning of the magnetic field rotation, while very often the (Huttunen et al., 2002) geomagnetic storm is caused by the sheath region of enhanced magnetic field, density and temperature preceding the magnetic field rotation.

Solar Differential Rotation and Magnetic Clouds

As magnetic clouds are so much more geoeffective than coronal mass ejections which are not magnetic clouds, of a great practical interest is with what probability a solar eruption will result in a CME or in a MC. It has been noted that occurrence frequency of CME's follows the sunspot cycle (Gopalswamy et al., 2003), while the occurrence frequency of MC's follows neither the sunspot cycle nor the occurrence frequency of CME's (Wu et al., 2003).

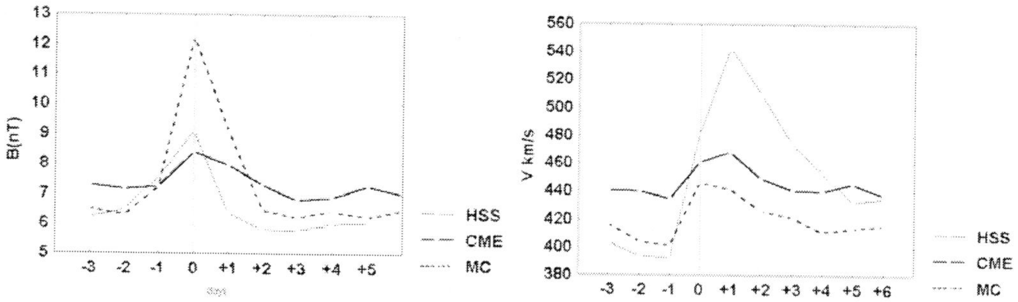

Figure 23. Daily averaged IMF magnitude (left) and solar wind speed (right) on days with CME's (solid line), HSS's (broken line) and MC's (dotted line) in the period 1997-2001.

Different estimates have been made about the percentage of CME's which are MC's: 30% (Gosling, 1990), 50% (Bothmer, 1996), 60-70% (Webb, 2000), until it was finally realized that this ratio varies with the sunspot cycle – from practically 100% at sunspot minimum though with poor statistics, to 15% at sunspot maximum (Richardson and Cane, 2003). Our results (Georgieva and Kirov, 2005) confirm this conclusion (Figure 24). This explains why from sunspot min to sunspot max, the intensity of storms associated with CME's increases, however the degree of association between CMEs and storms decreases (Webb, 2000) – because less CME's are MC's.

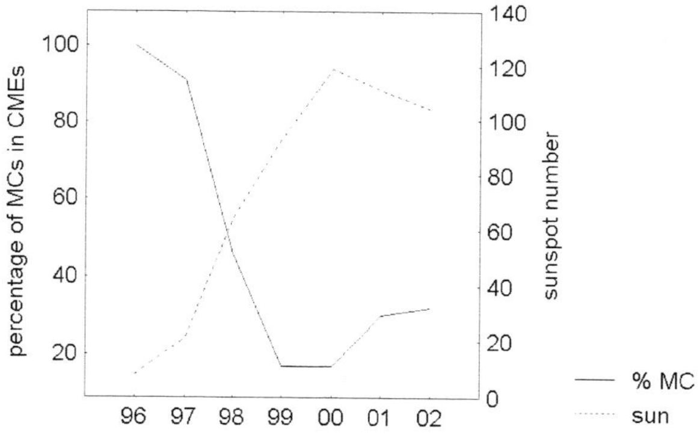

Figure 24. Percentage of CME's which are MC's (solid line) and the sunspot cycle (dotted line); from Georgieva and Kirov, 2005.

We should be reminded here that what distinguishes magnetic clouds from the other coronal mass ejections is the magnetic field rotation observed when a spacecraft crosses the cloud - which means that the magnetic field inside the magnetic clouds is twisted.

A quantitative measure of the twistedness of the magnetic field is the magnetic helicity. Magnetic helicity is constantly being generated in the Sun, but its dissipation is a very slow process (Berger, 1984), so the only way for the Sun to get rid of the helicity accumulated in the corona is to eject it in the coronal mass ejections which carry this helicity away into the interplanetary space (Low, 1997). Therefore, the more helicity is accumulated in the corona, the more of the CME's will contain helicity and will be registered as magnetic clouds.

The amount of helicity in the corona depends on 2 factors:

- the net helicity transferred from the solar interior into the southern and northern corona (Berger and Ruzmaikin, 2000);
- the surface differential rotation which can increase or reduce, or even reverse the twist of emerging magnetic tubes (De Vore, 2000).

The net helicity transferred from the solar interior into the northern corona is given by (Berger and Ruzmaikin, 2000)

$$\frac{dH_{CN}}{dt} = -\frac{\Omega_0}{2\pi}\Phi^2$$

Where H_{CN} is the net helicity from the solar interior into the northern solar hemisphere, Ω_0 is the solar equatorial rotation rate, and

$$\Phi = \pi R_{\odot}^2 B_1$$

is the net dipole flux through the northern photosphere. The expression for the southern hemisphere is with a positive sign. This means that the helicity transferred from the interior into the corona is positive in the southern hemisphere, and negative in the northern hemisphere (because $\Phi^2>0$ always). As the net dipole flux is maximum in sunspot minimum and minimum in sunspot maximum, the net transferred helicity has maximum positive values in the southern hemisphere and maximum negative values in the northern hemisphere around sunspot minimum, and is close to zero in both hemispheres around sunspot maximum (See Figure 3 in Berger and Ruzmaikin, 2000).

In the corona, the emerging flux tubes are subjected to differential rotation (de Vore, 2000):

$$\Omega(\varphi) = \Omega_o + b\sin^2\varphi$$

where $\Omega(\varphi)$ is the rotation rate at latitude φ, Ω_o is the equatorial rotation rate, and b is the latitudinal gradient of the rotation rate; $b<0$ so that the equator rotates faster than the higher latitudes. The generated helicity is negative in the north and positive in the south:

$$H \sim -\mathrm{sgn}(\varphi)\ \pi/32\ \cos(\varphi)$$

It twists initially untwisted or containing a finite amount of initial helicity flux tubes, and unwinds fields with opposite initial helicity.

The helicity generated by surface differential rotation is independent of the solar magnetic cycle and, as the helicity transferred from the solar interior, is also always negative in the north and positive in the south. This hemispheric helicity rule was first noted by Seehafer (1990). The magnetic clouds carry the helicity of their source regions (Kumar and Rust, 1994; Bothmer and Schwenn, 1996). Therefore, MC's originating from the northern solar hemisphere, should have negative helicity (counterclockwise rotation of the magnetic field), and from the southern hemisphere – positive helicity (clockwise rotation). However, this is true in only 70-80% of the cases (Hale, 1927; Richardson, 1941; Pevtsov et al., 1995).

Differential Rotation and Helicity

We have studied 39 cases of major geomagnetic storms (Dst<-100) caused by CME's in the period 1997-2001 for which the solar sources of the CME's have been identified (http://cdaw.gsfc.nasa.gov/geomag_cdaw/).

Out of them, in 27 cases (73%) a MC was observed at Earth's orbit with the expected chirality.

In 7 cases MC's originating from the northern solar hemisphere exhibited right-handed helicity (clockwise rotation of the magnetic field)

In 3 cases MC's originating from the southern solar hemisphere exhibited left-handed helicity (counterclockwise rotation of the magnetic field)

In 2 cases CME's originating from the southern solar hemisphere were not MC's (no magnetic field rotation).

In 10 out of the 12 cases when the hemispheric helicity rule was violated, the solar differential rotation in the source region of the CME was "reversed" (increasing with latitude). The dates of the CME's, their identified source regions, the magnetic field rotation in the structure at the Earth's orbit, and the differential rotation in the source region are summarized in the Table 2. Figure 25 illustrates the normal solar rotation as derived from Mt.Wilson Doppler shift measurements. The grid of the data used for the calculation of the rotation velocities averaged over one Carrington rotation is available online at *http://wso.stanford.edu/synoptic.html* and consists of 30 equal steps in sine latitude from 75.2 North to 75.2 South degrees and 5 degrees step in the heliographic longitude.The details of the calculations of the rotation velocity are given in Gavryuseva and Godoli, 2005). The calculated rotation data were kindly provided by Dr. E.Gavryuseva from Instituto Nazionale of Astrofisica, Osservatorio Astrofisico di Arcetri, Italy.

Table 2. CME's at the Earth's orbit with no magnetic field rotation or violating the hemispheric helicity rule.

date	Source region	CME at the Earth's orbit	Differential rotation in the source region
23 Nov 1997	N20E05	Right-handed MC	normal
04 May 1998	S18E20	No rotation	reversed
14 Nov 1998	N18W02	Right-handed MC	reversed
12 Feb 2000	N25E26	Right-handed MC	reversed
14 Oct 2000	N01W14	Right-handed MC	reversed
18 Apr 2000	S20W58	No rotation	reversed
17 Aug 2001	N16W36	Right-handed MC	normal
26 Sep 2001	S12E23	Left-handed MC	reversed
01 Oct 2001	N10E18	Right-handed MC	reversed
03 Oct 2001	S13E03	Left-handed MC	reversed
06 Nov 2001	N06W18	Right-handed MC	reversed
24 Nov 2001	S17W36	Left-handed MC	reversed

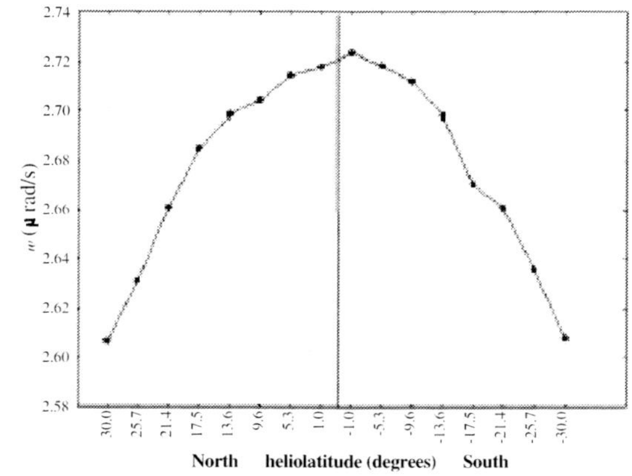

Figure 25. Dependence of the rotation rate on latitude averaged over 333 Carrington rotations from 1976 to 2001 (from Mt.Wilson Doppler shift measurements).

Figure 26 demonstrates a case of differential rotation reversed in the source region of a CME (marked by a circle).

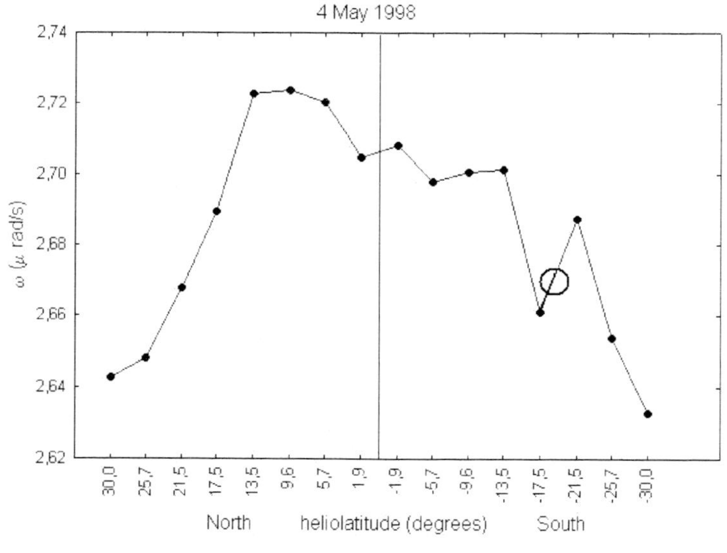

Figure 26. An example of reversed differential rotation in the source region of a CME.

Reversed (Anti-solar) Differential Rotation

Though the differential rotation of the Sun has been known for centuries, and in the last decades the rotation of dozens of stars has been measured, none of the numerous theories proposed to explain the origin and maintenance of the differential rotation is commonly accepted. The most popular one is probably the direct interaction between rotation and convection (Kitchatinov, 2005) which was first proposed by Lebedinsky (1941). Convective turbulence in rotating medium is subjected to the Coriolis force, and the feedback distorts rotation and makes it nonuniform. Models for differential rotation based on this theory provide rotation law for the Sun (Kitchatinov & Rűdiger, 1995) in close agreement with helioseismology (Shou et al., 1998). On the other hand, the theory of the reversed, or "anti-solar" differential rotation, is still in its initial phase. The suggestion is that this type of rotation can result from a fast meridional flow which in turn can be caused by deviation in spherical symmetry in gravity or temperature distributions. The reason can be large-scale thermal inhomogeneities or tidal forcing from a companion star. About ten stars have already been identified with anti-solar type of differential rotation (Strassmeier et al., 2003), six of them are close binaries, and one is a giant for which dark spots have been observed at low latitudes with temperature contrast of about 200^{0} K (Kitchatinov & Rűdiger, 2004).

In the case of the Sun, the anti-solar type of rotation is only observed in narrow latitudinal zones and in limited temporal intervals. In some of the cases listed in the table above, the planetary configurations hint at a possibility for tidal forcing. An example is shown in Figure 27. However, much additional study is needed to confirm this suggestion, completed by detailed data for the meridional circulation and the temperature distribution on the Sun.

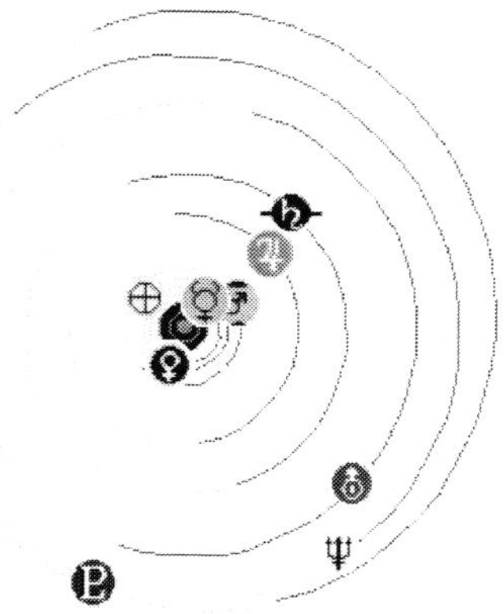

Figure 27. Planetary configuration for February 12, 2000.

Conclusion

The dynamics of the solar system is a factor which should be taken into account when studying the dynamics of the Sun and the solar activity. The dynamics of the Sun affects the way in which solar activity influences the Earth, with the solar wind and the interplanetary magnetic field playing the role of the mediator.

References

[1] Abarca del Rio R. ; Gambis D. ; Salstein D. ; Nelson P. ; Dai A., *J. Geodynamics* 2003, 36, 423–443.

[2] Abetti G., Sunspots from Galileo to Hale, 1965, Astronomical Society of the Pasific, *Leaflet* 432, 257-266.
[3] Antonucci, E. ; Hoeksema, J.T.; Scherrer, P.H, *Astrophysical J.*, 1990, 360, 296-304.
[4] Apostolov, E.M.; Letfus, V., *Acad. Bulg. Sci.* 1992. , 45 (2), 39-41.
[5] Arevalo, M. J.; Gomez, R.; Vazquez, M.; Balthasar, H., Woehl, H, *Astronomy and Astrophysics*, 1982,111 (2), 266-271.
[6] Arora, B. R. ; Rangarajan, G. K., *J. Geophys. Res.*, 1981, 86, 3369-3374.
[7] Babcock, H. W., *Astrophysical J.*, 1961, 133, 572-587.
[8] Balthasar, H.; Woehl, H., *Astronomy and Astrophysics*, 1980,92 (1-2), 111-116.
[9] Balthasar, H.; Vazquez, M. and Woehl, H., *Astronomy and Astrophysics*, 1986, 155 (1), 87-98.
[10] Berger, M. A., *Geophysical and Astrophysical Fluid Dynamics*, 1984, 30, 79-104.
[11] Berger, M.A.; Ruzmaikin, A., *J. Geophys. Res.*, 2000, 105 (A5), 10481-10490.
[12] Besprozvannaya, A. S.; Ohl, G. I.; Sazonov, B. I.; Scherba, I. A.; Schuka, T. I.; Troshichev, O. A., *J. Atm. Terr. Phys.*, 1997, 59, 1233-1244.
[13] Blizard, J.B., In M.R. Rampino, J.E.Sanders, W.S.Newman, L.K.Konigsson (eds.). *Climate*. Van Nonstrand Reinhold Co, New York, 1987, 415-420.
[14] Bothmer, V.; Schwenn, R., *Adv. Space Res.* 1996, 17, 319-322.
[15] Bruno, R.; Bavassano B., *Geophys.l Res. Lett*, 1997, 24 (18), 2,267-2,270.
[16] Buchha, V.; Jakubcova, I., Pick, Milos, *Geod.* 1985, 85 (2), 107-111.
[17] Chapman, S.; and J. Bartels, *Geomagnetism*, chap. 11, Oxford University Press, New York, 1940.
[18] Clayton, H.H., *World Weather*. MacMillan, New York, 1923.
[19] Clayton, H.H., *World Weather Records, Smithsonian Miscellaneous Collection*, V. 79, Washington, 1927.
[20] Cliver, E.W., Eos, Transactions, *American Geophysical Union*, 1994, 75 (49), 569, 574-569,575.
[21] DeVore, C. R., *Astrophysical J.*, 2000, 539 (2), 944-953.
[22] Djurovic, D., *Astronomy and Astrophysics*, 1983, 118(1), 26-28.
[23] Djurovic, D., Stajic, D., *Bull. Astron. Inst. Czechosl.*, 1985, 36, 246–250.
[24] Duane, G.S.; Webster, P.J., Weiss, J.B., *J. Atmos. Sci,.* 2000, 56 (24), 4183-4205.
[25] Emden, B R., *Astrophysical Journal*, 1902, 15, 38-42.
[26] Fairbridge, R.W., Sanders, J.E., In M.R. Rampino, J.E.Sanders, W.S.Newman, L.K.Konigsson (eds.). *Climate*. Van Nonstrand Reinhold Co, New York, 1987, 446.
[27] Feynman, J., *J. Geophys. Res,.* 1982, 87, 6153-6162.
[28] Fisk, L.A., *J. Geophys., Res.*, 2000, 106 (A5), 15,849-15,857.
[29] Dungey, J.W., In *Geophysics, The Earth's Environment*, ed. by C. DeWitt, J. Hieblot, A. Lebeau, Gordon and Breach, New York, 1963.
[30] Gavryuseva, E.; Godoli, G., *Phys. Chem. Earth* 2006 (In press).
[31] Georgieva, K.; Kirov, B., *Bulg. J. Phys.*, 2000, 27, 2, 28-31.
[32] Georgieva, K., *Phys. Chem. Earth*, 2002, 27, 433-440.
[33] Georgieva, K.; Javaraiah, J.; Kirov, B., Atanasov, D. In: *Proceedings of 10th European Solar Physics Meeting "Solar Variability: From Core to Outer Frontiers"*, ESA SP-506, 2002, 47-50.
[34] Georgieva, K.; Kirov, B., In: *Coronal and Stellar Mass Ejections*, IAU 226, Cambridge University Press, 2005.

[35] Giacalone, J., *J. Geophys. Res.*, 2001, 106 (A8), 15,881-15,887.

[36] Gigolashvili, M. Sh., Mdzinarishvili, T. G., Japaridze, D. R., Chargeishvili, B. B., *New Astronomy*, 2003, 8(6), 529-536.

[37] Girs, A.A. , *Meteorologia i hidrologia*, 1956, 10, 3-13 (In Russian).

[38] Girs, A.A.; Kondratovich, K.V., *Methods of long term weather forecasts*. Gidrometeoizdat, Leningrad, 1978. (In Russian).

[39] Golovkov, V.P., *Geomagn. Aeron,*. 1977, 17, 735–741 (in Russian).

[40] Gonzalez, W.D., Joselyn, J.A., Kamide, Y., Kroehl, H.W., Rostoker, G., Tsurutani, B.T., Vasyliunas, V.M., *J. Geophys. Res.*, 1994, 99, 5771-5792.

[41] Gopalswamy, Lara, A., Yashiro, S., Nunes, S., Howard, R. A., In *Solar variability as an input to the Earth's environment*, ISCS Symposium, 2003, 403.

[42] Gosling, J., *AGU Geophys.Monogr.*, 1990, 58, 343

[43] Greiner-Mai, H., *Astron. Nachr,*. 1987. 308 (3), 217-226.

[44] Gross, R.S.; Marcus, S.L.; Eubanks, T.M.; Dickey, J.O.; Keppenne, C.L., *Geophys. Res. Lett.*, 1996, 23 (23), 3373-3376.

[45] Gu, Zh., *Astrophy. Space Sci.*, 1998, 259 (4), 427 – 432.

[46] Hale, G. E., *Ap. J.*, 1908, 28, 315-348.

[47] Hale, G. E., *Publ. Astron. Soc. Pacific*, 24(144), 1912, 226-259

[48] Hale, G. E., *Nature*, 1927, 119, 708-711.

[49] Hale, G. E., Nicholson, S. B., *ApJ*, 1925, 62, 270-301.

[50] Hubrecht, J. B., *M.N.R.A.S.*, 1915, 75, 611-626.

[51] Hansen, J.; S. Lebedeff,. *Geophys. Res. Lett.*, 1988, 15, 323-326.

[52] Hathaway, D.; Wilson, R., *Ap J.*, 1990, 357, 271-274.

[53] Herman, J.R.; Goldberg R.A., *Sun, Weather, and Climate*. NASA Scientific and Technical Information Branch, Washington DC, 1969.

[54] Hide, R., *Geophys. Res. Lett,*. 1995, 22 (8), 961–964.

[55] Howard,B R.; Harvey,B J., *Solar Phys.*, 1970, 12, 23-51.

[56] Huttunen, K.E.J.; Koskinen, H.E.J.; Schwenn, R., *J. Geophys. Res,*. 2002, 107, (A7), doi: 10.1029/2001JA900171.

[57] Javaraiah, J., *ApA*, 2000, 21, 167-173.

[58] Javariah J., *Bull. Astr. Soc. India*, 1996, 24, 351-354.

[59] Javaraiah, J., *Solar Phys.*, 2003, 212 (1), 23-49.

[60] Javariah J.; Gokhale, M.H., *Solar Phys.*, 1997, 170, 389-410.

[61] Jin, R.-Sh., *J. Geophys. Res.* 1992, 97 (B12), 17,251-17,260.

[62] Jones, P.D.; Jónsson, T.; Wheeler, D., *Int. J. Climatol.*, 1997, 17, 1433-1450.

[63] Jones, P.D.;Moberg, A., *J. Climate*, 2003, 16, 206-223.

[64] Jones, P.D.; New, M.; Parker, D.E.; Martin, S.; Rigor, I.G., *Rev. Geophys.*, 1999, 37, 173-199.

[65] Jose, P.D., *Astron.J.*, 1965, 70 (3), 193-200.

[66] Kirov, B.; Georgieva, K.; Javaraiah, J., In: *Proceedings of 10th European Solar Physics Meeting "Solar Variability: From Core to Outer Frontiers"*, 2002, ESA SP-506, 149-152

[67] Kiselev, V.M.. *Astron. Zh.*, 1981, 58, 590-596 (In Russian).

[68] Kitchatinov, L.L., *Physics-Uspekhi*, 2005, 48, 475-494.

[69] Kitchatinov, L.L.; Rűdiger, G. *A&A*, 1995, 299, 446-451.

[70] Klyashtorin, L.B. In *The Possibility of Forecasting, FAO FISHERIES TECHNICAL PAPER,* 2001, 410, Rome.
[71] Kopecky, M., *BAICz.*, 1991, 42, 158-160.
[72] Köppen W., *Z.öst. Ges. Meteorel.*, 1873, 8, 241-250.
[73] Köppen W., *Meteorol.Zeitchrift.*, 1914, 31-36.
[74] Kosek, W. *Journal of Geodesy,* 1993 (67), 1– 9.
[75] Kumar, A.; Rust, D., *J. Geophys. Res.*, 1994, 101 (A7), 15,667-15,684.
[76] Krymskij, P. F., *Geomagn. Aeron.,* 1993, 33 (3), 7-13 (In Russian).
[77] Landscheidt, T., Swinging sun, *J. Interdiscipl. Cycle Res.* 1981, 12, 3-19.
[78] Laptukhov, A. I., *Geomagn. Aeron.*, 1980, 20, 670-673 (In Russian).
[79] Le Mouel, J.-L. ; Courtillot, V., *J. Geophys. Res.* 1982, 87, 4103–4108.
[80] Lin, Y., *Solar Physics*, 1998, 179(1), 179 - 188
[81] Loginov V.F., Prichiny I sledstvia klimaticeskih izmenenii. *Navuka I tehnika*, Minsk, 1992 (in Russian).
[82] Loginov, V. F.; Vysotskii, A. M.; Sherstiukov, B. G., In: *Some problems on the effect of solar activity on the lower atmosphere.* Moscow, Gidrometeoizdat, 1975, 43-49 (In Russian).
[83] Low, B.C., *Phys.Plasmas,* 1994, 1, 1684-1690.
[84] McFadden, P.L.; Merrill, R.T., *J. Geophys. Res.,* 1993, 98, 6189-6199.
[85] Miesch, M. S., *Living Rev. Solar Phys.,* 2005 2, (1), www.livingreviews.org/ lrsp-2005-1.
[86] Mörner, N.-A., In: *Abstracts from the AGU -EGS meeting,* 2003, abstract #10012.
[87] Munk, W., *Progress in Oceanography,* 1997, 40, 7-36.
[88] Mursula, K.; Hiltula, T.; Zieger, B., In: *Proceedings of the Tenth International Solar Wind Conference*, AIP Conference Proceedings, 2003, 679, 98-101.
[89] Okazaki, S., *Publ. Astron. Soc. Japan,* 1977, 29, 619-629.
[90] Parker, E. N., *Astrophysical J.*, 1958, 128, 664-676.
[91] Peterson, T. C.; Vose, R.S., *Bull. Am. Meteor. Soc.*, 1997, 78, 2837-2849.
[92] Pevtsov, A. A.; Canfield, R. C.; Metcalf, T. R., *ApJ,* 1995, 440, L109-L112.
[93] Piddington, J. H., *Solar Physics*, 1971, 21,.4-20.
[94] Pulkkinen, P.; Tuominen, I., *A&A*, 1998, 332, 748-754.
[95] Richardson, R. S., *ApJ,* 1941, 41, 24-31.
[96] Richardson, I.G.; Cliver, E.W.; Cane, H.V., *Geophys. Res. Lett,.* 2001, 28, 2569-2572.
[97] Rivin, Yu.I., *Tsikly Zemli I Solntsa*, Nauka, Moscow, 1989 (In Russian).
[98] Russell, C.T; McPherron, R.L., *J. Geophys. Res.*, 1973, 78(1), 92-98.
[99] Salstein D.A., In: Reid, M.J., Moran, J.M. (Eds.), *The impact of VLBI on astrophysics and geophysics.* IAU, 1988, 401-410.
[100] Salstein, D., Rosen, R., *J. Appl. Met.*, 1986, 25 (12), 1870-1877.
[101] Sazonov, B.I.; Loginov, V.F., *Solar-tropospheric relations*, Gidrometeorolo-gicheskoe izdatel'stvo, Leningrad, 1969 (In Russian).
[102] Schatten, K.H.; Wilcox, J.M.; Ness, N.F., *Solar Phys.*, 1969, 6, 442-447.
[103] Scherrer, P. H.; Wilcox, J. M.; Kotov, V.; Severnyi, A. B.; Howard, R., *Solar Physics*, 1977, 52, 3-12.
[104] Schou, J.; Antia, H.M.; Basu, S., et al., *ApJ,* 1998, 505, 390-417.
[105] Seehafer, N., *Solar Phys.*, 1990, 125, 219-232.
[106] Sidorenkov N.S., *Trudy Gidrometcentra SSSR,* 1978, 205, 48-66, (In Russian).

[107] Smirnov, R.V., *AN SSSR Dokl.*, 1967, 175 (1), 76-79 In Russian.
[108] Smirnov, R.V., *AN SSSR Dokl.*, 1968, 180 (11), 88-92. In Russian.
[109] Smith. C.W.; Bieber, J.W., *Astrophysical J.*, 1991, 370, 435-441.
[110] St. John, C., *Publ. Astr. Soc. Pacific*, 1918, 30(178), 318-325.
[111] Spangler, W. M. L.; Jenne, R. L., *World Monthly Surface Station Climatology (and Associated Data Sets).* National Center for Atmospheric Research, Boulder, Colorado, 1990.
[112] Starr V.P.; Oort, *Nature*, 1973, 242, 310-314.
[113] Stephenson, F.R.; Morrison, L.V., *Phil. Trans. Royal Soc. London*, Ser. A, 1984, 313, 47-70.
[114] StatSoft, Inc., *STATISTICA* (data analysis software system), version 6. 2001, www.statsoft.com.
[115] Strassmeier K.G.; Kratzwald L.; Weber M., *A&A*, 2003,408, 1103.
[116] Temmer M.; Veronig A.; Hanslmeier A., *ESA SP-505,* 2002, 587-590.
[117] Temmer, M.; Veronig, A.; Rybák, J., Hanslmeier, A., *Hvar Observatory Bulletin*, 2003, 27 (1), 59-66.
[118] Troup A.J., A., *Geofis.Pure Appl.*, 1962, 51, 184.
[119] Verma, V.K., *Ap J.*, Part 1, 1996, 403 (2), 797-800.
[120] Verma, V.K., 2000. Periodic variation of the North-South asymnmetry of solar activity phenomena. *J. Astrophys. Astr.* 21, 173-176.
[121] Vinnikov, K.Ya.; Groisman, P.Ya.; Lugina, K.M., *J. Clim.,* 1990, 3, 662 – 677.
[122] Vitinsky, Yu. I.; Kopecký, M.; Kuklin, G. V., Statistics of Sunspot Activity, *Nauka*, Moscow, 1986 (In Russian)
[123] Vose, R. S.; Schmoyer, R. L.; Steurer, P. M.; Peterson, T. C.; Heim, R.; Karl, T. R.; Eischeid J., *ORNL/CDIAC-53*, NDP-041. Carbon Dioxide Information Analysis Center, Oak Ridge National Laboratory, Oak Ridge, Tennessee, 1992.
[124] Waldmeier, M., *Zeitschr. Astrophts.* 1957, 43, 149.
[125] Walker, G.T., *Mem. India Meteorol. Dept,.* 1915, 21, 61-90.
[126] Wang, Y.-M.; Sheeley, N.R. Jr.; Andrews, M.D., *J. Geophys. Res.*, 107 (A12) doi:10.1029/2002JA009463,2002.
[127] Webb 2000. SOHO 11 Symp.on From Solar Min to Max: Half a Solar Cycle with SOHO, Noordwijk: *ESA Publications Division*, p. 409
[128] Wood, K.D., 1972. Sunspots and planets. *Nature*, 240, 91-93.
[129] Wood, R.M., 1975. Comparison of sunspot periods with planetary synodic period resonances. *Nature*, 255, 312-313.
[130] Yoshimura, H.. 1979. The solar-cycle period-amplitude relation as evidence of hysteresis of the solar-cycle nonlinear magnetic oscillation and the long-term /55 year/ cyclic modulation. *ApJ*, Part 1, 227, 1047-1058.
[131] Zieger, B., Mursula, K., 1998. Annual variation in near-Earth solar wind speed: Evidence for persistent north-south asymmetry related to solar magnetic polarity. *Geophys. Res. Lett.*, 25 (6), 841-844.
[132] Zurbuchen, T. H., Fisk, L. A., Hefti, S., Schwadron, N. A., 1998. *The new Heliospheric Magnetic Field: Observational Implications, Ninth International Solar Wind Conference*, ed. S.R.Habbal et al., American Institute of Physics, Melville, NY, USA. 471: 87-92.

In: Space Science: New Research
Editor: Nick S. Maravell, pp. 83-96
ISBN: 1-60021-005-8

Chapter 3

THE DYNAMICS OF THE PLASMASPHERE

Viviane Pierrard*

Belgian Institute for Space Aeronomy, 3 av. Circulaire, B-1180 Brussels, Belgium

Abstract

The plasmasphere is the high altitude extension of the ionosphere. To ensure accuracy and reliability of communications, navigation and other satellites stationed in this region, there is a considerable interest to understand the plasmaspheric environment and its dependence on external parameters.

Dynamical simulations have been developed at the Belgian Institute for Space Aeronomy to simulate the deformation of the plasmasphere during geomagnetic substorms and other variations in the level of geomagnetic activity. These simulations are based on the mechanism of plasma instability and on the E5D empirical electric field model. The results of these simulations are compared with satellite observations of the plasmasphere. Comparisons with observations of the IMAGE spacecraft launched in March 2000 are especially useful to understand the dynamics of the plasmasphere since the satellite provides the first global comprehensive images of the Earth's plasmasphere. The EUV (Extreme UltraViolet) instrument gives intensity maps of the 30,4 nm emissions of Helium ions integrated along the line of sight. It revealed the global behavior of the plasmasphere and its dynamics influenced by the solar wind and interplanetary magnetic field variations. New structures in the plasmasphere such as shoulders, channels, notches and the formation of plumes were discovered. The four Cluster spacecraft launched in 2000 also provide useful data to compare with the simulations.

Introduction

The plasmasphere is filled by cold (a few eV or less) plasma originating from the ionosphere distributed along geomagnetic field lines and co-rotating with the Earth. This toroidal region encircles the Earth at geomagnetic latitudes typically less than about 65 degree. This region where the number density of the particles is in the range of $10\text{-}10^4$ particles/cm^3 was discovered at the end of the fifties with the pioneering ground-based whistler wave

* E-mail address: viviane.pierrard@aeronomie.be; Tel: (+32) 2 3730418; Fax: (+32) 2 3748423

investigations and confirmed with the first in situ observations of plasma detectors on board the Soviet LUNIK 1 and 2 spacecraft launched in 1959. Besides ground-based whistler technique and direct measurements of electron and ion number density, other techniques to observe the plasmasphere were then developed, based on plasma wave experiments and remote sensing by optical means.

The observations showed that the number density of the particles decreases sometimes sharply at the limit of the plasmasphere, called the plasmapause. This discontinuity crosses the geomagnetic equatorial plane at radial distances ranging from 2 Re to 7 Re depending on the geomagnetic activity level. The plasmapause is closer to the Earth during geomagnetic substorms.

After four decades of study by ground-based and in situ measurements, many of the details of the plasma distribution remain unexplained and the processes in the inner magnetosphere poorly understood [Carpenter, 1995].

New progress in the plasmaspheric field of research is provided by the recent observations of EUV on board IMAGE [Sandel et al., 2003]. IMAGE provides indeed the first global views of the plasmasphere by imaging the distribution of He^+ in its 30.4 nm resonance line. The details of the plasmaspheric dynamics are then revealed: new structures such as shoulders, channels, notches and the formation of plumes, appear frequently in the observations. These are keys to understanding the ways that electric fields affect the plasma distribution and the mechanisms influencing the formation of the plasmapause.

Different processes have been proposed for the formation of the plasmapause [see Lemaire and Gringauz, 1998 for a review]. The role of the convection electric field and of the instability mechanism has been studied since 1974 at the Belgian Institute for Space Aeronomy [Lemaire, 1974, 1985]. Recently, dynamical simulations based on this instability mechanism have been developed [Pierrard and Lemaire, 2004] and compared with new observations of the plasmasphere by AURORAL PROBE, CLUSTER, and IMAGE [Bezrukikh et al., 2003; Pierrard, 2004; Dandouras et al. 2005; Pierrard and Cabrera, 2005; 2006].

AURORAL-PROBE shows clearly that the plasmasphere is eroded during geomagnetic substorms and that the new plasmapause is not the result of compression. Indeed, the number density inside the plasmasphere is not increased after the geomagnetic event.

This spacecraft also measured the temperature in the plasmasphere and shows large difference between dayside and nightside.

CLUSTER launched in 2000 revealed new density structures in the plasmasphere and the plasmapause [Darrouzet et al., 2004]. These four spacecraft flight on a polar orbit in a tetrahedral configuration to study the magnetosphere and its environment in three dimensions. The inter-satellite distance has been varied from 100 km to several thousand kilometers. The satellites penetrate periodically (57 h) inside the plasmasphere when they are at the perigee at 4 Re. The instrument WHISPER measures number density profiles of the electron by operating a relaxation sounder. CLUSTER crosses the plasmapause only when the limit of the plasmasphere is further from the Earth than 4 Rs, thus during quiet geomagnetic activity. The comparisons between the global view of IMAGE and the density profiles of CLUSTER gave interesting results about the plume formation and the ionic structures among others.

In the present chapter of this book, new study cases of the dynamical simulations based on the instability mechanism are presented and compared with the observations of EUV/IMAGE. These typical cases show the extended plasmasphere during quiet periods and the formation and evolution of plumes during geomagnetic substorms. The advances in plasmaspheric research as revealed by EUV and CLUSTER are also summarized.

Plasmapause Formation

The dynamical simulations presented in the present paper use a given Kp-dependent magnetospheric electric field model. This electric field is the E5D model determined from dynamical proton and electron spectra measured on board the geostationary satellites ATS-5 and 6 [McIlwain, 1986]. The Kp index observed during the analyzed period of time and the previous 24 hours is the only input of the time-dependent model. The convection electric field intensity is the largest in the post-midnight sector. Due to the increased convection velocity during periods of increased geomagnetic activity, the plasmasphere is peeled off in this region, according to the physical mechanism of plasma instability used in the numerical simulations. The instability mechanism and the simulation process are described in detail in Pierrard and Lemaire [2004]. Since plasma is detached in the post-midnight sector due to instability during geomagnetic substorms, the plasmapause is then closer to the Earth in this sector. Moreover, due to the differential rotation, a plume is formed later in the afternoon and dusk region.

The model predictions are compared with EUV observations from IMAGE. The instrument consists of three wide-field cameras that are tuned to the 30.4 nm resonance line of sunlight. The observations are intensity maps of these emissions of Helium ions integrated along the line of sight. They are projected in the geomagnetic equatorial plane to have the same view over the pole as in the simulations. In the Figures, sunlight is incident from the left. The plasmapause is assumed to be the sharp edge where the brightness of 30.4 nm He^+ emissions drops drastically. To better visualize the plasmapause, a white line is drawn corresponding to a threshold of a percentage of the maximum intensity of the image, where the intensity is the logarithm of the luminosity.

June 18, 2001

First, let us show the example of June 18, 2001. During this day, Kp increased up to 5+ and remained high during several hours. The upper panel of Fig. 1a shows the geomagnetic activity index Kp observed from June 16 to June 18, 2001. The Dst index that becomes negative during the geomagnetic substorm event is also illustrated.

The lower panel shows the result of the simulation on June 18, 2001 at 13h00 UT. Kp is then observed to be maximum. The colored dots represent successive positions of test plasma elements released every 10 minutes at 02:00 MLT in the post-midnight sector. These dots identify the geomagnetic equatorial position of the plasmapause, as predicted by the mechanism of interchange motion. The position of the plasmapause depends on the evolution of the geomagnetic index Kp during the previous hours, which controls the intensity of the convection electric field E5D. The different colors of the elements correspond to the epoch of their release. The dotted circles correspond to L=1, 2, 3, 4 and 5.

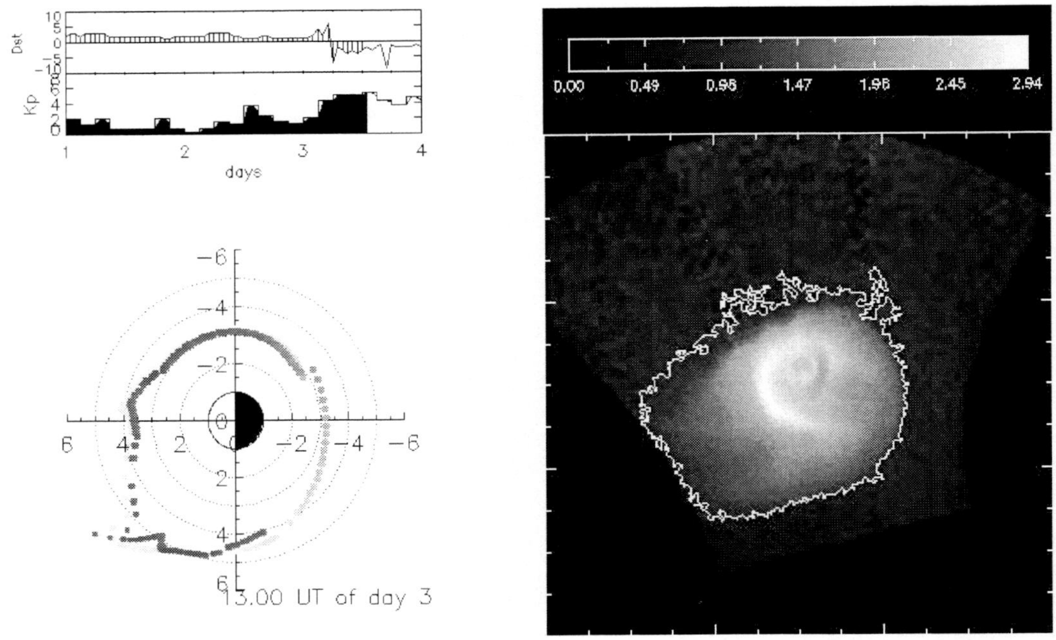

Fig. 1: Left panel (a): result of the simulation based on the instability mechanism, the E5D model and the value of Kp on 18 June 2001, 13h00. The dots indicate the position of plasmapause in the geomagnetic equatorial plane. The indexes Dst and Kp observed during the previous days are also displayed. The dotted circles correspond to L=1, 2, 3, 4 and 5.

Right panel (b): EUV observations on 18 June 2001 at 13h06, projected in the geomagnetic equatorial plane. The white line corresponds to 35% of the maximum intensity of the image and permits to visualize the plasmapause. The limit of the black square corresponds to L = 8.

The simulation predicts the development of a so-called plume in the afternoon region of the plasmasphere. Such plumes appear especially during periods of geomagnetic substorm events, i.e., after an increase of Kp.

The result of the simulation is compared with the plasmasphere observed by EUV/IMAGE on June 18, 2001 at 13h06 illustrated in Fig. 1b. The plasmasphere is viewed from a point of view about the North pole and projected in the geomagnetic equatorial plane. The edges of the plane extend up to 8 Rs. The observations of EUV/IMAGE show indeed the presence of a plume in this sector. The position of the plasmapause is about 3 Re in the midnight LT sector, i.e. close to the Earth, as in the simulations.

The white line corresponds to 35% of the maximum luminosity and illustrates the position of the plasmapause as observed by EUV. A smooth decrease in density is observed in the dayside sector. The dayside position the plasmapause would have been found at lower radial distances in the dayside sector if a higher threshold had been chosen. On the contrary, the gradient in density is sharper in the night sector. A sharp plasmapause knee is predicted by the mechanism of instability in this sector, since it is the region where the plasmasphere is peeled off during geomagnetic substorms.

Plumes are often observed by EUV/IMAGE and follows even moderate increases of Kp (above 3^{+}-4). Unfortunately, there are no observations in the present case study during the hours preceding its formation. But the EUV/IMAGE observations, which have generally a typical duration of 7 hours (with a time resolution of 10 minutes) out of each 14-hour orbit, allow us to follow the evolution of the plume during the next hours.

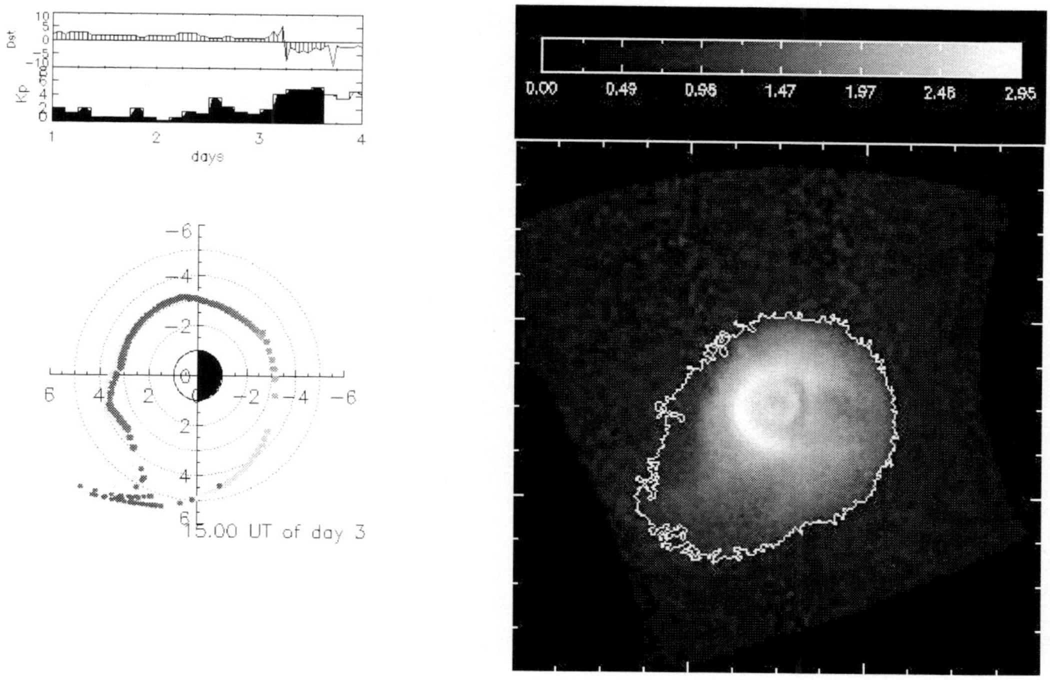

Fig. 2: Same as Fig. 1 for 18 June 2001 at 15h00 (left panel (a), simulation) and 14h59 UT (right panel (b), EUV observation).

Fig. 2 a and b present the plasmasphere two hours later than in Fig. 1, respectively as predicted by the model at 15h00 and as observed by EUV at 14h59. Since Kp remains high, the plasmapause remains very close to the Earth in the post-midnight sector. The base of the plume corotates with the Earth. But its top located at a large radial distance has a lower velocity, so that the plume becomes more and more bent and wraps. Moreover, it becomes thinner with time since the inner edge rotates faster than the outer edge. This evolution of the plume seems quite general, since it was observed also in other sequences [Spasojevic et al., 2003].

The evolution of the plume continues during the following hours and is illustrated in the next Figures. Fig. 3a shows the result of the simulation at 18h00 and 4a at 20h00. The corresponding EUV/IMAGE observations are displayed on Fig. 3b (18h03) and Fig. 4b (20h06). The comparison shows that the simulations reproduce quite well the formation of the plume due to the Kp increase and its evolution with time. This is confirmed also by other study cases obtained during geomagnetic substorms [Pierrard and Cabrera, 2005].

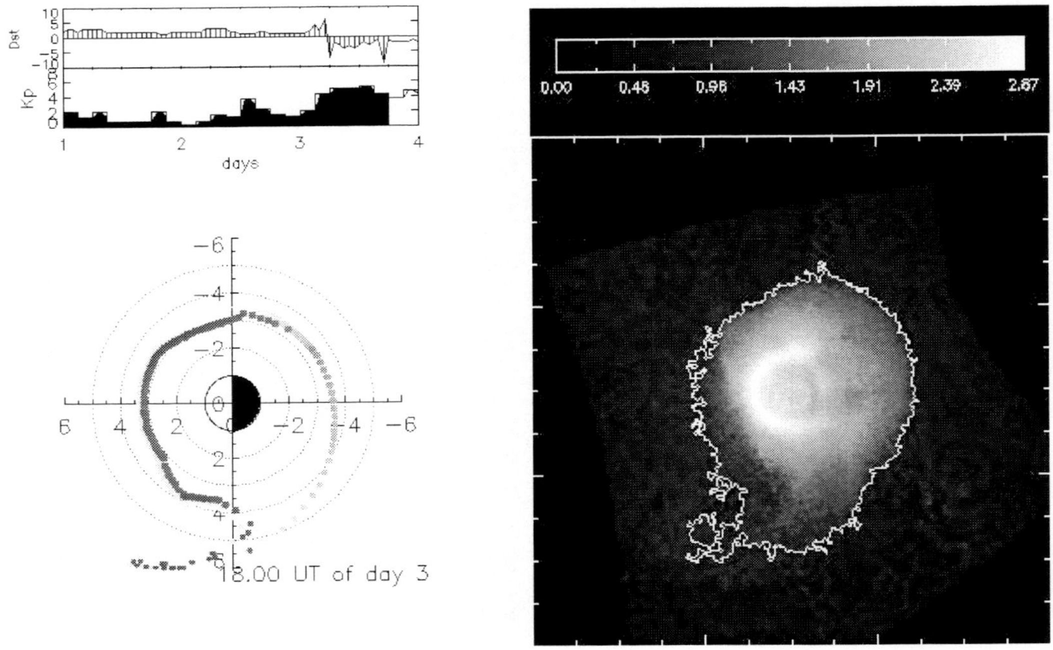

Fig. 3: Same as Fig. 1 for 18 June 2001, 18h00 (left panel (a), simulation) and 18h03 UT (right panel (b), EUV observation).

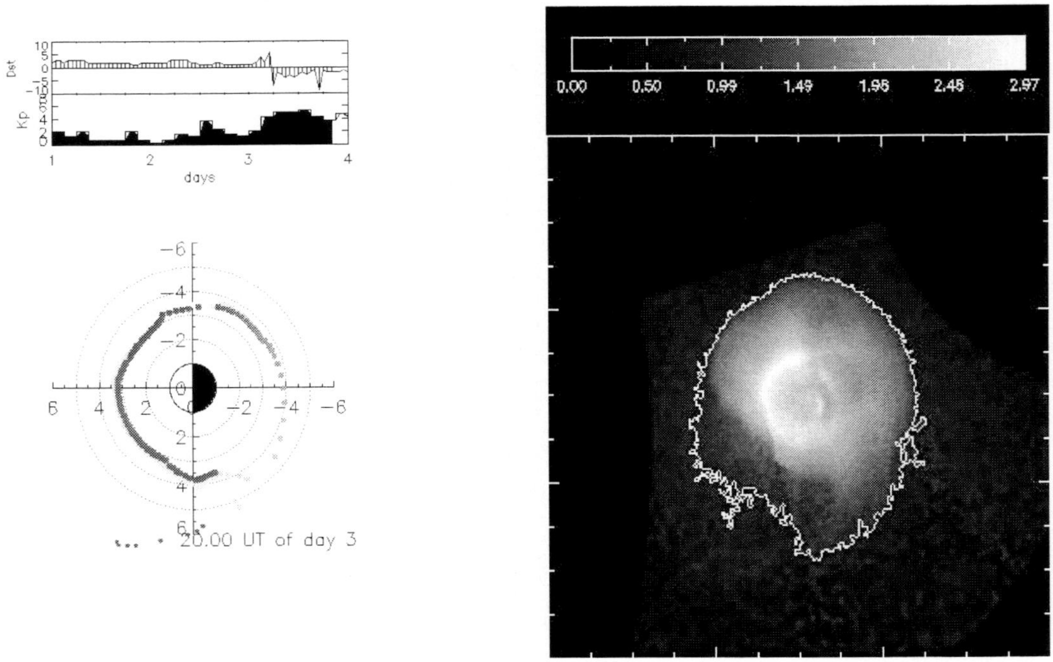

Fig. 4: Same as Fig. 1 for 18 June 2001, 20h00 (left panel (a), simulation) and 20h06 UT (right panel (b), EUV observation).

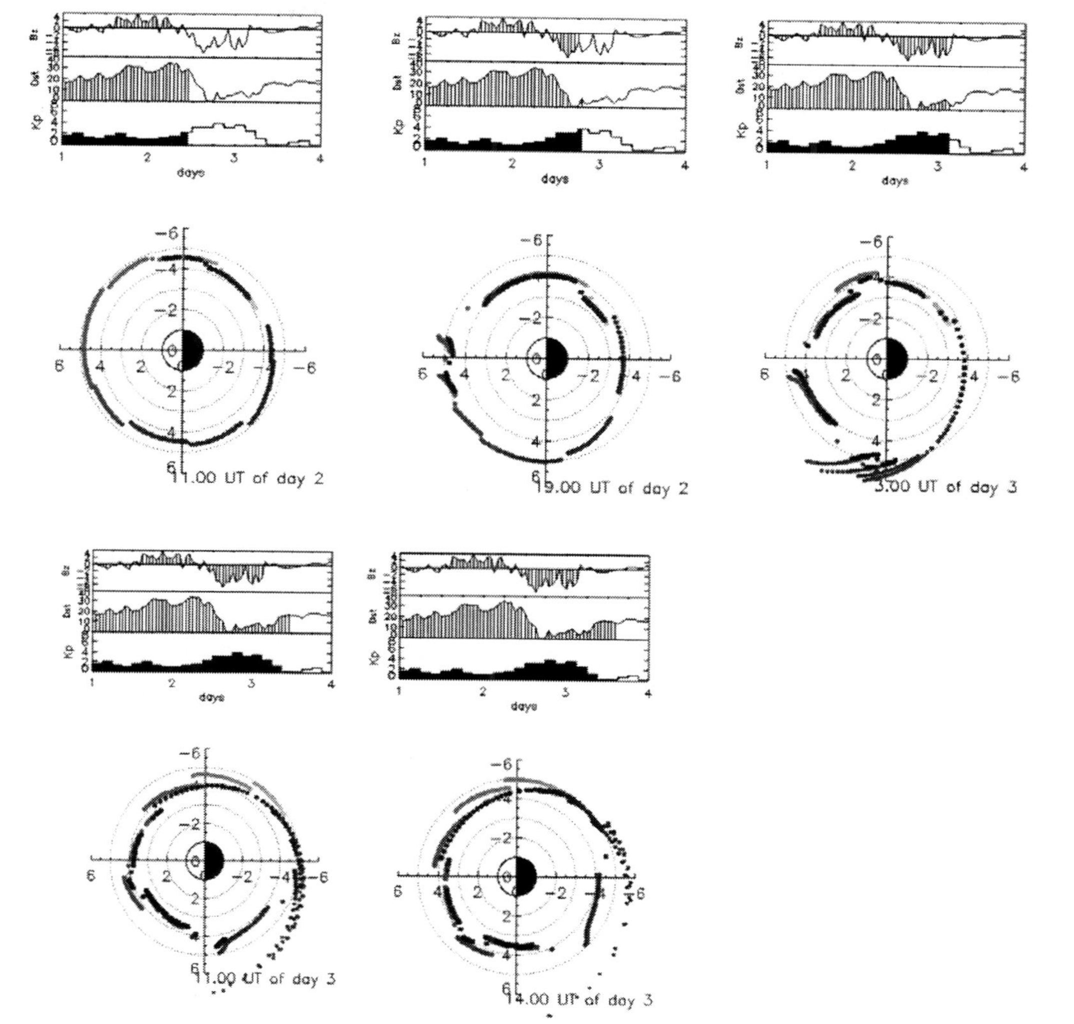

Fig. 5: Sequence showing the result of the simulation based on the instability mechanism, the E5D model and the value of Kp on 26 June, 2001, 11h00 (a); 19h00 (b); 27 June 2001, 3h00 (c); 11h00 (d) and 14h00 (e). The plasmapause in the geomagnetic equatorial plane corresponds to the dots. The indexes Bz, Dst and Kp observed during the previous and following days are also displayed. The dotted circles correspond to L=1, 2, 3, 4 and 5.

26-27 June, 2001

Another interesting study case corresponds to the relatively weak geomagnetic substorm of 26-27 June 2001. Kp increased gradually up to 4 and then gradually decreased. This event is weak but well isolated. In the upper panels of Fig. 5, the Kp index is illustrated, as well as Dst and Bz, the z component of the interplanetary magnetic field. Note that the geomagnetic substorm corresponds to an increase of Kp, to a decrease of Dst and to Bz becoming southward.

This sequence of IMAGE had been analyzed by Spasojevic et al [2003]. It is interesting because it is possible to follow the development of the plume and its evolution during two days. The EUV/IMAGE observations are interrupted during some hours but observations are available on June 26 from 4h31 to 14h35 and from 18h39 to 1h23 and from 9h39 to 18h31 on June 27. The rotation velocity is observed to be 80% of co-rotation during this period. This observed co-rotation lag has been accounted for in the model.

The result of the simulation is displayed on the sequence presented in Fig. 5.

The first panel corresponds to June 26, 2001 at 11h00. The plasmapause is then located between 4 and 5 Re following the LT sectors and is quite circular. Indeed, before the geomagnetic substorm, Kp was small and quite constant. Its irregular shape is due to the small variations of Kp during the previous 24 h. Note that the plasmapause is a little bit closer to the Earth in the post-midnight sector. This is due to the small increase of Kp at the beginning of the substorm. Indeed, the interchange mechanism predicts a plasmapause closer to the Earth in the post-midnight sector when Kp increases [Pierrard and Lemaire, 2004].

A little bit later, a plume begins to form in the morning sector. This is observed in the second panel showing the result of the simulation at 19h00. Later, the plume becomes thinner and more elongated. It rotates and is located in the afternoon sector at 3h00 (third panel) on 27 June 2001, in the dusk sector at 11h00 (fourth panel) and enters in the post-midnight sector at 14h00 (fifth panel).

Let us now compare with observations of EUV/IMAGE obtained during the same period of time. Global EUV images of the plasmasphere are available during three intervals of time on June 26 and 27, when the satellite was close to its apogee.

We use the threshold 0.45 of the maximum intensity in this sequence to visualize the plasmapause by a white line. The threshold cannot be chosen to be lower during this observation period, otherwise the white line corresponds in some cases to the limit of the camera view. This happens especially when the satellite is far from the apogee. The relative intensity threshold procedure is automatic and provides slightly different plasmapause positions than those obtained by the manual procedure used by Spasojevic et al. [2003]. The plasmapause is located at larger radial distances when a lower threshold is chosen. Note nevertheless that the modification of the threshold does not modify the position of the plasmapause in the postmidnight sector, since the gradients are generally sharper in this sector. White circles are also displayed on the observations to better evaluate the radial distance of the observed plasmapause: they correspond to L=1, 2, 4, 6 and 8.

The sequence shows indeed the formation of a plume, as predicted by the model. The plume begins its formation in the morning sector like in the simulations (see Fig. 6b). The formation of the plume is generally thought to be located in the afternoon sector, but one can see that in the present case, a small bulge is already visible in the morning sector. On Fig. 6c, d and e, the plume position corresponds quite well to what is obtained in the simulation. Nevertheless, the position of the plasmapause is generally located further in the simulations than what is observed. This effect can be due to the large value of the threshold chosen in the present case. Note also that the observations show an erosion at the western edge of the plume, with crenulations and irregularities. The plasmasphere is eroded in the dawn sector, while the plasmapause is more circular in the simulations.

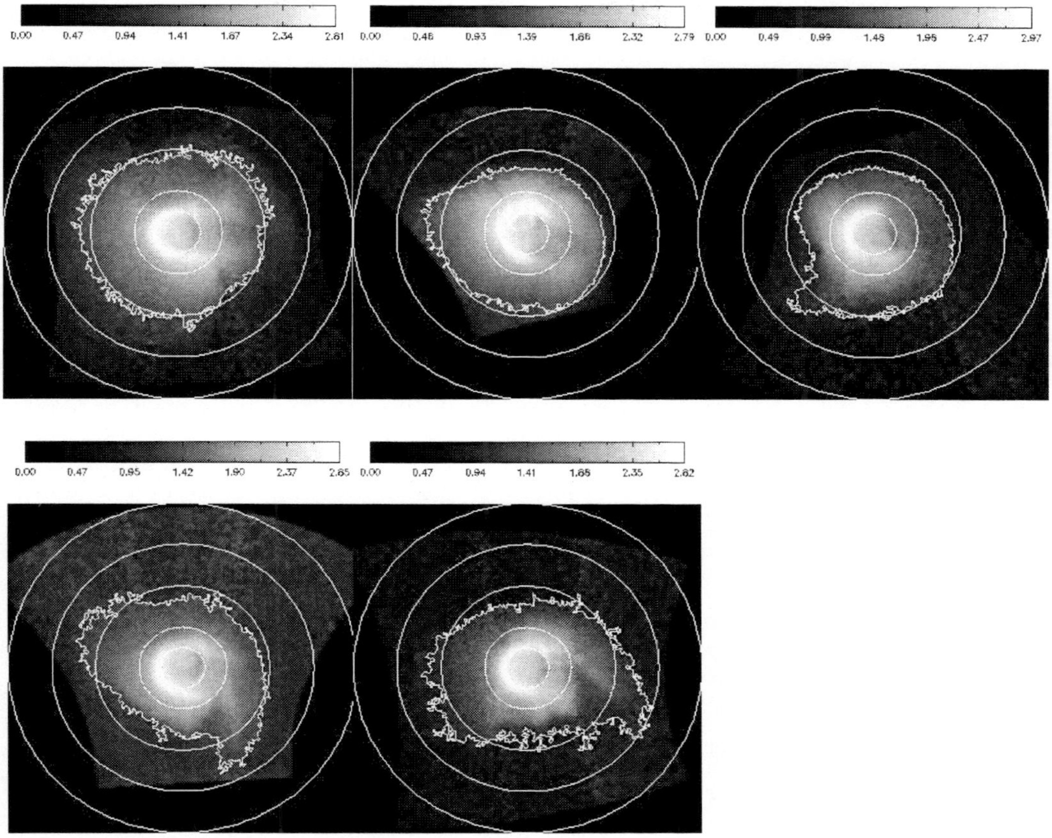

Fig. 6: Sequence showing the observations of EUV/IMAGE on 26 June, 2001, 11h00 (a); 19h00 (b); 27 June 2001, 3h00 (c); 11h00 (d) and 14h00 (e). The white line corresponds to a threshold of 45% of the maximum intensity and corresponds to the plasmapause in the geomagnetic equatorial plane. The white circles correspond to L=1, 2, 4, 6 and 8.

Other comparisons between EUV/IMAGE observations and results of simulation are also presented in Pierrard and Cabrera [2005], like for instance 9-10 June 2001 and 25-26 June, 2000. These events show an extended plasmasphere during periods of low geomagnetic activity level and the formation of a plume during geomagnetic substorms. Plumes begin to form in the morning LT sector and are well developed in the afternoon LT sector during geomagnetic substorms and storms. Subsequently, they tend to co-rotate with the Earth when the geomagnetic activity decreases. After a sharp decrease of Kp, the formation of shoulders has also been observed. The global evolution of the plume and the radial distance of the plasmapause in the different LT sectors are generally well reproduced by the simulations. But of course, the model has inherent limitations and is not expected to reproduce all observed features. The empiric 3-hourly Kp dependent quasi-static electric field model E5D is a global scale average: it can be quite different from the actual E-field during some periods of time.

IMAGE and CLUSTER Revelations about the Plasmasphere

The main observed characteristics and theories developed about the plasmasphere before 1998 are well summarized in Lemaire and Gringauz [1998]. Since then, many new results were obtained with the American IMAGE and the European CLUSTER missions, launched in 2000.

Scientific results about the whole magnetosphere were obtained by the IMAGE mission [Burch, 2003]. During the first two years of the IMAGE mission, most EUV observations have been from high latitudes with a favorable viewpoint for studying the distribution of plasma in the equatorial plane. For the first time, it was possible to observe the plasmapause in different MLT. During the extended mission of IMAGE between 2003-2005, the imaging perspective has changed to lower latitudes and the magnetospheric environment undergoes a transition from solar maximum to solar minimum.

Global imaging of the plasma distribution inside the plasmasphere led to new types of observations and of investigations summarized in this section.

1. Comparisons between EUV images and in situ measurements (RIP aboard IMAGE for instance) allow the determination of the particles number density inside the plasmasphere and show that the sharp outer boundary in EUV images (detecting He^+ ions) corresponds indeed to the plasmapause [Goldstein et al., 2003c].

2. Formation of plumes is a common consequence of increased convection. As they form, the connection is in the afternoon or dusk sector and the plumes extend sunward. The base of the plumes generally co-rotates [Spasojevic et al., 2003].

3. "Shoulders", asymmetric bulges in the plasmapause, were unexpected features of the plasmasphere seen for the first time in EUV images [Burch et al., 2001].

4. "Notches" were also discovered by EUV: they are localized density cavities in the plasmapause.

5. In times of increasing magnetospheric convection, the plasmasphere shrinks rapidly. The observations show that the plasmasphere is eroded, not compressed. Erosion occurs over the antisunward hemisphere, but not uniformly. The plasmapause moves at different times and at different rates at different local times. The plasmapause can move fast inward, for instance by almost 2 Re in less than 3 hours on 10 July 2000 [Goldstein et al., 2003a].

6. When convection returns to normal, flux tubes depleted by convection refill from the ionosphere. This process is slow but can be studied with EUV observations. During prolonged periods of geomagnetically quiet conditions, the plasmasphere extends to $L > 5$. The plasmaspheric refilling has been observed to be completed in less than 28 h (2 IMAGE orbital periods), which is less than theoretical expectations (2-3 days) [Reinisch et al., 2003].

7. "Channels" are regions of lower density extending roughly in the azimuthal direction. They usually appear first in the pre-midnight sector and allow the measurement of the rotation velocity. Some channels are formed by plume wrapping: the inner part of a plume nearly co-rotates while the outer part remains nearly fixed in local time.

8. Low-density regions (notches, channels) can serve as natural markers to track the angular velocity of the plasma over a range of L values. Some notches can persist for periods as long as 60 hours. Cold plasma in the range $2<L<4$ most frequently rotates at a rate 85-90 % of the co-rotation. But it may move at the co-rotation velocity for a time and then begin to lag.

The plasmasphere corotation lag is observed to be linked to a corresponding corotation lag in the ionosphere [Burch et al., 2004].

9. A comparison of how well three different electric field models can predict the plasmapause position as observed by EUV during the magnetic storm of 17 April 2002 is provided by Liemohn et al. [2004].

10. The multiple instruments on IMAGE give a comprehensive image of the entire ionosphere-plasmasphere-magnetosphere coupling [Goldstein et al., 2005]. Links between plasmaspheric plumes and sub-auroral proton arcs have been detected [Spasojevic et al., 2004]. There is an increasing interest on the role of the sub-auroral polarization stream (SAPS) in plume formation and in the sunward transport of plasma from dusk [Foster and Vo, 2002; Goldstein et al., 2003b].

Moreover, the four CLUSTER spacecraft launched in 2000 revealed also many important features in the magnetosphere. The local measurements of 4 spacecraft, combined with correlating global images of EUV, are useful to deconvolve spatial from temporal effects. The main new results concerning the plasmasphere obtained with these exceptional data are summarized in the next section:

1. The instrument WHISPER measures the electron number density profiles along the orbit of the spacecraft. Two different types of plasmasphere crossings are observed: a) smooth density variations with a sharp and well-defined knee corresponding to the plasmapause and b) small-scale density structures observed in the plasmasphere and the plasmapause. Due to the time resolution of 2 seconds, these thin structures and density irregularities (density variations of a factor 2 on 10 km) were for the first time observed by CLUSTER [Darrouzet et al., 2004].
2. Apparently detached plasma elements observed by both instruments are in many cases plasmaspheric plumes. The density values measured during the detached plasmasphere observations are by about an order of magnitude lower than the ones measured in the main plasmasphere. The thin structures and the density decrease are consistent with the density reduction for the detached plasma shells predicted by the peeling-off models due to plasma instability. These observations show that plumes have large latitudinal extent.
3. The Cluster Ion Spectrometry (CIS) instruments provide full three-dimensional distributions for the main ion species (H^+, He^+, O^+). It shows that H^+ and He^+ present mostly similar profiles with the He^+ densities being lower by a factor of ~15. On the contrary, O^+ is not observed as trapped plasmaspheric population at CLUSTER orbit altitude ($R > 4$ Re) [Dandouras et al., 2005].
4. Low-energy ($E < 25$ eV) O^+ is observed only as upwelling ions, escaping from the ionosphere along auroral field lines. O^{++} can be also observed in these upwelling ion populations.
5. In the vicinity of the plasmapause, around the geomagnetic equator, banded hiss-like electromagnetic emissions (2 to 10 kHz) are often observed [Masson et al., 2004].
6. EFW (Electric Field and Wave) instrument on Cluster give some simultaneous electric field and density observations [Gustafsson et al., 2001].

Conclusion

We presented here two detailed studies of the global dynamics of the plasmasphere during geomagnetically disturbed periods. A larger set of cases was studied and other sequences were presented in Pierrard and Cabrera [2005] and Dandouras et al. [2005].

From the direct comparison between the EUV observations from IMAGE and the numerical simulations of the plasmapause position determined from the value of Kp using the instability mechanism and the E5D electric field model, it appears that the formation of the plumes and their evolution with time are globally well reproduced. Plumes are formed after a moderate increase of Kp and rotate then with the plasmasphere during the quieting period following the enhanced magnetic activity. The observed shoulders are produced just at the end of magnetic substorms or storms and generally develop one or two hours before a sharp decrease of Kp. The plasmapause position obtained with the simulations corresponds in average to the position observed by EUV: it is not simply determined by the instantaneous Kp value, but it depends also on its past history.

The plasmasphere is highly structured and complex. Reproducible features evolve clearly during disturbed periods, but some differences appear also. The new observations of EUV/IMAGE will help to improve the theory for the formation of the plasmapause and the understanding of the new structures recently discovered in the plasmasphere.

Acknowledgments

V. Pierrard thanks Dr J. Cabrera (Center for Space Radiations, UCL) for the image processing and Prof. J. Lemaire and L. Dricot for their contributions in the development of the dynamical simulations. She thanks also Dr. B. Sandel, Lunar and Planetary Laboratory, University of Arizona, Tucson, USA and Dr. D.L. Gallagher, NASA, National Space Science and Technology Center, Huntsville, USA for the access to the EUV observations of the satellite IMAGE.

This project is financially supported by the Belgian Federal Science Policy Office (PPS).

References

[1] Bezrukikh V. V., Kotova G. A., Lezhen L. A., Lemaire J., Pierrard V. and Venediktov Yu. I. (2003). Dynamics of temperature and density of cold protons of the Earth's plasmasphere measured by the Auroral Probe/Alpha-3 experiment data during geomagnetic disturbances. *Cosmic Research*, Vol. 41, No. 4, 392-402.

[2] Burch J. L. (2003). The first two years of IMAGE. *Space Sci. Rev.*, 109, 1-24.

[3] Burch J. L., Goldstein J. and Sandel B. R. (2004). Cause of plasmasphere corotation lag. *Geophys. Res. Lett.*, 31, L05802, doi:10.1029/2003GL019164.

[4] Burch J. L., Mende S. B., Mitchell D. G., Moore T. E., Pollock C. J., Reinisch B. W., Sandel B. R., Fuselier S. A., Gallagher D. L., Green J. L., Perez J. D., and Reiff. P. H. (2001). Views of Earth's Magnetosphere with the IMAGE Satellite. *Science*, 291, 619-624.

[5] Carpenter D. L. (1995). Earth's plasmasphere awaits rediscovery. *EOS Trans. Am. Geophys. U.*, 76(9), 89.

[6] Dandouras I., Pierrard V., Goldstein J., Vallat C., Parks G. K., Rème H., Gouillart C., Sevestre F., McCarthy M., Kistler L. M., Klecker B., Korth A., Bavassano-Cattaneo M. B., Escoubet Ph., and Masson A. (2005). Multipoint observations of ionic structures in the Plasmasphere by CLUSTER-CIS and comparisons with IMAGE-EUV observations and with Model Simulations. Inner Magnetosphere Interactions: New Perspectives from Imaging, *AGU Geophysical Monograph*, 159, 23-54, doi: 10.1029/2004BK000104.

[7] Darrouzet F., Décréau P. M. E., De Keyser J., Masson A., Gallagher D. L., Santolik O., Sandel B. R., Trotignon J. G., Rauch J. L., Le Guirriec E., Canu P., Moullard O., Sedgemore F., André A. and Lemaire J. F. (2004) Density structures inside the plasmasphere: Cluster observations. *Annales Geophys.*, 22: 2577-2585.

[8] Foster, J. C., and Vo H. B. (2002). Average characteristics and activity dependence of the subauroral polarization stream. *J. Geophys. Res.*, 107(A12), 1475, doi: 10.1029/2002JA009409.

[9] Goldstein, J., Burch J. L., Sandel B. R., Mende S. B., son Brandt P. C., and Hairston M. R. (2005). Coupled response of the inner magnetosphere and ionosphere on 17 April 2002. *J. Geophys. Res.*, 110, A03205, doi:10.1029/ 2004JA010712.

[10] Goldstein J., Sandel B. R., Forrester W. T., and Reiff P. H. (2003a). IMF-driven plasmasphere erosion of 10 July 2000. *Geophys. Res. Lett.*, 30, 3, 1146, 10.1029/2002GL016478.

[11] Goldstein, J., Sandel, B. R., Hairston, M. R., and Reiff, P. H. (2003b). Control of plasmaspheric dynamics by both convection and sub-auroral polarization stream. *J. Geophys. Res.*, 30 (24), 2243, doi: 10.1029/2003GL018390.

[12] Goldstein J., Spasojevic M., Reiff P. H., Sandel B. R., Forrester W. T., Gallagher D. L., and Reinisch.B. W. (2003c). Identifying the plasmapause in IMAGE EUV data using IMAGE RPI in situ steep density gradients. *J. Geophys. Res.*, 108, A4, 1147, doi:10.1029/2002JA009475.

[13] Gustafsson G., André M., Carozzi T., Eriksson A. I., Falthammar C.-G., Grard R., Holmgren G., Holtet J. A., Ivchenko N., Karlsson T., Khotyaintsev Y., Klimov S., Laakso H., Lindqvist P.-A., Lybekk B., Marklund G., Mozer F., Mursula K., Pedersen A., Popielawska B., Savin S., Stasiewicz K., Tanskanen P., Vaivads A., and Wahlund J.-E. (2001). First results of electric field and density observations by Cluster EFW based on initial months of operation. *Annales Geophys.*, 19: 1219-1240.

[14] Lemaire J. (1974). The 'Roche limit' of ionospheric plasma and the formation of the plasmapause. *Planet. Space Sci.*, 22, 757-766.

[15] Lemaire J. F. (1985). *Frontiers of the plasmasphere* (Theoretical aspects). Université Catholique de Louvain, Faculté des Sciences, Editions Cabay, Louvain-La-Neuve, ISBN-2-87077-310-2.

[16] Lemaire J. F. and Gringauz K. I. *The Earth's plasmasphere*; Cambridge University Press, United Kingdom, 1998, 350 p.

[17] Liemohn M. W., Ridley, A. J., Gallagher, D. L., Ober, D. M. and Kozyra J. U. (2004). Dependence of plasmaspheric morphology on the electric field description during the recovery phase of the 17 April 2002 magnetic storm. *J. Geophys. Res.*, 109, A03209, doi:10.1029/2003JA010304.

[18] Masson A., Inan U. S., Laakso H., Santolik O., and Décréau P. (2004). Cluster observations of mid-latitude hiss near the plasmapause. *Annales Geophys.*, 22: 2565-2575.

[19] McIlwain C. E. (1986). A Kp dependent equatorial electric field model, The Physics of Thermal plasma in the magnetosphere. *Adv. in Space Res.*, 6 (3), 187-197.

[20] Pierrard V. (2004). La plasmasphère vue par IMAGE. *Ciel et Terre*, 120, 4, 114-117.

[21] Pierrard V. and Cabrera J. (2005). Comparisons between EUV/IMAGE observations and numerical simulations of the plasmapause formation. *Annales Geophysicae,* 23, 7, 2635, SRef-ID: 1432-0576/ag/2005-23-2635.

[22] Pierrard V. and Cabrera J. (2006). Dynamical simulations of plasmapause deformations. In Press in *Space Science Review.*

[23] Pierrard V. and Lemaire J. (2004). Development of shoulders and plumes in the frame of the interchange instability mechanism for plasmapause formation. *Geophys. Res. Lett.*, 31, 5, L05809, 10.1029/2003GL018919.

[24] Reinisch B. W., Huang X., Song P., Green J. L., Fung S. F., Vasyliunas V. M., Gallagher D. L., and Sandel B. R. (2003). Plasmaspheric mass loss and refilling as a result of a magnetic storm. *J. Geophys. Res.*, 10.1029/2003JA009948, 1-11.

[25] Sandel B. R., Goldstein J., Gallagher D. L.and Spasojevic M. (2003). Extreme ultraviolet imager observations of the structure and dynamics of the plasmasphere. *Space Sci. Rev.*, 109, 25-46.

[26] Spasojevic M., Frey H. U., Thomsen M. F., Fuselier S. A, Gary S. P., Sandel B. R., and Inan U. S. (2004). The link between a detached subauroral proton arc and a plasmaspheric plume. *Geophys. Res. Lett.*, 31, L04803, doi:10.1029/2003GL018389.

[27] Spasojevic M., Goldstein J., Carpenter D. L., Inan U. S., Moldwin M. B., and Reinisch B. W. (2003). Global response of the plasmasphere to a geomagnetic disturbance. *J. Geophys. Res.*, 108, 1340.

In: Space Science: New Research
Editor: Nick S. Maravell, pp. 97-139
ISBN: 1-60021-005-8

Chapter 4

FLUTE AND BALLOONING MODES IN THE INNER MAGNETOSPHERE OF THE EARTH: STABILITY AND INFLUENCE OF THE IONOSPHERIC CONDUCTIVITY

O.K. Cheremnykh and A.S. Parnowski
Space Plasma Department, Space Research Institute of NASU and NSAU, 40, prosp. Akad. Glushkova, Kyiv-187, MSP, 03680, Ukraine

Abstract

In this article we represent a survey of the present state and analytical methods of the theory of transversally small-scale standing MHD perturbations in the inner magnetosphere of the Earth, as well as our own views on this matter. We restrict our consideration by two important types of such perturbations: flute and ballooning modes.

In the most general case arbitrary three-dimensional transversally small-scale standing MHD perturbations in ideal plasmas with nested magnetic surfaces are described by a pair of Dewar-Glasser equations [34]. In some papers such equations are derived from the MHD equations by application of differential operators. This casts certain doubts upon the correctness of the resulting spectra of perturbations. We choose a different path, using just the condition of transversally small-scaleness and longitudinal elongation of perturbations, which, nevertheless, appeared to be sufficient to derive mentioned equations.

In addition we used ideal MHD approximation, neglected the convection and considered the equilibrium to be static. Obtained equations were applied to a dipolar magnetic configuration, approximately representing the inner magnetosphere of the Earth. This made the equations much simpler and after introducing dimensionless variables they reduced to linear homogeneous ordinary differential equations of second order.

When hydrodynamic pressure is considerable flute and ballooning perturbations are generated. The most unstable of them are the perturbations with a transversal to the magnetic surfaces polarization of the magnetic field.

Obtained equations were supplemented with ionospheric boundary conditions derived in the same approximation. Considered perturbations appeared to be affected only by integral Pedersen conductivity of the ionosphere, which was approximated by a thin spherical layer. Moreover, when Pedersen conductivity of the ionosphere is finite flute modes can appear in the magnetosphere, determining its stability in this case. Using a modified energetic principle we derived the corresponding stability criterion, which sets stronger restrictions on the stability than well known Gold criterion.

Introduction

Plasma confined by the geomagnetic field is thermodynamically non-equilibrium like any other plasma with pressure. Therefore, according to general concepts of theory of plasma instabilities, it can spontaneously acquire collective degrees of freedom, becoming unstable. Magnetospheric plasma never occupies a low-energy state due to constant flow of energy from outside. Trying to reach the minimum energy state magnetosphere constantly generates perturbations, which change its configuration and transport properties.

Instabilities in plasma can be divided by two main classes — magnetohydrodynamic (MHD) and kinetic ones. MHD instabilities are usually associated with motions of macroscopic volumes of plasma. They can be described with MHD equations [1]. The other instabilities, which essentially depend on difference in motions of different groups of particles in the same volume, are called kinetic instabilities. These are microscopic in comparison to large-scale slow MHD motions. The greatest threat to plasma stability is known to be posed by MHD instabilities with large growth rates, which lead to a rapid reconstruction of the initial equilibrium.

As magnetospheric plasma pressure grows, the main role goes to MHD instabilities, which are powered by plasma's thermal energy. These are flute and ballooning modes, which are driven by pressure and magnetic field line curvature. These instabilities were first investigated in papers [2, 3]. Kruskal and Schwarzschild [2] were first to demonstrate the development of instabilities shaped in a form of flutes elongated along the field lines (hence the name "flute modes") in plasma with sharp boundaries. The criterion of stability of magnetospheric plasma against flute perturbations was derived for the first time by Gold [3]. This criterion is best understood in terms of virtual flux tube interchange calculations made by Chandrasekhar [4], hence the second name of these modes, "interchange modes". Later on this criterion acquired an apt physical interpretation [5]: among all motions of flux tubes across the magnetic field lines, only those, which do not perturb the magnetic field, are allowed. Note that this criterion is analogous to the convective stability condition of gas, compressed by a gravity field. Gold's assumption of large-scale interchange motions of magnetic flux tubes being responsible for realignment of mass of planetary magnetosphere was further developed in papers [6 — 11].

In 1966 Furth et al [12] considered plasma stability in a flat geometry replacing field line curvature with variable gravity and discovered ballooning instability. Flute modes in this case appeared to be not constant, increasing in unfavourable curvature area and decreasing in favourable curvature area. Ballooning instability occurs when plasma pressure exceeds a certain critical pressure, which depends on ratio of lengths of favourable and unfavourable curvature sections of the field line.

Ballooning modes are well known in the magnetic fusion literature as pressure driven instabilities [13, 14], analogous to the instability developing at a weak spot of a pressurized elastic container. Ballooning perturbations are also sometimes called transversally small-scale modes. The assumption of smallness of transversal scale of perturbations has a profound physical meaning: (fast) magnetosonic waves, exhibiting themselves in compression and depression of field lines and being suppressed by stationary magnetic field of the Earth at low plasma pressure, are thrown away and only Alfvén and slow magnetosonic (ionosonic) waves, which are the most dangerous to the stability, are considered. So, perturbations of plasma, submersed in a stationary magnetic field, are a hybrid of Alfvén and slow magnetosonic waves. It's just these waves, which form perturbations of ballooning type. In the case of the magnetosphere such stationary field is the geomagnetic one, and in the case of fusion research — the stationary field of a device. This is one of the reasons for ballooning modes to be so important in magnetospheric and fusion research.

Many theoreticians were attracted by the problem of ballooning modes in the magnetosphere [15 — 29]. These investigations were motivated by the responsibility of ballooning modes for some observed geomagnetic pulsations [30]. Besides, interchange instability can explain strong ejections of plasma at the time of magnetospheric substorms [31]. An interesting result was obtained in paper [17], where flute perturbations were shown to be a special class of ballooning perturbations being described by the same equations. Note that the equation for ballooning perturbations, as it was shown in [17], generalizes differential operators for Alfvén waves derived by several authors [32, 33]. It was specified ibid that the stability is mostly determined by large-scale perturbations, whose structure essentially depends on the distribution of plasma parameters and ionospheric boundary conditions.

At the same time intensive research of ballooning stability of toroidal fusion devices were carried out [13, 14, 34 — 36]. In particular, paper [34] introduced equations for ballooning modes in arbitrary magnetic geometry with nested magnetic surfaces. For this reason, the approach of paper [34] was later successfully applied to magnetospheric plasma.

All attempts of deriving an analytical stability criterion against such perturbations had no success until recently, so large-scale ballooning modes were usually studied numerically. In last years an interest for generation of flute and ballooning perturbations in the Earth's magnetosphere was also prompted by an investigation of mechanisms of self-organisation of non-linear structures in the magnetosphere [37] and their influence on the redistribution of mass of magnetospheric plasma.

Hitherto the matter concerned the stability of magnetospheric plasma without detailed elaboration of ionospheric boundary conditions. Real ionospheric plasmas, which bound magnetospheric plasma, are always of finite conductivity and thus are not completely tied to the magnetic field lines in the ionosphere. This causes dissipative instabilities, which develop much easier than ideal ones.

The first articles dealing with derivation of ionospheric boundary conditions were also the first ones investigating the influence of dissipation caused by finite ionospheric conductivity on ballooning (and interchange) plasma stability. The boundary problem for magnetospheric perturbations is discussed for half a century, starting from pioneer articles by Dungey and Nishida [38, 39]. Earlier investigators considered either isolating boundary conditions, or no boundary conditions at all (local perturbations, see e.g. [5]). More complicated boundary conditions, alternative to [38, 39], were later derived by Maltsev [40]. It was only at the beginning of nineties, when boundary conditions for magnetospheric perturbations were thoroughly analyzed in paper [41] as part of an investigation of magnetospheric pulsations excitement mechanisms [42]. Different attempts to describe the influence of the ionosphere on magnetospheric waves using the boundary conditions are given out in details in paper [43] and references therein. Boundary conditions for ballooning perturbations were derived in paper [25].

The aim of this work is to derive a stability criterion of magnetospheric plasma against flute and ballooning instabilities. This research was motivated by three reasons. The first one is the result of paper [25], which states that ballooning stability threshold essentially depends on ionospheric conductivity. The second one is our earlier research of magnetospheric plasma's ballooning stability with perfect ionospheric conductivity, which showed that stability is determined by ballooning modes, and interchange modes are absent [27]. This reminded us of paper [44] by Furth, Killen and Rosenbluth about interchange plasma stability with dissipation caused by finite conductivity, which demonstrated for the first time that finite conductivity leads to flute instability of plasma, considered to be interchange stable in the framework of ideal MHD. We think that this situation applies to magnetospheric plasma stability research as well. The third reason is that scientific publications on ballooning modes do not always present consistent and thorough descriptions of mathematical techniques, involved in ballooning perturbations research. We decided to fill this gap and try to give out in fine detail the mathematical part of the problem. For the same reason we give a detailed description of a plasma equilibrium, dipole model of the geomagnetic field and ionospheric boundary conditions, as well as a brief overview of the energetic principle (δW) and its generalisations for the case of a non-Hermitian operator.

1. Main Equations

The Earth's magnetosphere represents an apt illustration of a system, which is describable in terms of ideal MHD. The simplest estimates [45] show that MHD approximation is applicable for describing large-scale structures with spatial scales above 100 km and temporal scales above 1 sec. These include the long-period magnetospheric perturbations of interest, which are observed both from the ground and in the space. We shall investigate their stability in the framework of linear perturbation theory.

The description of wave processes in inhomogeneous medium with magnetic fields, such as magnetospheric plasma, even in the framework of linear theory of MHD perturbations is a complicated mathematical problem. Further we shall derive equations of small perturbations for a special class of perturbations called the ballooning ones in scientific literature. When deriving these equations we shall use the approaches developed in papers [34, 35]. For all this, we shall not use variational methods, as in paper [34], or apply differential operators to the initial equations, as in paper [35], because it can distort the perturbations spectrum. Instead, we derive these equations using only the approximation of transversal localization of perturbations and their elongation along the magnetic field lines. We shall use the derived equations to analyze plasma stability in the inner magnetosphere of the Earth.

1.1. Initial Equations

In the framework of MHD plasma is usually considered as an ideally conductive isotropic quasineutral fluid described by equations

$$\rho\frac{\partial\vec{v}}{\partial t}=-\nabla p+\frac{1}{c}\left[\vec{J}\times\vec{H}\right],\ \frac{\partial\vec{H}}{\partial t}=\mathrm{rot}\left[\vec{v}\times\vec{H}\right],$$

$$\frac{\partial\rho}{\partial t}+\mathrm{div}\,\rho\vec{v}=0\,,\ \left(\frac{\partial}{\partial t}+\vec{v}\cdot\nabla\right)\frac{p}{\rho^{\gamma}}=0\,, \tag{1.1}$$

$$\mathrm{rot}\,\vec{H}=\frac{4\pi}{c}\vec{J}\,,\ \mathrm{div}\,\vec{H}=0\,,\ \vec{\varepsilon}=-\frac{1}{c}\left[\vec{v}\times\vec{H}\right],$$

where ρ is plasma density, $\vec{v}$ is a velocity of an elementary plasma volume, γ is a ratio of specific heats. Symbols p, $\vec{J}$, $\vec{\varepsilon}$ and $\vec{H}$ denote plasma pressure, current density, and electric and magnetic intensities. For the sake of brevity we introduce new unit-independent variables $\vec{j}$, $\vec{B}$ and $\vec{E}$ as follows

$$\vec{H}=\sqrt{4\pi}\vec{B}\,,\ \vec{J}=\frac{c}{\sqrt{4\pi}}\vec{j}\,,\ \vec{\varepsilon}=\frac{\sqrt{4\pi}}{c}\vec{E}\,. \tag{1.2}$$

One can easily express the same variables in MKS units. As a result, we obtain from (1.1) a set of equations, which we shall use for further analysis

$$\rho\frac{\partial\vec{v}}{\partial t}=-\nabla p+\left[\vec{j}\times\vec{B}\right],\ \frac{\partial\vec{B}}{\partial t}=\mathrm{rot}\left[\vec{v}\times\vec{B}\right],$$

$$\frac{\partial\rho}{\partial t}+\mathrm{div}\,\rho\vec{v}=0\,,\ \left(\frac{\partial}{\partial t}+\vec{v}\cdot\nabla\right)\frac{p}{\rho^{\gamma}}=0\,, \tag{1.3}$$

$$\mathrm{rot}\,\vec{B}=\vec{j}\,,\ \mathrm{div}\,\vec{B}=0\,,\ \vec{E}=-\left[\vec{v}\times\vec{B}\right].$$

Taking into account linearized versions of Faraday induction law, combined with the condition of perfect conductivity

$$\tilde{\vec{B}}=\mathrm{rot}\left[\vec{\xi}\times\vec{B}\right],$$

adiabatic equation

$$\tilde{p}=-\left(\vec{\xi}\cdot\nabla\right)p-\gamma p\,\mathrm{div}\,\vec{\xi}\,,$$

and Ampere's law

$$\tilde{\vec{j}} = \operatorname{rot} \tilde{\vec{B}},$$

and introducing a plasma displacement vector $\vec{\xi}$, $\vec{v} = \dfrac{\partial \vec{\xi}}{\partial t}$, we obtain that all perturbed quantities in a linearized Euler equation (the first of (1.3)) can be expressed through $\vec{\xi}$ and the whole set (1.3) reduces to a single vector equation

$$\rho \frac{\partial^2 \vec{\xi}}{\partial t^2} = \nabla\left(\vec{\xi} \cdot \nabla p\right) + \nabla\left(\gamma p \operatorname{div} \vec{\xi}\right) + \frac{1}{c}\left[\tilde{\vec{j}} \times B\right] + \frac{1}{c}\left[\vec{j} \times \tilde{\vec{B}}\right]. \tag{1.4}$$

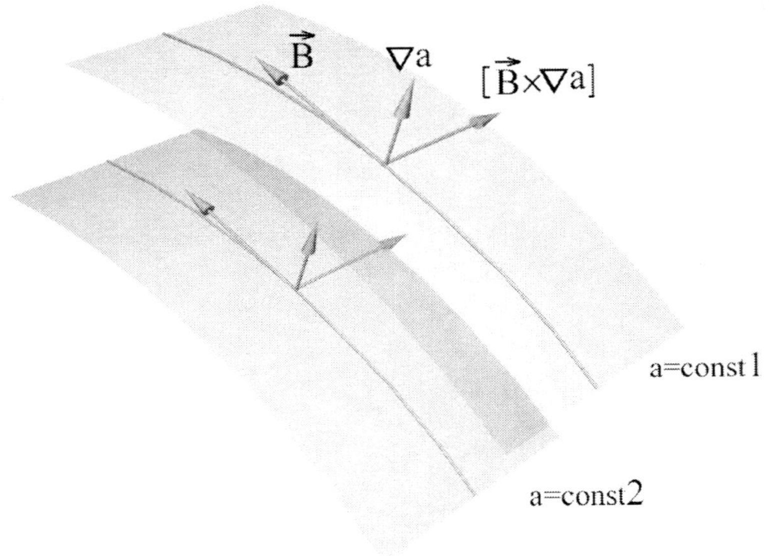

Fig. 1. A label of magnetic surface.

1.2. Magnetic Label

Further we shall assume that equilibrium magnetic field has magnetic surfaces, i.e. surfaces, containing magnetic field lines and current lines. In a magnetic field with flux surfaces the following useful geometrical relations that do not depend on plasma pressure hold

$$\vec{B} \cdot \nabla a = 0, \ \operatorname{div} \vec{B} = 0, \ \vec{j} \cdot \nabla a = \operatorname{div}\left[\vec{B} \times \nabla a\right] = 0, \tag{1.5}$$

where a is a value, traditionally called as magnetic label, which is constant on any given magnetic surface, see Fig. 1. It was introduced by Kruskal and Kulsrud in paper [46]. A magnetic label can be arbitrary chosen. In particular, it can be associated with toroidal magnetic flux or with a volume of plasma, confined inside this surface. We shall return to this matter in section 2. The knowledge of the precise value of a is not necessary for deriving the equations of small perturbations. In this sense (1.5) is independent from the equilibrium equation

$$p'|\nabla a|^2 = \left[\operatorname{rot}\vec{B} \times \vec{B}\right] \cdot \nabla a, \tag{1.6}$$

where $p' = dp/da$ and which is needed to find $a(\vec{r})$. Note that ∇a, $\left[\vec{B} \times \nabla a\right]$ and $\vec{B}$ directions are orthogonal and can be chosen as a local flux coordinate system.

1.3. Equations of Small Oscillations

In this subsection we shall follow a procedure set out in [35]. It is convenient to express the values of perturbed magnetic field $\tilde{\vec{B}}$ and current density $\tilde{\vec{j}}$ in (1.4) through a vector $\vec{T}$ defined as

$$\vec{T} \equiv \operatorname{rot}\left[\vec{\xi} \times \vec{B}\right] + \frac{\left[\vec{j} \times \nabla a\right]}{|\nabla a|^2} \vec{\xi} \cdot \nabla a. \tag{1.7}$$

Introducing (1.7) into (1.4) transforms the latter into

$$\begin{aligned} \frac{\rho \partial^2 \vec{\xi}}{\partial t^2} = &\left[\operatorname{rot}\vec{T} \times \vec{B}\right] + \left[\vec{j} \times \left(\vec{T} - (\vec{T} \cdot \nabla a)\frac{\nabla a}{|\nabla a|^2}\right)\right] + \nabla(\gamma p \operatorname{div}\vec{\xi}) + \\ &+ \vec{\xi} \cdot \nabla \left\{ \nabla p' + \frac{|\vec{j}|^2}{|\nabla a|^2}\nabla a - \left[\operatorname{rot}\frac{\vec{j} \times \nabla a}{|\nabla a|^2} \times \vec{B}\right]\right\}. \end{aligned} \tag{1.8}$$

When obtaining (1.8) we used the equation of hydrostatic equilibrium (1.6) and vector equality

$$\vec{T} \cdot \nabla a = \vec{B} \cdot \nabla(\vec{\xi} \cdot \nabla a), \tag{1.9}$$

which follows from (1.7).

Let us decompose vectors $\vec{T}$ i $\vec{\xi}$ into three orthogonal vectors ∇a, $\left[\vec{B} \times \nabla a\right]$ and $\vec{B}$:

$$\vec{T} = T_1 \nabla a + T_2 \frac{\left[\vec{B} \times \nabla a\right]}{|\nabla a|^2} - T_3 \vec{B},$$
$$\vec{\xi} = \xi \frac{\nabla a}{|\nabla a|^2} + \eta \frac{\left[\vec{B} \times \nabla a\right]}{|\vec{B}|^2} + \tau \frac{\vec{B}}{|\vec{B}|^2}. \tag{1.10}$$

The components of (1.10) have the form

$$T_1 = \frac{\vec{B} \cdot \nabla \xi}{|\nabla a|^2}, \quad T_2 = \frac{1}{\alpha_s}\left[\vec{B} \cdot \nabla \eta + (\gamma_s - S)\xi\right],$$
$$T_3 = \frac{\nabla a \cdot \nabla \xi}{|\nabla a|^2} + \frac{\left[\vec{B} \times \nabla a\right]}{|\vec{B}|^2} \cdot \nabla \eta + \frac{\xi}{|\vec{B}|^2}\left[\operatorname{div}(\alpha_s \nabla a) + 2p'\right], \tag{1.11}$$

where

$$S = \frac{\left[\vec{B} \times \nabla a\right]}{|\nabla a|^2} \cdot \operatorname{rot} \frac{\left[\vec{B} \times \nabla a\right]}{|\nabla a|^2}, \ \alpha_s = \frac{|\vec{B}|^2}{|\nabla a|^2}, \ \gamma_s = \frac{\vec{j} \cdot \vec{B}}{|\nabla a|^2}. \tag{1.12}$$

Substituting (1.10) — (1.12) into (1.8) we get

$$\rho \frac{\partial^2 \vec{\xi}}{\partial t^2} = \nabla a \left[\alpha_s \nabla a \cdot \nabla T_3 + \frac{\nabla a \cdot \nabla T_0}{|\nabla a|^2} - 2p' T_3 + T_2 (S - \gamma_s) + \vec{B} \cdot \nabla T_1 + K\xi \right] +$$
$$+ \frac{\left[\vec{B} \times \nabla a\right]}{|\nabla a|^2} \left[\frac{\left[\vec{B} \times \nabla a\right] \cdot \nabla T_0}{|\vec{B}|^2} + \vec{B} \cdot \nabla T_2 + \left[\vec{B} \times \nabla a\right] \nabla T_3 \right] + \frac{\vec{B}}{|\vec{B}|^2} \left(\vec{B} \cdot \nabla\right) T_0. \tag{1.13}$$

Here the following denotations are used:

$$T_0 = \gamma p \operatorname{div} \vec{\xi}, \ K = \frac{\gamma_s}{\alpha_s}(\gamma_s - S) + \frac{p'}{|\vec{B}|^2} \frac{\nabla a \cdot \nabla\left(2p + |\vec{B}|^2\right)}{|\nabla a|^2}. \tag{1.14}$$

Equation (1.13) does not depend on the coordinate system, because it is based only on the general properties of differential operators in an arbitrary flux coordinate system. For this reason it is exact in such a way that it describes arbitrary low-frequency plasma perturbations.

In review [35] it was noted that equation (1.13) is too difficult for further analysis, because it contains high-order derivatives of T_3. Really, only this quantity contains transversal derivatives of ξ and η, which are large for transversally small-scale perturbations. This complicates the analysis of (1.13). For this reason, authors of [34] used different equations. They decomposed (1.13) into ∇a and $\left[\vec{B}\times\nabla a\right]$ directions, and then subtracted one component from another. As a result, they got an equation, which did not contain transversal derivatives of T_3. So, their set of equations consisted from this new equation and two equations, following from (1.13) and proportional to $\left[\vec{B}\times\nabla a\right]$ and $\vec{B}$. In our opinion, such procedure may significantly distort the spectrum of perturbations. Thus we shall use the original equation (1.13) and shall not apply any differential operators to it.

1.4. Ballooning Approximation

Equation (1.13) is still cumbersome because it describes arbitrary MHD perturbations. Here and later we restrict to considering pressure driven transversal small-scale perturbations. Further for the description of the perturbations we turn to the approximation $\left|\vec{k}_\perp\right| >> b^{-1}, \left|\vec{k}_\parallel\right|$, where $\vec{k}_\parallel$ and $\vec{k}_\perp$ are the longitudinal and the transversal wave vectors, and b is a characteristic spatial scale of variation of equilibrium quantities. This approximation, which is usually called ballooning (see [35] and references therein), will significantly simplify the analysis of equation (1.13). In our problem's geometry ballooning approximation has a simple physical meaning: any component X of the displacement vector satisfies an inequality

$$\frac{|\nabla a\cdot\nabla X|}{|\nabla a|}, \frac{\left|\left[\vec{B}\times\nabla a\right]\cdot\nabla X\right|}{|\vec{B}||\nabla a|} >> \frac{1}{b}, \frac{|\vec{B}\cdot\nabla X|}{|\vec{B}|}, \tag{1.15}$$

where b is a characteristic spatial scale of the system.

For description of perturbations (1.15) in an inhomogeneous medium it seems natural to use eikonal approximation. However, the wavelengths of perturbations of interest are of the same order as the spatial scale of the magnetosphere. Thus, usual eikonal approximation is inapplicable. However, it is possible to apply it to transversal components [34], leaving the spatial part of the longitudinal components unchanged. In this case magnetospheric plasma perturbations are described in terms of transversal rays and longitudinal MHD waves. Following Dewar and Glasser [34], we search for solution of (1.13) in the transversal eikonal form

$$\vec{\xi}(\vec{r},t) = \hat{\vec{\xi}}(\vec{r})\exp(-i\omega t + i\chi/\varepsilon),\; T_i(\vec{r},t) = \hat{T}_i(\vec{r})\exp(-i\omega t + i\chi/\varepsilon), \tag{1.16}$$

where $\varepsilon << 1$ is a characteristic transversal spatial scale, and χ is a transversal eikonal, which satisfies an equation

$$\vec{k}_\perp \cdot \vec{B} = 0\,,\ \vec{k}_\perp = \nabla \chi\,. \tag{1.17}$$

We assume the amplitudes $\hat{\vec{\xi}}(\vec{r})$ and $\hat{\vec{T}}(\vec{r})$ to be smooth coordinate functions. One can easily see that the relation (1.15) holds due to smallness of ε.

Perturbed components T_i from (1.16) are equal to (see (1.11))

$$\hat{T}_0 = \gamma p\left(\operatorname{div}\hat{\vec{\xi}} + \frac{i}{\varepsilon}\vec{k}_\perp \cdot \hat{\vec{\xi}}\right),\ \hat{T}_1 = \frac{\vec{B}\cdot\nabla\hat{\xi}}{|\nabla a|^2},$$

$$\hat{T}_2 = \frac{1}{\alpha_s}\left[\vec{B}\cdot\nabla\hat{\eta} + (\gamma_s - S)\hat{\xi}\right],\ \hat{T}_3 = \frac{\hat{T}_0}{\gamma p} - \vec{B}\cdot\nabla\left(\frac{\hat{\tau}}{|\vec{B}|^2}\right) + 2\vec{\kappa}\cdot\hat{\vec{\xi}}\,, \tag{1.18}$$

where $\vec{\kappa} = \left(\frac{\vec{B}}{|\vec{B}|}\cdot\nabla\right)\frac{\vec{B}}{|\vec{B}|}$ is a magnetic field line curvature. Taking (1.18) and (1.13) into account and omitting terms of order ε, we obtain

$$\hat{T}_0 + |\vec{B}|^2\hat{T}_3 = 0\,. \tag{1.19}$$

Taking into account the relation between $\hat{T}_3$ and $\hat{T}_0$ in (1.18) we find from (1.19)

$$\hat{T}_0 = \frac{\gamma p|\vec{B}|^2}{\gamma p + |\vec{B}|^2}\left[\vec{B}\cdot\nabla\left(\frac{\hat{\tau}}{|\vec{B}|^2}\right) - 2\vec{\kappa}\cdot\hat{\vec{\xi}}\right]. \tag{1.20}$$

Comparing the obtained expression for $\hat{T}_0$ to (1.18) we see that the following relation holds to within ε

$$\vec{k}_\perp \cdot \hat{\vec{\xi}} = 0\,. \tag{1.21}$$

Substituting $\hat{\vec{\xi}}$ for its expression given by (1.16) we obtain

$$\hat{\eta} = -\hat{\xi}\frac{\vec{k}_\perp\cdot\nabla a}{k_{\perp\beta}|\nabla a|^2}, \tag{1.22}$$

where

$$k_{\perp\beta} = \frac{\vec{k}_\perp \cdot \left[\vec{B} \times \nabla a\right]}{\left|\vec{B}\right|^2}. \tag{1.23}$$

From (1.10), (1.16), and (1.22) we get

$$\hat{\vec{\xi}} = \hat{\xi}\frac{\left[\vec{k}_\perp \times \vec{B}\right]}{k_{\perp\beta}\left|\vec{B}\right|^2} + \hat{\tau}\frac{\vec{B}}{\left|\vec{B}\right|^2}. \tag{1.24}$$

Then

$$\vec{\kappa} \cdot \hat{\vec{\xi}} = \hat{\xi}\frac{\vec{k}_\perp \cdot \left[\vec{B} \times \vec{\kappa}\right]}{k_{\perp\beta}\left|\vec{B}\right|^2}. \tag{1.25}$$

As a result, the expression (1.20) for $\hat{T}_0$ can be rewritten as

$$\hat{T}_0 = \frac{\gamma p\left|\vec{B}\right|^2}{\gamma p + \left|\vec{B}\right|^2}\left[\vec{B} \cdot \nabla\left(\frac{\hat{\tau}}{\left|\vec{B}\right|^2}\right) - 2\hat{\xi}\frac{\vec{k}_\perp \cdot \left[\vec{B} \times \vec{\kappa}\right]}{k_{\perp\beta}\left|\vec{B}\right|^2}\right] = -\hat{T}_3\left|\vec{B}\right|^2. \tag{1.26}$$

Using two easily verifiable equalities

$$\vec{B} \cdot \nabla k_{\perp\beta} = 0, \; \vec{B} \cdot \nabla\left(\frac{\vec{k}_\perp \cdot \nabla a}{\left|\nabla a\right|^2}\right) = -Sk_{\perp\beta} \tag{1.27}$$

and equation (1.22) we obtain

$$\vec{B} \cdot \nabla \hat{\eta} = S\hat{\xi} - \frac{\vec{k}_\perp \cdot \nabla a}{k_{\perp\beta}\left|\nabla a\right|^2}\,\vec{B} \cdot \nabla \hat{\xi}. \tag{1.28}$$

Taking into account (1.28) we find from (1.18)

$$\hat{T}_2 = \frac{\gamma_s}{\alpha_s}\hat{\xi} - \frac{\vec{k}_\perp \cdot \nabla a}{k_{\perp\beta}\left|\vec{B}\right|^2}\frac{\vec{B} \cdot \nabla \hat{\xi}}{\alpha_s}. \tag{1.29}$$

Casting (1.16) and (1.24) in (1.13) we get

$$\rho\left(\frac{\left[\vec{k}_\perp\times\vec{B}\right]}{k_{\perp\beta}\left|\vec{B}\right|^2}\frac{\partial^2\hat{\xi}}{\partial t^2}+\frac{\vec{B}}{\left|\vec{B}\right|^2}\frac{\partial^2\hat{\tau}}{\partial t^2}\right)=\frac{\vec{B}}{\left|\vec{B}\right|^2}\vec{B}\cdot\nabla\hat{T}_0+$$
$$+\nabla a\left[\frac{2\vec{\kappa}\cdot\nabla a}{|\nabla a|^2}\left(\hat{T}_0+p'\hat{\xi}\right)-\frac{\vec{k}_\perp\cdot\nabla a}{k_{\perp\beta}\left|\vec{B}\right|^2}\left(\vec{B}\cdot\nabla\hat{\xi}\right)\left(S-\gamma_s\right)+\vec{B}\cdot\nabla\left(\frac{\vec{B}\cdot\nabla\hat{\xi}}{|\nabla a|^2}\right)\right]+ \tag{1.30}$$
$$+\frac{\left[\vec{B}\times\nabla a\right]}{|\nabla a|^2}\left[\frac{\left(\vec{B}\cdot\nabla\hat{\xi}\right)}{\alpha_s}\left(\gamma_s+S\right)+2\frac{\left[\vec{B}\times\nabla a\right]}{\left|\vec{B}\right|^2}\vec{\kappa}\left(\hat{T}_0+p'\hat{\xi}\right)-\frac{\vec{k}_\perp\cdot\nabla a}{k_{\perp\beta}|\nabla a|^2}\vec{B}\cdot\nabla\left(\frac{\vec{B}\cdot\nabla\hat{\xi}}{\alpha_s}\right)\right].$$

Now we shall decompose this equation. First, we obtain the longitudinal component (along $\vec{B}$)

$$\rho\frac{\partial^2\tau}{\partial t^2}=\vec{B}\cdot\nabla\hat{T}_0. \tag{1.31}$$

Then we obtain the transversal component (along $\frac{\left[\vec{k}_\perp\times\vec{B}\right]}{k_{\perp\beta}\left|\vec{B}\right|^2}$)

$$\rho\frac{\left|\vec{k}_\perp\right|^2}{k_{\perp\beta}^2\left|\vec{B}\right|^2}\frac{\partial^2\hat{\xi}}{\partial t^2}=2\left(\hat{T}_0+p'\hat{\xi}\right)\frac{\vec{\kappa}\cdot\left[\vec{k}_\perp\times\vec{B}\right]}{\vec{k}_\perp\cdot\left[\vec{B}\times\nabla a\right]}+\vec{B}\cdot\nabla\left[\left(1+\frac{\left(\vec{k}_\perp\cdot\nabla a\right)^2}{k_{\perp\beta}^2\left|\vec{B}\right|^2}\right)\frac{\vec{B}\cdot\nabla\hat{\xi}}{|\nabla a|^2}\right]. \tag{1.32}$$

Using an obvious relation

$$k_{\perp\beta}^2\left|\vec{B}\right|^2+\left(\vec{k}_\perp\cdot\nabla a\right)^2=\left|\vec{k}_\perp\right|^2|\nabla a|^2 \tag{1.33}$$

we rewrite (1.32) in the form

$$\rho\frac{\left|\vec{k}_\perp\right|^2}{k_{\perp\beta}^2\left|\vec{B}\right|^2}\frac{\partial^2\hat{\xi}}{\partial t^2}=2\left(\hat{T}_0+p'\hat{\xi}\right)\frac{\vec{\kappa}\cdot\left[\vec{k}_\perp\times\vec{B}\right]}{\vec{k}_\perp\cdot\left[\vec{B}\times\nabla a\right]}+\vec{B}\cdot\nabla\left(\frac{\left|\vec{k}_\perp\right|^2}{k_{\perp\beta}^2\left|\vec{B}\right|^2}\vec{B}\cdot\nabla\hat{\xi}\right). \tag{1.34}$$

Equations (1.26) and (1.34) coincide with those obtained by Dewar and Glasser [34]. They are principal for further analysis. Equation (1.8), supplied with corresponding boundary conditions, can be used to obtain a spectrum of ballooning perturbations.

2. Dipole Model of the Geomagnetic Field

The obtained equations of small oscillations (1.26) and (1.34) should be adjusted with the equilibrium plasma configuration. Therefore, prior to analyzing them we shall briefly discuss the magnetospheric equilibrium.

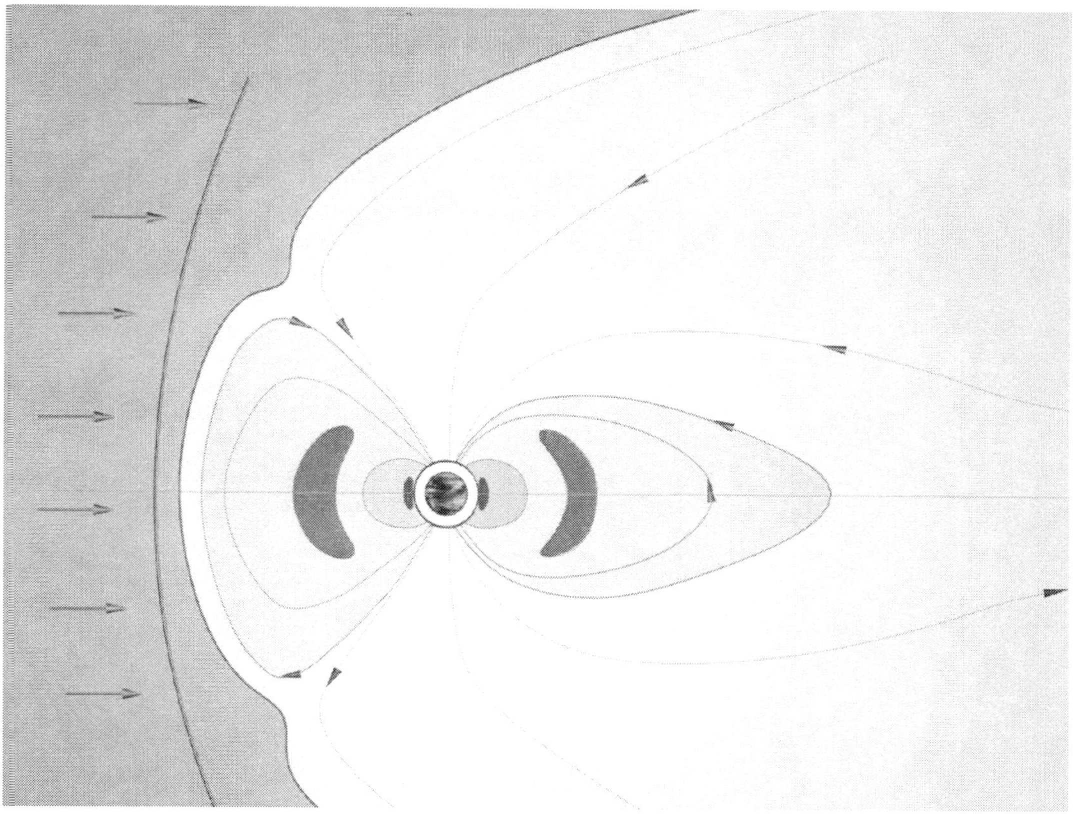

Fig. 2. Plasma structure of the Earth's magnetosphere.

At present time, the configuration of large-scale magnetic field in the Earth's magnetosphere (see Fig. 2) is already studied well enough to create its global empirical model [47, 48]. Nevertheless, this model is very unwieldy for analytical description of magnetospheric perturbations. To illustrate the influence of electromagnetic forces and pressure gradient on the stability of MHD perturbations in the inner magnetosphere of the Earth and not to complicate the consideration we shall use the simplest three-dimensional model of the geomagnetic field — axially symmetric magnetic dipolar field [49] with nested magnetic surfaces (Fig. 3). This model is very convenient for the analysis of geophysical phenomena and approximates the real geomagnetic field well enough up to 6 R_E. At greater distances this model becomes less and less accurate due to the influence of magnetospheric currents, becoming inapplicable at 10 R_E.

2.1. Magnetic Field

The Earth's magnetic field is mostly determined by internal currents and can be approximated by a multipole expansion with a dominant dipole term. For this reason this field can be assumed to be provided by a point dipole $\vec{M}$ located at origin and pointed southwards, i.e.

$$\vec{B} = \operatorname{rot}\vec{A}\,,\ \vec{A} = \frac{\left[\vec{M}\times\vec{r}\right]}{r^3}, \tag{2.1}$$

where $\vec{A}$ is a vector potential, and $\vec{r}$ is radius vector. In spherical coordinates r, θ, φ with $\theta = 0$ being an equatorial plane equation (2.1) has the form

$$\vec{B} = \left[\nabla\psi\times\nabla\varphi\right], \tag{2.2}$$

where $\psi = \dfrac{M\cos^2\theta}{r}$ is a poloidal magnetic flux, and $M = \left|\vec{M}\right|$ is the norm of the magnetic dipole momentum of the Earth. Note that in spherical coordinates $\nabla\varphi = (r\cos\theta)^{-1}\vec{e}_\varphi$, and the components of the magnetic field have the form

$$\vec{B}_r = -\frac{2M}{r^3}\sin\theta\,\vec{e}_r\,,\ \vec{B}_\theta = \frac{M\cos\theta}{r^3}\vec{e}_\theta\,, \tag{2.3}$$

and its norm $\left|\vec{B}\right|$ equals to

$$\left|\vec{B}\right| = \frac{M\sqrt{b(\theta)}}{r^3}, \tag{2.4}$$

where $b(\theta) = 1 + 3\sin^2\theta$. The field lines of such a magnetic field are shown on Fig. 3.

Further we will also need a field line equation. The arc element $\overrightarrow{ds}$ of a field line satisfies a condition

$$\left[\overrightarrow{ds}\times\vec{B}\right] = 0\,.$$

For the considered axially symmetric field this condition reduces to an equation

$$\frac{dr}{B_r} = \frac{rd\theta}{B_\theta},$$

whence

$$r = L\cos^2\theta, \tag{2.5}$$

where L is a radial distance to the field line in the equatorial plane, measured in R_E. This value can be expressed through the magnetic latitude θ_0 of the field line at the Earth's surface as

$$L = R_E \cos^{-2}\theta_0,$$

where R_E is the Earth's radius. The length element of the field line is given by

$$(dl)^2 = (dr)^2 + r^2(d\theta)^2 = L^2\cos^2\theta b(\theta)(d\theta)^2.$$

Hence

$$dl = L\sqrt{b(\theta)}\cos\theta d\theta. \tag{2.6}$$

Integrating this equation gives us the field line length at any given value of McIlwain parameter.

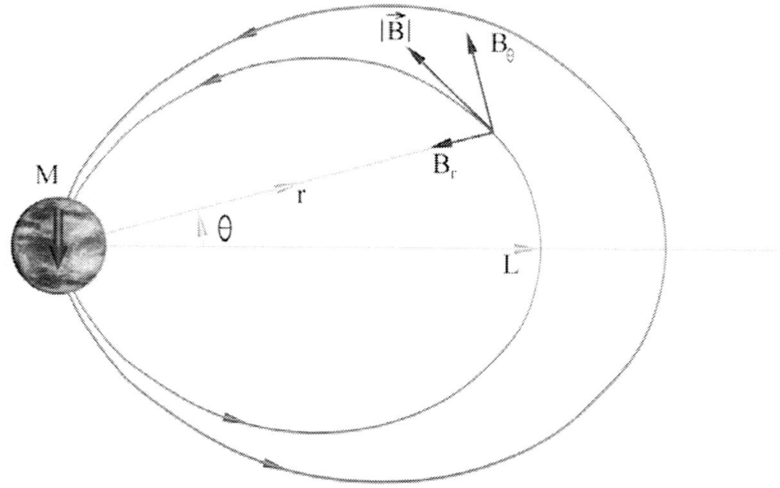

Fig. 3. Dipolar magnetic field lines.

2.2. Equilibrium

We assume magnetospheric plasma convection to be small ($\vec{v} = 0$), so that magnetospheric equilibrium can be approximated by a static MHD equilibrium

$$\nabla p = \left[\vec{j} \times \vec{B}\right]. \tag{2.7}$$

It follows from (2.7) that there is a transversal current with density

$$\vec{j}_\perp = \frac{\left[\vec{B} \times \nabla p\right]}{\left|\vec{B}\right|^2}. \tag{2.8}$$

Besides the transversal current (2.8) there can be a longitudinal current

$$\vec{j}_\parallel = \lambda \vec{B}.$$

It should be of such value that the total current is divergenceless

$$\mathrm{div}\left(\vec{j}_\parallel + \vec{j}_\perp\right) = 0. \tag{2.9}$$

As before, we assume that equilibrium magnetic field has no curls and can be approximated by vacuum magnetic dipolar field $\vec{B}_D$

$$\mathrm{rot}\,\vec{B}_D = 0. \tag{2.10}$$

Then we obtain from (2.9)

$$\vec{B} \cdot \nabla \lambda = 2\nabla\left|\vec{B}\right| \cdot \frac{\left[\vec{B} \times \nabla p\right]}{\left|\vec{B}\right|^3}. \tag{2.11}$$

Following [50], we shall assume that equilibrium field-aligned current is absent ($\lambda = 0$), and the magnetospheric equilibrium is provided mostly by the toroidal current

$$\vec{j} = j_\varphi(r, \theta)\vec{e}_\varphi. \tag{2.12}$$

Then from (2.1), (2.10) and (2.11) one can derive the following useful relations

$$\vec{j} \cdot \vec{B} = 0, \quad \left[\vec{B} \times \nabla \psi\right] \cdot \nabla\left|\vec{B}\right| = 0, \tag{2.13}$$

which will be used later. When deriving (2.12) we used that

$$p = p(\psi), \tag{2.14}$$

which follows from (2.1) i (2.7). One can see from (2.7), (2.8) and (2.13) that equilibrium toroidal ring current is equal to

$$j_\varphi = \frac{dp}{d\psi} r\cos\theta \tag{2.15}$$

and depends on particle distribution in the magnetosphere.

In particular, since $\vec{B}\cdot\nabla\psi = \vec{B}\cdot\vec{e}_\varphi = 0$, magnetic field lines and the lines of constant ψ coincide. Thus it follows from (2.15) that the quantity $dp/d\psi$ is constant along the field line, and, therefore, current (2.15) reaches its maximum value at the equator, and, besides, equation (2.15) correctly describes eastward and westward ring currents, flowing in the magnetosphere [50]. All stated witnesses that the model of magnetospheric equilibrium, which we based upon, correctly describes main observed phenomena.

Note that in the considered model of magnetospheric plasma with isotropic pressure the density of plasma is constant along the magnetic field lines. Really, from the equation of equilibrium (2.7) it follows that pressure is constant along the magnetic field line ($\vec{B}\cdot\nabla p = 0$, see (2.7)) and then from the relation $\frac{d}{dt}\left(\frac{p}{\rho^\gamma}\right) = 0$ follows the mentioned statement.

At this point we would like to discuss the accuracy of our calculations. We based our derivation of the static equilibrium upon (2.7) and

$$\vec{j} = \mathrm{rot}\,\vec{B}. \tag{2.16}$$

Setting

$$\vec{B} = \vec{B}_D + \vec{B}_1, \tag{2.17}$$

where $\vec{B}_1$ and $\vec{j}_1$ are respectively the magnetic field and current density regarded to pressure gradient. They satisfy the equations

$$\begin{aligned} \nabla p &= \vec{j}_1\times\vec{B}_D + \vec{j}_1\times\vec{B}_1, \\ \vec{j}_1 &= \mathrm{rot}\,\vec{B}_1. \end{aligned} \tag{2.18}$$

As distinct from Chan et al. [20], we consider only the low-pressure case ($\beta = \gamma p/B_0^2 < 1$, where $B_0 = M/L^3$ is a value of the magnetic intensity at equatorial plane) and thus cancel all terms of order β^2. In that way, plasma equilibrium is described by an equation

$$\nabla p = \vec{j}_1 \times \vec{B}_D . \tag{2.19}$$

It follows from (2.18) and (2.19) that $j_1 \sim B_1 \sim \beta < 1$.

Obviously, one should carry out all the calculations with the same accuracy. That is why we should linearize all equilibrium quantities in pressure.

2.3. Equations of Small Oscillations in Dipole Geometry

While discussing the distribution of pressure in the magnetosphere, we noted that $\vec{B} \cdot \nabla \psi = 0$. This condition, combined with (2.2) allows us to set the poloidal magnetic flux ψ equal to a magnetic label

$$a = \psi . \tag{2.20}$$

In this case (1.6) describes the toroidal current (2.15), which provides plasma equilibrium. Then due to (2.13), coefficients (1.11), (1.12) and (1.14) of the equation of small oscillations become much simpler, because both the local magnetic shear and the longitudinal current vanish

$$S = 0 , \; \gamma_s = 0 . \tag{2.21}$$

In this case eikonal χ satisfies the equation (1.17)

$$\vec{B} \cdot \vec{k}_\perp = [\nabla \psi \times \nabla \varphi] \cdot \nabla \chi = 0 . \tag{2.22}$$

Since eikonal is constant on a given field line, $\vec{k}_\perp$ can be represented without the loss of generality as

$$\vec{k}_\perp = \nabla \varphi + \hat{\lambda} \nabla \psi , \tag{2.23}$$

where $\hat{\lambda}$ is a constant quantity along a field line, whose meaning will be explained further.

From (1.22) and (2.23) we obtain

$$\hat{\eta} + \hat{\lambda} \hat{\xi} = 0 \tag{2.24}$$

and

$$\frac{\vec{k}_\perp \cdot \nabla a}{k_{\perp \beta}} = \hat{\lambda} \frac{|\nabla \psi|}{|\nabla \varphi|} . \tag{2.25}$$

Then the equations (1.26) and (1.34) have the form

$$\frac{\rho}{|\nabla\psi|^2}\left(1+\hat{\lambda}^2\frac{|\nabla\psi|^2}{|\nabla\varphi|^2}\right)\frac{\partial^2\hat{\xi}}{\partial t^2}=\frac{2(\vec{\kappa}\cdot\nabla\psi)}{|\nabla\psi|^2}\left(\hat{T}_0+p'\hat{\xi}\right)+\vec{B}\cdot\nabla\left[\left(1+\hat{\lambda}^2\frac{|\nabla\psi|^2}{|\nabla\varphi|^2}\right)\frac{\vec{B}\cdot\nabla\hat{\xi}}{|\nabla\psi|^2}\right],$$

$$\rho\frac{\partial^2\hat{\tau}}{\partial t^2}=\vec{B}\cdot\nabla\hat{T}_0,\quad \hat{T}_0=\gamma p\left[\vec{B}\cdot\nabla\left(\frac{\hat{\tau}}{|\vec{B}|^2}\right)-\frac{2(\vec{\kappa}\cdot\nabla\psi)}{|\nabla\psi|^2}\hat{\xi}\right]. \qquad (2.26)$$

Note that this equation contains only the $(\vec{B}\cdot\nabla)$ differential operator. Thus it seems convenient to replace this operator with the longitudinal derivatives, i.e. write that

$$(\vec{B}\cdot\nabla)=|\vec{B}|\frac{d}{dl}, \qquad (2.27)$$

where $\frac{d}{dl}$ is a derivative along the field line. Taking into account that one should cast into the expression (2.4) for the norm of dipole field $|\vec{B}|$ the value of r on a magnetic field line (2.5), and that the arc element of the magnetic field line has the form (2.6), we finally obtain

$$(\vec{B}\cdot\nabla)=\frac{M}{L^4\cos^7\theta}\frac{d}{d\theta}. \qquad (2.28)$$

Let us also note that along the field line

$$|\nabla\psi|^2=\frac{M^2b}{L^4\cos^6\theta},\quad |\nabla\varphi|^2=\frac{1}{L^2\cos^6\theta},$$

$$|\vec{B}|^2=\frac{M^2b}{L^6\cos^{12}\theta},\quad \frac{2\vec{\kappa}\cdot\nabla\psi}{|\nabla\psi|^2}=\frac{4L\cos^2\theta}{Mb^2}. \qquad (2.29)$$

Introducing dimensionless quantities

$$\Omega^2=\frac{\omega^2}{\omega_A^2},\quad \omega_A^2=\frac{M^2}{L^8\rho},\quad \alpha=-\frac{L}{p}\frac{dp}{dL},\quad \beta=\frac{\gamma pL^6}{M^2},$$

$$\xi=\hat{\xi}\frac{L}{M},\quad \eta=\hat{\eta},\quad \tau=\hat{\tau}\frac{L^2}{M}, \qquad (2.30)$$

$$\lambda = \hat{\lambda}\frac{M}{L},\ T_0 = \hat{T}_0\frac{L^6}{M^2},$$

we transform (2.24) and (2.26) into

$$\Omega^2\xi\left(1+\lambda^2 b(\theta)\right)+\frac{4}{b(\theta)\cos^4\theta}\left(T_0+\frac{\alpha\beta}{\gamma}\xi\right)+\frac{b(\theta)}{\cos^{13}\theta}\frac{\partial}{\partial\theta}\left[\frac{1+\lambda^2 b(\theta)}{b(\theta)\cos\theta}\frac{\partial\xi}{\partial\theta}\right]=0,$$

$$\eta+\lambda\xi=0,\ \Omega^2\tau+\frac{1}{\cos^7\theta}\frac{\partial}{\partial\theta}T_0=0, \tag{2.32}$$

$$T_0=\beta\left[\frac{1}{\cos^7\theta}\frac{\partial}{\partial\theta}\left(\frac{\tau\cos^{12}\theta}{b(\theta)}\right)-\frac{4\cos^2\theta}{(b(\theta))^2}\xi\right].$$

According to subsection 2.3, we dropped all terms of order less than β in (2.32). Alfvén frequency ω_A is assumed constant along the field line (see note at the end of subsection 2.2).

One can easily see that (2.32) contains only even powers of $\sin\theta$ and $\cos\theta$ and the derivatives have the form $\frac{1}{\cos\theta}\frac{\partial}{\partial\theta}=\frac{\partial}{\partial(\sin\theta)}$. Thus it seems logical to introduce a new variable $x=\sin\theta$ and relieve (2.32) of trigonometric functions:

$$\Omega^2\xi\left(1+\lambda^2 b\right)+\frac{4}{bc^2}\left(T_0+\frac{\alpha\beta}{\gamma}\xi\right)+\frac{b}{c^6}\left[\frac{1+\lambda^2 b}{b}\xi'\right]'=0,$$

$$\eta+\lambda\xi=0,\ \Omega^2\tau+\frac{T_0'}{c^3}=0, \tag{2.33}$$

$$T_0=\beta\left[\frac{1}{c^3}\left(\frac{c^6}{b}\tau\right)'-\frac{4c}{b^2}\xi\right],$$

where $b=1+3x^2$, $c=1-x^2$, and a prime indicates a derivative with respect to x.

2.4. Polarization of Ballooning Modes

Equations (2.33) can be expressed as

$$\hat{L}_P(\Omega,\xi)+\lambda^2\hat{L}_T(\Omega,\xi)+\hat{F}(\Omega,\beta,\xi,\tau)=0,$$

$$\eta+\lambda\xi=0, \tag{2.34}$$

$$\hat{L}_S(\Omega,\beta,\xi,\tau)=0,$$

where

$$\hat{L}_P(\Omega,\xi)=\Omega^2\xi+\frac{b}{c^6}\left(\frac{\xi'}{b}\right)',\ \hat{L}_T(\Omega,\xi)=\Omega^2\xi+\frac{\xi''}{c^6},$$

$$\hat{L}_S(\Omega,\beta,\xi,\tau)=\Omega^2\tau+\frac{T_0'}{c^3},\ \hat{F}(\beta,\xi,\tau)=\frac{4}{bc^2}\left(T_0+\frac{\alpha\beta}{\gamma}\xi\right). \tag{2.35}$$

This set contains two parameters: $0<\beta<1$ and $-\infty<\lambda<\infty$. While the physical meaning of parameter β is obvious, the latter one requires some explanations. The last of equations (1.3) yields

$$\vec{E}=i\omega\left[\vec{\xi}_\perp\times\vec{B}\right].$$

Hence

$$\vec{\xi}_\perp=\frac{i}{\omega}\frac{\left[\vec{E}\times\vec{B}\right]}{\left|\vec{B}\right|^2}. \tag{2.36}$$

From (1.17), (1.22) and (2.36) we obtain to within ε

$$\vec{k}_\perp\cdot\hat{\vec{\xi}}=\frac{i}{\omega}\frac{\vec{B}\cdot\left[\vec{k}_\perp\times\vec{E}\right]}{\left|\vec{B}\right|^2}=0. \tag{2.37}$$

Since $\vec{k}_\perp\perp\vec{B}$, (2.37) can hold only when $\left[\vec{k}_\perp\times\vec{E}\right]=0$, so

$$\vec{E}\parallel\vec{k}_\perp=\nabla\varphi+\lambda\nabla\psi. \tag{2.38}$$

In such a manner, when $\lambda = 0$ we have $\vec{E} \parallel \nabla\varphi$, $\hat{\vec{\xi}}_\perp \parallel \nabla\psi$, and when $\lambda \to \pm\infty$ we have $\vec{E} \parallel \pm\nabla\psi$, $\hat{\vec{\xi}}_\perp \parallel \mp\nabla\varphi$. At intermediate values of λ vector $\vec{E}$ lies in the same plane as vectors $\nabla\psi$ and $\nabla\varphi$. Thus, parameter λ describes the polarization of perturbations.

When $\lambda = 0$, as follows from (2.34), $\eta = 0$ and equations (2.33) reduce to a set of equations

$$\hat{L}_P(\xi,\tau) = \Omega^2\xi + \frac{4}{bc^2}\left(T_0 + \frac{\alpha\beta}{\gamma}\xi\right) + \frac{b}{c^6}\left(\frac{\xi'}{b}\right)' = 0,$$

$$\eta = 0, \tag{2.39}$$

$$\Omega^2\tau + \frac{T_0'}{c^3} = 0,$$

obtained earlier in paper [26] and describing coupled slow magnetosonic and poloidal Alfvén wave, modified by pressure and field line curvature.

The case $\lambda \to \infty$ yields $\xi = 0$, $\tau = 0$, but $\lambda\xi = -\eta \neq 0$, and equation (2.33) reduces to an equation

$$\hat{L}_T(\eta) = \Omega^2\eta + \frac{\eta''}{c^6} = 0, \tag{2.40}$$

obtained earlier in papers [32, 51] and describing toroidal Alfvén waves.

3. Ionospheric Boundary Conditions

Due to a common magnetic field and large longitudinal (Spitzer) conductivity, the magnetosphere and the ionosphere constitute a joint electrodynamical system. A perturbed electrical field created in the magnetosphere propagates along the field lines and provides magnetosphere-ionosphere coupling. Magnetospheric events usually affect ionospheric processes weakly. This is due to characteristic temporal scale of magnetospheric perturbations being much larger than that of ionospheric ones. For this reason, the magnetosphere can be affected by the ionosphere, whereas reverse influence is negligible. That is why it is necessary to take into account electrodynamical properties of ionospheric plasma, first of all, its conductivity, when considering the generation of magnetospheric MHD perturbations.

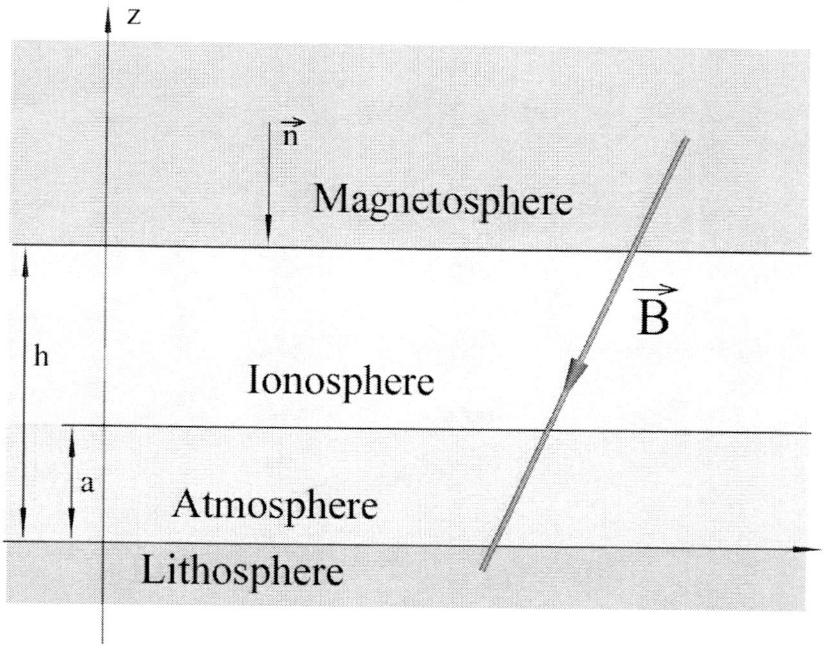

Fig. 4. A sketch of bounding layers.

Further we shall take this influence into account in the form of boundary conditions or equations, describing the propagation of ballooning perturbations in the magnetosphere. We shall derive these boundary conditions taking into account the existence of two near-Earth layers: insulating atmosphere and partially ionized ionosphere (see Fig. 4). We can speak about the ionospheric layer because for the real ionosphere the transversal conductivity essentially differs from zero only in a layer on heights from 150 km to 350 km. Below this layer the ionosphere can be regarded weakly different from neutral atmosphere.

3.1. Boundary Conditions

The magnetosphere-ionosphere coupling is determined by properties of magnetospheric and ionospheric plasmas. The magnetosphere is filled with totally ionized plasma with low concentration, and the ionosphere is a layer of weakly ionized plasma with higher concentration with thickness of order 100 km. Ohm's law in the ionosphere can be written as [1]

$$\vec{j} = \ddot{\sigma}\vec{E}\,, \tag{3.0}$$

where $\ddot{\sigma}$ is a unit-independent tensor of conductivity given by

$$\vec{\vec{\sigma}} = \begin{pmatrix} \sigma_P & \sigma_H & 0 \\ -\sigma_H & \sigma_P & 0 \\ 0 & 0 & \sigma_{\parallel} \end{pmatrix}. \qquad (3.0)$$

Here $\sigma_{\parallel}$ is a longitudinal (Spitzer) conductivity along the magnetic field, σ_P is Pedersen conductivity along the electric field, and σ_H is Hall conductivity along the vector product of electric and magnetic fields. Note that in accordance to (1.2) conductivity $\vec{\vec{\sigma}}$ is scaled as follows

$$\frac{4\pi}{c^2}\vec{\vec{\sigma}} \to \vec{\vec{\sigma}}.$$

Spitzer conductivity significantly exceeds the transversal conductivity $\vec{\vec{\sigma}}_{\perp}$, which equals to

$$\vec{\vec{\sigma}}_{\perp} = \begin{pmatrix} \sigma_P & \sigma_H \\ -\sigma_H & \sigma_P \end{pmatrix} \qquad (3.0)$$

and is of essence only in the ionospheric layer. Further we will also need a tensor of integral transversal conductivity, equal to

$$\vec{\vec{\Sigma}}_{\perp} = \int_a^h \vec{\vec{\sigma}}_{\perp} dz = \begin{pmatrix} \Sigma_P & \Sigma_H \\ -\Sigma_H & \Sigma_P \end{pmatrix}, \qquad (3.0)$$

where a and h are the heights of respectively the lower and the upper boundaries of the ionosphere, $\Sigma_P = \int_a^h \sigma_P dz$ and $\Sigma_H = \int_a^h \sigma_H dz$ are integral Pedersen and Hall conductivities of the ionosphere.

The problem of propagation of perturbations in the ionosphere in the general case is too complex for direct solution. Nevertheless, as shown in paper [41], one could significantly simplify it if he notifies that it contains two small parameters — a ratio of transversal components of the conductivity to a longitudinal one ($\sim 10^{-2}$ — 10^{-4}) and the ratio of the height of the ionosphere to the magnetospheric wavelength ($\sim 10^{-1}$). Decomposing Maxwell equations by these two parameters leads to exclusion of longitudinal conductivity and the condition of conservation of horizontal relatively to ground surface component of the electric field in the ionosphere. This allows us to represent ionosphere as an infinitely thin layer, which is characterized only by its integral transversal conductivity. Using also the conditions of continuity of perturbed magnetic field and the horizontal component of the electric field, one can obtain that ionospheric currents, which flow horizontally in this layer, close the magnetospheric currents, which flow to the surface of the ionosphere. These estimates were obtained in papers [25, 41]. We will use them in further considerations to avoid unwieldy calculations.

After choosing the appropriate boundary conditions, we will formulate them mathematically. We start from the "non-penetration" boundary condition. Its most logical notation would be

$$\vec{\xi} \cdot \vec{n}\Big|_b = 0, \tag{3.0}$$

where $\vec{n}$ is an external (earthward) unit normal to the ionospheric layer, subscript b indicates that value should be taken at the boundary, more precisely, at $z = a$. This condition controls vertical motions of plasma and holds due to large neutrals population in the ionosphere and to the presence of the insulating atmosphere.

The second boundary condition requires that total current is divergenceless, or, in different way, that electric charge of the ionospheric layer is constant. To derive this boundary condition we expand a $\operatorname{div} \vec{j} = 0$ condition

$$\frac{\partial j_z}{\partial z} + \nabla_s \cdot \left(\vec{\sigma}_s \cdot \vec{E}_s\right) = 0, \tag{3.0}$$

where subscript s indicates a horizontal component, and subscript z stands for a vertical one. Height-integrating this equation over the ionospheric layer, we derive that upward component of the ionospheric current, assuming the longitudinal current to vanish at $z = a$,

$$j_z\Big|_{z=h} = -\int_a^h \nabla_s \cdot \left(\vec{\sigma}_s \cdot \vec{E}_s\right) dz \approx -\nabla_s \cdot \left(\vec{\Sigma}_s \vec{E}_s\right), \tag{3.0}$$

where $\vec{\Sigma}_S = \int_a^h \vec{\sigma}_S dz$ is an integral horizontal conductivity of the ionosphere. Since $\vec{n} = -\vec{e}_z$, we obtain the second boundary condition in a form, given in [41]

$$\vec{j}_M \cdot \vec{n}\Big|_b = \nabla_S \left(\vec{\Sigma}_S \cdot \vec{E}_S\right)\Big|_b. \tag{3.0}$$

Subscript b in (3.8) is due to thinness of the ionospheric layer, and j_M is a boundary value of the magnetospheric current.

Note that in addition to boundary conditions (3.1) and (3.8) there is also a boundary condition $\vec{B} \cdot \vec{n}\Big|_b = 0$, connected with the "freezing" of the magnetic flux. In paper [17] it was shown that this condition corresponds to fast magnetosonic waves and thus can be rejected in the ballooning approximation.

3.2. Expression for $\dot{\overleftrightarrow{\Sigma}}_S$

Despite its simple form, boundary condition (3.8) is quite unwieldy. This unwieldiness is caused by the presence of an unknown tensor $\overleftrightarrow{\Sigma}_S$. However, the tensor $\overleftrightarrow{\Sigma}_\perp$ is known to us and we shall try to express the former through the latter.

To find the relation between $\overleftrightarrow{\Sigma}_S$ and $\overleftrightarrow{\Sigma}_\perp$ we should find a similar relation for the corresponding currents in the ionosphere ($\vec{I}_S \perp \vec{n}$) and in the plasma layer, perpendicular to the magnetic field ($\vec{I}_\perp \perp \vec{b}$)

$$\vec{I}_S = \overleftrightarrow{\Sigma}_S \vec{E}_S,$$

$$\vec{I}_\perp = \overleftrightarrow{\Sigma}_\perp \vec{E}_\perp. \tag{3.0}$$

One can write the relation between currents using the projector $\overleftrightarrow{Q}$, which transforms the horizontal ionospheric current into a transversal magnetospheric one and likewise for the electric intensity:

$$\overleftrightarrow{Q}\vec{I}_S = \vec{I}_\perp,\ \overleftrightarrow{Q}\vec{E}_S = \vec{E}_\perp. \tag{3.0}$$

Casting (3.8) and (3.9) into (3.10) yields

$$\overleftrightarrow{Q}\overleftrightarrow{\Sigma}_S \vec{E}_S = \overleftrightarrow{\Sigma}_\perp \vec{E}_\perp = \overleftrightarrow{\Sigma}_\perp \overleftrightarrow{Q}\vec{E}_S, \tag{3.0}$$

whence

$$\overleftrightarrow{Q}\overleftrightarrow{\Sigma}_S = \overleftrightarrow{\Sigma}_\perp \overleftrightarrow{Q}. \tag{3.0}$$

Since projector $\overleftrightarrow{Q}$ has an inverse operator $\overleftrightarrow{Q}^{-1}$, the sought-for tensor has the form

$$\overleftrightarrow{\Sigma}_S = \overleftrightarrow{Q}^{-1}\overleftrightarrow{\Sigma}_\perp \overleftrightarrow{Q}. \tag{3.0}$$

Now all that remains is to find $\overleftrightarrow{Q}$.

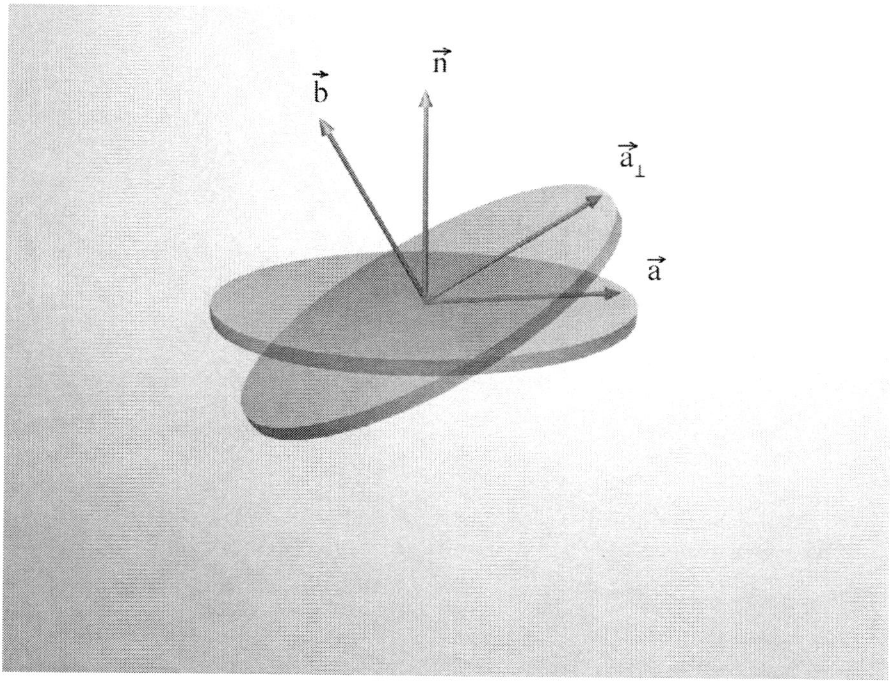

Fig. 5. To derivation of projector $\overleftrightarrow{Q}$.

3.3. Derivation of $\overleftrightarrow{Q}$

To find the projector $\overleftrightarrow{Q}$ we consider the following problem. Let an arbitrary vector $\vec{a}$ lie in a horizontal plane, perpendicular to the unit normal $\vec{n}$. We seek the operator $\overleftrightarrow{Q}$, which uniformly projects this vector on the transversal plane, perpendicular to the unit vector $\vec{b} = \vec{B}/|\vec{B}|$ (see Fig. 5.). Obviously, this would be the operator we seek. To define the vector $\vec{a}_\perp$, satisfying the relation $\overleftrightarrow{Q}\vec{a} = \vec{a}_\perp$, we introduce the vector $\vec{c} = \alpha\vec{n}$, where α is an arbitrary scalar, and vector $\vec{d}(\alpha) = \vec{a} + \vec{c}$. Let us find such a value $\alpha = \alpha_\perp$ that $\vec{d} \perp \vec{b}$. For that we set their scalar product equal to zero:

$$\vec{d} \cdot \vec{b} = \vec{a} \cdot \vec{b} + \alpha_\perp (\vec{n} \cdot \vec{b}) = 0, \tag{3.0}$$

whence

$$\alpha_\perp = -\frac{(\vec{a} \cdot \vec{b})}{(\vec{n} \cdot \vec{b})}. \tag{3.0}$$

Thus, the sought-for vector $\vec{a}_\perp = \vec{d}(\alpha_\perp)$ has the form

$$\vec{a}_\perp = \vec{a} - \vec{n}\frac{\vec{b}\cdot\vec{a}}{\vec{b}\cdot\vec{n}} = \vec{Q}\vec{a}, \tag{3.0}$$

therefore,

$$\vec{Q} = \vec{I} - \frac{\vec{n}\otimes\vec{b}}{\left(\vec{n}\cdot\vec{b}\right)}, \tag{3.0}$$

where $\vec{I}$ is an identity tensor, and $\vec{n}\otimes\vec{b}$ is a dyadic. Equation (3.17) coincides with the analogous expression from paper [41].

3.4. Boundary Conditions for Ballooning Modes in Dipole Geometry

Now we shall combine (3.5), (3.8), (3.9), and (3.17) to find the boundary conditions for our specific problem.

Equation (3.5) yields

$$\hat{\xi}\frac{\partial\psi}{\partial r} + \hat{\tau}\frac{B_r}{\alpha_s} = 0 \quad . \tag{3.0}$$

Casting (1.3) into (3.7), we obtain

$$\mathrm{rot}\,\tilde{\vec{B}}\cdot\vec{n} = \nabla_s\left(\vec{\Sigma}_s\cdot\vec{E}_s\right)\Big|_b . \tag{3.0}$$

Considering the ballooning approximation (1.16) and dropping terms of order ε we get from (3.19)

$$\mathrm{rot}\,\vec{B}\cdot\vec{n} = \frac{i}{\varepsilon}\left[-\frac{\lambda B_r}{\alpha_s}\vec{B}\cdot\nabla\hat{\eta} + \frac{B_r}{|\nabla\psi|^2}\vec{B}\cdot\nabla\hat{\xi} + \frac{\partial\psi}{\partial r}\frac{1}{r^2\cos^2\theta}\left\{\frac{\nabla\hat{\xi}\cdot\nabla\psi}{|\nabla\psi|^2} + \right.\right.$$
$$\left.\left. + \frac{\left[\vec{B}\times\nabla\psi\right]\cdot\nabla\hat{\eta}}{|\vec{B}|^2} + \frac{\hat{\xi}}{|B|^2}\left[\mathrm{div}(\alpha_s\nabla\psi) + p'\right]\right\}\right], \tag{3.0}$$

$$\nabla_s\cdot\left(\vec{\Sigma}_s\cdot\vec{E}_s\right) = \frac{\omega}{\varepsilon}\left[\hat{\lambda}\hat{\xi}\nabla_s\psi\cdot\left(\vec{\Sigma}_s\cdot\nabla_s\varphi\right) - \left[\hat{\lambda}\hat{\eta}\nabla_s\psi\cdot\left(\vec{\Sigma}_s\cdot\nabla_s\psi\right) + \right.\right.$$
$$\left.\left. + \hat{\xi}\nabla_s\varphi\cdot\left(\vec{\Sigma}_s\cdot\nabla_s\varphi\right) - \hat{\eta}\nabla_s\varphi\cdot\left(\vec{\Sigma}_s\cdot\nabla_s\psi\right)\right]\right]. \tag{3.0}$$

A term in (3.21), proportional to $\nabla_s\psi\cdot\left(\vec{\vec{\Sigma}}_s\cdot\nabla_s\psi\right)$, can be transformed as

$$\nabla_s\psi\cdot\left(\vec{\vec{\Sigma}}_s\cdot\nabla_s\psi\right)=\nabla_s\psi\left(\vec{\vec{Q}}^{-1}\vec{\vec{\Sigma}}_\perp\cdot\vec{\vec{Q}}\cdot\nabla\psi\right)=\vec{\vec{Q}}\cdot\nabla\psi\cdot\left(\vec{\vec{\Sigma}}_\perp\cdot\vec{\vec{Q}}\cdot\nabla\psi\right) \quad (3.0)$$

When obtaining (3.22) we took into account that $\vec{\vec{Q}}\cdot\vec{n}=0$, therefore, $\vec{\vec{Q}}\cdot\nabla_s\psi=\vec{\vec{Q}}\cdot\nabla\psi$. Since

$$\vec{\vec{Q}}\cdot\nabla\psi=\nabla\psi, \quad (3.0)$$

it is finally rewritten in the form

$$\nabla_s\psi\cdot\left(\vec{\vec{\Sigma}}_s\cdot\nabla_s\psi\right)=\nabla\psi\cdot\left(\vec{\vec{\Sigma}}_\perp\cdot\nabla\psi\right). \quad (3.0)$$

Taking into account (3.3), we have from (3.22)

$$\nabla\psi\cdot\left(\vec{\vec{\Sigma}}_s\cdot\nabla_s\psi\right)=\Sigma_P\left|\nabla\psi\right|^2. \quad (3.0)$$

A term, proportional to $\nabla_s\varphi\cdot\left(\vec{\vec{\Sigma}}_s\cdot\nabla_s\varphi\right)$ transforms in a similar way to the form

$$\nabla_s\varphi\cdot\left(\vec{\vec{\Sigma}}_s\cdot\nabla_s\varphi\right)=\Sigma_P\left|\nabla\varphi\right|^2. \quad (3.0)$$

The other two terms in (3.21) vanish, because $\left(\nabla\varphi\cdot\nabla\psi\right)=0$. As a result, we get from (3.21) — (3.26)

$$\nabla_s\cdot\left(\vec{\vec{\Sigma}}\vec{E}_s\right)=-\frac{\omega}{\varepsilon}\lambda\hat{\eta}\Sigma_P\left|\nabla\psi\right|^2+\frac{\omega}{\varepsilon}\lambda\hat{\xi}\Sigma_P\left|\nabla\varphi\right|^2. \quad (3.0)$$

Equating (3.20) and (3.27), we get

$$\hat{\lambda}\hat{\eta}\omega\left|\nabla\psi\right|^2-\frac{\omega\hat{\xi}}{r^2\cos^2\theta}+\frac{i}{\Sigma_P}\left(-\frac{\hat{\lambda}}{\alpha_s}B_r\vec{B}\cdot\nabla\hat{\eta}+B_r\frac{\vec{B}\cdot\nabla\hat{\xi}}{\left|\nabla\psi\right|^2}+\right.$$
$$\left.\left.+\frac{\partial\psi}{\partial r}\frac{1}{r^2\cos^2\theta}\left\{\frac{\nabla\hat{\xi}\cdot\nabla\psi}{\left|\nabla\psi\right|^2}+\frac{\left[\vec{B}\times\nabla\psi\right]\cdot\nabla\hat{\eta}}{\left|\vec{B}\right|^2}+\frac{\hat{\xi}}{\left|\vec{B}\right|^2}\left[\mathrm{div}(\alpha_s\nabla\psi)+p'\right]\right\}\right)\right|_{-\theta_0}^{\theta_0}=0. \quad (3.0)$$

Casting (2.24) into (3.28), we derive the second boundary condition in the form

$$\frac{\omega\hat{\xi}}{r^2\cos^2\theta}\left(1+\hat{\lambda}^2|\nabla\psi|^2\cos^2\theta\right)-\frac{i}{\Sigma_P}\times\left(\frac{B_r\left(\vec{B}\cdot\nabla\hat{\xi}\right)}{|\nabla\psi|^2}\left(1+\hat{\lambda}^2|\nabla\psi|^2 r^2\cos^2\theta\right)+\right.$$
$$\left.+\frac{\partial\psi}{\partial r}\frac{1}{r^2\cos^2\theta}\left[\frac{\gamma p}{|\vec{B}|^2}\frac{4r\hat{\xi}}{Mb^2}-\frac{\gamma p}{|\vec{B}|^2}\vec{B}\cdot\nabla\left(\frac{\hat{\tau}}{|\vec{B}|^2}\right)-\frac{p'\hat{\xi}}{|\vec{B}|^2}\right]\right)\Bigg|_{-\theta_0}^{\theta_0}=0. \tag{3.0}$$

The boundary conditions (3.18) and (3.29) should be made dimensionless. Following the same procedure, as in section 2, considering (2.28) — (2.30) and

$$B_r=-\frac{2M}{L^3c^3}x,\ \frac{\partial\psi}{\partial r}=\frac{M}{L^2c}, \tag{3.0}$$

we finally obtain the boundary conditions in the form

$$\Omega\xi\left(1+\lambda^2b\right)+\frac{i\delta}{b}\left[2x\left(1+\lambda^2b\right)\xi'-c^5\left(T_0+\frac{\alpha\beta}{\gamma}\xi\right)\right]\Bigg|_{x=\pm x_0}=0,$$

$$\xi+2xc\tau\Big|_{x=\pm x_0}=0, \tag{3.0}$$

where $\delta=\dfrac{1}{\Sigma_P\omega_A}$ is a unit-independent squared skin depth.

Linear homogenous ordinary differential equations (2.33) with boundary conditions (3.31) are an eigenvalue problem. Its eigenvalues are the perturbation frequencies Ω, and eigenfunctions are the corresponding amplitudes ξ and τ. Thus the formulated problem describes a spectrum of MHD perturbations of plasma equilibrium in dipole geometry of the magnetic field with a resistive ionosphere.

In the case of a conducting ionosphere ($\Sigma_P\to\infty$, $\delta\to0$), equations (2.33) hold true, and the boundary conditions (3.31) acquire the form

$$\xi\Big|_{x=\pm x_0}=0,$$

$$\tau\Big|_{x=\pm x_0}=0. \tag{3.0}$$

For an insulating ionosphere ($\Sigma_P\to0$, $\delta\to\infty$), boundary conditions can be written as

$$2x\left(1+\lambda^2 b\right)\xi' - c^5\left(T_0 + \frac{\alpha\beta}{\gamma}\xi\right)\Bigg|_{x=\pm x_0} = 0,$$

$$\xi + 2xc\tau\Big|_{x=\pm x_0} = 0. \tag{3.0}$$

In conclusion of this section we would like to mention that equations (2.33) with boundary conditions (3.31) represent an eigenvalue problem for a non-Hermitian operator, while the same equations with boundary conditions (3.32) or (3.33) — for a Hermitian one [52].

4. Magnetospheric Plasma Stability

From the mathematic point of view, the stability problem of MHD perturbations comes to investigation of small oscillations in the vicinity of the equilibrium. Only in several simplest cases the corresponding eigenvalue problem can be completely solved. These solutions contain full information about low-frequency oscillations of plasma and its stability. For more complicated cases, like ours, this problem is too complex for direct solving. For this reason we would like to investigate its stability without solving the complete problem. This can be accomplished with an energetic principle.

4.1. Energetic Analysis

In the common form the energetic method [53] lies in studying the potential energy of perturbations

$$\rho\frac{\partial^2\vec{\xi}}{\partial t^2} = \vec{F}\left(\vec{\xi}\right) = -\hat{K}\vec{\xi}, \tag{4.0}$$

where $\hat{K}$ is an operator, which can be identified as a coefficient of resilience when plasma displacement is small. The quantity $\vec{F}\left(\vec{\xi}\right)$ is the force, affecting the plasma elementary volume. For "good" boundary conditions without dissipation (i.e., without imaginary quantities) the operator $\hat{K}$ is self-adjoint [5, 52], and equation (4.1) allows to investigate the stability of the system in any geometry. Truly, after choosing the time dependence of all quantities in the form $\exp(-i\omega t)$, equation of small oscillations (4.1) acquires the form

$$\omega^2 = \frac{\int\vec{\xi}\hat{K}\vec{\xi}d\vec{r}}{\int\rho\vec{\xi}^2 d\vec{r}}. \tag{4.0}$$

This means that ω^2 is purely real due to self-adjointness of $\hat{K}$ and the condition of stability reads

$$\delta W = \int \vec{\xi}\hat{K}\vec{\xi}d\vec{r} > 0\,. \qquad (4.0)$$

Since we are not concerned with the value of the growth rate, we can now simply minimize δW . Naturally, we can minimize δW only partially, because its full minimization is equivalent to finding an exact solution of (4.1) for all three components.

In our case, equations of small oscillations (2.33) have "bad" boundary conditions (3.21) and we can not directly use (4.3). Fortunately, the structure of the equations and boundary conditions allows us to make certain deductions about plasma stability.

For energetic analysis we multiply the first equation of (2.33) by ξ^* (an asterisk indicates a complex conjugate), and the second one — by $\frac{c^6}{b}\tau^*$, add them up and integrate with respect to x over the field line. This yields

$$K\Omega^2 + iS\Omega - W = 0\,, \qquad (4.0)$$

where

$$K = \int_{-x_0}^{+x_0} c^3\left[\left(1+\lambda^2 b\right)\left|\xi\right|^2 + \frac{c^6}{b}\left|\tau\right|^2\right]dx\,, \qquad (4.0)$$

$$S = \left\| \frac{\left|\xi\right|^2\left(1+\lambda^2 b\right)}{2x\delta} \right\|, \qquad (4.0)$$

$$W = \int_{-x_0}^{+x_0} \frac{\left(1+\lambda^2 b\right)}{b}\left|\xi'\right|^2 dx + \int_{-x_0}^{+x_0} c^3\left|T_0\right|^2 dx - \frac{\alpha\beta}{\gamma}\int_{-x_0}^{+x_0} \frac{4c^4}{b^2}\left|\xi\right|^2 dx - \frac{\alpha\beta}{\gamma}\left(\frac{c^5}{2xb}\left|\xi\right|^2\right)\Bigg|_{-x_0}^{+x_0} = 0\,. \qquad (4.0)$$

Here we used boundary conditions (3.31).

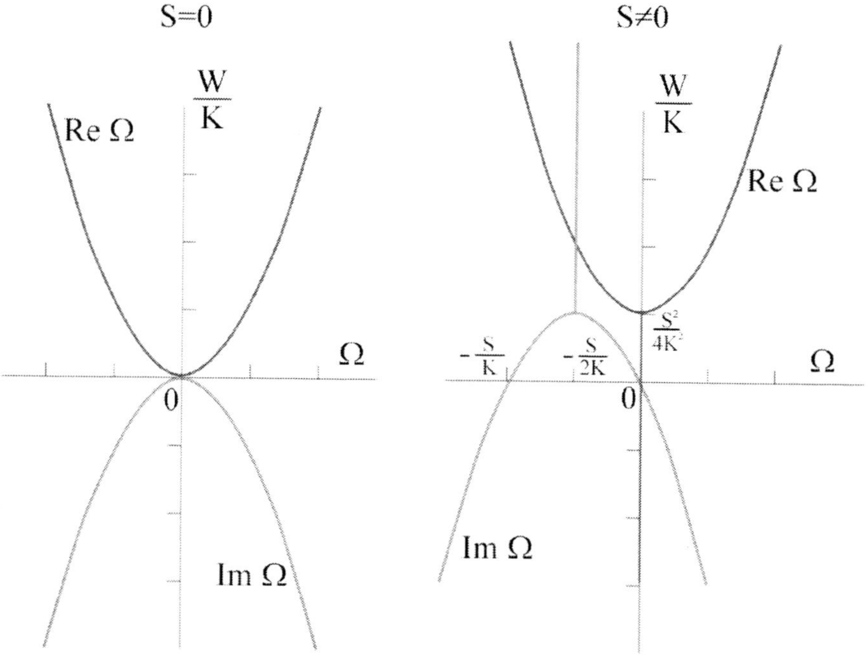

Fig. 6. A solution of the equation (4.4).

Notice that all coefficients in (4.4) are purely real and have simple physical meanings. K is a kinetic energy of the perturbation, S is a normal Poynting vector and describes energy losses in the ionospheric layer due to its finite conductivity, W is a potential energy of the perturbation.

A graphical solution of (4.4) is plotted on Fig. 6. One can see that $W > 0$ yields $\mathrm{Im}\,\Omega < 0$ and the perturbations are stable. When $W > \dfrac{S^2}{4A}$, all oscillations decay. The case $W < 0$ leads to $\mathrm{Im}\,\Omega > 0$ and the perturbations become unstable. Thus, an equation

$$W(\xi, \tau) = 0 \tag{4.0}$$

describes a stability threshold regardless of the ionospheric conductivity.

Usually [17, 28], considered perturbations become unstable due to the common influence of unfavourable curvature of magnetic field lines and of outward pressure gradient. Now, in addition to that source of instability, we have a new one, caused by pressure gradient in the ionospheric layer.

4.2. Boundary Problem for the Stability Threshold

One can see from Fig. 6, that growing perturbations have zero frequencies, so functional W is self-adjoint in the region of instability [52]. For this reason, we can replace $|\ldots|^2$ with $(\ldots)^2$, and assume ξ and τ to be purely real. In that way, the stability threshold is determined by an equation

$$\int_{-x_0}^{+x_0}\frac{(1+\lambda^2 b)}{b}(\xi')^2\,dx+\int_{-x_0}^{+x_0}c^3T_0^2\,dx-\frac{\alpha\beta}{\gamma}\int_{-x_0}^{+x_0}\frac{4c^4}{b^2}\xi^2\,dx-$$
$$-\frac{\alpha\beta}{\gamma}\frac{c_0^5}{2bx_0}\left(\xi^2\Big|_{+x_0}+\xi^2\Big|_{-x_0}\right)=0, \tag{4.0}$$

where $c_0=1-x_0^2$, x_0 is a boundary value of x.

With the help of boundary conditions (3.32) one can easily persuade himself that the last destabilizing term is absent when the ionosphere is perfectly conductive ($\delta\to 0$). When its conductivity is finite ($\delta\neq 0$), the two last terms are destabilizing. In addition, the most unstable situation corresponds to $\lambda=0$, which conforms to the result of paper [54]. Thus we should minimize the positive summands in the functional

$$W=\int_{-x_0}^{+x_0}\frac{(\xi')^2}{b}dx+\int_{-x_0}^{+x_0}c^3\left\{\frac{\beta b}{b+\beta c^6}\left[\frac{1}{c^3}\left(\frac{c^6}{b}\tau\right)'-\frac{4c}{b^2}\xi\right]'\right\}^2dx-$$
$$-\frac{\alpha\beta}{\gamma}\int_{-x_0}^{+x_0}\frac{4c^4}{b^2}\xi^2dx-\frac{\alpha\beta}{\gamma}\frac{c_0^5}{2b_0x_0}\left(\xi^2\Big|_{+x_0}+\xi^2\Big|_{-x_0}\right)=0. \tag{4.0}$$

To minimize this functional we shall follow paper [53] and cancel all strictly positive terms by proper selection of ξ and τ. Setting the minimized functional equal to zero, we get a stability threshold and amplitudes ξ and τ of the most unstable perturbations. Here we consider ξ and τ to be independent.

Firstly, we minimize (4.10) over τ. Since τ is present only in the second term of (4.10), proportional to T_0^2, the functional W can be made minimal over τ regardless of ξ. Varying W over τ, we obtain

$$T_0'=\beta\left[\frac{1}{c^3}\left(\frac{c^6}{b}\tau\right)'-\frac{4c}{b^2}\xi\right]'=0. \tag{4.0}$$

Hence

$$\frac{1}{c^3}\left(\frac{c^6}{b}\tau\right)' - \frac{4c}{b^2}\xi = D, \tag{4.0}$$

where D is a constant of integration, equal to

$$D = -\frac{\dfrac{c_0^5}{2b_0x_0}\left(\xi\big|_{+x_0} + \xi\big|_{-x_0}\right) + \int_{-x_0}^{x_0}\dfrac{4c^4}{b^2}\xi dx}{\int_{-x_0}^{x_0} c^3 dx}. \tag{4.0}$$

Casting (4.12) into (4.10), we get the following expression for W

$$W = \int_{-x_0}^{+x_0}\left[\frac{(\xi')^2}{b} - \frac{\alpha\beta}{\gamma}\frac{4c^4}{b^2}\xi^2\right]dx +$$
$$+\frac{\beta}{\int_{-x_0}^{x_0} c^3 dx}\left(\frac{c_0^5}{2b_0x_0}\left(\xi\big|_{+x_0} + \xi\big|_{-x_0}\right) + \int_{-x_0}^{x_0}\frac{4c^4}{b^2}\xi dx\right)^2 - \tag{4.0}$$
$$-\frac{\alpha\beta}{\gamma}\frac{c_0^5}{2b_0x_0}\left(\xi^2\big|_{+x_0} + \xi^2\big|_{-x_0}\right).$$

The stability threshold of MHD perturbations is now given by

$$W(\xi) = 0. \tag{4.0}$$

In paper [25] the stability threshold of ballooning perturbations was proved to be independent from the value of ionospheric conductivity. Equation (4.15) confirms this result. Further minimization of (4.14) depends on ξ, which should satisfy a boundary condition

$$\left. 2x\xi' - \frac{\alpha\beta}{\gamma}c^5\xi + \beta c^5\frac{\dfrac{c_0^5}{2b_0x_0}\left(\xi\big|_{+x_0} + \xi\big|_{-x_0}\right)}{\int_{-x_0}^{x_0} c^3 dx} + \beta c^5\frac{\int_{-x_0}^{x_0}\dfrac{4c^4}{b^2}\xi dx}{\int_{-x_0}^{x_0} c^3 dx}\right|_{x=\pm x_0} = 0. \tag{4.0}$$

Now we should derive a differential equation for ξ, casting (4.14) into (4.15) with boundary condition (4.16). This is a variational problem, so we must ensure that (4.15) holds not only at given ξ, but also at its small fluctuations.

To find ξ we calculate the first variational derivative, which we set equal to zero. This gives us a differential equation for ξ on the stability threshold

$$\left(\frac{\xi'}{b}\right)' + \frac{\alpha\beta}{\gamma}\frac{4c^4}{b^2}\xi - \beta\frac{4c^4}{b^2}\frac{\int_{-x_0}^{x_0}\frac{4c^4}{b^2}\xi dx}{\int_{-x_0}^{x_0}c^3 dx} - \beta\frac{4c^4}{b^2}\frac{\frac{c_0^5}{2x_0 b_0}\left(\xi\big|_{+x_0} + \xi\big|_{-x_0}\right)}{\int_{-x_0}^{x_0}c^3 dx} = 0, \quad (4.0)$$

which together with boundary condition (4.16) is equivalent to the mentioned variational problem. Equation (4.17) with boundary condition (4.16) has solutions only at certain relation between α and β. This relation determines the stability threshold.

It is easy to see from (2.33), (3.33) that equations (4.16), (4.17) hold for the insulating boundary ($\Sigma_P = 0$, $\delta \to \infty$) as well. For this reason the stability threshold in cases $0 < \Sigma_P < \infty$ and $\Sigma_P = 0$ will be the same. This result conforms to Theorem 2 from paper [25], which states that a ballooning perturbation occurs for resistive bounding ends if, and only if it occurs when bounding ends are insulators.

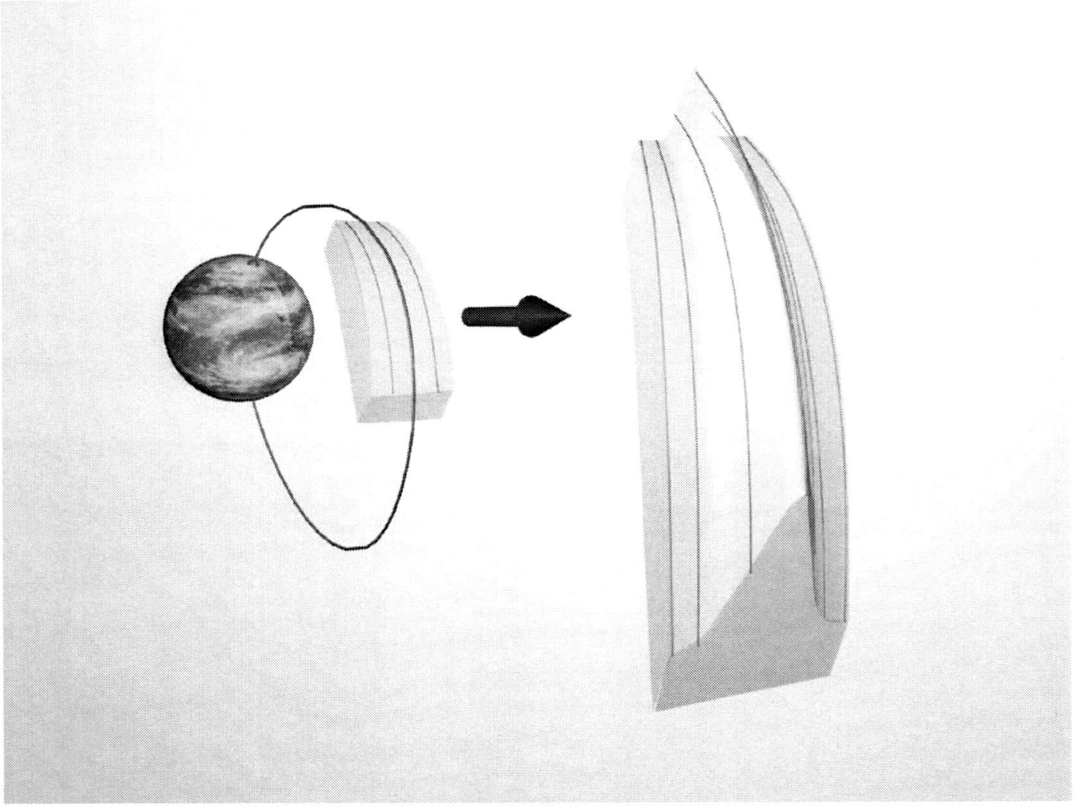

Fig. 7. A flute perturbation.

4.3. Stability Threshold

Prior to solving the boundary problem (4.16), (4.17), let us analyze functional (4.14). It contains two stabilizing terms. The first one, proportional to $(\xi')^2$, describes stabilization by boundary conductivity. The last summand describes stabilization by compressibility. These two terms cannot vanish simultaneously. We shall consider two cases when one of these terms vanishes. Of course, these cases do not cover all possibilities, but one of them proves to be the most unstable.

4.3.1. Flute Perturbations

The first stabilizing term in (4.14) will be minimal (equal to zero) when $\frac{\partial \xi}{\partial l} \sim \xi' = 0$. For this reason one should expect first to become unstable will be perturbations of the form

$$\xi = const(x), \tag{4.0}$$

which remind of flutes (see Fig. 7), stretched along the field lines, hence the name "flute modes". For these modes elasticity of plasma is caused only by its compressibility. Thus, they will grow at high pressure profile, pointed towards the unfavourable field line curvature.

Usually perturbations (4.18) exist in magnetic systems with closed field lines, for example, in fusion devices, but are not always suitable for magnetospheric equilibrium, because they do not satisfy the boundary conditions in several cases. At the same time, they can be used to obtain useful estimates.

The first stability criterion for flute modes was obtained in a pioneering paper [3] in the form

$$\frac{\alpha}{\gamma} = 4 . \tag{4.0}$$

Although being the main interchange criterion for a long time, this criterion became a very controversial subject, enlisting both supporters and opponents. A very similar criterion was obtained in paper [8], which dealt with finite pressure plasma. The latter was then criticized in paper [6]; its author disputed the integrability of the energetic principle along the field lines, which would led to a completely different criterion than (4.19). This statement was later supported by Liu [11], who also criticized the confirmation of (4.19) by the authors of paper [7], stating that it was only due to an incidental property of their toroidal model. We tend to agree with Liu [11], because, as it was shown in [20], the dipole flux surface is pushed outward by the finite pressure and taking this effect into account would lead, in our opinion, to a significant distortion of stability criterion.

Functions (4.18) satisfy equations (4.17), (4.18) when

$$\frac{\alpha}{\gamma} = \frac{\int_{-x_0}^{x_0} \frac{4c^4}{b^2} dx + \frac{c_0^5}{x_0 b_0}}{\int_{-x_0}^{x_0} c^3 dx}. \tag{4.0}$$

Equation (4.20) essentially differs from (4.19) and describes less restricting conditions of the development of instability.

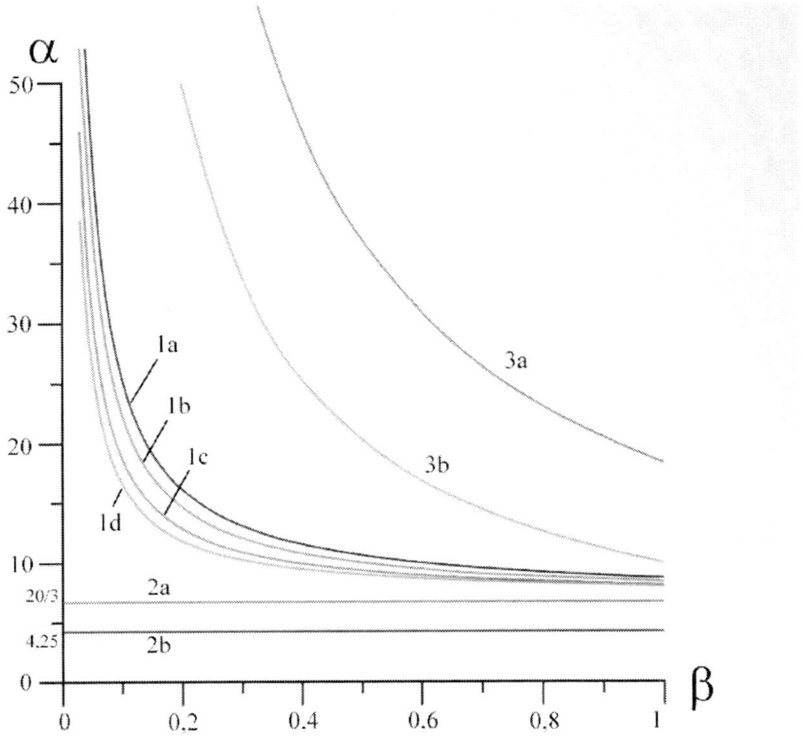

Fig. 8. Stability thresholds for different perturbations. 1 — ballooning perturbations with perfectly conductive ionosphere (a: L=2, b: L=2.5, c: L=4, d: L=7); 2 — flute perturbations (a: Gold criterion (4.19), b: criterion (4.20)); 3 — ballooning modes with resistive or insulating ionosphere (a: compressible, b: incompressible).

4.3.2. Incompressible Perturbations

Now we consider incompressible ballooning perturbations which induce

$$\operatorname{div}\vec{\xi} \sim T_0 \sim C = 0 \tag{4.0}$$

and cancel the term, proportional to T_0^2 in W. It means that

$$\frac{c_0^5}{2x_0b_0}\left(\xi\big|_{+x_0}+\xi\big|_{-x_0}\right)+\frac{\int_{-x_0}^{x_0}\frac{4c^4}{b^2}\xi dx}{\int_{-x_0}^{x_0}c^3dx}=0\,. \tag{4.0}$$

For such perturbations (4.17) takes the form

$$\left(\frac{\xi'}{b}\right)'+\frac{\alpha\beta}{\gamma}\frac{4c^4}{b^2}\xi=0\,, \tag{4.0}$$

and boundary conditions (4.16) become

$$\left.2x\xi'-\frac{\alpha\beta}{\gamma}c^5\xi\right|_{x=\pm x_0}=0\,. \tag{4.0}$$

One could think that equation (4.23) has an excessive number of boundary conditions. Nonetheless, equation (4.22) directly follows from (4.23) and (4.24), and the problem is self-consistent.

4.3.3. Numerical Analysis

Quite obviously, the two considered types of MHD perturbations are not the only ones possible. Their stability threshold can be apparently calculated only numerically from (4.16) and (4.17). The results of these calculations are plotted on Fig. 8. The flute modes are the easiest to exceed their stability threshold (4.20) and become unstable. They are absent when the ionosphere is perfectly conductive. In this case, stability is determined by ballooning modes [28] (see Fig. 9). Less prone to instability are incompressible modes, and the most stable are compressible modes.

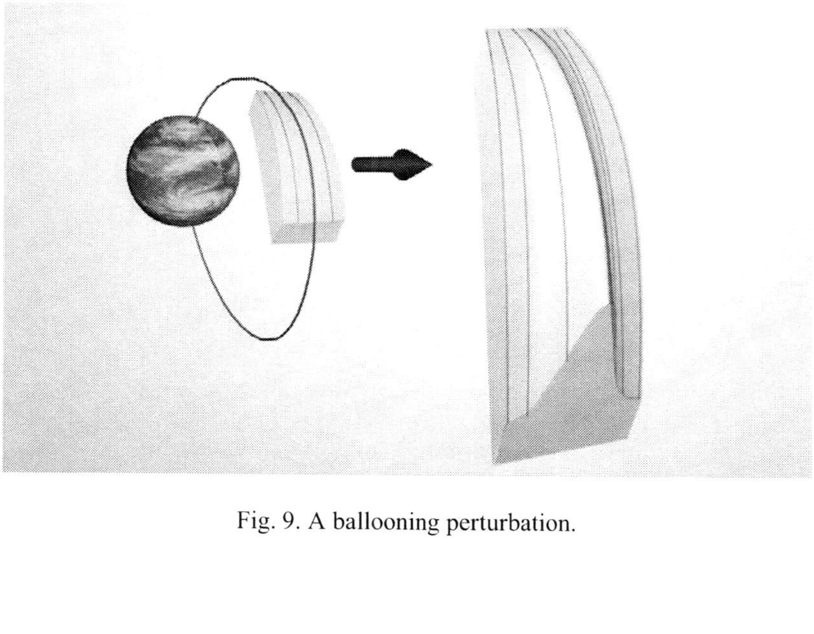

Fig. 9. A ballooning perturbation.

Thus we have obtained that general MHD stability of the magnetospheric plasma is determined by flute modes at arbitrary finite ionospheric conductivity with threshold (4.20). The latter lies lower than Gold criterion (4.19). As it was shown in our previous works [28], the same stability criterion (4.20) applies to an insulating boundary case, which fully agrees with the analytical analysis of paper [25].

4.4. Stability Criterion of Ballooning Modes with Perfectly Conductive Boundary

In general case stability analysis of MHD perturbations is a complex problem. However, in some special case it can be solved without involvement of the energetic principle. One of such special cases is ballooning modes with perfectly conductive ionosphere. It corresponds to boundary conditions (3.32) and all quantities in (2.33) are purely real. Therefore, the stability threshold corresponds to a condition $\Omega^2 = 0$. Numerical calculations show [28] that the instability occurs first on the lowest even eigenmode at all considered values of α and β. Equations (2.33) on the stability threshold have the form

$$\left(\frac{1+\lambda^2 b}{b}\xi'\right)' + \frac{4c^4}{a^2}\left(T_0 + \frac{\alpha\beta}{\gamma}\xi\right) = 0,$$

$$T_0' = 0, \tag{4.0}$$

and can be analytically solved by reducing to an integral equation. Integrating (4.25) with respect to x twice taking into account boundary conditions (3.32) we derive a final expression for the stability threshold

$$\xi\big|_{x=0} = \frac{\alpha\beta}{\gamma}\int_0^{x_0}\frac{4c^4}{b^2}\xi\int_x^{x_0}\frac{b}{1+\lambda^2 b}dtdx - \frac{\int_0^{x_0}\frac{4c^4}{b^2}\xi\int_x^{x_0}\frac{b}{c^6}dtdx}{\int_0^{x_0}\frac{c^3\left(a+\beta c^6\right)}{\beta a}\int_x^{x_0}\frac{a}{c^6}dtdx}\int_0^{x_0}\frac{4c^4}{b^2}\int_x^{x_0}\frac{b}{1+\lambda^2 b}dtdx. \tag{4.0}$$

Substituting a test function $\xi = 1 - \frac{x^2}{x_0^2}$, which approximates the numerical results to within a few percent with normalization $\xi\big|_{x=0} = 1$, into equation (4.26) gives a following expression for the stability threshold:

$$\alpha = \frac{K_1}{\beta} + K_2. \tag{4.0}$$

The exact form of coefficients K_1 and K_2 is too cumbersome to publish, and we will restrict by presenting their values for some magnetic shells in Table 4.1. At latitudes around 70 degrees (6 — 10 R_E) they are roughly the same. The plots of stability thresholds calculated analytically after (4.27) and numerically after (2.33), (3.32) coincide.

Table 4.1. Values of coefficients in (4.27).

θ_0, °	L	K_1	K_2
45	2.00	1.83	5.2
50	2.42	1.54	5.0
60	4.00	1.17	4.8
72	10.47	0.95	4.7

Conclusion

In this chapter we investigated the influence of interchange and ballooning modes on plasma stability in the inner magnetosphere of the Earth. Staying in the framework of linear MHD we tried to find the stability threshold of these perturbations in dependence of ionospheric conductivity.

We applied both analytical and numerical methods to solve this problem. They allowed to understand the main mechanisms of destabilization of these perturbations. Firstly, we discovered that finite ionospheric conductivity permits the interchange modes to exist, in which case they define the general MHD stability of the magnetosphere. Secondly, their threshold does not depend on the ionospheric conductivity. We derived a stability criterion (5.20) for this threshold. This threshold appeared to be lower than predicted by Gold [3] and Kadomtsev [5]. The stability threshold of ballooning modes lies higher than that of interchange modes.

It is hard to tell which role interchange and ballooning modes play in the magnetospheric processes. Only by investigating a non-linear stage of development of instability and interaction of perturbations with hot particles, this question could be answered. Only these effects and, naturally, space experiments can provide some information on macroscopic effects. Certain information can be gained from papers [55 — 57]. They conclude that interaction of interchange and ballooning modes with high-energy particles could lead to a substorm onset. Multi-satellite observations in the magnetosphere [58] witness that ballooning instability in the field-reversal region of the magnetotail [21] is connected to substorm detonations [19, 59]. The picture of collective processes, caused by these perturbations, can become much more complicated in presence of convection.

References

[1] Callenrode, M.-B. *Space Physics*; Springer-Verlag: Berlin, 2004.

[2] Kruskal, M.; Schwarzschild, M. *Proc. Roy. Soc.* London 1954, A233, 348-360.

[3] Gold, T. *J. Geophys. Res.* 1959, 64, 1219-1226.

[4] Chandrasekhar, S. *Plasma Physics*; Univ. of Chicago Press: Chicago, IL, 1960; pp. 85-96.
[5] Kadomtsev, B. B. In *"Reviews of Plasma Physics"*; Leontovich M. A.; Ed.; Consultants Bureau: New York, NY, 1966; Vol. 2, pp. 132-167.
[6] Cheng, A. F. *J. Geophys. Res.* 1974, 80, 9900.
[7] Rogers, B.; Sonnerup, B. U. O. *J. Geophys. Res.* 1986, 91, 8837-8845.
[8] Southwood, D. J.; Kivelson, M. G. *J. Geophys. Res.* 1987, 92, 109.
[9] Ivanov, V. N.; Pokhotelov, O. A. *Fizika Plazmy* 1987, 13, 1446-1454, in Russian.
[10] Southwood, D. J.; Kivelson, M. G. *J. Geophys. Res.* 1989, 94, 299-308.
[11] Liu, W. W. *J. Geophys. Res.* 1996, 101, 27443-27447.
[12] Furth, H. P.; Killen, J.; Rosenbluth, M. N.; Coppi, B. In *Proceedings of International Conference on Plasma Physics and Controlled Nuclear Fusion Research*. Culham, 6-10 September 1965; IAEA: Vienna, 1966; Vol.1; pp. 103-126.
[13] Coppi, B. *Phys. Rev. Lett.* 1977, 39, 939-942.
[14] Connor, J. W.; Hastie, R. J.; Taylor, J. B. *Phys. Rev. Lett.* 1978, 40, 396-399.
[15] Lakhina, G. S.; Mond, M.; Hameiri, E. *J. Geophys. Res.* 1990, 95, 4007-4016.
[16] Lakhina, G. S.; Hameiri, E.; Mond, M. *J. Geophys. Res.* 1990, 95, 10441-10448.
[17] Hameiri, E.; Laurence, P.; Mond, M. *J. Geophys. Res.* 1991, 96, 1513-1526.
[18] Lee, D.-Y.; Wolf, R. A. *J. Geophys. Res.* 1992, 97, 19251-19257.
[19] Ohtani, S.-I.; Tamao, T. *J. Geophys. Res.* 1993, 98, 19369-19379.
[20] Chan, A. A.; Xia, M.; Chen, L. *J. Geophys. Res.* 1994., 99, 17351-17366.
[21] Lee, D.-Y.; Min, K. W. *J. Geophys. Res.* 1996, 101, 17347-17354.
[22] Liu, W. W. *J. Geophys. Res.* 1997, 102, 4927-4931.
[23] Hurricane, O. A. *J. Geophys. Res.* 1997, 102, 19903-19911.
[24] Lee, D.-Y. *Geophys. Res. Lett.* 1998, 25, 4095-4098.
[25] Hameiri, E. *Phys. Plasmas* 1999, 6, 674-685.
[26] Burdo, O. S.; Cheremnykh, O. K.; Verkhoglyadova, O. P. *Izv. Akad. Nauk Fiz.* 2000, 64, 1896-1900.
[27] Cheremnykh, O. K.; Parnowski, A. S. *Adv. Space Res.* 2004, 33, 769-773.
[28] Cheremnykh, O. K.; Parnowski, A. S.; Burdo, O. S. *Planet. Space Sci.* 2004, 55, 1217-1229.
[29] Cheremnykh, O. K.; Parnowski, A. S. *Adv. Space Res.* 2005, article in press.
[30] Jacobs, J. A. *Geomagnetism Micropulsations*; Springer-Verlag: New York, NY, 1970.
[31] Swift, D. W. *Planet. Space Sci.* 1967, 15, 1225-1237.
[32] Cummings, W. D.; O'Sullivan, R. J.; Coleman, P. J. *J. Geophys. Res.* 1969, 74,778-789.
[33] Singer, H. J.; Southwood, D. J.; Walker, R. J.; Kivelson, M. G. *J. Geophys. Res.* 1981, 86, 4589-4596.
[34] Dewar, R. L.; Glasser, A. H. *Phys. Fluids* 1983, 26, 3038-3052.
[35] Pustovitov, V. D.; Shafranov, V. D. In *"Reviews of Plasma Physics"*; Kadomtsev, B. B.; Ed.; Consultants Bureau: New York, NY, 1990; Vol. 15; p.163.
[36] Cheremnykh, O. K.; Revenchuk, S. M.; Omelchenko, A. Ya.; Burdo, O. S. *Phys. Scripta* 1995, 51, 264-276.
[37] Pokhotelov, O. A.; Stenflo, L.; Shukla, P. N. *Fizika Plazmy* 1996, 22, 941-953, in Russian.

[38] Dungey, J. W. In *Proceedings of the International Conference on the Ionosphere; Institute of Physics*: London, 1963; p. 230.
[39] Nishida, A. *J. Geophys. Res.* 1964, 69, 1861.
[40] Maltsev, Yu. P. *Geomagn. Aeron.* 1977, 17, 1008.
[41] Hameiri, E.; Kivelson, M. G. *J. Geophys. Res.* 1991, 96, 21125-21134.
[42] Kivelson, M. G. In: *Introduction to Space Physics*; Kivelson, M. G.; Russel, C. T.; Ed.; Cambridge University Press: Cambridge, 1995; pp. 330-355.
[43] Kivelson, M. G.; Southwood, D. J. *Geophys. Res. Lett.* 1988, 15, 1271-1274.
[44] Furth, H. P.; Killen, J.; Rosenbluth, M. N. *Phys. Fluids* 1963, 6, 459.
[45] Saunders, M. In: *Advances in Solar System Magnetohydrodynamics*; Priest, E. R.; Hood, A. W.; Ed.; Cambridge University Press: Cambridge, 1991; pp. 357-397.
[46] Kruskal, M. D.; Kulsrud, R. M. *Phys. Fluids* 1958, 1, 265-274.
[47] Tsyganenko, N. A. *J. Geophys. Res.* 2002, 107, doi:10.1029/2001JA000219.
[48] Tsyganenko, N. A. *J. Geophys. Res.* 2002, 107, doi:10.1029/2001JA000220.
[49] Lyons, L. R.; Williams, D. J. D. Reidel Publishing Company: Dordrecht, 1984.
[50] Cheng, C. Z. *J. Geophys. Res.* 1992, 97, 1497-1510.
[51] Cheng, C. Z.; Chang, T. C.; Lin, C. A.; Tsai, W. H. *J. Geophys. Res.* 1993, 98, 11339-11347.
[52] Reed, M.; Simon, B. *Methods of modern mathematical physics*; Academic Press: New York, NY, 1978; Vol.4.
[53] Bernstein, I. B.; Frieman, E. A.; Kruskal, M. D.; Kulsrud, R. M. *Proc. Roy. Soc.* London 1958, A244, 17-40.
[54] Hameiri, E. *Phys. Fluids* 1983, 26, 230-241.
[55] Zaharia, S.; Cheng, C. Z.; Johnson, J. R. *J. Geophys. Res.* 2000, 105, 18741-18752.
[56] Cheng, C. Z.; Lui, A. T. Y. *Geophys. Res. Lett.* 1998, 25, 4091-4094.
[57] Levitt, B.; Maslovsky, D.; Mauel, M. E. *Phys. Plasmas* 2002, 9, 2507-2516.
[58] Petrukovich, A. A. et al. *J. Geophys. Res.* 1988, 103, 47-49.
[59] Hurricane, O. A.; Fong, B. H.; Cowley, S. C.; Coronity, F. V.; Kennel, C. F.; Pellat P. *J. Geophys. Res.* 1999, 104, P. 10221-10231.

In: Space Science: New Research
Editor: Nick S. Maravell, pp. 141-164

ISBN: 1-60021-005-8

Chapter 5

PALEOSHORELINES AND THE EVOLUTION OF THE LITHOSPHERE OF MARS

Javier Ruiz
Departamento de Geodinámica, Facultad de Ciencias Geológicas, Universidad Complutense de Madrid, 28040 Madrid, Spain
Rosa Tejero
Departamento de Geodinámica, Facultad de Ciencias Geológicas, Universidad Complutense de Madrid, 28040 Madrid, Spain
David Gómez-Ortiz
Escuela Superior de Ciencias Experimentales y Técnicas-Área de Geología, Universidad Rey Juan Carlos, 28933 Móstoles, Madrid, Spain
Valle López
Servicio de Sistemas de Información Geográfica, Instituto Geológico y Minero de España, Ríos Rosas 23, 28003 Madrid, Spain

Abstract

The existence of features indicative of shorelines of ancient oceans on Mars has been proposed for several authors. In this chapter we revise the topography of possible Martian paleoshorelines, and their consequences for the amount of water infilling ocean basins when water load is considered. We show that a re-evaluation of paleoshorelines is need. For example, the putative Meridiani shoreline could be the same feature as some portions of the Arabia shoreline. Indeed, elevations in the Meridiani shoreline are roughly similar to that of the Arabia shoreline in northeast Arabia, Utopia (not taken into account the Isidis basin), Elysium, and Amazonis regions. This is still far of an equipotential surface, but a paleoshoreline through these regions and the Meridiani shoreline would be better candidate to represent a paleoequipotential surface than the Arabia shoreline *sensu strito*. Moreover, the elevation of the Arabia shoreline in northern Arabia Terra after is intriguingly close to the mean elevation of the Deuteronilus shoreline, and it cannot be discarded a "mixed" Arabia/Deuteronilus shoreline, which would include the Arabia shoreline in northern Arabia Terra, and the Deuteronilus shoreline elsewhere.

If reality of global shorelines is accepted, as increasing evidence suggests, then present-day topographic variations in these features postdate shorelines formation. So, their topographic range should provide information on large-scale vertical movement of the lithosphere, which in turn provides information on the thermal evolution of Mars. We describe the application of thermal isostasy concept to constraint the ancient thermal state of the lithosphere from present-day paleoshoreline topography. For the ~1.1 km total elevation range of the Deuteronilus shoreline, the relative amplitude of heat flow variations (the ratio between maximum and minimum heat flow) is ≤1.6. This value is clearly lower than that presently observed on continental areas on the Earth. If heat flow variations on Mars are currently greatly disappeared, then the obtained heat flow variations upper limits must be mostly related to the paleoshoreline formation time: the present-day elevation range along Deuteronilus shoreline suggests that differences in the thermal state of the lithosphere in regions along this putative paleoshoreline have been relatively small since the feature was formed, and therefore the absence of lithospheric tectonothermal events by the latest ~3 Gyr, at least. If the Deuteronilus shoreline is a combination of portions of several paleoshorelines, then the total elevation range, and the implied heat flow variations, would be lower, and the lithosphere stability higher.

Introduction

A plenty of studies about the characteristics, thermal state, and evolution of the lithosphere of Mars has been previously performed from several lines of work.

On a hand, numerous efforts have focused on the mechanical properties of the lithosphere, which serve to calculate its effective elastic thickness and thermal structure (e.g., Comer et al., 1985; Solomon and Head, 1990; Anderson and Grimm, 1998; Zuber et al., 2000; Nimmo, 2002; McGovern et al., 2002, 2004; McKenzie et al., 2002; Kieffer, 2004; Ruiz et al., 2006), and the topography and geometry or greath faults, and hence the brittle-ductile transition depth (e.g., Schultz and Lin, 2001; Schultz and Watters, 2001; Vidal et al., 2005). These studies inform about the thermal structure and heat flow in the time when the structures were formed.

On the other hand, thermal history models make predictions about the evolution of surface heat flow and lithospheric thickness, or even crustal growth (e.g., Stevenson et al., 1983; Schubert and Spohn, 1990; Schubert et al., 1992; Grasset and Parmentier, 1998; Nimmo and Stevenson, 2000; Choblet and Sotin, 2001; Spohn et al., 2001; Hauck and Phillips, 2002; Breuer and Spohn, 2003). Obviously, results from thermal history models should be consistent with constraints imposed by the studies of the mechanical properties of the lithosphere.

A third line of information about the evolution of the Martian lithosphere is the analysis of present-day topography of features interpreted as paleoshorelines. Indeed, the presence of features indicative of shorelines of ancient oceans on Mars has been proposed (Parker et al., 1989, 1993; Edgett and Parker, 1997; Clifford and Parker, 2001). If reality of global paleoshorelines is accepted, then present-day topographic variations in these features postdate shorelines formation. So, their topographic range should provide information on large-scale vertical movement of the lithosphere, which in turn would provide information on the thermal evolution of Mars (Ruiz, 2003): total elevation differences along paleoshorelines impose constraints to the differences in the evolution of the thermal structure of the lithosphere in shoreline-crossed regions. In fact, it is possible to make an approximate calculation of the amplitude of the ancient heat flow variations necessary to compensate, through thermal isostasy, elevation differences, and transform the paleoshorelines into equipotential surfaces. Like other geological processes could have produced vertical movements, the results so obtained suppose an upper limit.

Diverse efforts have been carried out to test reality of the proposed paleoshorelines. Specifically targeted MOC images have been interpreted as not supporting the shoreline hypothesis (Malin and Edgett, 1999, 2001), although these results have been disputed (Parker et al., 2001; Clifford and Parker, 2001; Fairén et al., 2003). Moreover, recent works using high resolution images have found clear evidences of erosion in places located in putative paleoshorelines (Webb and McGill, 2003; Webb, 2004), which support its formation in relation to coastal processes. The observed present-day Martian topography (Smith et al., 1999, 2001) has also been used to analyze elevations along of the main proposed paleoshorelines, in order to test their chance to represent true paleoshorelines (Head et al., 1998, 1999; Carr and Head, 2003, Webb, 2004). These analyses obtain that the Late Hesperian Deuteronilus shoreline is a viable paleoshoreline, since this feature slightly deviates from an equipotential surface. Otherwise, the putative older and higher-standing Arabia shoreline deviates substantially from an equipotential surface, indicating that it may not be representative of a true shoreline.

The evaluation of possible paleoshorelines through assessment of present-day topography must be made cautiously, because it is not necessarily true that a paleoequipotential surface must fit well a present-day equipotential surface. Lithosphere rebound due to water unloading associated with the disappearance of an ocean with irregularly shaped margins could result in deviations of equipotentiality of up to several hundreds of meters (Leverington et al., 2003; Leverington and Ghent, 2004). Different thermal histories among regions may have appreciably contributed to the deformation of the original large wavelength topography of putative paleoshorelines: variations in Martian heat flow, similar in relative amplitude to those observed in terrestrial continental tectonothermally stable areas, could result in large wavelength elevation differences of kilometric scale through differential thermal isostasy (Ruiz, 2003, Ruiz et al., 2003, 2004), an important amount of deformation for any possible paleoshoreline. Tectonic processes could also produce vertical movements, and erosional or sedimentary activity could affect the original paleoshoreline signatures (Clifford and Parker, 2001). Moreover, lateral continuity of paleoshorelines is not well established, and diverse division and mixing of the originally proposed features are likely required (Ruiz et al, 2003; Webb, 2004; Ruiz, 2005).

In this chapter we revise and re-evaluate the topography of possible Martian paleoshorelines, and the consequences of the water load for the amount of water infilling ocean basins. We also describe the application of thermal isostasy concept to constraint the ancient thermal state of the lithosphere from present-day paleoshoreline topography, and we present the results obtained and their implications for the evolution of the lithosphere of Mars.

Paleoshorelines

In this section we first revise elevation ranges in the paleoshorelines originally proposed by Parker et al. (1989, 1993) and Edgett and Parker (1997), which were revised, redrawn and renamed (as Deuteronilus, Arabia and Meridiani) by Clifford and Parker (2001) (Figure 1 shows a map of Mars with some of the geographical features mentioned in the text, and

Figure 2 shows the paleoshorelines after Clifford and Parker (2001) represented on the Martian topography). Later, we use MOLA topography (Smith et al., 1999, 2001) to refine calculations of the water volume infilling the ancient oceanic basins, by taking into account both present-day topography and the maximum possible effect caused by the water weight on the oceanic floors topography. Finally, we re-evaluate these features through of comparing their respective elevations (geomorphologic revisions or re-evaluations are beyond the scope of this chapter), using the mapping of Clifford and Parker (2001) and MOLA topography.

In this point it is necessary recall that Martian elevations are given with respect to an arbitrary zero elevation level, defined as the equipotential surface (gravitational + rotational) whose average value at the equator is equal to the mean radius (see Smith et al., 2001).

Table 1. Martian chronology after Hartmann and Neukum (2001) and Hartmann (2005).

Period	Age (Gyr)
Late Amazonian	
	0.2-0.6
Middle Amazonian	
	1.4-2.1
Early Amazonian	
	2.9-3.2
Late Hesperian	
	3.2-3.6
Early Hesperian	
	3.5-3.7
Late Noachian	
	3.6-3.9
Middle Noachian	
	3.8-4.1
Early Noachian	

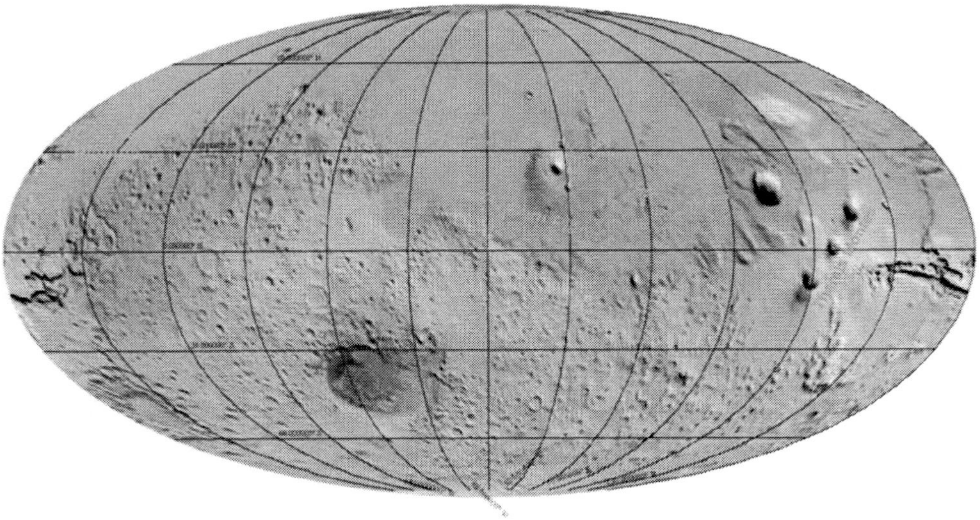

Figure 1. Map of Mars with some of the geographical features mentioned in the text, and other representative features indicated on the Martian topography.

Elevations Range along Deuteronilus, Arabia, and Meridiani Shorelines

The better evidence for a Martian paleoshoreline (dating from the Late Hesperian; temporal equivalences, as obtained from crater counts statistical, are given in Table 1) is the presence in the northern lowlands of a mapping contact (originally named Contact 2) marking the outer boundary of the northern plains, which was interpreted to be the shoreline of an ancient Martian ocean (Parker et al., 1989, 1993). This "contact" was later redrawn and renamed Deuteronilus shoreline by Clifford and Parker (2001). Head et al. (1998, 1999) and Carr and Head (2003), using MOLA data, have shown that this putative shoreline represents a relatively good approximation to an equipotential surface: its mean altitude is −3.792 ± 0.236 km, and its whole topographic range is ~1.1 km, from −3.2 to −4.3 km (Carr and Head, 2003). The elevation of the base levels of the Chryse outflow channels, −3.742 ± 0.153 km (Ivanov and Head, 2001), is close to the mean level of Deuteronilus shoreline, could also indicate that they debouched into a large standing body of water (Head et al., 1999; Ivanov and Head, 2001); this seem also be the case for other channels termini (Salamuniccar, 2004). Additionally, there is clear evidence for erosion along the Deuteronilus shoreline (Webb, 2004).

Alternatively (or complementarily), Carr and Head (2003) considered that the Late Hesperian Vastitas Borealis Formation, which extends for a great part of the northern lowlands, represents better support for the past existence of a large standing body of water on Mars. The Vastitas Borealis Formation has been interpreted as a sedimentary veneer at least 100 m thick on the East Hesperian ridged plains (Head et al., 2002), which could have originated as a sublimation residue from a large (probably frozen) water body (Kreslavsky and Head, 2002). The outer contact of the Vastitas Borealis Formation is coincident with the trace of the Deuteronilus shoreline in the Deuteronilus, Nilosyrtis, Isidis, Tempe, and Chryse regions, but not in Elysium or the Olympus Mons aureole. If the outer contact of the Vastitas Borealis Formation in the Utopia basin is ignored (where it is covered by younger Amazonian Elysium materials, and therefore, the original contact trace is not visible), the outer contact of the Vastitas Borealis Formation has a mean altitude of −3.658 ± 0.282 km, with a whole elevation range of ~1.0 km, from −3.3 to −4.3 km (Carr and Head, 2003; see their Figure 12).

Figure 2. The Deuteronilus (yellow), Arabia (green) and Meridiani (red) shorelines after Clifford and Parker (2001), represented on the Martian topography.

Parker et al. (1989, 1993) also proposed an older, higher-standing Contact 1, later on renamed Arabia shoreline (Clifford and Parker, 2001). This shoreline, which would be of Noachian age (see Clifford and Parker, 2001), is roughly coincident with the Martian dichotomy separating the lowlands from the highlands, and the elevation along its outline highly deviates from an equipotential surface (Head et al., 1998, 1999), and thus it is not a good candidate to paleoshoreline. The topography along the Arabia shoreline is characterized (Carr and Head, 2003) by a mean altitude of –2.090 ± 1.400 km, and a total elevation range of ~5.6 km (from 1.6 to –4.0 km).

Finally, the existence of a pre-Arabia shoreline has been proposed in northern Sinus Meridiani and western Arabia Terra (Edgett and Parker, 1997; Clifford and Parker, 2001); precisely, the MER Opportunity has recently found evidences for an aqueous, maybe sea-related, environment at Meridiani Planum, close to this possible paleoshoreline (Squyres et al., 2004). The mean elevation of this Meridiani shoreline (as named by Clifford and Parker, 2001), would be about –1.5 km (Parker et al., 2000), although its topography has not been examined in previous works. Figure 3 shows constant elevation contours (0.5 km spaced) crossed by the Meridiani shoreline. It can be seen that elevations along the mapped paleoshoreline mostly range between 0 and –2 km. If Hesperian chaos materials (see Tanaka et al., 1992), impact craters, and an isolated peak are not taken into account (see also Figure 6), then total elevation range is ~1 km, from –0.5 to –1.5 km, a reasonable amount for a very old paleoshoreline.

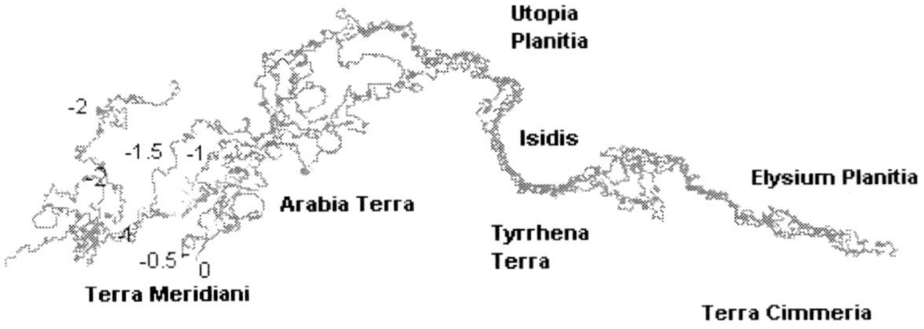

Figure 3. Constant elevation contours in km (black) crossed by the Meridiani shoreline (red). Elevation contours are spaced 0.5 km.

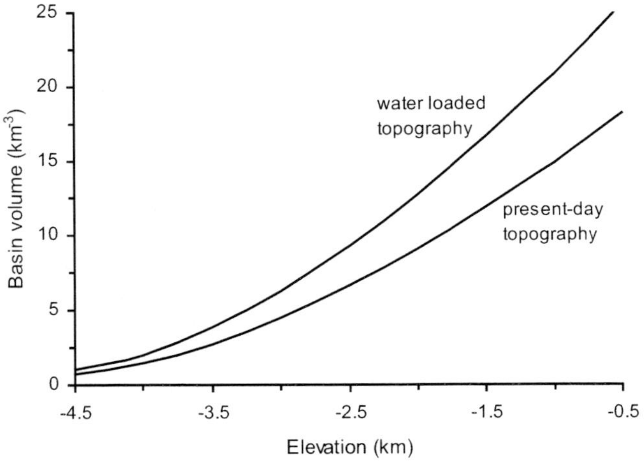

Figure 4. Basin volume as a function of the elevation level, for both, present-day topography and water loaded, Airy compensated, topography (below the -4.35 km level volume includes the contribution from both North Polar and Utopia basins).

Water Volumes in the Ancient Oceans

MOLA topography and mean elevations of proposed paleoshorelines has been previously used to calculate the water volumes contained in the basin related to these putative coastal limits (Head et al., 1998, 1999; Carr and Head, 2003; Öner et al., 2004). Similar calculations are presented in Table 2, and Figure 4 shows basin volume as a function of the elevation level. For the Meridiani shoreline Results in Table 2 are very lower than that in Carr and Head (2003) due to the fact that these authors used a value of 0 for the mean paleoshoreline elevation, whereas we use a mean elevation of −1.5 km following Parker et al. (2000).

Table 2. Basin characteristics at the mean elevations of the proposed Deuteronilus, Arabia and Meridiani shorelines. North polar cap contribution is included in every case.

	Deuteronilus shoreline	Arabia shoreline	Meridiani shoreline
Mean elevation (km) [a]	−3.792	−2.090	−1.5
Basin area enclosed	2.47	4.66	5.34
Present-day basin volume [b] (10^7 km^3)	1.93	8.66	11.88
Total water volume [c,d] (10^7 km^3)	2.00-2.80	8.77-12.28	11.99-16.79
GEL (km) [c,d]	0.14-0.19	0.61-0.85	0.83-1.16
Mean depth (km) [c,d]	0.81-1.13	1.88-2.63	2.25-3.14
Maximum depth (km) [c]	1.46-2.04	3.16-4.42	3.75-5.25

[a] Mean elevations for Deuteronilus and Arabia shoreline after Carr and Head (2003), and for Meridiani Shoreline after Parker et al. (2000),

[b] North polar cap contribution not included.

[c] Lower limit: present-day topography; Upper limit: water loaded, Airy compensated, topography.

[d] North polar cap contribution included.

These calculations assume that the topography of the northern plains has changed little since the putative shorelines were formed. Irrespective of the accuracy of this assumption, estimations based on observed present-day topography can only provide lower limits to the basins volume. Indeed, the existence of oceans in the northern lowlands would imply that the water column provided an additional load over the lithosphere in the regions covered by water (Leverington et al., 2003; Leverington and Ghent, 2004). The effects of variations of the water load on topography are well known for Earth (for a review see Watts, 2001). Therefore, the weight of the water column would have produced the subsidence of the sea floor during the possible periods in which an ocean occupied the lowlands, increasing the basins volume; the subsequent desiccation of the ocean would results in the opposite effect (Hiesinger and Head, 2000; Thomson and Head, 2001; Kreslavsky and Head, 2002; Leverington et al., 2003; Leverington and Ghent, 2004).

We take into account the effect of water load by assuming Airy compensation to calculate upper limits to the basins volume for Meridiani, Arabia and Deuteronilus shoreline. To assume Airy compensation achieved in the mantle implies that elevation variation in the sea floor due to changes in the height of the overlying water column is

$$\Delta y_{floor} = \frac{y_w \rho_w}{\rho_m},$$

where y_w is the height of the water column (i.e., the ocean depth), and ρ_w and ρ_m are the ocean and mantle densities, respectively. If y_w is taken as the depth of an ancient ocean, and y is the altitude difference between the ocean basin floor for the empty basin and the sea level when the basin was filled, obviously $\Delta y_{floor} = y_w - y$, and the isostasy principle requires

$$\frac{y_w}{y} = \frac{\rho_m}{(\rho_m - \rho_w)}.$$

So, if y is taken as the present-day depth of the basin under the sea level, then this implies that estimations for the water volume enclosed in the Martian oceans should be increase by a factor y_w/y (with higher ocean density and lower mantle density increasing this factor). If we assume $\rho_w = 1000$ kg m^{-3} and $\rho_m = 3500$ kg m^{-3}, it is obtained $y_w/y = 1.4$. Figure 4 shows basins volume as a function of the elevation level for water loaded, Airy-compensated, topography.

It is important to note that, although water load of putative ancient oceans would result in a substantial increase in the basin volume with respect to calculations based on present-day topography, the assumption of Airy isostasy would imply that the lithosphere has not rigidity, and therefore the increasing factor y_w/y is an upper limit. So, values presented in Table 2 for water volume, GEL (the Global Equivalent Layer if the water is homogeneously distributed on the surface of Mars), mean depth and maximum depth are given as intervals between estimated results for present-day topography and results for water load compensated by Airy isostasy.

In the calculations of total water volumes the north polar cap volume under the appropriate mean shoreline elevation is taken into account (the total volume of the north polar cap is 1.14×10^6 km^3; Smith et al., 2001), since this contributes to the present-day topography. Lithospheric flexure due to north polar cap loading is not considered here, since that the assumption of not flexure represents the lower limit for north polar contribution to the basins volumes, and upper limits are here calculate assuming Airy compensation.

Re-evaluation of Paleoshorelines

The original global mapping of the putative paleoshorelines was limited by resolution of Viking images. Besides this, it is fairly evident that diverse degradational processes could have affected the original morphology (and topography) of any putative paleoshoreline. Thus, reevaluations of paleoshorelines mapping are probably guaranteed. These revaluations would be of great interest to improve the knowledge the hydrogeological history, but also for the tectonothermal history of Mars, because they could affect the elevation range attributed to a given paleoshoreline, and hence the information derived from elevation ranges.

The possibility that the putative Meridiani shoreline could be the same feature as some portions of the Arabia shoreline was first suggested by Ruiz et al. (2003) related to preliminary work about thermal isostasy applied to Mars. Indeed, the mean elevation in the Meridiani shoreline (–1.5 km following Parker et al. (2000)) is roughly similar to that of the Arabia shoreline in northeastern Arabia, Utopia (not taken into account the Isidis impact basin), Elysium, and Amazonis regions. Thus, a possible paleoshoreline might follow the outline of the Arabia shoreline in these regions, but including the outline of Meridiani Shoreline in western Arabia Terra and Sinus Meridiani. The elevation range of this "mixed" Meridiani/Arabia shoreline, although not examined, would be mostly about 2 km, from –1 to –3 km after Ruiz et al. (2004) on the basis of the topography analysis of Arabia shoreline in Carr and Head (2003). This is still far of an equipotential surface, but this Meridiani/Arabia shoreline would be better candidate to represent a paleoequipotential surface than the Arabia shoreline *sensu strito*: for that reason, it was incorporated to the hypothesis for the Martian hydrogeological history suggested by Fairén et al. (2003), in order to represent the boundary of a putative Noachian ocean.

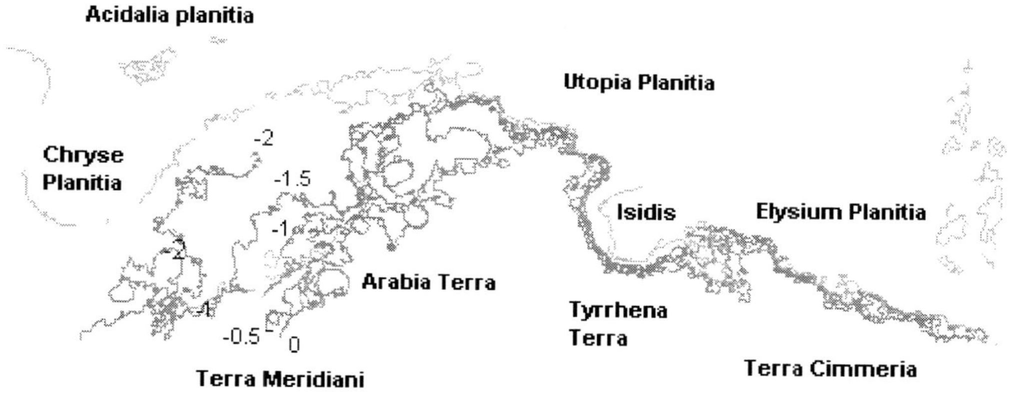

Figure 5. Meridiani (red) and Arabia (green) shorelines, and constant elevation contours in km (black) crossed by the Meridiani shoreline, which are spaced each 0.5 km..

Figure 5 shows Meridiani and Arabia shorelines, as mapped by Clifford and Parker (2001), and constant elevation contours (0.5 km spaced) crossed by the Meridiani shoreline. It can be seen that elevations of the Arabia shoreline at northeastern Arabia, Utopia, and Elysium regions are similar to these along the Meridiani Shoreline, which support the "mixed" Meridiani/Arabia shoreline as a true paleoshoreline. Also is evident that elevations in the Arabia shoreline are much lower in northwestern Arabia Terra, as well as further to the west. Figure 6a shows Meridiani and Arabia shorelines and the contour of the –1.5 km elevation level (the mean level of the Meridiani shoreline after Parker et al. (2000)) superimposed on MOLA topography. The Arabia shoreline at northeastern Arabia, Utopia, Elysium, and Amazonis regions is well close to the –1.5 km elevation level at the majority of places, further supporting the "mixed" Meridiani/Arabia shoreline as a true paleoshoreline. Figure 6b is similar, but showing the contour of the –2.09 kn elevation, corresponding to the mean elevation of the Arabia shoreline *sensu strito*: the Arabia shoreline at Arabia, Utopia, Elysium, and Amazonis regions is at, or generally above, the –2.09 kn elevation level (with the exception of Isidis impact basin, which probably postdates paleoshoreline formation). Thus, the whole topographic range in the Meridiani/Arabia shoreline would be ~1.6 km, between –0.5 and –2.1 km. Moreover, if the Arabia shoreline *sensu strito* is not a true paleoshoreline, then areas, volumes, mean depths and GELs obtained for this "feature" from MOLA topography are not representative of any Martian oceanic stage, but estimations for the –1.5 km elevation level would be roughly appropriate for the Meridiani/Arabia shoreline.

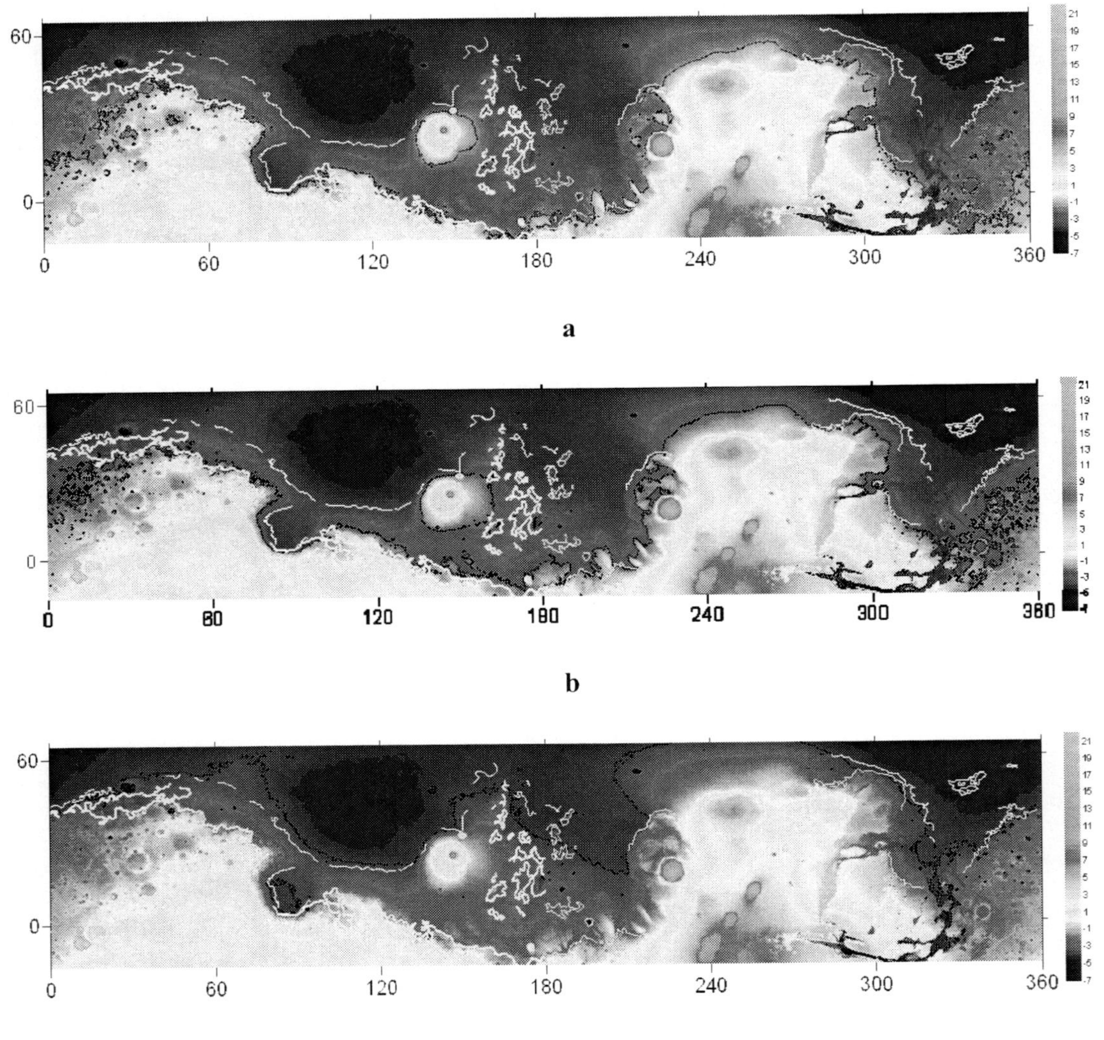

Figure 6. The Deuteronilus (yellow), Arabia (green) and Meridiani (red) shorelines after Clifford and Parker (2001), represented on the Martian topography (scale in km). Also represented (black) are the contour of the (a) -1.5 km, (b) -2.09 km, and (c) -3.792 km elevation levels.

The elevations along putative shorelines on northern Arabia Terra has been recently analyzed in higher resolution (Webb, 2004), finding a elevation of 3707 ± 21 m, for the Arabia shoreline, and two different elevations, 4000 ± 14 m and 4200 ± 12 m, for two separate portions of the Deuteronilus shoreline, which could therefore represent two distinct shorelines. Elevation difference between the two separate portions of the Deuteronilus shoreline is difficultly due to post-formation processes, because elevation in the two portions is nearly constant along distances of about 500 km, and to the clear bimodality in the elevation values. On the other hand, the elevation of the Arabia shoreline in northern Arabia Terra after (Webb, 2004) is intriguingly close to the mean elevation of the Deuteronilus shoreline, and it the possibility of a "mixed" Arabia/Deuteronilus shoreline, which would include the Arabia shoreline in northern Arabia Terra and the Deuteronilus shoreline elsewhere, has been mentioned (Ruiz, 2005).

Figure 6c shows the Arabia and Deuteronilus shorelines, and the contour of the –3.792 km elevation level, the mean elevation of the Deuteronilus shoreline, superimposed on MOLA topography. It is obvious that the Arabia shoreline is very close to –3.792 km elevation level at northwestern Arabia Terra and northeastern Tempe Terra. In turn, Deuteronilus shoreline elevation at Tempe Terra is close to –4 km (see also Carr and Head, 2003), a similar elevation to that found for this feature by Webb (2004) for northern Arabia Terra. Thus, locally at northern Arabia Terra and northeastern Tempe Terra putative shorelines fit well equipotential surfaces, but they are suggesting a complex scenario for the possible evolution of Martian oceans.

Elevation Ranges along Paleoshorelines and Vertical Movements of the Lithosphere

Implications of paleoshoreline reevaluations are evident: the lower the true elevation range of a paleoshoreline the lower the magnitude of vertical movements postdating its formation. For example, elevation range of 1.1 km along the Deuteronilus shoreline is an upper limit, because this range probably includes portions of several different paleoshorelines. Vertical movements deduced from the differences of this feature with respect to an equipotential surface should also be an upper limit.

The fit, although rough, of paleoshorelines to equipotential surfaces would imply a relatively calmed history for the Martian lithosphere, at least since these features were formed. This is clearly the case for the Deuteronilus shoreline *sensu strito*, if this feature is a true paleoshoreline, and moreover for the "revised" Deuteronilus shorelines, which are dating from the Late Hesperian, ~3 Gyr ago (see Table 1). The evidences for a reasonably equipotential Noachian paleoshoreline are attractive, but they must be taken more carefully. If the Meridiani/Arabia shoreline represents a true paleoshoreline, then its whole elevation range of 1.6 km would imply a quite stable lithosphere since 3.5 Gyr ago at least.

By contrast, Gratton et al. (2003) find that the Meridiani, Arabia and Deuteronilus shorelines could fit equipotential surfaces if the reference ellipsoid for the planet has changed with the time, maybe due to the dichotomy formation or Tharsis volcanism. Although this possibility is attractive, it is based on the interpretation of the Meridiani, Arabia, and Deuteronilus *sensu strito* shorelines as true paleoshorelines, which is unlikely (see above).

Thermal isostasy and Thermal Evolution of the Lithosphere

A significant relation exists between Earth's surface elevation and the thermal state of the lithosphere: the warmer the lithosphere, the lower its mean density, and the higher its buoyancy with respect to the underlying fluid materials. This principle (known as thermal isostasy) has been broadly applied to the thermal subsidence of the cooling oceanic lithosphere (e.g., Turcotte and Schubert, 2002), but it can be also applied in a general way to the continental lithosphere (Lachenbruch and Morgan, 1990), including a tectonothermally stable one. This allows the use of topography and surface heat flow data to constrain Earth's continental lithospheric thermal structure (e.g., Tejero and Ruiz, 2002; Lewis et al., 2003). This relation between surface elevation and thermal state of the lithosphere has been also applied to Mars (Ruiz, 2003; Ruiz et al., 2003, 2004).

In this section we present a complete description of the application to Mars of the thermal isostasy concept. We also revise and extend the results obtained about the evolution of the Martian lithosphere.

Thermal Isostasy

Because of thermal expansion and contraction, the elevation of the surface, referenced to the free height of the asthenosphere, depends on the thermal state of the lithosphere and has contributions from the lithospheric mantle and crust,

$$H = H_{\mathrm{m}} + H_{\mathrm{c}}, \tag{1}$$

where H_{m} and H_{c} are the lithospheric mantle and crust contribution to the elevation of the surface respectively. (The term lithosphere is used here to define a thermally conductive layer, in which base isostatic compensation can be achieved.) The contribution due to the lithospheric mantle is given by (Lachenbruch and Morgan, 1990)

$$H_{\mathrm{m}} = \alpha(\overline{T}_{\mathrm{m}} - T_{\mathrm{a}})b_{\mathrm{m}}, \tag{2}$$

where α is the volumetric thermal expansion coefficient, T_{a} is the temperature of the asthenosphere, $\overline{T}_{\mathrm{m}}$ is the mean temperature of the lithospheric mantle, and b_{m} is the lithospheric mantle thickness; density differences between asthenospheric and lithospheric mantle are taken as solely due to temperature differences, which is a very reasonable approximation. A similar equation can be written to describe the crustal contribution, but taking into account a correction factor for the lesser crustal density,

$$H_{\mathrm{c}} = \frac{\alpha\rho_{\mathrm{c}}}{\rho_{\mathrm{a}}}(\overline{T}_{\mathrm{c}} - T)b_{\mathrm{c}}, \tag{3}$$

where ρ_{c} and ρ_{a} are respectively a reference crustal density and the density of the asthenosphere, and b_{c} is the crustal thickness. In turn, the mean temperatures of each lithospheric layer is given by

$$\overline{T} = \frac{1}{z_2 - z_1}\int_{z_1}^{z_2} T(z)\mathrm{d}z. \tag{4}$$

where z, z_1 and z_2 are the depth, the depth in the layer top, and the depth at the layer base, respectively. The component of the topography due to thermal isostasy, expressed in terms of heat flow, is (Ruiz, 2003)

$$h = H_m(F_h) + H_c(F_h) - H_m(F_o) - H_c(F_o)\,, \tag{5}$$

where h is the local elevation with respect to a reference elevation, F_h is the local surface heat flow, and F_o is the heat flow for the reference elevation.

Temperature Profiles

Temperature profiles in this chapter are calculated by assuming radioactive heat sources homogeneously distributed in the crust, and linear thermal gradients for the lithospheric mantle mantle. The temperature at a depth z within the crust is given by (Roy et al., 1968)

$$T_z = T_s + \frac{Fz}{k_c} - \frac{Az^2}{2k_c}\,, \tag{6}$$

where T_s is the surface temperature, F is the surface heat flow, k_c is the thermal conductivity of the crust, and A is the volumetric heat production rate. Ruiz (2003) calculate b_m, $\overline{T}_m$, and $\overline{T}_c$ in terms of the surface heat flow, and the proportion f of the heat flow originated from crustal heat sources, assuming that crustal heat sources are homogeneously distributed. The factor f can be formally defined as

$$f = \frac{Ab_c}{F}\,; \tag{7}$$

so, within the crust, the temperature at a depth z is

$$T_z = T_s + \frac{Fz}{k_c}\left(1 - \frac{fz}{2b_c}\right). \tag{8}$$

The existence of heat sources in the lithospheric mantle is not take into account, because on Earth radiogenic sources are sparse beneath the near-surface radioactive element-rich layer, and within the lithospheric mantle, the heat flow can be assumed constant (e.g., Turcotte and Schubert, 2002); so, within the mantle lithosphere, the temperature at a depth z is

$$T_z = T_c + \frac{F(1-f)(z-b_c)}{k_m}\,, \tag{9}$$

where T_c is the temperature at the crust base, calculated taking $z = b_c$ in equation (8); in turn, b_m is calculated from

$$b_m = b_c + \frac{k_m(T_a - T_c)}{F(1-f)} \quad (10)$$

Mean lithospheric mantle and crust temperatures are respectively given by

$$\overline{T}_m = \frac{T_a + T_c}{2}, \quad (11)$$

and

$$\overline{T}_c = T_s + \frac{F b_c (1 - f/3)}{2k_c}. \quad (12)$$

Parameter Values

The calculations have been performed using $\alpha = 3 \times 10^{-5}$ °C^{-1}, $k_c = 2.5$ W m^{-1} °C^{-1}, $k_m = 3.5$ W m^{-1} °C^{-1}, ρ_c = 2900 kg m^{-3}, and ρ_a = 3500 kg m^{-3} for material properties. The surface temperature is taken as 0°C, maybe more appropriate for a time in which an ocean is assumed than the current mean surface temperature of about –50°C (although the results are relatively insensitive to the election of this parameter). The asthenosphere temperature is taken as 1300°C, which is a value typically used for the Earth's asthenosphere (e.g., Ranalli, 1997). Crustal thickness is assumed to be 40 km, in accordance with the typical mean crustal thickness below the northern lowlands derived from topography and gravity data (Zuber et al., 2000). Two possibilities have been taken for the value of f in equation (8), although we note that this value could locally vary: $f = 0$ (corresponding to a linear thermal gradient through the crust) and $f = 0.5$. The latter value is in accordance with the proposal (made from geochemical arguments drawn from the materials on Mars' surface) that perhaps over 50% (or even 75%) of radioactive heat sources in this planet are placed in its crust (McLennan, 2001, 2003); similarly, in the Earth, the 40-60% of the heat flow lost in continental areas originates from crustal heat sources (Pollack and Chapman, 1977; Turcotte and Schubert, 2002).

Deformation of Paleoshorelines

Ruiz et al. (2004) have used the thermal isostasy concept to show that it is not necessarily true that a paleoequipotential surface implies a good fit to a present-day equipotential surface. A similar conclusion has been obtained taking into account the lithosphere rebound due to water uploading associated to the disappearance of an ocean of irregularly shaped margins (Leverington et al., 2003; Leverington and Ghent, 2004). Here we present the first-order calculations of the magnitude of the relation between possible variations in the thermal state of the lithosphere and elevation differences, which could cause concrete deviations in equipotentiallity along the proposed paleoshorelines. Note that for the purposes of this chapter, only large wavelength topographic differences are relevant, since that the rigidity of the Martian lithosphere could prevent small-scale isostatic adjustment.

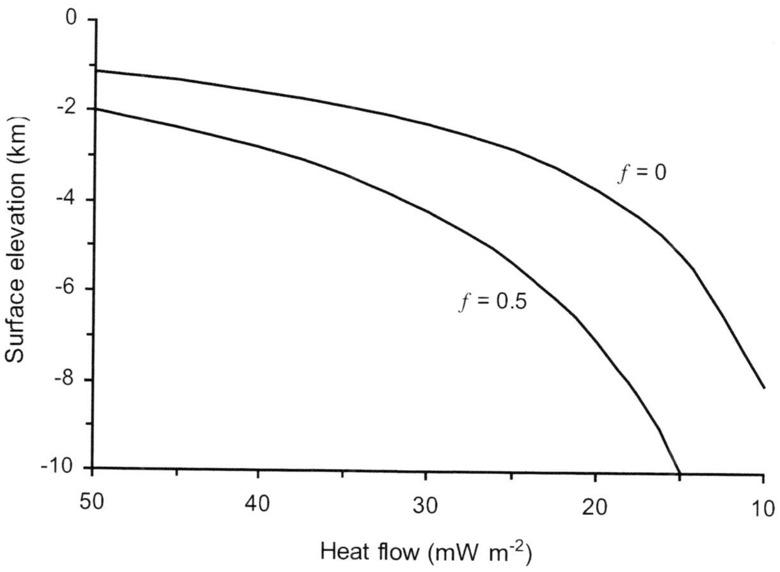

Figure 7. Elevation over the free height of the asthenosphere in terms of surface heat flow for $f = 0$ and $f = 0.5$. Surface heat flow is represented in reverse order (describing the topographic evolution of a cooling region isostatically compensated). Adapted from Ruiz et al. (2004).

Figure 7 shows H in terms of surface heat flow for $f = 0$ and $f = 0.5$. Calculations have been made for a range of F values between 10 and 50 mW m^{-2}, which roughly correspond to the whole range of surface heat flows proposed for diverse regions and times using estimates of the elastic thickness of the lithosphere (McGovern et al., 2002, 2004). It is important to note that the f value can change along a possible paleoshoreline (for example, due to local variations in crustal heat sources, mantle heat flow, or both), or with time (due to waning of radiogenic dissipation intensity or to changes in the efficiency of convective heat transfer). In any case, these possibilities are not important for the purpose of this first-order calculation, which is show the feasibility of differential thermal isostasy histories to affect the large wavelength topography of possible Martian paleoshorelines. For the purposes of this analysis the interesting point is the relative differences of H, and not the absolute values obtained for this parameter (planetary topographies are referred to arbitrary datum). Figure 7 indicate that variations in thermal state of the lithosphere can result in differential thermal isostasy, which in turn can result in important elevation differences, even of kilometric scale as occur on the Earth (Lewis et al., 2003).

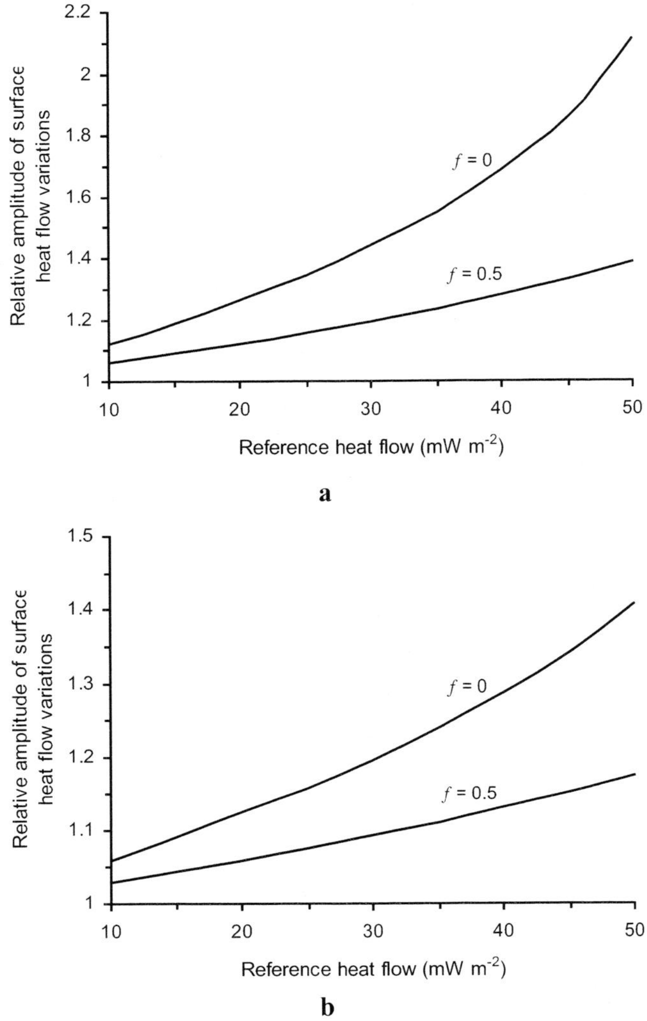

Figure 8. Relative amplitude of surface heat flow variations that can produce elevation ranges of 1 (a) and 0.5 (b) km centered on the *H* value corresponding to a reference heat flow in the range from 10 to 50 mW m^{-2}. As in Figure 1 surface heat flow is represented in reverse order. Adapted from Ruiz et al. (2004).

Assuming a Martian paleoshoreline, the posterior attenuation, disappearance (as is expected with the waning of internal heat sources), or formation (if reheating of the lithosphere postdating the shoreline formation occurred) of heat flow variations must result in the deformation of the paleoshoreline topography, deviating it from an equipotential surface. Figure 8 shows the relative amplitude of surface heat flow variations that can produce elevation ranges of 1 and 0.5 km centered on the *H* value corresponding to a reference heat flow in the range from 10 to 50 mW m^{-2}. The relative amplitude of heat flow variations is obtained as the quotient between the maximum and minimum heat flow that can produce positive and negative elevations, respectively, of 0.5 and 0.25 km with respect to the reference *H* value. In Figure 8 it can be seen that ancient surface heat flow variations less than a factor ~2 may account for differences of elevation of 1 km. An elevation range of 0.5 km could be produced by surface heat flow variations less than a factor of ~1.5. These values are further lowered if a substantial amount of the Martian heat sources are located within the crust.

On the Earth, present-day surface heat flow variations expand by more than one order of magnitude (e.g. Pollack et al., 1993). The major part of these variations is due to plate tectonics, but for Mars an early phase of plate tectonics is controversial. Variations in surface heat flow in Earth's continental regions, from contoured maps (Cermak, 1993; Pollack et al., 1993), can be higher than a factor of 2 or 3, sufficient for important deformation of paleoshorelines (see Figure 7). Those areas include terrains of different ages, and it is known for continental areas that an inverse relation exists between surface heat flow and age of the last tectonothermal stabilization (e.g., Hamza, 1979; Vitorello and Pollack, 1980; Cermak, 1993). Moreover, continental heat flow depends also on a wide array of factors, as for example radioactive heat production in the crust, local mantle heat flow, or tectonic or erosive redistribution of crustal heat producing elements (for reviews see Beardsmore and Cull, 2001; Sandiford and McLaren, 2002).

In any case, heat flow variations on old and tectonothermally stable terrestrial continental areas can be as high as a factor ~1.5-2 (e.g., Cermak, 1993; Roy and Rao, 2000; Rolandone et al., 2002). If local variations of surface heat flow of at least similar amplitude existed in Mars during any moment of its history, then these results indicate that differential thermal isostasy should result in important deformation, and deviation of equipotentiallity, along putative shorelines. Moreover, it is significant that if the half of the surface heat flow was originated from crustal heat sources (as it is the case on the Earth) when the paleoshorelines were formed, then heat flow variations lower than a modest factor of ~1.2-1.4 may account for present-day elevation ranges of 0.5-1 km (if, as it seem reasonable, these heat flow variations are currently greatly attenuated). These elevation ranges are respectively similar to the ±1 standard deviation and whole elevation ranges in the Deuteronilus shoreline, but they represent an important amount of deformation along any possible paleoshoreline.

It is important remind that thermal isostasy is only a contributor to the topography. Though it is not our intentions to discuss the many other factors that may have contributed to the modification of an equipotential surface, we highlight that other possibilities may be degradation by wind, water, tectonic, and volcanic modification (e.g., Clifford and Parker, 2001; Fairén et al., 2003), rebound of the lithosphere due to dissipation of a water body (Leverington et al., 2003; Leverington and Ghent, 2004; for a review of isostatic and flexural effects related to changes in sea level see Watts, 2001), flexure (non-thermal) isostasy due to surface loading, erosion, or subsurface magmatic intrusions. In fact, endogenic-driven geologic activity (probably implying vertical movements) and exogenic activity clearly postdates the possible paleoshorelines in Arabia Terra and Tharsis and Elysium (e.g., Head et al., 1999; Anderson et al., 2001). Together, all those possibilities make more pressing the main argument in this section: any paleoequipotential surface dating of the ancient Mars must be importantly deformed at present, even in a range of elevations of a kilometric scale.

Ancient Heat Flow Variations

If the existence of paleoshorelines is accepted, then elevation differences along these paleoshorelines impose constraints to the differences in the evolution of the thermal structure of the lithosphere in shoreline-crossed regions (Ruiz, 2003). Indeed, it is possible to calculate the amplitude of heat flow variations necessary to compensate present-day topography and transform a paleoshoreline into an equipotential surface. If surface heat flow variations on Mars are currently almost disappeared (as it would expectable for an efficiently cooled planet), then the heat flow variations deduced from shoreline topography must be mostly related to the time when this feature was formed. Like other than thermal isostasy processes could have produced vertical movements in the shoreline-crossed regions, the results obtained in this way suppose an upper limit to the amplitude of heat flow variations. Diverse processes, including geomorphologic evolution, can affect small-scale topographic variations, and the rigidity of the Martian lithosphere could also prevent small-scale isostatic adjustment; therefore, large wavelength topographic variations, in which isostatic adjustment can work, are more relevant again for the purposes of this argument.

The calculations have been performed for the Deuteronilus shoreline, but the argument and results are also valid for the outer contact of the Vastitas Borealis Formation (if this is considered to be an ancient oceanic limit) since the ranges of elevations are similar in both features. So, h is taken as ±0.55 km, for a total elevation range of 1.1 km for the Deuteronilus shoreline. This corresponds to a reference elevation (for which $h = 0$ and $F_h = F_o$) of –3.75 km, close to the mean elevation of the Deuteronilus shoreline and to the mean elevation of the termini of the Chryse outflow channels. Like elevation differences along the shoreline must be compensated by heat flow variations, h should have sign minus in equation (5) for mathematical consistency. The F_o value is not known, and for that reason, the calculations have been performed for a range of F_o values between 15 and 35 mW m^{-2}, a range based on estimates (uncertainty included) of the late Hesperian/early Amazonian elastic lithosphere thickness (Zuber et al., 2000; McGovern et al., 2002, 2004).

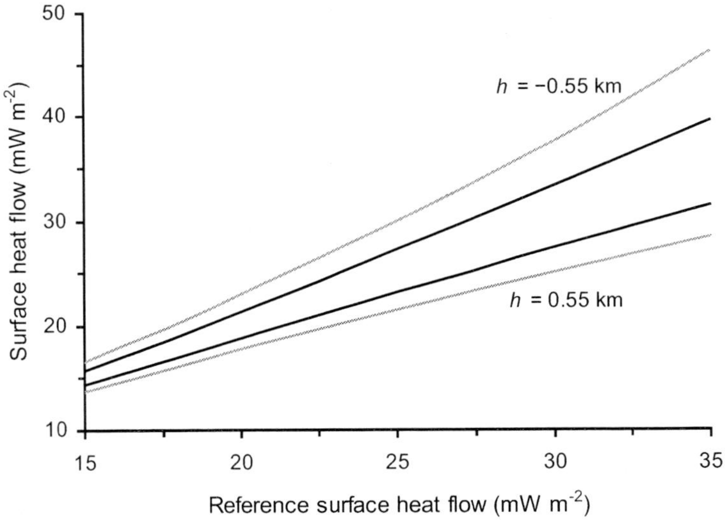

Figure 9. F_h values for h = 550 m (lower curves) and h = –550 m (upper curves) in terms of F_o. Gray and black lines indicate $f = 0$ and $f = 0.5$, respectively. In each case, the difference between the upper and lower curves gives the maximum surface heat flow variations permitted assuming the Deuteronilus shoreline as a paleo-equipotential surface. Adapted from Ruiz (2003).

Figure 9 shows F_h values for h = ±0.55 km in terms of F_o. In each case, the difference between the upper and lower curves gives the maximum heat flow variations allowed, taking into account the Deuteronilus shoreline topography. Figure 10 shows upper limits to the relative amplitude of surface heat flow variations on Deuteronilus shoreline locations; these relative amplitude upper limits are the ratio between the maximum and minimum values shown in Figure 9.

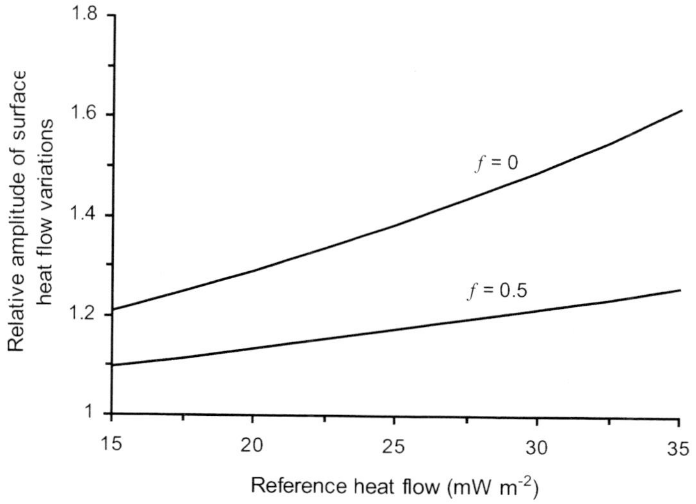

Figure 10. Upper limits to the relative amplitude of surface heat flow variations on the Deuteronilus shoreline locations. These relative amplitude upper limits are given by the ratio between upper and lower values. Adapted from Ruiz (2003).

It can be seen in Figures 9 and 10 that variations in heat flow in regions through the Deuteronilus shoreline were small in the Late Hesperian. In fact, the obtained upper limits for the relative amplitude of these variations are, at most, a factor of 1.6. If crustal heat sources are taken into account, the magnitude of the relative amplitude variations decreases for each given value of F_0. For the F_0 range used here, the depth to the 1300ºC isotherm is ~100-300 km for $f = 0$ and ~200-600 km for $f = 0.5$. Since the small radius of Mars the calculations for the case $f = 0.5$ (at low heat flows values) should take in account the spherical shape of Mars. In addition, the range of F_0 values used here is based on works assuming linear thermal gradients. Calculation of surface heat flows from elastic thicknesses would result in higher values if heat sources are present in the crust (Solomon and Head, 1990; Ruiz et al., 2005). This, in turn, decreases the depth to the 1300ºC isotherm and also increases the relative amplitude of variations in F_h for the case $f = 0.5$. As the relative amplitude of F_h variations is clearly lower in the $f = 0.5$ case than in the $f = 0$ one, the main conclusions so obtained are not altered.

The upper limits for the relative amplitude of heat flow variations obtained here are clearly lower than those presently observed on Earth. On our planet, the higher heat flows are associated with sea floor spreading centers, but there is not clear evidence for a phase of plate tectonics in the Mars's history (and, in any case, not for the late Hesperian or later on), and so, those very high heat flows are not relevant for the purposed of is chapter. As above mentioned, heat flow show variations in continental areas can be higher than a factor 2-3, and as high as a factor ~1.5-2 in tectonothermally stable terrestrial continental areas. Those areas include terrains of different ages, and it is known for continental areas that an inverse relation exists between surface heat flow and age of the last tectonothermal stabilization (e.g., Hamza, 1979; Vitorello and Pollack, 1980; Cermak, 1993).

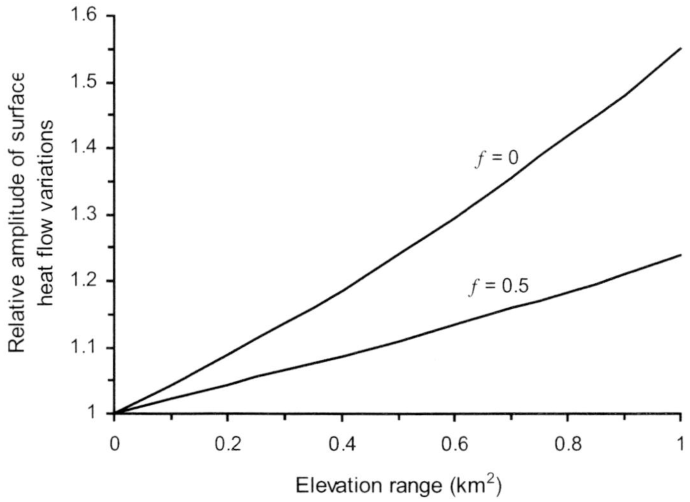

Figure 11. Upper limits to the relative amplitude of surface heat flow variations in terms of the total elevation range of a paleoshoreline, calculated for $F_0 = 35$ mW m^{-2} (the upper limits in the range of heat flows assumed for analyze the Deuteronilus shoreline case) at the H value corresponding to the central value in the elevation range.

If surface heat flow variations on Mars are currently almost disappeared, then the upper limits to the heat flows variations deduced of Deuteronilus shoreline topography are related to the time when this feature was formed (i.e., the Late Hesperian, ~3 Gyr ago; Hartmann and Neukum, 2001). In this case, the present-day elevation range along Deuteronilus shoreline suggests that differences in the thermal state of the lithosphere in the "Deuteronilus shoreline regions" have been relatively small since the feature was formed, and therefore, that very large areas of the Martian lithosphere has been tectonothermally stable since the Late Hesperian. This is consistent with near complete building of the Tharsis rise by the end of the Noachian (Phillips et al., 2001), with a significant decrease in volcanic resurfacing rates following the Hesperian's end (e.g., Hartmann and Neukum, 2001) and with the localization of the waning Amazonian magmatic and tectonic activity at areas in Tharsis and Elysium (e.g., Anderson et al., 2001; Dohm et al., 2001; Head et al., 2001).

Alternatively, reheating of the lithosphere postdating the Deuteronilus shoreline could have caused, or contributed to, the distortion of the topography. In this case, the reheating should have been maintained (at least partially) until the present time, since the dissipation of the thermal anomalies should lead to the disappearance of their effect on the topography. But, as above mentioned, Amazonian geological activity represents the waning and localized magmatic and tectonic activity on Mars, and for that reason, the obtained upper limits to the heat flow variations more probably refer to the thermal state of the Martian lithosphere when the Deuteronilus shoreline was formed.

In summary, if the Deuteronilus shoreline (or equivalently, the outer contact of the Vastitas Borealis Formation) represents a Late Hesperian paleo-equipotential surface, then three conclusions can be deduced from these calculations. First, the relative variations in Late Hesperian surface heat flow in shoreline regions were lower than relative variations in present-day surface heat flow in continental areas on Earth. Second, if substantial amounts of radiogenic heat sources are located in the Martian crust, those relative variations are likely lower. Finally, very large areas of the Martian lithosphere have been tectonothermally stable since (at least) the Late Hesperian.

If the Deuteronilus shoreline is a feature combining portions of several paleoshorelines, then the total elevation range would be lower, and the arguments above more pressing, as clearly illustrated by Figure 11. This figure shows upper limits to the relative amplitude of heat variations, for $f = 0$ and $f = 0.5$, as functions of the total elevation range of a paleoshoreline. As only upper limits for a given elevation range are represented, the calculations have been performed for a F_0 value of 35 mW m^{-2} (the upper limits in the range of heat flows assumed for analyze the Deuteronilus shoreline case) at the H value corresponding to the central value in the elevation range, and therefore $h = \pm\Delta H/2$.

Conclusion

The range of elevations of the Deuteronilus shoreline is strongly indicating the great tectonothermal stability of the Martian lithosphere since the Late Hesperian, at least. If an important proportion of radioactive heat sources are located in the crust, as geochemical evidences suggest, the lithosphere should have been greatly thermally homogeneus in "Deuteronilus shoreline" regions when coastal processes creating this paleoshoreline were working.

There are evidences suggesting that the lateral continuity of the originally proposed paleoshorelines is not well established: diverse division and mixing of the originally proposed paleoshorelines seem be required. These revaluations are of great interest for the understanding of the evolution of the Martian lithosphere, because they would modify the elevation range attributed to a given paleoshoreline. Thus, it is necessary a careful geomorphologic reassessment of the diverse features interpreted as paleoshorelines and of the relation among them. The Deuteronilus shoreline (*sensu strito*) seems integrate portions of several separate paleoshorelines, which would imply a still higher thermal stability and homogeneity of the lithosphere by the latest ~3 Gyr. The confirmation of the Meridiani/Arabia shoreline would be indicative of a longer lithospheric stability.

Acknowledgments

The authors thank Frank Columbus for your invitation to prepare this chapter. JR was supported by a grant of the Spanish Secretaría de Estado de Educación y Universidades.

References

[1] Anderson, S.; Grimm, R. E. *J. Geophys. Res.* 1998, 103, 11,113-11,124.

[2] Anderson, R. C.; Dohm, J. M.; Golombek, M. P.; Haldemann, A. F. C.; Franklin, B. J.; Tanaka, K. L.; Lias, J.; Peer, B. *J. Geophys. Res.* 2001, 106, 20,563-20,585.

[3] Beardsmore, G. R.; Cull, J. P. *Crustal Heat Flow. A Guide to Measurement and Modelling*; University Press: Cambridge, UK, 2001; pp 1-324.

[4] Breuer, D.; Spohn, T. *J. Geophys. Res.* 2003, 108 (E7), 8.1-8.13.

[5] Carr, M. H.; Head, J.W. *J. Geophys. Res.* 2003, 108 (E5), 8.1-8.28.

[6] Cermak, V. *Phys. Earth Planet. Inter.*1993, 79, 179-193.

[7] Choblet, G.; Sotin, C. *Geophys. Res. Lett.* 2001, 28, 3035-3038.

[8] Clifford, S. M.; Parker, T. J. *Icarus* 2001, 154, 40–79.

[9] Comer, R. P.; Solomon, S. C.; Head, J. W. *Rev. Geophys.* 1985, 23, 61-92.

[10] Dohm, J. M.; Ferris, J. C.; Baker, V. R.; Anderson, R. C.; Hare, T. M.; Strom R. G.; Barlow N. G.; Tanaka, K. L.; Klemaszewski, J. E.; Scott D. H. *J. Geophys. Res.* 2001, 106, 32,943-32,958.

[11] Edgett, K. S.; Parker, T. J. *Geophys. Res. Lett.* 1997, 24, 2897-2900.

[12] Fairén, A. G.; Dohm, J. M., Baker V. R.; de Pablo M. A.; Ruiz J.; Ferris J. C.; Anderson, R. C. *Icarus* 2003, 165, 53-67, 2003.

[13] Grasset, O.; Parmentier, E. M. *J. Geophys. Res.* 1998, 103, 18,171-18.181.

[14] Gratton, R.; Mareschal, J.; Sotin, C. *Eos Trans. AGU Fall Meet. Suppl.* 2003, 84(46), abstract G41B-0033.

[15] Hamza, V. M. *Pure Appl. Geophys.* 1979, 117, 65-74.

[16] Hartmann, W. K. *Icarus* 2005, 174, 294-320.

[17] Hartmann, W. K.; Neukum, G. *Space Sci. Rev.* 2001, 96, 165-194.

[18] Hauck, S. A.; Phillips, R. J. *J. Geophys. Res.* 2002, 107 (E7), 6.1-6.19.

[19] Head, J. W.; Kreslavsky, M.; Hiesinger, H.; Ivanov, M. A.; Pratt, S.; Seibert, N.; Smith, D. E.; Zuber, M. T. *Geophys. Res. Lett.* 1998, 25, 4401-4404.

[20] Head, J. W.; Hiesinger, H.; Ivanov, M. A.; Kreslavsky, M. A.; Pratt, S.; Thomson, B. J. *Science* 1999, 286, 2134-2137.

[21] Head, J. W.; Greeley R.; Golombek, M. P.; Hartmann, W. K.; Hauer, E.; Jaumann, R.; Masson, P.; Neukum, G.; Nyquist, L. E.; Carr, M. H. *Space Sci. Rev.* 2001, 96, 263-292.

[22] Head, J. W.; Kreslavsky, M. A.; Pratt, S. *J. Geophys. Res.* 2002, 106 (E1), 3.1-3.29.

[23] Hiesinger, H.; Head, J. W. *J. Geophys. Res.* 2000, 105, 11,999-12,022.

[24] Ivanov, M. A.; Head, J. W. *J. Geophys. Res., 106*, 3275-3295, 2001.

[25] Kiefer, W. S. *Earth Planet. Sci. Lett.* 2004, 222, 349-361.

[26] Kreslavsky, M. A.; Head, J. W. *J. Geophys. Res.* 2002, 107 (E12), 4.1-4.25.

[27] Lachenbruch, A. H.; Morgan, P. *Tectonophysics* 1990, 174, 39–62.

[28] Leverington, D. W.; Ghent R. R. *J. Geophys. Res.* 2004, 109 (E01005), 1-10.

[29] Leverington, D. W.; Ghent, R. R.; Irwin R. P.; Craddock R. A.; Maxwell T. A. *Lunar Planet. Sci.* 2003, 34, abstract 1282.

[30] Lewis, T. J.; Hyndman, R. D.; Flück P. *J. Geophys. Res.* 2003, 108 (B6), 16.1-16.18.

[31] Malin, M. C.; Edgett, K.C. *Geophys. Res. Lett.* 1999, 26, 3049-3052.

[32] Malin, M. C.; Edgett, K.C. *J. Geophys. Res.* 2001, 106, 23,429-23,570.

[33] McGovern, P. J.; Solomon, S. C.; Smith, D. E.; Zuber, M. T.; Simons, M.; Wieczorek, M. A.; Phillips, R. J.; Neumann, G. A.; Aharonson, O.; Head, J. W. *J. Geophys. Res.* 2002, 107 (E12), 19.1-19.25.

[34] McGovern, P. J.; Solomon, S. C.; Smith, D. E.; Zuber, M. T.; Simons, M.; Wieczorek, M. A.; Phillips, R. J.; Neumann, G. A.; Aharonson, O.; Head, J. W. *J. Geophys. Res.* 2004, 109 (E07007), 1-5.

[35] McKenzie, D.; Barnett, D. N.; Yuan, D. L. *Earth Planet. Sci. Lett.* 2002, 195, 1-16.

[36] McLennan, S. M. *Geophys. Res. Lett.* 2001, 28, 4019-4022.

[37] McLennan, S. M. *Sixth International Conference on Mars* 2003, abstract 3099.

[38] Nimmo, F.; *J. Geophys. Res.* 2002, 107 (E11), 27.1-27.16.

[39] Nimmo, F.; Stevenson, D. J. *J. Geophys. Res.* 2000, 105, 11,969-11.979.

[40] Öner, A. T.; Ruiz, J.; Fairén, A. G.; Tejero, R.; Dohm J. M. *Lunar Planet. Sci.* 2004, 35, abstract 1319.

[41] Parker, T. J.; Saunder, R. S.; Schneeberger, D. M. *Icarus* 1989, 82, 111-145.

[42] Parker, T. J.; Gorsline, D. S.: Saunders, R. S.; Pieri, D. C.; Schneeberger, D. M. *J. Geophys. Res.* 1993, 98, 11,061-11,078.

[43] Parker, T. J.; Clifford, S. M.; Banerdt, W. B. *Lunar Planet. Sci.* 2000, 31, abstract 2033.

[44] Parker, T. J.; Grant, J. A.; Franklin B. J.; Rice J. W. *Lunar Planet. Sci.* 2001, 32, abstract 2051.

[45] Phillips R. J. et al. *Science* 2001, 291, 2587-2591.

[46] Pollack, H. N.; Chapman, D. S. *Earth Planet. Sci. Lett.* 1977, 34, 174-184.

[47] Pollack, H. N.; Hunter, S. J.; Johnson, J. R. *Rev. Geophys.* 1993, 31, 267-280.

[48] Ranalli, G., *Geol. Soc. Spec. Publ.* 1997, 121, 19-37.

[49] Rolandone, F.; Jaupart, C.; Mareschal, J. C.; Gariépy, C.; Bienfait, G.; Carbonne, C., Lapointe, R. *J. Geophys. Res.* 2002, 107 (B12), 7.1-7.19.

[50] Roy, R. F.; Blackwell, D. D.; Birch, F. *Earth Planet. Sci. Lett.* 1968, 5, 1-12.

[51] Roy, S.; Rao, R. U. M. *J. Geophys. Res.* 2000, 105, 25,587-25,604.

[52] Ruiz, J. *J. Geophys. Res.* 2003, 108 (E11), 8.1-8.5.

[53] Ruiz, J. *Lunar Planet. Sci.* 2005, 36, abstract 1135.

[54] Ruiz, J.; Fairén, A. G.; de Pablo, M. A. *Lunar Planet. Sci.* 2003, 34, abstract 1090.

[55] Ruiz, J.; Fairén, A. G.; Dohm, J. M.; Tejero, R. *Planet. Space Sci.* 2004, 52, 1297-1301.

[56] Ruiz, J.; McGovern, P. J.; Tejero, R. *Earth Planet. Sci. Lett.* 2006, 241, 2-10.

[57] Salamuniccar, G. *Lunar Planet. Sci.* 2004, 35, abstract 1992.

[58] Sandiford, M.; McLaren, S. *Earth Planet. Sci. Lett.* 2002, 204, 133-150.

[59] Schultz, R. A.; Lin, J. *J. Geophys. Res.* 2001, 106, 16,549-16,566.

[60] Schultz, R. A.; Watters, T. R. *Geophys. Res. Lett.* 2001, 28, 4659-4662.

[61] Solomon, S. C.; Head, J. W. J. Geophys. Res. 1990, 95, 11,073-11,083.

[62] Schubert, G.; Spohn, T. *J. Geophys. Res.* 1990, 95, 14,095-14,104.

[63] Schubert, G.; Solomon, S. C., Turcotte, D. L.; Drake, M. J., Sleep, N. H. In *Mars*; Kieffer H.; Jakosky, B. M.; Snyder, C. W.; Matthews, M. S.; Eds.; University of Arizona Press: Tucson, AZ, 1992; pp 249-297.

[64] Smith, D. E. et al. *Science* 1999, 284, 1495-1503.

[65] Smith, D. E. et al. *J. Geophys. Res.* 2001, 106, 23,689-23,722.

[66] Spohn, T.; Acuña, M. H.; Breuer, D.; Golombek, M.; Greeley, R., Halliday, A.; Hauber, E.; Jaumann, R.; Sohl, F. *Space Sci. Rev.* 2001, 96, 231-262.

[67] Squyres, S. W. et al. *Science* 2004, 306, 1709-1714.

[68] Stevenson, D. J.; Spohn, T.; Schubert, G. *Icarus* 1983, 54, 466-489.
[69] Tanaka, K. N.; Scott, D.H.; Greeley, R. 1992. In *Mars*; Kieffer H.; Jakosky, B. M.; Snyder, C. W.; Matthews, M. S.; Eds.; University of Arizona Press: Tucson, AZ, 1992; pp 345-383
[70] Tejero, R.; Ruiz, J. *Tectonophysics* 2002, 350, 49-62.
[71] Thomson, B. J.; Head, J. W. *J. Geophys. Res.* 2001, 106, 23,209-23,230.
[72] Turcotte, D. L.; Schubert, G. *Geodynamics. Second edition*; Cambridge University Press: Cambridge, UK, 2002; pp 1-456.
[73] Vidal, A.; Mueller, K. M.; Golombek, M. P. *Lunar Planet. Sci.* 2005, 36, abstract 2333.
[74] Vitorello, I.; Pollack, H. N. *J. Geophys. Res.* 1980, 85, 983-995.
[75] Watts, A. B. *Isostasy and flexure of the lithosphere*; Cambridge University Press: Cambridge, UK, 2001; pp 1-458.
[76] Webb, V. E. *J. Geophys. Res.* 2004, 109 (E09010), 1-12.
[77] Webb, V. E.; McGill, G. E. *Lunar Planet. Sci.* 2003, 34, abstract 1132.
[78] Zuber, M. T. et al. *Science* 2000, 287, 1788-1793.

In: Space Science: New Research
Editor: Nick S. Maravell, pp. 165-195
ISBN: 1-60021-005-8

Chapter 6

THERMAL PROPERTIES AND TEMPERATURE VARIATIONS IN MARTIAN SOIL ANALOGUES

F. Gori*[*] *and S. Corasaniti
Department of Mechanical Engineering, University of Rome "Tor Vergata",
Via del Politecnico 1, 00133 Rome, Italy

Abstract

The present work reviews the information on mineralogy and chemistry of Martian soils from the data of Landers and remote sensing. Compositional information from Mars Pathfinder Landers indicates a mixture of weathered local rocks of mafic composition. If the Martian soil is composed by sediments only, porosity can be 50% at the surface. In other areas of interest a surface porosity of 20% can be present for lavas.

Martian soil analogues, in dry and frozen conditions, are investigated in the present work as far as thermal conductivity and temperature variation, along its depth, are concerned. The thermal conductivity is predicted theoretically with the cubic cell model, which considers the dry soil as a cubic cell, with a solid cubic particle at the center, and gas around it. The thermal conductivity of the porous medium is evaluated from the solution of the one-dimensional Fourier conduction equation with the assumption of isothermal lines. The model needs to know the thermal conductivities of the solid particle and of the materials present, i.e. atmospheric gas and/or frozen ice, and the porosity of the soil analogue. Martian soil analogue is simulated with a kind of olivine tested in laboratory. The soil mineral composition allows to evaluate experimentally the thermal conductivity of the olivine particle, with the help of the theoretical method, which results equal to $K_s = 2.94$ W/m K. The thermal conductivity of dry Martian soil analogue, evaluated with the theoretical model in the porosity range $p = 0 - 1$ and at several temperatures, increases with the temperature and decreases with the porosity. The Martian atmospheric pressure is about 6 mbar and the Martian atmospheric gas is composed of CO_2 (95%), N_2 (3%), Ar (1.5%) and traces of water vapour. Heat capacity of soil analogue is evaluated with the knowledge of its physical properties, the porosity and the specific heats of the materials present.

Thermal diffusivity, calculated as the ratio of thermal conductivity and heat capacity, is a function of porosity and ice mass content of the soil analogue.

The temperature of Mars surface is assumed variable during the day in the range: 148 K – 298 K. Temperature variations in dry and partially frozen soil analogues are predicted during

---

[*] E-mail address: gori@uniroma2.it (Corresponding author)

a Martian day. The temperature variation at different depth is attenuated, as compared to the surface variation, and a phase delay is present depending on the soil thermal properties. The temperature variation, as well as the derivative of the temperature variation with the depth, is dependent on the thermal diffusivity of the soil analogue. In conclusion, the temperature measurement along the depth of a Martian soil analogue can be used to verify its physical status, i.e. dry or partially frozen.

Keywords: Martian soil analogue, thermal conductivity prediction, cubic cell model, temperature variation, thermal diffusivity.

Introduction

Images of Mars surface, taken by different spacecraft (Viking, Pathfinder), suggest the upper few kilometres of the crust consists in a thick porous megaregolith, mainly composed by ejecta of impact craters. The lithology varies from loose fine-grained regolith to meter-sized block of ejecta, all with a significant amount of volcanics. If a large quantity of volatiles still resides on Mars, these volatiles may be concentrated in the Martian regolith and megaregolith as clay mineral, and as ground and underground condensates such as brines on the surface and ice in the underground. With a dry periglacial-type climate, Martian ground ice should be stable at some depth over virtually the entire planet while in the first meters the soil is able to exchange volatiles with the thin atmosphere of the planet. In this soil layer of ice free regolith there will be fluxes of volatiles in and out and the deposition or the evaporation of volatile constituents influence the physical parameters of the soil.

Direct measurement of the bulk thermal conductivity of the Martian regolith and megaregolith is up to now lacking, likely range of values can be inferred only by considering terrestrial analogues. Thermal properties of the regolith and megaregolith should be closely approximate by those of terrestrial frozen soil and basalt. Laboratory measurements of thermal conductivity of 144 samples of frozen soil and 301 samples of basalt are reported in [1]. The samples vary in composition, lithology, porosity, and degree of pore saturation, but all represent conditions that are likely to exist somewhere within the top crust of Mars. The thermal conductivity of the soil samples ranges from 0.08-4 W/m/K, with an average value of about 1.66 W/m/K, while the basalt conductivity exhibits a greater spread: 0.09-5.40 W/m/K, with an average value of about 2.06 W/m/K. If the assumption that the outer crust of Mars contains a significant amount of volcanics is correct, then the thermal properties will be most strongly influenced by basalt. A column averaged thermal conductivity of about 2.0 + 0.1 W/m/K, with lower values likely near the surface, where the fine-grained materials may be present in abundance, is suggested in [2]. In contrast, previous estimates are often based on the thermal properties of terrestrial analogues. For example, in [3] it is assumed that the properties of the megaregolith are well represented by a hard-frozen limonitic soil (0.8 W/m/K), while in [4] it is assumed a purely basaltic composition (2.09-2.5 W/m/K).

Thermal diffusivity is also uncertain, being unknown the value of density and specific heat, but in different environments values of about 10^{-6} m^2/s have been assumed [5].

Like thermal conductivity and diffusivity, estimates of the Martian thermal heat fluxes varied over a wide range: 15 to 45 mW/m^2. It has been noted [6] that if Mars loose heat at the same rate per unit mass as the Earth, then it should have a mean global heat flux of about 31 mW/m^2. For comparison purposes, the current best estimates of the average heat flux for the other planetary bodies for which some data exist, are 82 mW/m^2 for the Earth and 16 mW/m^2 for the Moon. In the case of Mars, using a globally averaged value of 30 mW/m^2 and a thermal conductivity of 2 W/m/K it is found a thermal gradient of 15 K/km.

Measurements from the newest orbital mission Mars Odyssey appear to confirm the existence of, perhaps, large quantities of shallow subsurface ice in certain parts of the planet. The study uses measurements of the neutron flux emitted from Mars, in several different energy regimes, and spectra of gamma-ray emissions induced by neutron capture reactions to map the global distribution of near-surface hydrogen on the planet. Details on the thickness of ice-rich lower layers are limited by the ~1 m sensing depth of the neutron instruments and the total quantity of subsurface ice cannot be determined. However, if the modeling is correct, the inferred ice concentration implies an extremely porous, nearly ice-filled regolith (the layer of rocky debris and dust resulting from repeated meteoritic impacts) at high latitudes.

The present investigation is aimed to evaluate theoretically the thermal properties of Martian soil analogues. The theoretical prediction of the thermal conductivity of terrestrial soils, subjected to permafrost conditions, has been carried out in [7] with reference to a four-phase soil, where solid particle, ice, unfrozen water and air are present. The model is based on the physical assumption that the unit cell is made of a cubic solid particle surrounded by the other three phases, disposed in a prescribed shape according to the phenomena of adsorption and capillarity. The prediction of the thermal conductivity has been extended to non-spherical particles in [8], with special emphasis to soils with large values of porosity. The cubic cell model has been used to study other kinds of porous media [9]. The hypothesis of parallel isotherms has been used to solve the one dimensional conduction equation in [7-9], while the parallel heat flux lines hypothesis have been employed in [10]. Different models have been proposed in [11-12] with approaches which take into account two-phases porous media. Radiation, coupled to conduction, has been introduced in the model in [13], while, the importance of mass transfer has been studied in [14-15]. The application of the cubic cell model to extraterrestrial conditions, taking into account thermal radiation, has been done in [16-18]. Comparisons with the data of the experiments carried out in the same laboratory have been done in [19-24]. The predictions of the cubic cell model have been compared to those of other models proposed in the literature, in [25-28]. The investigation on Mars soil analogues has been supported by ESA [29].

In conclusion, the cubic cell model can be considered as a reliable theoretical model, capable of investigating the thermal conductivity of soils in extraterrestrial conditions, like those present in other planets or comets, where the physical conditions can be different from the Earth, as far as pressure, temperature, and fluids in the atmosphere are concerned. The theoretical model can provide a relationship between the structure of the soil and the materials present (in solid or vapor phases). For example, the soil in contact with the atmosphere, in the hottest period of the day, has the lowest value of thermal conductivity because the fluid is present only in gas phase. The presence of frozen fluids within the soil increases its thermal conductivity as compared to the value on the surface. The theoretical models can assist in the prediction of higher values of thermal conductivity and thermal diffusivity, and, finally, in the temperature variation along its depth, as well as in the prediction of higher values of the thermal conductivity due to the presence of solid materials, e.g. ice layers. The variation of the soil temperature at the atmospheric level is assumed in order to reproduce the boundary condition of the soil as a function of time (day and night variation).

I. Characterization of Martian Soils

1. Martian Soil Compositions and Stratifications

Information on mineralogy and chemistry of sediments and soil of Mars (surface mineralogy) can be obtained from the data of Landers and remote sensing. The Viking Landers (Lander 1 "Chryse Planitia "-lat.22N long.48W-, Lander 2 "Utopia Planitia" -lat.46N long.150E, Figure 1) have analyzed the soil composition. The rocks have not been compositionally analyzed, whereas the Pathfinder Landers (area in the mouth of Ares Vallis -lat.19.33N 33.55W, Figure 1) have analyzed both rocks and soil. The alpha proton x-ray spectrometer (APXS), on board the rover of the Mars Pathfinder mission, have measured the chemical composition of six soils and five rocks in the Ares Vallis landing sites.

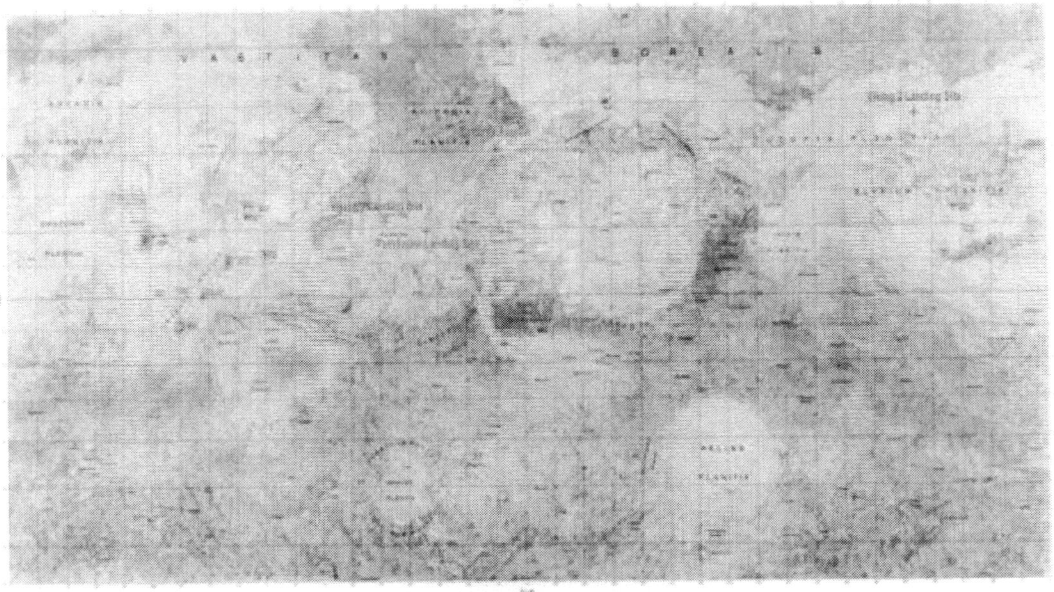

Figure 1. Map of Mars and landing sites of Viking and Pathfinder mission (http://mars.jpl.nasa.gov/mgs/target/marsmap1b.jpg) [30].

The composition of the soils analyzed has been interpreted to represent the weathering products of mafic igneous rocks (i.e., rocks composed of minerals rich in iron and magnesium, dense, and typically dark in color. The term comes from the words magnesium and ferric. Common mafic minerals are olivine and pyroxene. Basalt is a mafic igneous rock). Chemical analyses returned by Mars Pathfinder indicate that some rocks may be high in silica, implying differentiated parent materials. Moreover the roundness of pebbles and cobbles and the presence in the area of a possible conglomerate suggest fluvial processes that imply liquid water in equilibrium with the atmosphere and thus a warmer and wetter past.

On this basis, the attention is focused on two examples of Mars soils. These two sites are characterized by mafic igneous rocks representing a large fraction of Mars's surface which, from a sediment point of view, are associated with sediments or plains (i.e. extensive flows of lava). Then they can be considered as end-members as regard as porosity.

2. Site 1

Sediments and Lava Flows: Area in the Mouth of Ares Vallis (lat 19.33 N; Long 33.55 W).

From the Geological Map of Mars, [31], the surface material is referred to as Anch, that is, channel material that occurs on floors of sinuous to linear depressions resembling terrestrial stream beds. These channel materials have been interpreted in [31] as fluvial and Aeolian deposits. Compositional information in this area from Mars Pathfinder Landers indicates a mixture of weathered local rocks of mafic composition (Figure 2). The chemical composition of the five measured Pathfinder soils A-2, A-4, A-5, A-10, and A-15 has been normalized to an SiO_2 content of 44% by weight, and to the Viking soil data. The box for the Viking data reflects the range of the individual element concentrations, which are normalized to an SiO_2 content of 44% after being recalculated to 100% by weight [32]. The quartz content in these soils is about 44 %, following the indication given in [32].

The variety of rocks in the landing sites of Pathfinder is shown in Figure 3. The red arrows point rounded boulders and rocks. The blue arrows indicate rocks with sharp edges and points, most probably ejected from nearby impact craters and/or ancient volcanic activity. The white areas (of which Scooby Doo is one) are believed to be deposits left behind by evaporating water, or aggregates of materials fused together by the action of water, [33].

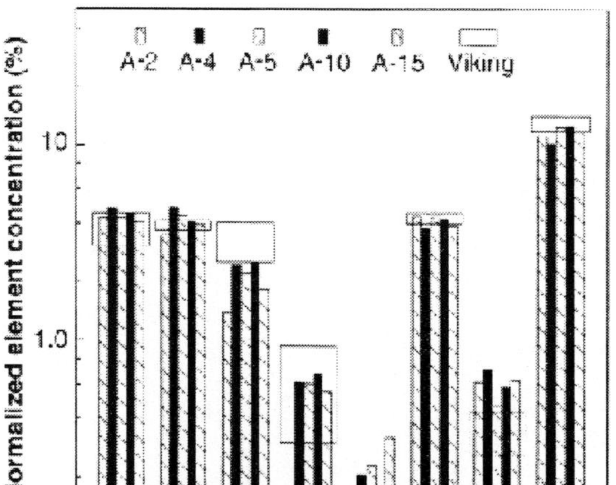

Figure 2.Chemical composition of the five measured Pathfinder soils A-2, A-4, A-5, A-10, A-15, [32].

The porosity can be obtained by similarity with the Northern plains, near the Alba Patera volcano [34]. Even if the genesis of these two soils is quite different they are both sediments. In the geological map of [31] the formation of Northern plains is indicated as Aps, i.e. Aeolian and volcanic deposits, tents to hundreds of meters thick. If the soil is composed of sediments only, porosity would be very high and can be assumed as about 50% at the surface.

Figure 3. View to the northeast of the Lander, (http://mars.jpl.nasa.gov/MPF/ops/NEcolor_annot.jpg), [33].

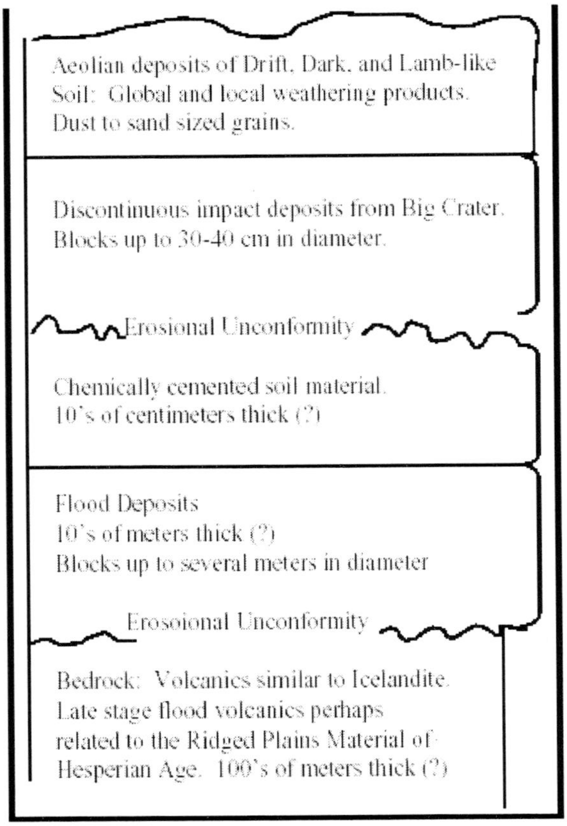

Figure 4. Preliminary Stratification column for the Pathfinder Landing Site [36].

The preliminary reconstruction of the stratification in the Pathfinder area is shown in Figure 4, [36]. The local bedrock is represented by samples brought up by the "Big Crater" impact event in rocks like Barnacle Bill and Shark. The contact of this unit is probably an erosion unconformity (i.e. a surface that separates younger strata from eroded, dipping, older strata and represents a gap in the geologic record), caused by catastrophic floods.

Evidence of local erosion is seen in the structure of the Twin Peaks and regional erosion is evident in Viking Orbiter imagery of nearby streamlined islands. The flood(s) deposited a layer of sedimentary debris including the boulders seen in the site. The thickness of this layer is not known but may be at least 10's meters since only the hills like Twin Peaks appear to remain unburied. These boulders are likely derived from the nearby upstream streamlined islands.

From all the former information, the hypothetical stratification column down to 5 m for the Ares Vallis site, is the following, reported in Figure 5.

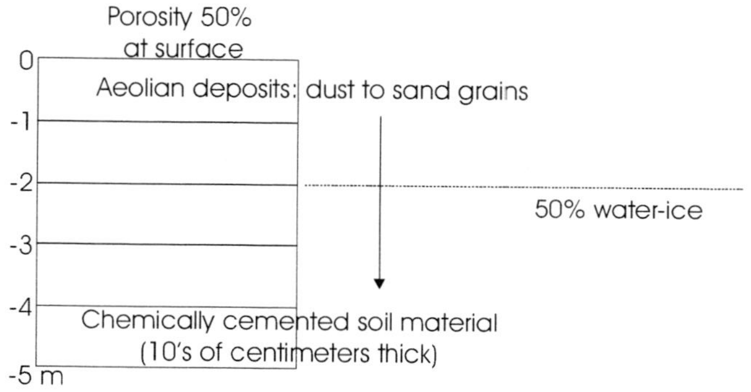

Figure 5. Hypothetical stratification column down to 5 m for the Ares Vallis site

In the first two meters of Figure 5 is possible to assume the presence of a variable quantity of ice. Theoretical models of the atmospheric water vapour deposition within the porous regolith indicate that the first ten's meters of the Martian crust could contain a seasonally variable quantity of ice. This data is confirmed in this area by the presence of the white areas already discussed.

3. Site 2

Hesperian Volcanic Plains: Lunae Planum(lat 10-20 N, long 70-60 W), Sinai Planum, Solis Planum, Syrtis Major Planitia and Hesperia Planitia

From the Geological Map of Mars [31] the surface material is referred to as Hprg, i.e. basaltic lava flow or andesitic dust-free rocks. Information on the mineral composition of these rocks could come from the Alpha Proton X-Ray Spectrometer on the Pathfinder rover that measured the compositions of nine rocks. The silicon content of some rocks is much higher than the Martian meteorites, our only other samples of Mars. The rocks analyzed by Pathfinder, however, are not basalts. If they are volcanic - as suggested by their pitted surface texture, presumably formed when gases trapped during cooling left small holes in the rock -

their silicon content classifies them as andesites (Figure 6). The SiO_2 content is about 50% for the basaltic rocks and 60 % for the andesitic rocks (Figure 6).

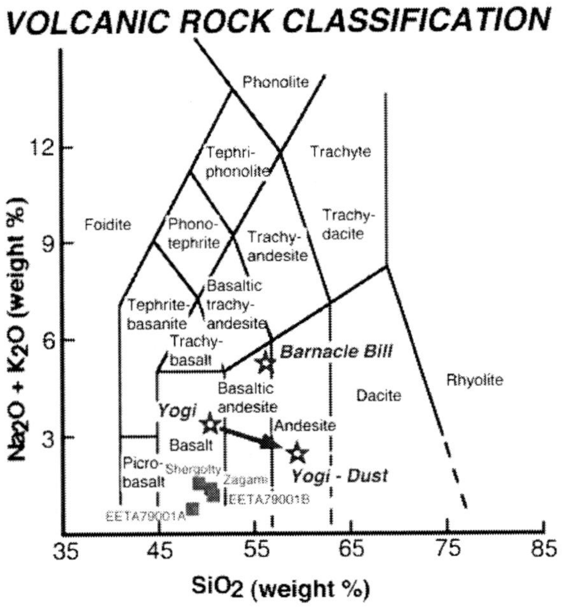

Figure 6. Volcanic rock classification, (http://mars.jpl.nasa.gov/MPF/ops/hap_5.jpg) [37].

The commonly used chemical classification for lavas, reported in Figure 6, shows that Barnacle Bill and Yogi (corrected for adhering dust) are distinct from basaltic Martian meteorites. The Pathfinder APXS analyses have been corrected for the presence of a small amount of salt, and sulfur is assumed to be present as sulfide. These rocks plot in or near the field of andesites, a type of lava common at continental margins on the Earth. The preliminary data for alkalis are likely to represent upper limits, so refinement of these analyses could shift them to slightly lower $Na_2O + K_2O$ and higher SiO_2 [37].

In this case it is possible to assume a surface porosity of 20 % for lavas [35]. Lavas will overlain porous and possibly ice-rich sediments or mega-regolith.

From all the former information, the hypothetical stratification column down to 5 m for the site 2, is the following, reported in Figure 7.

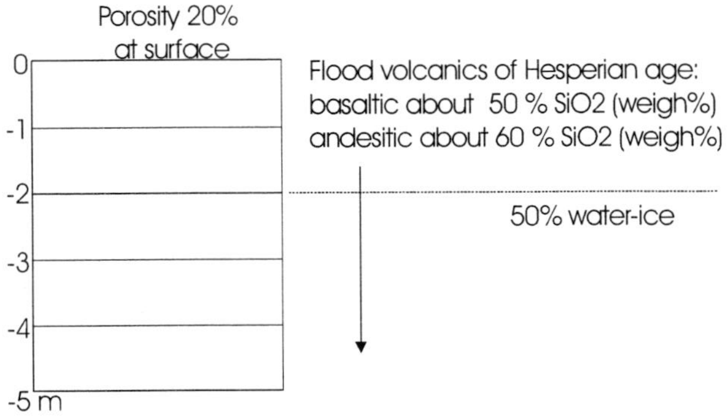

Figure 7. Hypothetical stratigraphic column down to 5m for the Hesperian volcanic plains.

II. Thermal Properties of Martian Soil Analogues

1. Theoretical Prediction of Thermal Conductivity and Diffusivity

1.1. Two Phase Martian Soil Analogue (Solid and Gas)

The theoretical model considers the dry soil analogue (two phase medium) as composed of a cubic cell, with a solid cubic particle at the centre and gas around it, as sketched in Figure 8a.

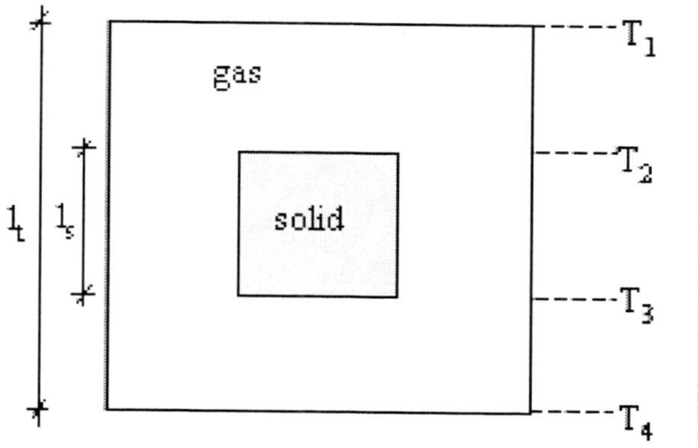

Figure 8a. Cubic cell model for dry soil analogue [38].

The thermal conductivity of the soil is evaluated from the solution of the one-dimensional heat conduction equation, under the assumption of parallel isotherms [7]. The effective thermal conductivity of the dry soil analogue, two-phase medium, is given by:

$$\frac{1}{k_{eff}}=\frac{\beta-1}{k_{atm}\cdot\beta}+\frac{\beta}{k_{atm}\cdot\left(\beta^2-1\right)+k_s} \tag{1}$$

where: k_{eff} is the effective thermal conductivity of the soil analogue, k_{atm} the thermal conductivity of the atmospheric gas, k_s the thermal conductivity of the solid particle. The parameter β is given by:

$$\beta=\frac{l_t}{l_s}=\sqrt[3]{\frac{\rho_s}{\rho_d}}=\sqrt[3]{\frac{1}{1-p}} \tag{2}$$

where p is the porosity, ρ_s the density of the solid particle, ρ_d the dry density of the soil, and l_t, l_s are the dimensions indicated in Figure 8a.

The first term in Eq. 1 corresponds to the thermal resistance of the gas in the cross section l_t^2, of length $(l_t - l_s)$, while the second term is the thermal resistance of the materials gas and solid in the cross section l_t^2, of length l_s.

1.2. Three phase Martian Soil Analogues (Solid, Gas and Ice)

If the Martian soil analogue is not dry and the volume of ice content, V_i, is low, as compared to the volume of the void, V_v, the ice is assumed adsorbed around the solid particle, as represented in Figure 8b. In terrestrial soils the amount of liquid water adsorbed around the solid particle is a fraction of the water content at the permanent wilting point [15]. The value of this fraction can been found on the base of the experimental data on several terrestrial soils. Because of the absence of reliable data on Martian soil analogue, the present paper assumes the ratio of adsorbed ice to void volume, or ice volume content equal to $f_a = V_i / V_v = 0.083$, as done in [7]. It must be noted that this is the only empirical constant of the theoretical model that can be removed with better knowledge of the Martian regolith The upper limit of f is 1 when ice occupies all voids and the soil is saturated with ice.

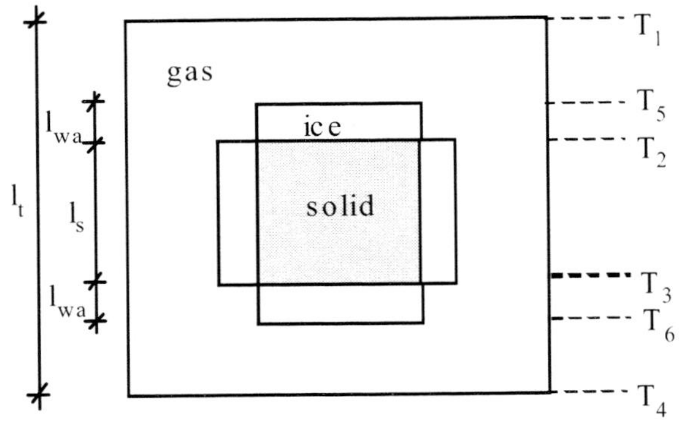

Figure 8b. Cubic cell model for a soil analogue with adsorbed ice [38].

The effective thermal conductivity is then given by:

$$\frac{1}{k_{eff}} = \frac{\beta - 1 - \delta/3}{\beta \cdot k_{atm}} + \frac{\beta \cdot \delta}{3\left[k_{atm}\left(\beta^2 - 1\right) + k_i\right]} + \frac{\beta}{k_s + \frac{2}{3} \cdot \delta \cdot k_i + k_{atm}\left(\beta^2 - 1 - \frac{2}{3} \cdot \delta\right)} \tag{3}$$

where k_i is the thermal conductivity of ice and the parameter δ is given by

$$\delta = \frac{Vi}{1-p} \tag{4}$$

with V_i the volume of ice.

The first term of Eq. (3) is the thermal resistance of the gas in the cross section l_t^2, of length $(l_t - l_s - 2\, l_{wa})$. The second term is the thermal resistance of the materials ice and gas in the cross section l_t^2, of length $(2\, l_{wa})$. The third term is the thermal resistance of the materials ice, gas and solid, in the cross section l_t^2, of length l_s.

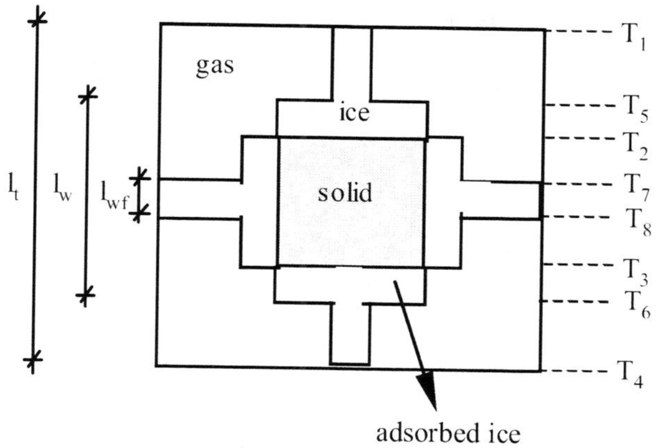

Figure 8c. Cubic cell model for a soil analogue with ice adsorbed around the solid particle and disposed among adjacent particles, for $\gamma_f < 1$ [38].

If the volume content of ice is higher than the adsorbed value, $f_a = 0.083$, but the amount of ice, present among the adjacent solid particles, is enough low, such that $\gamma_f < 1$, which is defined in the following Eq. (5), the distribution of ice and gas in the soil analogue is represented in Figure 8c.

The following parameters are then defined:

$$\gamma = \sqrt[3]{\frac{V_i}{V_s} - \frac{V_{if}}{V_s} + 1}, \quad \gamma_f = \sqrt{\frac{V_{if}/V_s}{3 \cdot (\beta - \gamma)}}. \tag{5}$$

where V_i is the ice volume, V_s the volume of the solid particle and V_{if} the ice volume accumulated among adjacent solid particles and called funicular, according to [39]. In order to have a simple model, V_{if} / V_s is assumed linearly variable with the porosity of the soil analogue, between 0.183, for $p = 0.4764$, and 0.226, for $p = 0.2595$. The resulting expression is then:

$$\frac{V_{if}}{V_s} = \frac{V_{if}}{V_v} \cdot \left(\beta^3 - 1\right) = \left[0.183 + \frac{0.226 - 0.183}{0.4764 - 0.2595}(0.4764 - p)\right] \cdot \left(\beta^3 - 1\right) \tag{6}$$

According to the configuration of Figure 8c, for $\gamma_f < 1$, k_{eff} is given by:

$$\frac{1}{k_{eff}} = \frac{\beta^2 - \beta \cdot \gamma}{k_{am} \cdot \left(\beta^2 - \gamma_f^2\right) + k_i \cdot \gamma_f^2} + \frac{\beta \cdot \gamma - \beta}{k_{am} \cdot \left(\beta^2 - \gamma^2\right) + k_i \cdot \gamma^2} + \frac{\beta - \beta \cdot \gamma_f}{k_{am} \cdot \left(\beta^2 - \gamma^2\right) + k_i \cdot \left(\gamma^2 - 1\right) + k_s} + \frac{\beta \cdot \gamma_f}{k_s + k_i \cdot \left(\gamma^2 - 1 + 2 \cdot \beta \cdot \gamma_f - 2 \cdot \gamma \cdot \gamma_f\right) + A} \tag{7}$$

where

$$A = k_{atm}\left(\beta^2 - \gamma^2 - 2\cdot\beta\cdot\gamma_f + 2\cdot\gamma\cdot\gamma_f\right) \tag{8}$$

The first term of Equation (7) is the thermal resistance of the materials gas and ice in the section $(l_t^{\,2})$, of length $(l_t - l_w)$. The second term is the thermal resistance of the materials gas and ice in the section $(l_t^{\,2})$, of length $(l_w - l_s)$. The third term is the thermal resistance of the materials gas, ice and solid in the section $(l_t^{\,2})$, of length $(l_s - l_{wf})$. The fourth term is the thermal resistance of the materials gas, ice and solid in the section $(l_t^{\,2})$, of length (l_{wf}).

The distribution of ice and gas in the soil analogue for a large volume of ice, such that $\gamma_f > 1$, is presented in Figure 8d.

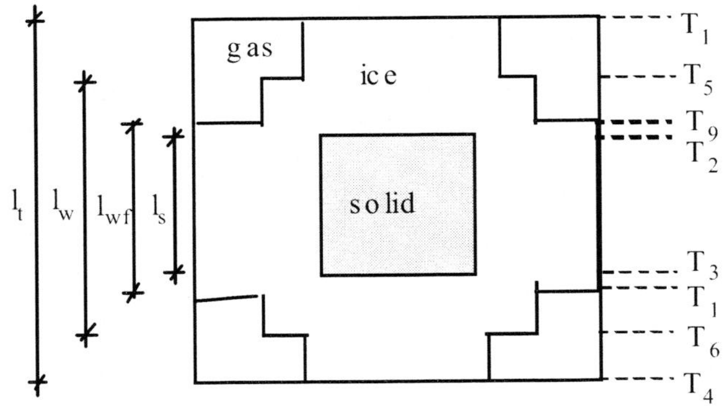

Figure 8d. Cubic cell model for a soil analogue with ice disposed among adjacent particles, for $\gamma_f > 1$ [38].

The effective thermal conductivity, k_{eff}, is then given by:

$$\frac{1}{k_{eff}} = \frac{\beta^2 - \beta\cdot\gamma}{k_{atm}\cdot\left(\beta^2 - \gamma_f^2\right) + k_i\cdot\gamma_f^2} + \frac{\beta\cdot\gamma - \beta\cdot\gamma_f}{k_{atm}\cdot\left(\beta^2 - \gamma^2\right) + k_i\cdot\gamma^2} + \frac{\beta\cdot\gamma_f - \beta}{A + k_i\cdot\left(\gamma^2 + 2\cdot\beta\cdot\gamma_f - 2\cdot\gamma\cdot\gamma_f\right)} + \frac{\beta}{k_s + k_i\cdot\left(\gamma^2 - 1 + 2\cdot\beta\cdot\gamma_f - 2\cdot\gamma\cdot\gamma_f\right) + A} \tag{9}$$

where A is given by Eq. (8).

The first term of Eq. (9) is the thermal resistance of the materials gas and ice in the section $(l_t^{\,2})$, of length $(l_t - l_w)$. The second term is the thermal resistance of the materials gas and ice in the section $(l_t^{\,2})$, of length $(l_w - l_{wf})$. The third term is the thermal resistance of the materials gas and ice in the section (l_t^{2}), of length $(l_{wf} - l_s)$. The fourth term is the thermal resistance of the materials gas, ice and solid in the section $(l_t^{\,2})$, of length (l_s)

2. Martian Analogues

The Martian soil analogue investigated in the present work is the olivine tested in the laboratory [21]. The thermal conductivity of the solid particle has been evaluated experimentally resulting equal to k_s= 2.94 W/m/K [21]. The evaluation of the thermal conductivity of the solid particle is based on two experimental measurements, i.e. in dry and water saturated conditions. The two experimental values are then used in Eq. (1) as k_{eff} to give two equations which are then solved to calculate the values of β and k_s .

The Martian atmospheric pressure is about 6 mbar and the atmospheric gas is composed of CO_2 (95%), N_2 (3%), Ar (1.5%) and traces of water vapour. The thermal conductivity of the Martian atmospheric gas is evaluated as 92.8% of the value at 1 bar and as a weighed average. Table 1 reports the values at several temperatures.

Table 1. Thermal conductivity of the Martian atmospheric gas, k_{atm} [38].

T (K)	*10^3(W/m/K)*
148	6.86
173	8.39
223	11.87
273	14.49
298	16.01

A relation between ice volume content, $f = V_i / V_v$, ice mass content, a = M_i / M_t and soil porosity, p, can be found, where M_i is the mass of ice in the porous medium, M_t the total mass of the porous medium, V_i the ice volume in the porous medium, and V_v the void volume of the cubic cell. The ice mass content, a, is given by:

$$a = \frac{1}{1 + \frac{\rho_{atm}}{\rho_{ice}} \cdot \frac{1-f}{f} + \frac{\rho_S \cdot (1-p)}{\rho_{ice} \cdot f \cdot p}} \tag{10}$$

For a given soil porosity, p, the ice mass content a = M_i / M_t, corresponding to saturated frozen soil, is reported in Table 2.

Table 2. Porosity and ice mass content, a = M_i / M_t, at T=223 K, for saturated soil [38].

p	**a**
0.20	0.074
0.30	0.12
0.40	0.17
0.50	0.24

3. Numerical Evaluation of Two Phase Martian Soil Analogue

Figure 9 presents the thermal conductivity of a dry Martian soil analogue in the porosity range p = 0 – 1 at several temperatures. The thermal conductivity increases with the temperature and decreases with the increase of porosity. At p = 0 the thermal conductivity is equal to the thermal conductivity of the solid particle (2.94 W/m/K) and at p = 1 is equal to the thermal conductivity of the Martian atmospheric gas.

Figure 10 presents the thermal diffusivity of a dry Martian soil analogue, in the porosity range p = 0 - 1 and at several temperatures, evaluated assuming the thermal conductivity given by Figure 9 and the value of $(\rho{\cdot}c_p)_d$ measured experimentally in [21].

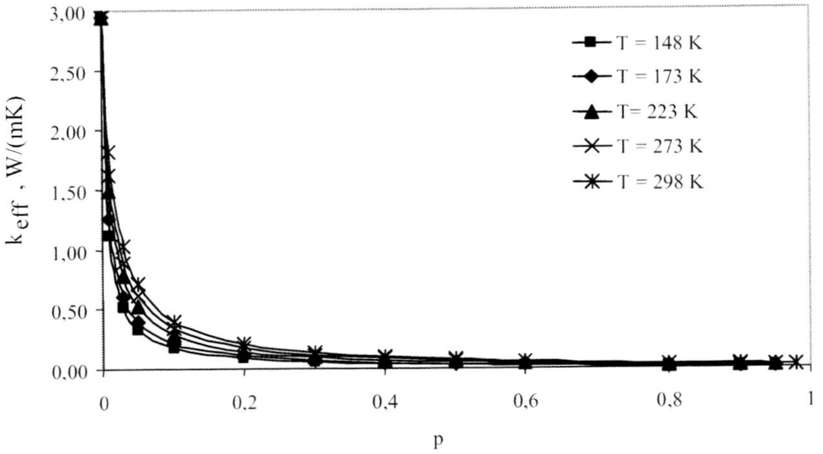

Figure 9. Thermal conductivity of dry Martian soil analogue versus porosity [29].

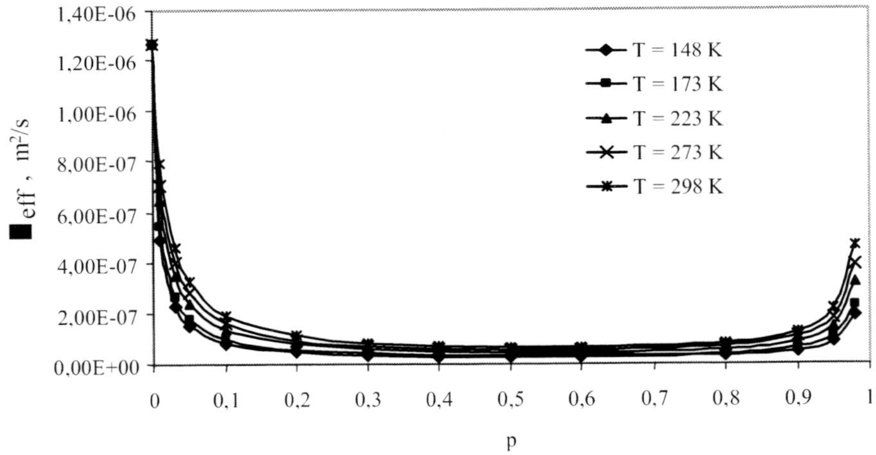

Figure 10. Thermal diffusivity of dry Martian soil analogue versus porosity [29].

4. Numerical Evaluation of Three Phase Partially Frozen Martian Soil Analogue

Figure 11 presents the thermal conductivity of a partially frozen Martian soil analogue versus ice volume content at several porosities and T = 223 K. The abrupt increase in the theoretical effective thermal conductivity is due to the step change from the adsorbed ice configuration of Figure 8b to the configuration of Figure 8c, where ice is disposed among adjacent soil particles. The experimental step increase is smooth as confirmed by unpublished results.

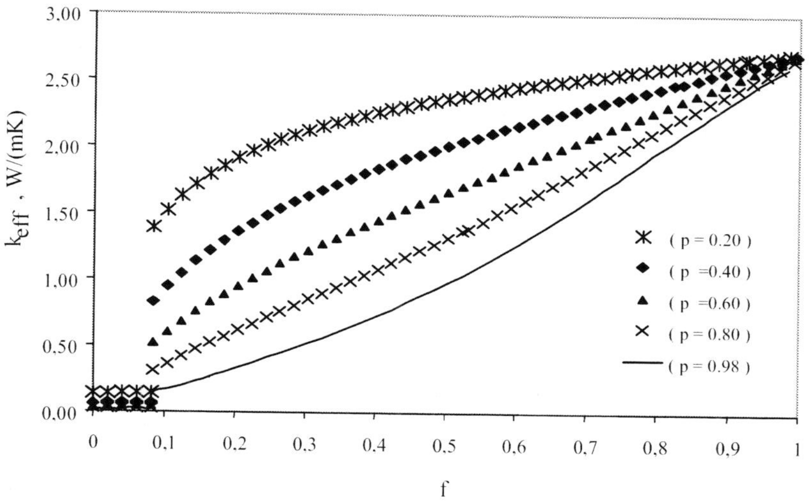

Figure 11. Thermal conductivity versus ice volume content at 223 K.

Figure 12 presents the thermal conductivity of a partially frozen Martian soil analogue versus the ice mass content, a, at several porosities and $T = 223$ K. For a porosity equal to 0.2, the thermal conductivity spans from the minimum value in dry condition, 0.144 W/m/K, to the maximum one in saturated conditions, 2.7 W/m/K. For a porosity equal to 0.5, the thermal conductivity spans from the minimum value in dry conditions, 0.05 W/m/K, to the maximum one, in saturated conditions, 2.7 W/m/K. Figure 12 reports also the lines with the same ice volume content *f*. The thermal conductivity, at the same *f*, decreases with the increase of the ice mass content, because the thermal conductivity of the solid particle is higher than the ice value.

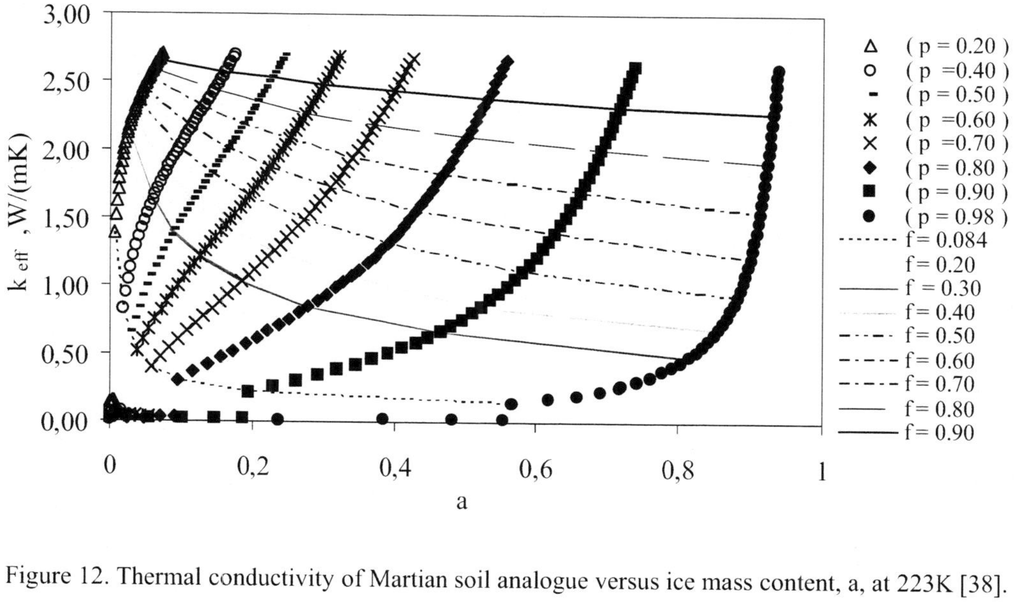

Figure 12. Thermal conductivity of Martian soil analogue versus ice mass content, a, at 223K [38].

Figure 13 presents the thermal conductivity of the Martian soil analogue versus porosity at several ice mass contents, a, and 223 K. In dry condition (a = 0) the thermal conductivity decreases from the value of the solid particle (2.94 W/m/K), p=0, to the atmospheric gas (0.012 W/m/K), p=1. At each ice mass content, $a = M_i/M_t$, the thermal conductivity decreases from the maximum value, predicted at low porosities, when the cubic cell is full of ice, to the minimum ones, at high porosity, when the void space inside the cell is occupied by the atmospheric gas. It can be noted that the maximum thermal conductivity is around 2.7 W/m/K, almost independently of the ice mass content.

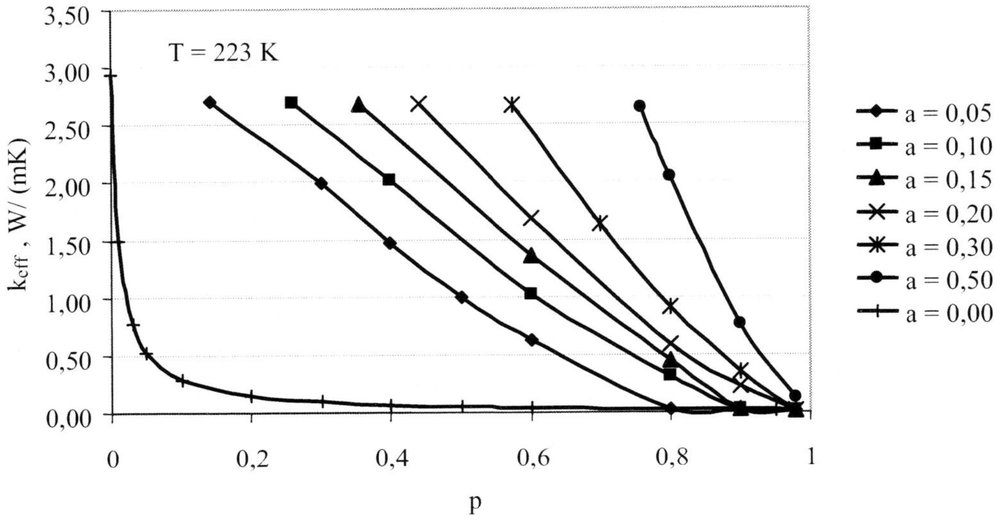

Figure 13. Thermal conductivity of Martian soil analogue versus porosity, at 223 K [38].

Figure 14 present the thermal conductivity of the Martian soil analogue versus porosity at several ice mass contents and 273 K.

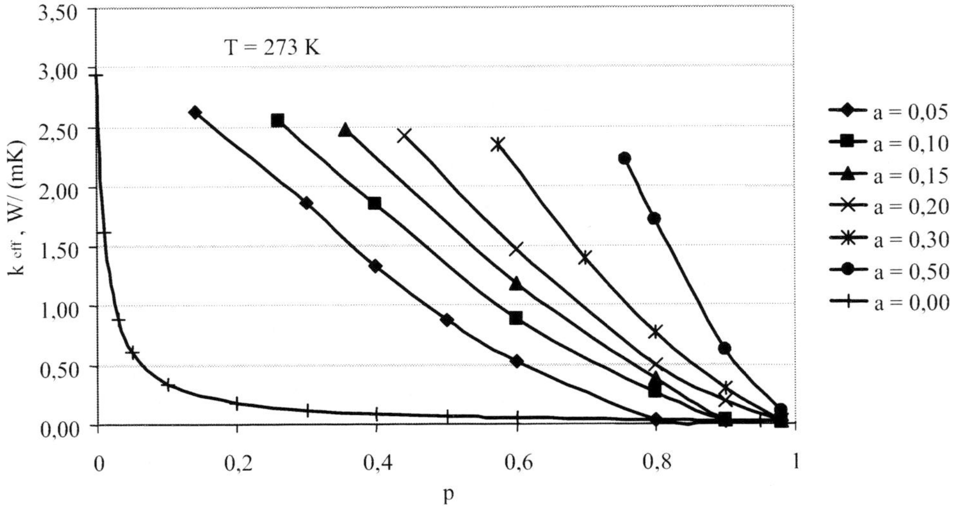

Figure 14. Thermal conductivity of Martian soil analogue versus porosity, at 273 K, [38].

The heat capacity of the Martian soil analogue is evaluated with the following Eq.11, and the values are reported in Figure 15. The heat capacity increases, for a fixed porosity, with the ice volume content, with a higher increase at higher porosities.

$$(\rho \cdot c_p)_{eff} = (\rho \cdot c_p)_s \cdot (1-p) + (\rho \cdot c_p)_{ice} \cdot f \cdot p + (\rho \cdot c_p)_{atm} \cdot (1-f) \cdot p \tag{11}$$

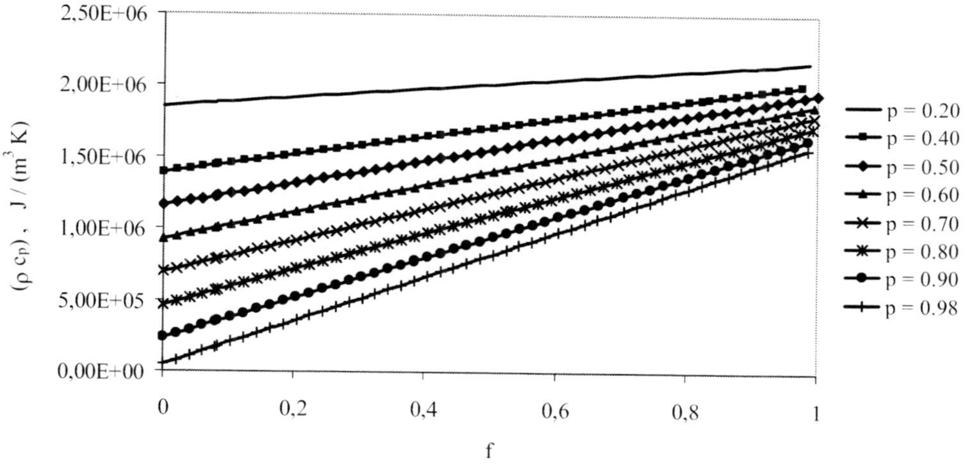

Figure 15. Heat capacity of Martian soil analogue versus ice volume content, a, at 223 K [38].

The thermal diffusivity of the Martian soil analogue is evaluated by the ratio of k_{eff} and ρ c_p. Figure 16 presents the thermal diffusivity versus the volume ice content, f, at several porosities and 223 K. The theoretical thermal diffusivity has a step increase from the configuration of ice adsorbed around the solid particle, Figure 8b, for $f < 0.083$, to the configuration of ice disposed among the solid particles, for $f > 0.083$, Figure 8c, because of the ice bridges present among the solid particles. The thermal diffusivity increases with f with a higher increase at higher porosities.

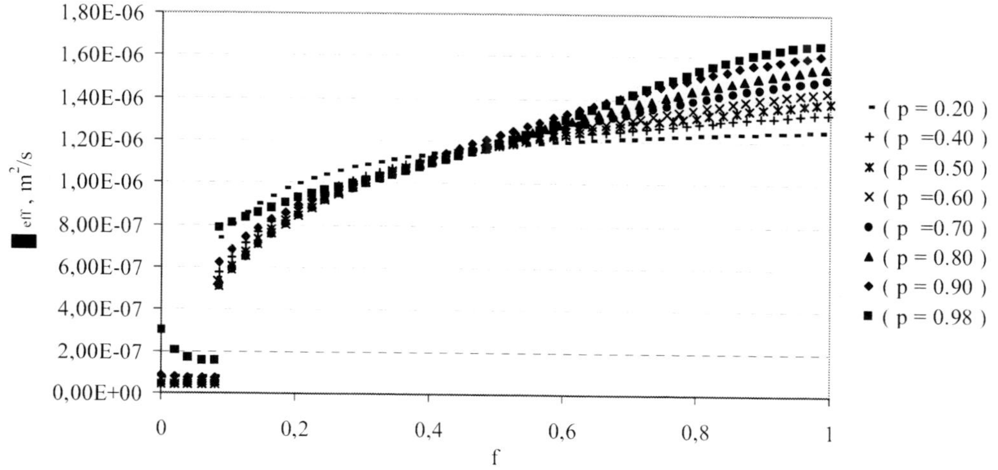

Figure 16. Thermal diffusivity of Martian soil analogue versus ice volume content, f, at 223 K [38].

Figures 17-19 report the thermal diffusivity of the Martian soil analogue, versus porosity, at several ice mass contents, a, and three temperatures, i.e. 148 K, 173 K and 223 K. The results show that the thermal diffusivity of the Martian soil analogue changes of more than an order of magnitude from dry to partially frozen conditions.

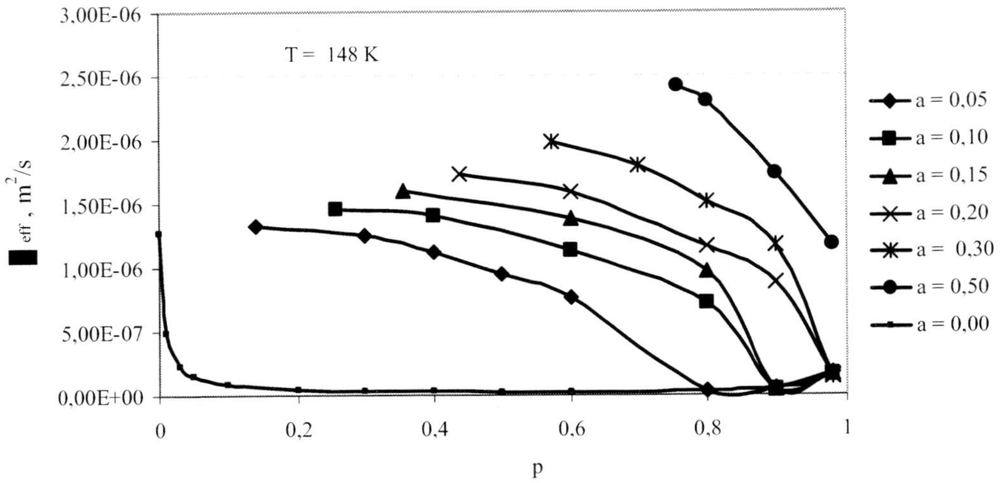

Figure 17. Thermal diffusivity of Martian soil analogue versus porosity, p, at 148 K [38].

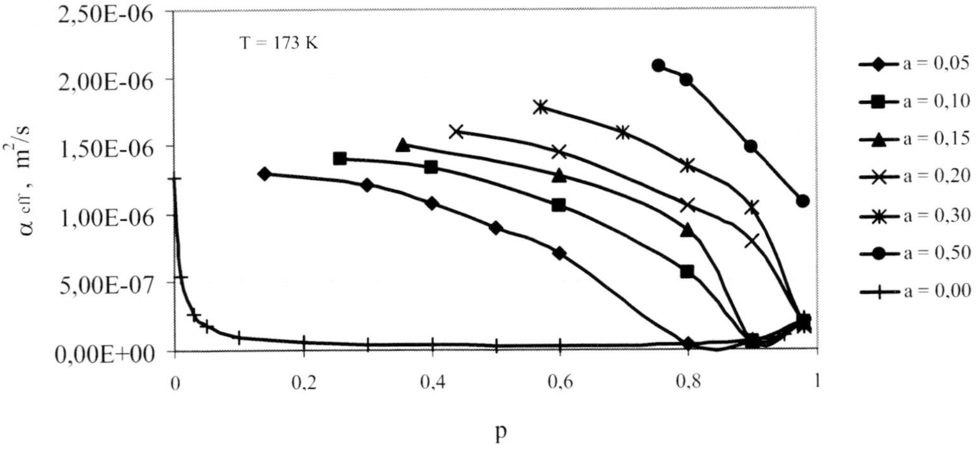

Figure 18. Thermal diffusivity of Martian soil analogue versus porosity, p, at 173 K [29].

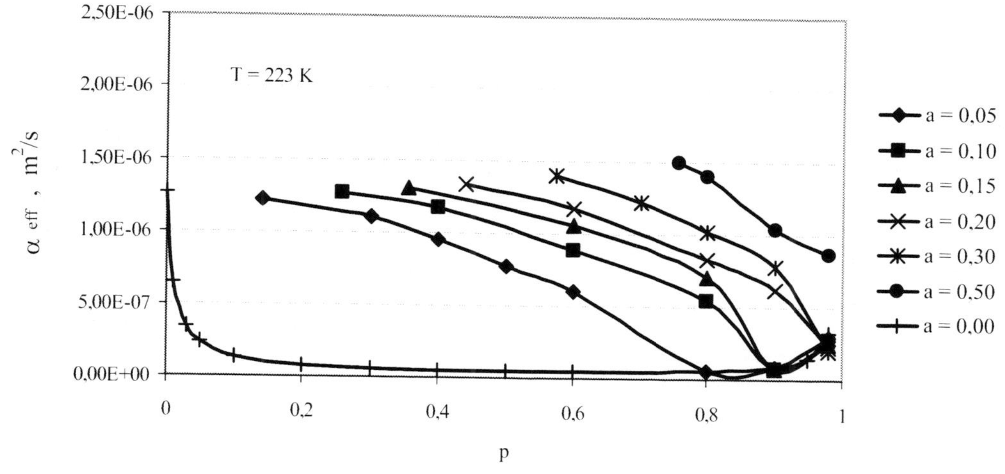

Figure 19. Thermal diffusivity of Martian soil analogue versus porosity, p, at 223 K [29].

Figure 20 presents the thermal diffusivity versus the ice mass contents, a, along to the lines of constant ice volume content, *f*.

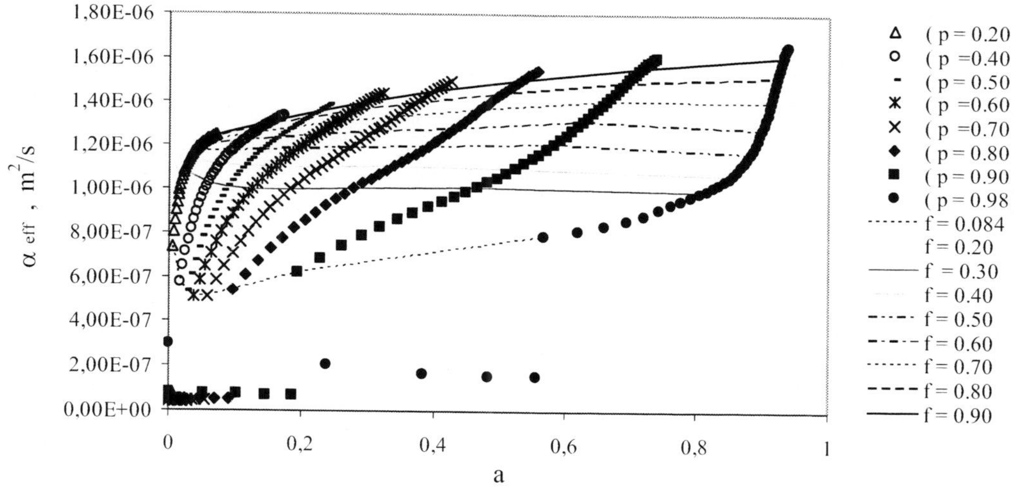

Figure 20. Thermal diffusivity of Martian soil analogue versus ice mass content, a, at 223 K [29].

III. Temperature Variation in Martian Soil Analogues

1. Temperature Variation

The differential equation of heat conduction in the soil

$$\frac{\partial}{\partial z}\left(k\frac{\partial \vartheta}{\partial z}\right) = C \cdot \frac{\partial \vartheta}{\partial t} \tag{12}$$

has the solution

$$\vartheta(z,t)=\vartheta_a+\vartheta_s \exp\left(-\frac{z}{D}\right)\cdot\sin\left(\omega t-\frac{z}{D}+\varphi\right) \quad (13)$$

under the following boundary conditions:
- on the surface, z = 0, the temperature variation is

$$\vartheta(0,t)=\vartheta_a+\vartheta_s\cdot\sin(\omega t+\varphi) \quad (14)$$

- at an infinite depth the temperature variation is zero

$$\lim_{z\to\infty}\vartheta=\vartheta_a \quad (15)$$

where ϑ_a is the average temperature in the soil, ϑ_s the amplitude of the temperature variation on the surface, t the time, ω is the radial frequency

$$\omega=\frac{2\pi}{T}=\frac{2\pi}{86400}=7.27\cdot10^{-5}\ \sec^{-1} \quad (16)$$

The Martian day is assumed equal to 24 h and φ is the phase delay.
Further on:

$$D=\sqrt{\frac{2\cdot k}{C\ \cdot\omega}}, \quad (17)$$

where k is the thermal conductivity of the soil, C the heat capacity and $\alpha = k/C$ the thermal diffusivity.

1.1. Martian Soil Analogues, Type A

The first Martian soil analogue investigated, type A, has the porosity p = 0.20. The temperature variation on the surface is variable from the minimum value of 148 K to the maximum one of 298 K.

1.1.1. Dry Martian Soil Analogue, Type AD.

The dry soil analogue, AD, has a thermal conductivity equal to k = 0.144 W/m/K, as evaluated by the theoretical model at 223 K, Figure 13, and a thermal diffusivity equal $\alpha = 10^{-6}\ m^2/s$, as suggested by the literature review. Figures 21-22 show the daily temperature variations (24 hours) in the soil AD. Figure 21 reports the temperatures of the soil up to a depth of 0.60 m and Figure 22 below 0.60 m. At 0.30 m the total temperature variation, during the day, is about 25 K and at 0.60 m is 4 K, Figure 21. At 1 m the temperature variation is 0.4 K, while at 1.30 m is 0.06 K, Figure 22.

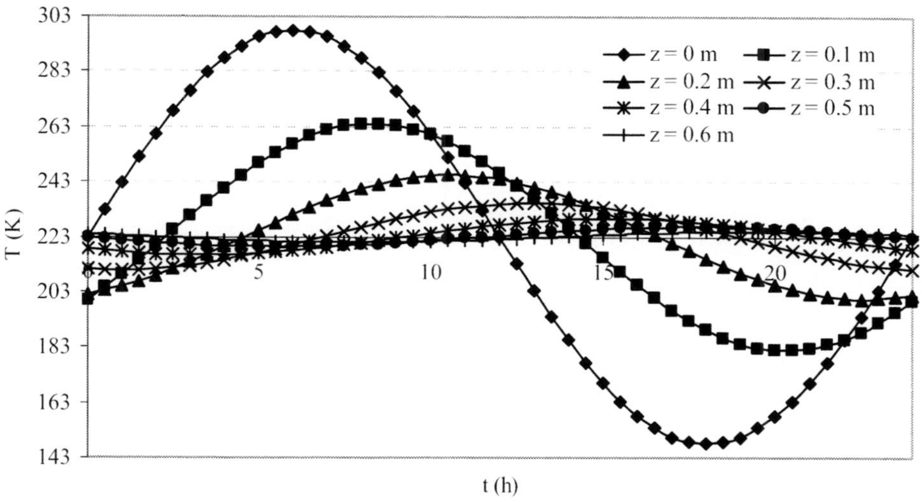

Figure 21. Temperature variations in the dry soil analogue, type AD, up to 0.60 m [29].

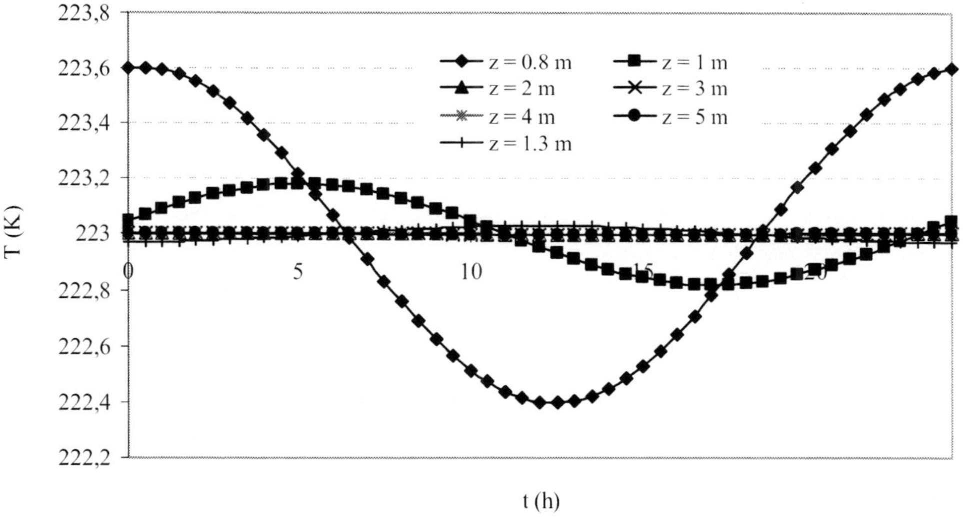

Figure 22. Temperature variations in the dry soil analogue, type AD, below 0.60 m [29].

The temperature distributions are presented versus the depth of the soil analogue, in Figure 23, at several hours during the day. Figure 23 shows that the temperature variations are detectable inside the dry soil analogue, type AD, up to a depth of about 1 m.

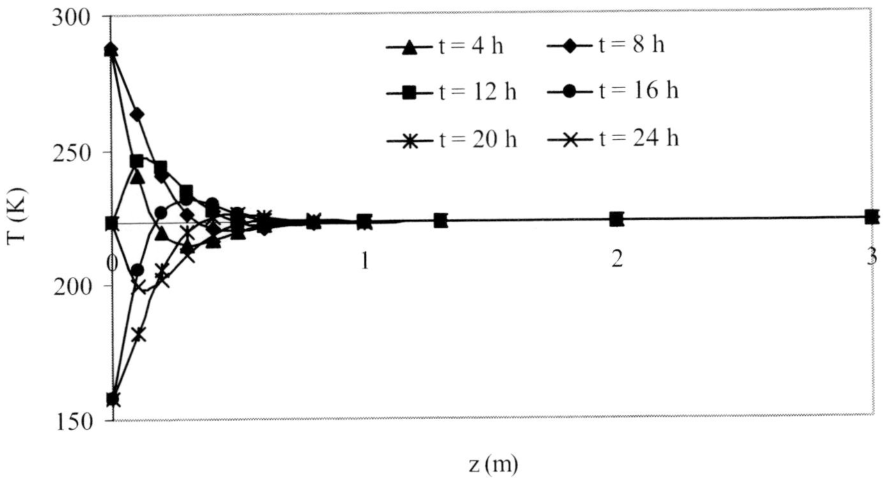

Figure 23. Temperature distribution in the dry soil analogue, type AD, versus depth, during the Martian day, [38].

1.1.2. Frozen Martian Soil Analogues, Type AF

The frozen Martian soil analogue, type AF, has the following characteristics: $p = 0.20$, $f = 1$, $a = M_{ice} / M_t = 0.0728$, thermal conductivity equal to $k = 2.70$ W/m/K according to the theoretical model at 223 K, Figure 13, and thermal diffusivity equal to $\alpha = 10^{-5}$ m^2/s, one order of magnitude higher than the dry soil, as suggested by the literature review. The temperature variation on the surface is variable from the minimum value of 148 K to the maximum one 298 K, with an average value of 223 K.

Figures 24-25 show the daily temperature variations in the soil analogue, type AF. Figure 24 reports the temperature of the soil up to a depth of 0.60 m. At 0.30 m, the temperature variation, during the day, is about 85 K and at 0.60 m is about 48 K, Figure 24. At 1 m the temperature variation is about 22 K, at 1.30 m about 13 K and at 2 m about 3.3 K, Figure 25.

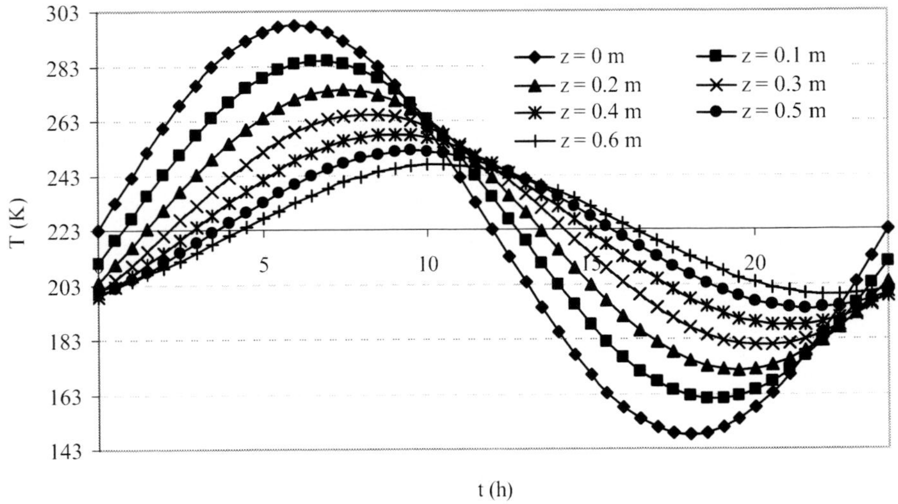

Figure 24. Temperature variations in the frozen soil analogue, type AF, up to 0.60 m [29].

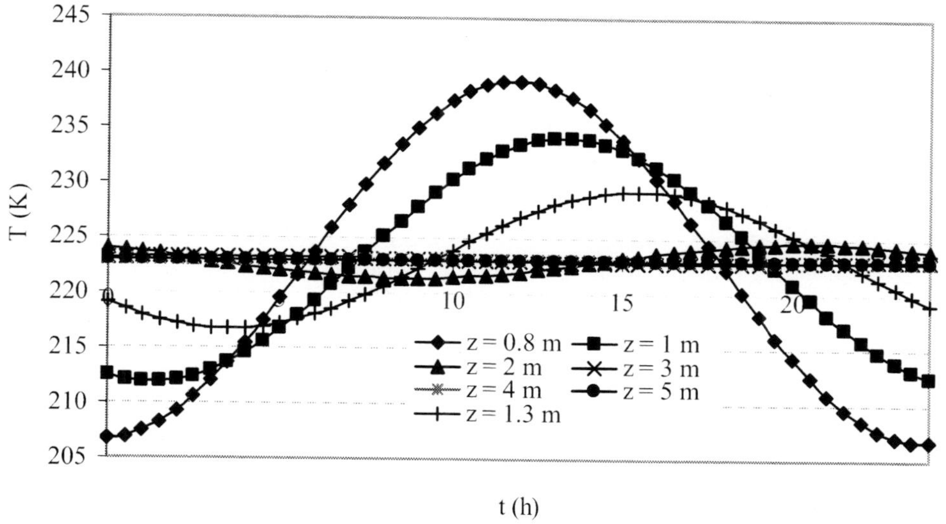

Figure 25. Temperature variations in the frozen soil analogue, type AF, below 0.60 m [29].

The temperature distributions are presented versus the depth of the soil analogue, type AF, in Figure 26, at several hours during the Martian day. Figure 26 shows that the temperature variations are detectable, inside the frozen soil analogue, up to a depth of about 3 m. In conclusion the temperature variation is more penetrating in the frozen soil, because the thermal diffusivity is one order of magnitude higher than in the dry soil.

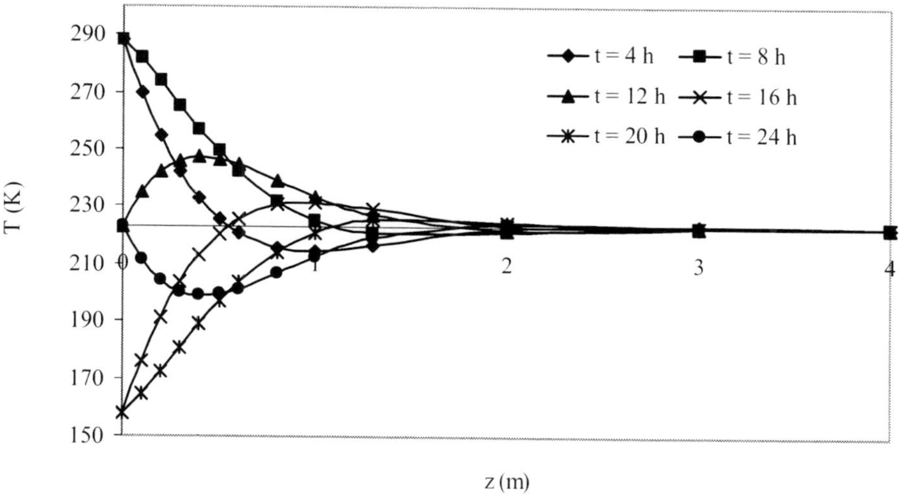

Figure 26. Temperature distribution in the frozen soil analogue, type AF, versus depth, during the Martian day [38].

1.2. Martian Soil Analogue, Type B

The second Martian soil analogue investigated, type B, has a porosity equal to $p = 0.50$. The temperature variation on the surface is variable from the minimum value of 148 K to the maximum one of 298 K.

1.2.1 Dry Martian Soil Analogue, Type BD

The dry soil analogue, type BD, has a thermal conductivity equal to k = 0.05 W/m/K, as evaluated by the theoretical model at 223 K, Figure 13, and a thermal diffusivity equal to $\alpha = 0.45 \cdot 10^{-7}$ m^2/s, as evaluated by the theoretical model, Figure 19. Figures 27-28 show the daily temperature variations (24 hours) in the soil BD. Figure 27 reports the temperatures of the soil up to a depth of 0.60 m and Figure 28 below 0.60 m. At 0.30 m the total temperature variation, during the day, is about 0.03 K and at 0.60 m is negligible, Figure 27. At 0.80 m and at 1 m the temperature variation is negligible, Figure 28.

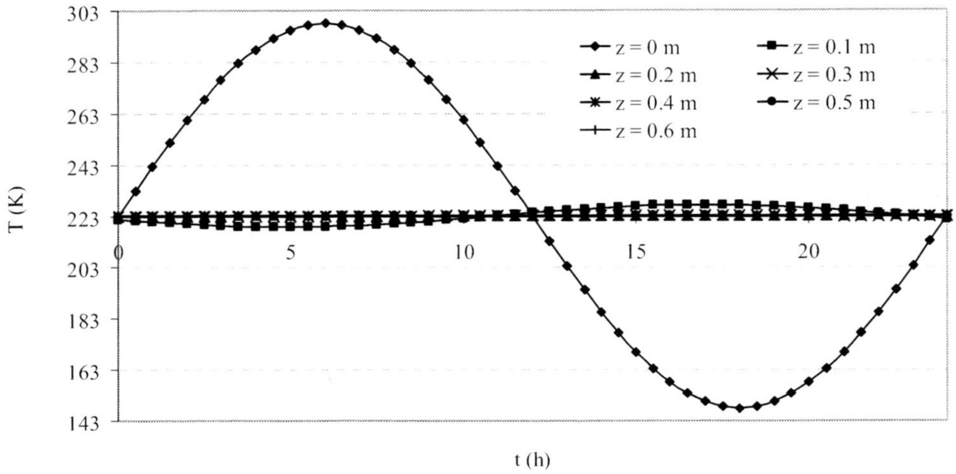

Figure 27. Temperature variations in the dry soil analogue, type BD, up to 0.60 m [29].

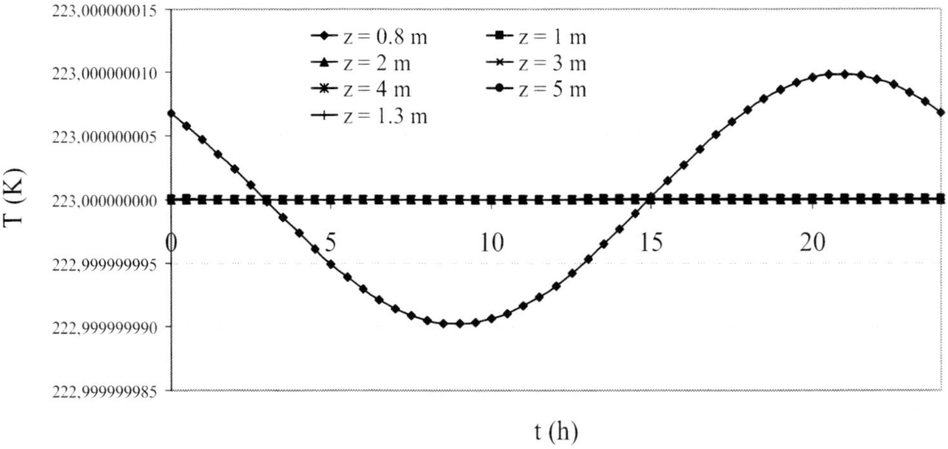

Figure 28. Temperature variations in the dry soil analogue, type BD, below 0.60 m [29].

The temperature distributions are presented versus the depth of the soil analogue, type BD, in Figure 29, at several hours during the Martian day. Figure 29 shows that the temperature variations are detectable, inside the dry soil analogue, up to a depth of about 0.20 m.

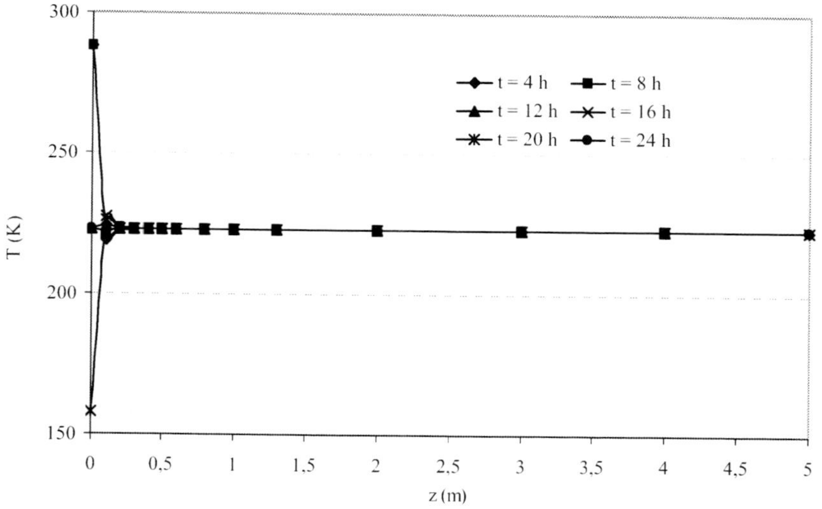

Figure 29. Temperature distribution in the dry soil analogue, type BD, versus depth, during the day [29].

1.2.2. Frozen Martian Soil Analogue, Type BF

The frozen Martian soil analogue, type BF, has the following characteristics: $p = 0.50, f = 1, a = M_{ice} / M_t = 0.2413$, thermal conductivity equal to k = 2.70 W/m/K, according to the evaluation of the theoretical model at 223 K, Figure 13, and thermal diffusivity equal to $\alpha = 1.36 \cdot 10^{-6}$ m^2/s, as evaluated by the theoretical model, Figure 19. The temperature variation on the surface is variable from the minimum value 148 K to the maximum one 298 K, with an average value of 223 K.

Figures 30-31 show the daily temperature variations in the soil analogue, type BF. Figure 30 reports the temperature of the soil up to a depth of 0.60 m. At 0.30 m the temperature variation, during the day, is 31.8 K and at 0.60 m is 6.8 K, Figure 30. Figure 31 reports the temperature variations below 0.60 m. At 1 m the temperature variation is 0.8 K, at 1. 30 m is 0.2 K while is negligible at 2 m, Figure 31.

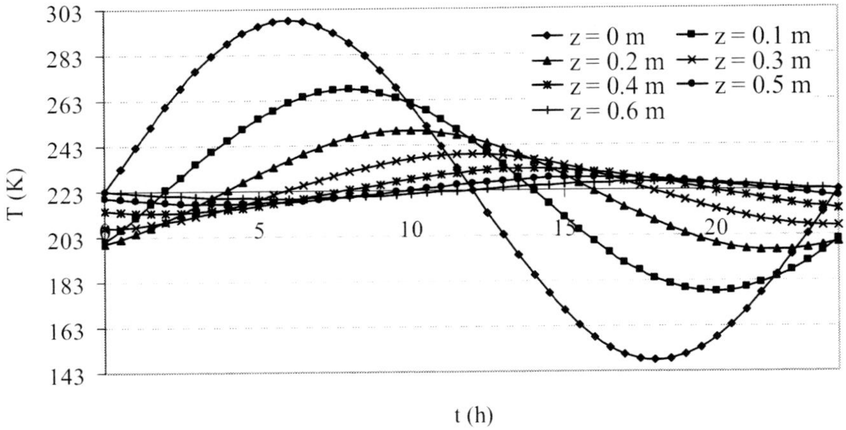

Figure 30. Temperature variations in the frozen soil analogue, type BF, up to 0.60 m [29].

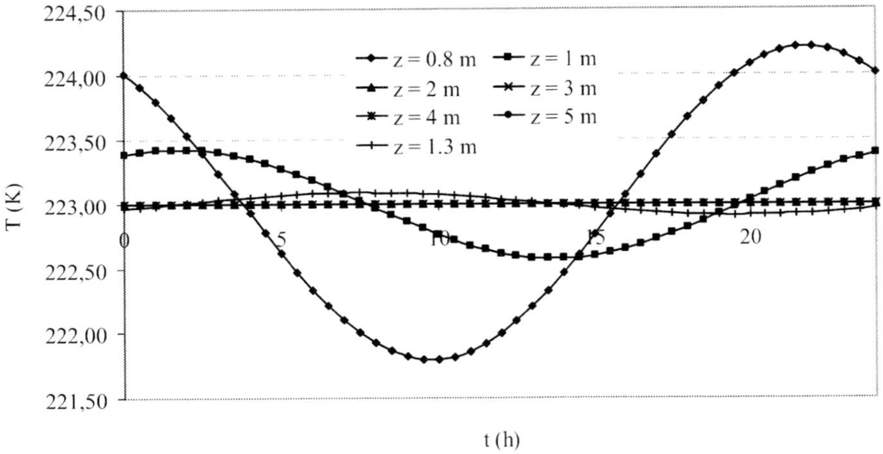

Figure 31. Temperature variations in the frozen soil analogue, type BF, below 0.60 m [29].

The temperature variations are presented versus the depth of the soil analogue, type BF, in Figure 32, at several hours during the Martian day. Figure 32 shows that the temperature variations are detectable, inside the frozen soil analogue, up to a depth of about 1.3 0 m. In conclusion, the temperature variation is more penetrating in the frozen soil because the thermal diffusivity is more than one order of magnitude higher than in the dry soil.

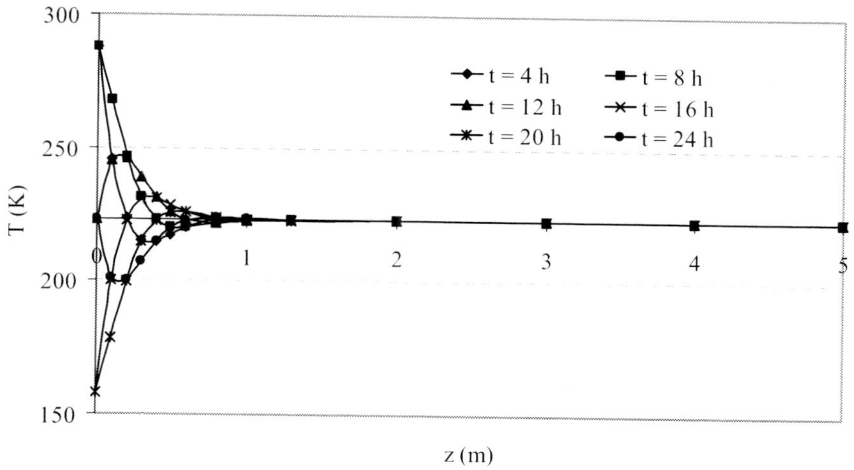

Figure 32. Temperature distribution in the frozen soil analogue, type BF, versus depth at several hours [29].

2. Relation between Temperature Distribution and Physical Properties

The temperature variations in the soil analogues, type A, are reported in Table 3, at several depths.

Table 3. Temperature variation, $\Delta T = T_{max} - T_{min}$, in the soil analogues, type A, at several depths [38].

Depth	Type AF	Type AD
z(m)	*ΔT* (K)	*ΔT* (K)
0.10	123.7	82.0
0.20	102.4	44.9
0.30	84.6	24.6
0.40	69.9	13.4
0.50	57.8	7.4
0.60	47.7	4.0
0.80	32.6	1.2
1.00	22.2	0.4
1.30	12.6	0.1
2.00	3.3	0.0
3.00	0.5	0.0
4.00	0.1	0.0
5.00	0.0	0.0

Table 3 shows the comparison between the temperature variations in the dry and frozen soil analogues, type AD and AF. At small depths, i.e. up to 0.50 m, the temperature variation is of the same order of magnitude, although higher in the frozen soil, but, at larger depths, i.e. below 0.60 m, Table 3 shows that the temperature variations are higher of one or two orders of magnitudes, in the frozen soil as compared to the dry soil.

Table 4. Temperature variation, $\Delta T = T_{max} - T_{min}$, in the soil analogues, type B, at several depths [29].

Depth	Type BF	Type BD
z(m)	**ΔT (K)**	**ΔT (K)**
0.10	89.5	8.7
0.20	53.4	0.5
0.30	31.9	0.03
0.40	19.0	0.0
0.50	11.4	0.0
0.60	6.8	0.0
0.80	2.4	0.0
1.00	0.9	0.0
1.30	0.2	0.0
2.00	0.0	0.0
3.00	0.0	0.0
4.00	0.0	0.0
5.00	0.0	0.0

Table 4 reports the temperature variations in the soil analogue, type B, at several depths. The temperature variations are lower than those of Table 3 because of the lower thermal diffusivity but they are detectable in the frozen soil analogue, type BF, up to 1.30 m while in the type BD they are detectable only up to 0.20 m.

Table 5. Derivative of the temperature variation, $\Delta T/\Delta z$ (K/m), for the soil analogues, type A, at several depths [38].

Depth	Type AF	Type AD
z(m)	**ΔT/Δz (K/m)**	**ΔT/Δz (K/m)**
0.15	213.0	371.0
0.25	178.0	203.0
0.35	147.0	112.0
0.45	121.0	60.0
0.55	101.0	34.0
0.70	75.5	14.0
0.90	52.0	4.0
1.15	32.0	1.1
1.65	13.3	0.1
2.50	2.8	0.0
3.50	0.4	0.0
4.50	0.1	0.0

The data of Table 3 have been used to calculate the variation of $\Delta T = (T_{max} - T_{min})$ with the depth or, in other words, the derivative of ΔT with z, $\partial T/\partial z = \Delta T/\Delta z$ (K/m). The results for the soil analogues, types A and B, are reported in Table 5-6.

Table 5 shows the comparison between the derivative of the temperature variation, $\Delta T/\Delta z$, in the two soil analogues, types AF and AD. At small depths, i.e. up to 0.25 m, $\Delta T/\Delta z$

is higher for the dry soil than for the frozen soil. At greater depths, i.e. below 0.35 m, Table 5 shows that $\Delta T/\Delta z$ are higher for the frozen soil. Below 0.70 m, $\Delta T/\Delta z$ are higher for the frozen soil of one order of magnitude and are detectable up to a depth of 3.5 m while in the type AD they are detectable only up to 1.15 m.

Table 6. Derivative of the temperature variation, $\Delta T/\Delta z$ (K/m), for the soil analogues, type B, at several depths [29].

Depth	**Type BF**	**Type BD**
z(m)	**$\Delta T/\Delta z$ (K/m)**	**$\Delta T/\Delta z$ (K/m)**
0.05	604.9	1412.8
0.15	360.9	82.1
0.25	215.5	4.8
0.35	128.5	0.3
0.45	76.8	0.02
0.55	45.8	0.0
0.70	21.8	0.0
0.90	7.8	0.0
1.15	2.2	0.0
1.65	0.3	0.0
2.50	0.0	0.0
3.50	0.0	0.0
4.50	0.0	0.0

Table 6 reports the derivative of the temperature variation, $\Delta T/\Delta z$, in the soil analogue, type B, at several depths. Table 6 shows that at very small depths, i.e. up to 0.05 m, $\Delta T/\Delta z$ is higher for the dry soil than for the frozen soil. At greater depths, i.e. below 0.15 m, $\Delta T/\Delta z$ is higher for the frozen soil. For the type BF $\Delta T/\Delta z$ are detectable up to a depth of 1.65 m while in the type AD they are detectable only up to 0.35 m.

3. Conclusions

The temperature variation along the depth of a soil analogue during a Martian day, and the knowledge of its porosity, can be employed to evaluate the physical conditions, i.e. either in dry or frozen state. The temperature variations in a frozen Martian soil analogue with porosity equal to 0.2 are detectable up to a depth of about 3 m while in a dry soil analogue they are detectable only up to 1 m. The derivative of the temperature variation with the depth is detectable in a frozen soil analogue up to a depth of 3.5m while in a dry soil analogue it is detectable only up to 1.15 m. The temperature variations in a frozen Martian soil analogue, with porosity equal to 0.5, are detectable up to a depth of about 1.3 m while in a dry soil analogue they are detectable only up to 0.2 m. The derivative of the temperature variation with the depth is detectable in a frozen soil analogue up to a depth of 1.65 m while in a dry soil analogue it is detectable only up to 0.35 m.

References

[1] Clifford, S.M. and S.P. Fanale, The Thermal Conductivity Of The Martian Crust, *Lunar Planet Sci. Conf. XVI*, 144-145, 1985.

[2] Clifford, S.M., A Model For The Hydrologic And Climate Behavior Of Water On Mars, *Jour. Geophys. Res.* 98 (E6), 10973-11016, 1993.

[3] Fanale F.P., Martian Volatiles: Their Degassing History And Geochemical Fate, *Icarus*,28,179-202, 1976.

[4] Crescenti G.H., Analysis Of Permafrost Depths On Mars, *NASA Tech.Mem. TM* 79730, 1984.

[5] Fanale, F.P., J.R. Salvail, A.P. Zent, and S.E. Postawko, Global Distribution And Migration Of Subsurface On Mars, *Icarus*, 67,1-18, 1986.

[6] Solomon, S.C. and J. W. Head, Heterogeneities In The Elastic Lithosphere Of Mars: Constraints In The Heat Flow And Internal Dynamics, *Jour. Geophys. Res.* 95,11073-11083, 1990.

[7] Gori F., A Theoretical Model for Predicting the Effective Thermal Conductivity of Unsaturated Frozen Soils, *The 4th International Conference on Permafrost*, Fairbanks, (U.S.A.), 363-368, 1983.

[8] Gori F., Prediction of Effective Thermal Conductivity of Two-Phase Media (including Saturated Soils), *XVI International Congress of Refrigeration*, Paris, B1-26, 525-531, 1983.

[9] Gori F., On the Theoretical Prediction of the Effective Thermal Conductivity of Bricks, *The Eight International Heat Transfer Conference*, S. Francisco (U.S.A.), II, 627-632, 1986.

[10] Gori F. and Corasaniti S., Considerations On A Theoretical Method For The Prediction Of The Thermal Conductivity Of Porous Media, *17th UIT National Heat Transfer Conference*, Ferrara (Italy), vol. II, pp. 683-694, 1999.

[11] Pande R.N. and Gori F., Effective Media Formation and Conduction Through Unsaturated Granular Materials, *International Journal of Heat and Mass Transfer*, 30, 5, 993-1000, 1987.

[12] Pande R.N., Chaudhary D.R. and Gori F., Effective Medium in Dispersed Systems, *Pramana - J. Phys.*, 29, 2, 217-223, 1987.

[13] Gori F. and Corasaniti S., A Theoretical Model for Combined Radiation Conduction Heat Transfer in Porous Media, *19th UIT National Heat Transfer Conference*, Modena (Italy), 371-376, 2001.

[14] Gori F. and Corasaniti S., Theoretical Thermal Conductivity of Soils at High Temperatures, *IMECE, International Mechanical Engineering Congress and Exposition, HTD-24152, ASME*, new York, NY, pp.1-8, 2001.

[15] Gori F. and Corasaniti S., Theoretical Prediction of the Thermal Conductivity of Soils at Moderately High Temperatures, *Journal of Heat Transfer, ASME*,126 (6), 1001-1008, 2002.

[16] Gori F. and Corasaniti S., Theoretical Prediction of the Effective Thermal Conductivity of Particulate Materials in Extraterrestrial Conditions and of Foams at Low Density, *Microgravity and Space Station Utilization*, Vol. 2, Nr. 2-3-4, pp 23-24, June-December 2001.

[17] Gori F. and Corasaniti S., Temperature Variations Inside Dry And Partially Frozen Mars Soils, *IEEE Aerospace Conference*, March 8-15, Big Sky, Montana,2003.

[18] Gori F. and Corasaniti S., Theoretical Prediction Of The Thermal Conductivity And Temperature Variation Inside Mars Soil Analogues, *Planetary and Space Science*, 52, 91-99, 2004.

[19] Gori F., Marino C. and Pietrafesa M., Experimental Measurements and Theoretical Predictions of the Effective Thermal Conductivity of Porous Saturated Two Phases Media, *International Communications in Heat and Mass Transfer*, Vol. 28, N. 8 (01), pp. 1091-1102, 2001.

[20] Gori F. and Corasaniti S., On the Effective Thermal Conductivity of Dry Olivine, *5th World Conference on Experimental Heat Transfer, Fluid Mechanics and Thermodynamics*, vol.2,1257-1262, 2001.

[21] Gori F. and Corasaniti S., Experimental Measurement and Theoretical Prediction of the Thermal Conductivity of Two and Three Phases Water Olivine Systems, *European Conference on Thermophysical Properties, ECTP*, London, 2002.

[22] Gori F. and Corasaniti S., Experimental Measurement of the thermal conductivity of porous materials, *57 th ATI Congress*, pp. II, 93-98, 2002.

[23] Gori F. and Corasaniti S., Experimental Measurements and Theoretical Prediction of the Thermal Conductivity of Two-and Three-Phase Water/Olivine Systems, *International Journal of Thermophysics*, Vol. 24, No. 5, pp. 1339-1353, September 2003.

[24] Gori F. and Corasaniti S., The Measurement Of The Thermal Diffusivity To Detect The Amount Of Water In Olivine, *Proceedings of the 14th International Symposium on Transport Phenomena*, Bali, Indonesia, pp. 581-585, 6-10 July, 2003.

[25] Tarnawski V.R., Gori F., Wagner B. and Buchan G.D., Modelling Approaches To Predicting Thermal Conductivity Of Soils At High Temperatures, *International Journal of Energy Research*, n 24, pp. 403-423, 2000.

[26] Tarnawski V.R. and Gori F., Enhancement of the Cubic Cell Soil Thermal Conductivity Model, *International Journal of Energy Research*, n. 26, 143,157, 2002.

[27] Tarnawski V.R., Leong W.H., Gori F., Buchan G.D. and Sundberg J., Inter-Particle Contact Heat Transfer In Soil Systems At Moderate Temperatures, *International Journal of Energy Research*, 26:1345-1358, 2002.

[28] Leong W. H., Tarnawski V. R., Gori F., Buchan G. D. and Sundberg J., Inter-Particle Contact Heat Transfer Models: An Extension To Soils At Elevated Temperatures, *International Journal of Energy Research*, 2005; 29:131-144.

[29] Gori F. and Corasaniti S., Extraterrestrial Soil Thermophysical- Properties, EST-PRO, *Final Report, European Space Agency*, 2002.

[30] Map of Mars and landing sites of Viking and Pathfinder mission *(http://mars.jpl.nasa.gov/mgs/target/marsmap1b.jpg)*

[31] Scott D.H. and M.H. Carr, Geological Map Of Mars, 1:25.000.000, Miscellaneous investigations series, *U.S. Geological Survey*, 1978.

[32] Rieder R., T. Economou, H. Wanke, A. Turkevich, J. Crisp, J. Bruckner, G. Dreibus and H.Y. McSween Jr., The Chemical Composition Of Martian Soil And Rocks Returned By The Mobile Alpha Proton X-Ray Spectrometer: Preliminary Results From The X-Ray Mode, *Science*, 278, 1771-1774, 1997.

[33] *(http://mars.jpl.nasa.gov/MPF/ops/NEcolor_annot.jpg)*.

[34] Withers P. and G.A. Neumann, Enigmatic Northern Plains Of Mars, *Nature*, 410, 651, 2001.

[35] Mangold N., P. Allemand, P. Duval, Y. Geraud and P. Thomas, Experimental And Theoretical Deformation Of Ice-Rock Mixtures: Implications On Rheology And Ice Content Of Martian Permafrost, *Planetary and Space Science*, 50, 385-401, 2002.

[36] Britt D.T., R. Anderson, J.F. Bell III, J. Crisp, T. Economou, K.E. Herkenhoff, M.B. Madsen, H.Y. Mc Sween, S. Murchie, R. Reid, R. Rieder, R.B. Singer and L. Soderblom, The Mineralogy Of The Mars Pathfinder Landing Site, 29th *Lunar and Planetary Science conference*, 1998.

[37] *(http://mars.jpl.nasa.gov/MPF/ops/hap_5.jpg).*

[38] Reprinted from [18] with permission from Elsevier.

[39] Luikov A.V., Heat And Mass Transfer In Capillary-Porous Bodies. *Pergamon Press*, 1966.

In: Space Science: New Research
Editor: Nick S. Maravell, pp. 197-243

ISBN: 1-60021-005-8

Chapter 7

DEALING WITH POTENTIALLY HAZARDOUS ASTEROIDS

Eric W. Elst
Royal Observatory at Uccle & Holbach-Foundation
Mortsel, Belgium

Abstract

An excurse is made through the main belt of asteroids and beyond. Starting with the discovery of the asteroid Ceres, during the first night of January 1801, by Piazzi in Palermo (Sicily), the discipline of asteroid-discovery evolved quickly into large searches of similar bodies in the solar system. At this very moment, more than130.000 new objects have been numbered and the more of them classified into *families*. One of these groups of asteroids, the NEO's (Near Earth Objects), subdivided in three substantially different *subfamilies* (Aten-, Amor- and Apollo-objects) have presently become search objects of primary importance; since they may collide with Earth, resulting in disastrous situations on our planet and (posing) serious risks for the preservation of life on its surface.

Introduction

Ceres, the first and largest asteroid of the main belt of asteroids, was discovered visually at the 5 feet vertical circle by G. Piazzi (1746-1826), during the first night of 1801 at the Observatory of Palermo (Sicily).

J. Palisa (1848-1925), with the 15cm astrograph at the Pola Observatory (Yougoslavia) and afterwards with the 62cm astrograph at the Vienna Observatory (Austria) found visually 122 asteroids during the period of 1874-1923, promoting himself with these discoveries to the most succesful visual discoverer of asteroids. By introducing the photographical method at the Bruce-twin 40cm astrograph for hunting asteroids, M. Wolf (1863-1932) at the Heidelberg Observatory was honoured with 248 discoveries. One of his discoveries was the earth approaching asteroid (887) Alinda, in 1920. Before that remarkable event he had already discovered the first Trojan (588) Achilles, in 1906.

At the same observatory K. Reinmuth (1892-1979) would even discover 395 asteroids, holding for a long time the record of asteroid discovery. He was removed from the first place by the team of the van Houten's at Leiden and T. Gehrels (1970) at Palomar, with the Palomar-Leiden surveys during 1960-1977, with 3428 discoveries, and individually by E.W. Elst, during 1986-1999, with 3161 discoveries (current situation). However Reinmuth was the first to discover a PHA (Potentially Hazardous Asteroid), on April 24th of 1932. It was named *(1862) Apollo* and it became the prototype of the Apollos from which several objects have been classified as *potentially hazardous*. With the discovery of the prototype of the Amor-asteroids, *(1221) Amor* by E. Delporte in 1932, at Uccle Observatory and in 1976 of the firstAten asteroid *(2062) Aten* by E. Helin at Palomar, more PHAs have been added to the list of potentially hazardous objects. Introducing CCDs (charge-coupled devices) into astronomical imaging in 1987, CCD-discovery by larger teams, such as LINEAR, LONEOS, Spacewatch end NEAT asteroid discovery, went far beyond all earlier visual and photographical discoveries. Above mentioned teams account for 65.113 CCD-discoveries from a total of 99947 numbered asteroids (by all means; July 2005). Among these discoveries we distinguish 286 Aten-, 1663 Apollo- and 1472 Amor asteroids, including 663 PHAs, of which 120 objects have so far been definitively numbered and 48 of them have received an appropriate (mythological) name (MPC, July 2005). Table 1 lists the total amount of discoveries by individuals and by teams. The heading of the table speaks for itself including the symbols Co for *codiscoveries* and * for *+I. van Houten-Groeneveld, T. Gehrels*. Table 2 lists the most prolific sites where asteroids have been observed very intensively. Table 3 deserves our special attention, since it reviews the almost incredible effort that has been put into observation of asteroids, during the last 5 years. The table lists the evolution of the observation of numbered and unnumbered asteroids (cumulative numbers). Under *Grand total* all individual observations of comets have also been included. It is interesting to note from this table, that in the beginning of 2000 the larger part of the observations still deal with the unnumbered objects (17995549 observations for unnumbered objects versus against 1509331 observations for numbered asteroids). Around May 2001 we notice a turning point, where the numbered objects take over from the unnumbered objects, pointing to the fact that statistically more numbered objects than unnumbered objects show up in the sky. This evolution is undoubtly confirmed after June 2001, where the recovery of the numbered objects is steadily moving higher. At present (21 July 2005) a total of 21609020 observations of numbered objects have been collected (recovery of numbered objects) against 9254137 observations of unnumbered objects (discovery of new objects and recovery of unnumbered objects). Are we approaching the end of asteroid discovery? This question will be answered in next paragraph. Table 4_a,b,c lists the Apollos and numbered PHAs. For the meaning of the symbols in these tables we refer to Annex1.

1. The End of Asteroid Discovery?

Table 3 suggests that in a near future the only observations may be of numbered asteroids. Elst (2001) has made a statistical investigation, in order to elucidate the problem. From the enormous increase of recent asteroid discoveries by LINEAR, LONEOS, NEAT, Catalina Sky Survey and Spacewatch, there is indeed an indication that the number of truly new discoveries is slowly decreasing. What is the reason for this? Is there a cutoff in the distribution of fainter and faintest asteroids or is it the consequence of the fact that most instruments/devices are not reaching far enough?

In the past many searches for new asteroids have been performed. Among the photographical searches we notice the survey done during 1951-1952 at the Observatory of McDonald, the Palomar-Leiden Asteroid Surveys (1960-1973) and the Palomar-Comet-Asteroid Survey (1970-1995). At the European Southern Observatory (ESO) succesful asteroid-searches were performed during 1976-1998. Intensive CCD-searches for asteroids were initiated in 1997 by LINEAR at Socorro (New Mexico), taking very quickly a leading role in the discovery of asteroids (and comets), with the discovery of 50.113 new asteroids (Table 1, MPC, July 2005). However, all these discoveries remain below visual magnitude v=19.5. Only the CCD-surveys done at Spacewatch (Arizona, 1m reflector) and NEAT (Palomar,1m Schmidt) are reaching further and have discovered several thousands of new and fainter asteroids (Table 1, MPC, July 2005), including all photographical discoveries during the past.

In 2001, in the Annual Report (1999) of the Minor Planet Center, Brian Marsden, director of MPC wrote the following: "As anticipated, the year 1999 handsomely broke all previous records for Minor Planet Center activity. The number of pages of actual observations in the *Minor Planet Circulars Supplement (MPSs),* 4998, each exceeded the total for the previous year by more than 65 percent. The 3145 new numberings of minor planets were 85 percent more than in 1998 and more that *four times* as many as in 1997. A milestone was reached in March 1999 with the numbering of minor planet (10000). Although it had taken more than 198 years, following the discovery of Ceres, to attain this milestone, the milestone rather loses significance when one realizes that it will less than *two years* more to reach the numbering of minor planet (20000)"!

From this quote we could get the impression that discovering and numbering minor planets will continue for ever. But the figures from table 3 remain and therefore it is necessary that we investigate the problem more thoroughly.

During September/October 1960 Gehrels took 130 photographical plates at Palomar, with the 48 inch Schmidt telescope, which were afterwards measured by van Houten et al (1970). More than 2000 faint asteroids were discovered. Statistical investigation made clear that, at least for asteroids as faint as magnitude v=19.5, there was surely no cutoff to expect for asteroids beyond the limiting magnitude of the instrument.

2: Asteroid Families and Completeness of Searches

In 1866 Kirkwood discovered the presence of gaps in the main belt of asteroids and found that for particular distances from the sun corresponding to the value of the semimajor axis of the objects involved- asteroids were accumulating (Figure1), revealing the existence of *families*, i.e. asteroids with almost the same orbital elements (Kirkwood, 1866). We want to mention here that if we want to study *families in* a more appropriate way, then it will be necessary to use *proper elements* (orbital elements, freed from perturbation)(Knezevic and Milani, 1993).

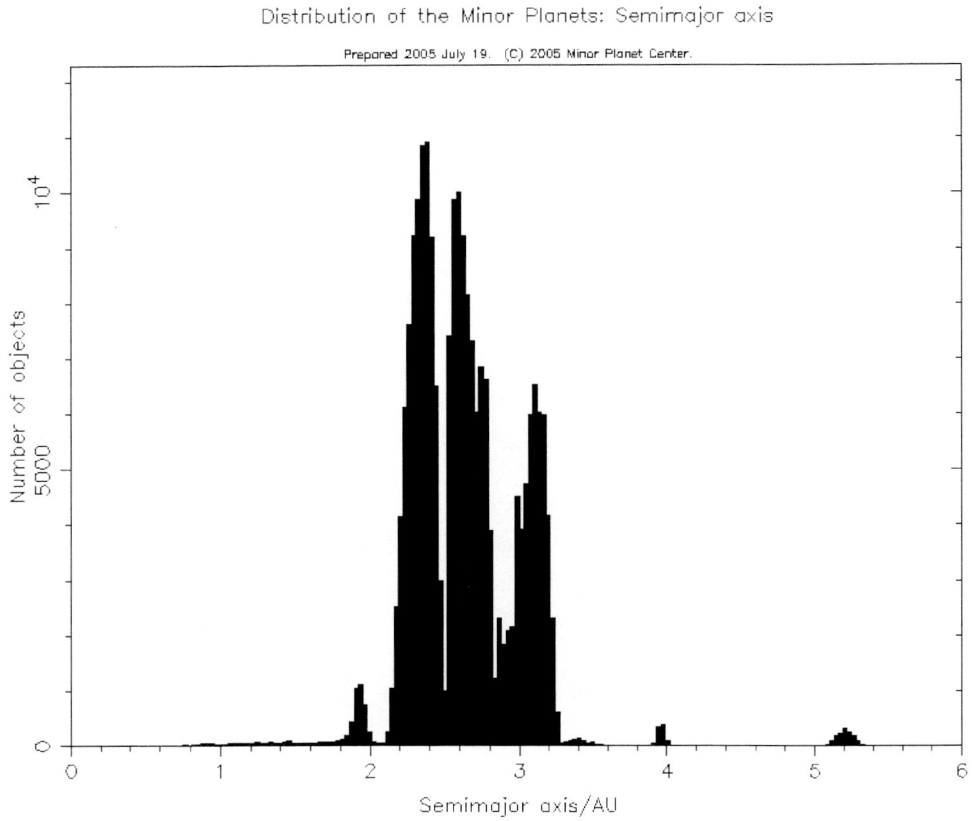

Figure 1. Number of asteroids versus semimajor axis.

Flora asteroids are accumulating at 2.3 A.U., Themis asteroids at 3.15 A.U., Hilda asteroids at 3.9 A.U. and Trojans at 5.2 A.U. At present we deal with about 20 well distinguished families, but there seems to exist a lot more (e.g. *Hirayama asteroid families*). It is clear that the height of a peak may be interpreted as a measure for relative completeness of a particular family, assuming that below the instrumental limiting magnitude v all their members have been discovered. Extending the limiting magnitude, by using larger instruments and more sensitive CCDs, will result in higher peaks, since we are reaching the fainter asteroids of that particular family. Since we know the distance a of the family from the sun, it is possible to calculate the absolute magnitude H of the newly discovered asteroids, by using the following approximate formula (Cunningham, 1988)

$$H = v - 5*\log(a(a-1)) \tag{1}$$

where v is the limiting magnitude, H the absolute magnitude of the asteroid and a its semimajor axis. With v=19.5 and a=2.3 we obtain a value H=17.1. By means of the formula (Meeus J., 1991)

$$\log D(\text{km}) = 3.6 - 0.2*H \tag{2}$$

we obtain the metric dimension of the asteroid (i.e. D=1.5km).

We may now plot the number of all discovered asteroids (intermixing all families), for a particular limiting magnitude (i.e. v=19.5), versus absolute magnitude H (Figure2).

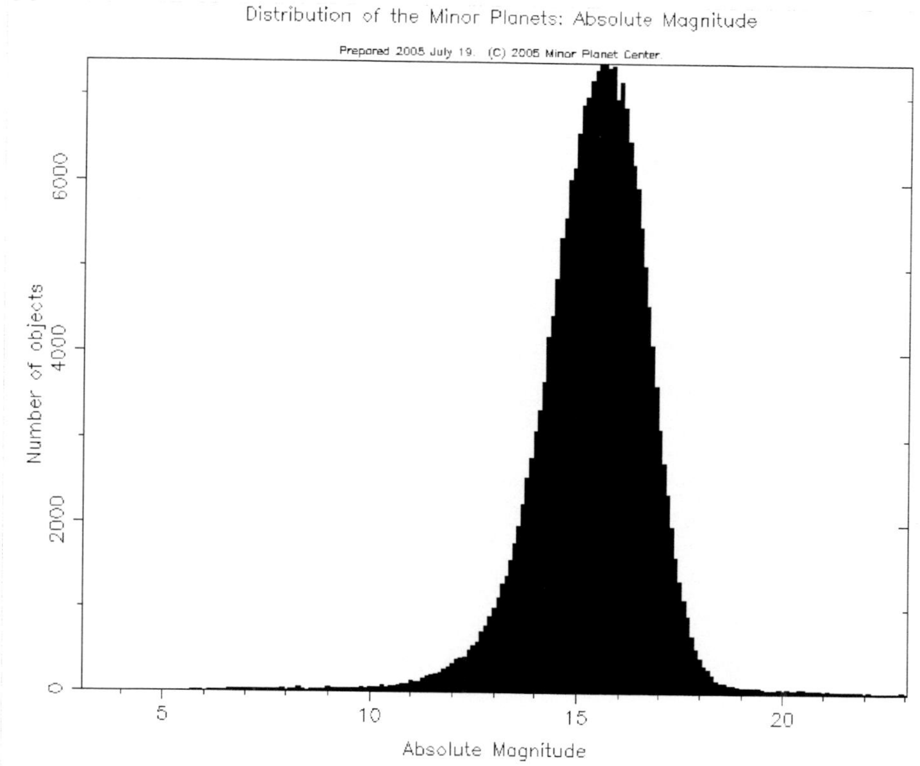

Figure 2. Total number of asteroids versus absolute magnitude H.

Figure 2 provides a reliable measure for completeness, assuming that full sky coverage has been performed for observations up to the limiting magnitude v. In the interval H=15-15.5 (step=0.25 magnitude) about 15000 numbered asteroids have been discovered. This implies also that no more asteroids with H=15-15.5 will be discovered in that particular interval. Only, by pushing the observations beyond the limiting magnitude v new fainter objects will be sighted and eventually numbered, resulting in a higher peak which will move at the same time to fainter values of the absolute magnitude. It goes without saying that we can construct analogue plots for all families separately (e.g.: Figure3).

3. Hazardous Asteroids

What is a hazardous asteroid? It is a celestial body that during its revolution around the sun crosses the orbit of the earth (e.g. Alinda) and may therefore collide (impact) with our home planet, with all possible disastrous consequences. Figure 3 shows the total number of numbered and unnumbered NEAs _Near Earth Asteroids_ (Apollos, Amors and Atens, including all PHAs) versus absolute magnitude. From inspection of this figure it follows that we have nearly completeness for an absolute magnitude of H=20.0 (corresponding with a metric dimension of about 400 meter), to be compared with an absolute magnitude of 18.5

(850 meter) in 2000. It is clear that NEAs, and therefore also PHAs, with the same absolute magnitude may be still discovered in the near future; but they will generally show large inclinations, a signature for objects that arediscovered more seldomly, since they are moving for the largest part of their existence far outside the ecliptic, where during the past searches for new asteroids have been more scarce or even completely been neglected. Only by pushing the limiting instrumental magnitude v to fainter absolute magnitudes new asteroids will be discovered. It is interesting to note that with *completeness* of H=20.0, and even with H=18.5, we have reached the objects with dimension smaller than 1km, and therefore the theoretical limit for *non global destruction effects*, if one of these objects should hit the earth.

The main belt remains the principal *reservoir* of new objects that, due to internal collisions could find their way towards the inner parts of the solar system (Ipatov, 1995). However, the collisions do not lead directly to the generation of new potentially hazardous asteroids, but they may supply new objects that are partially injected into resonant, chaotic orbits undergoing large variations of the eccentricity and becoming therefore amenable to planetary encounters (Menichella et al.,1996). Farinella et al. (1993) estimated the fraction of main belt fragments that may reach the 1:3 mean motion resonance with Jupiter or the v6 secular resonance, which are the two most effective dynamical routes from the main belt to the inner planet zone (Elst, 2001)

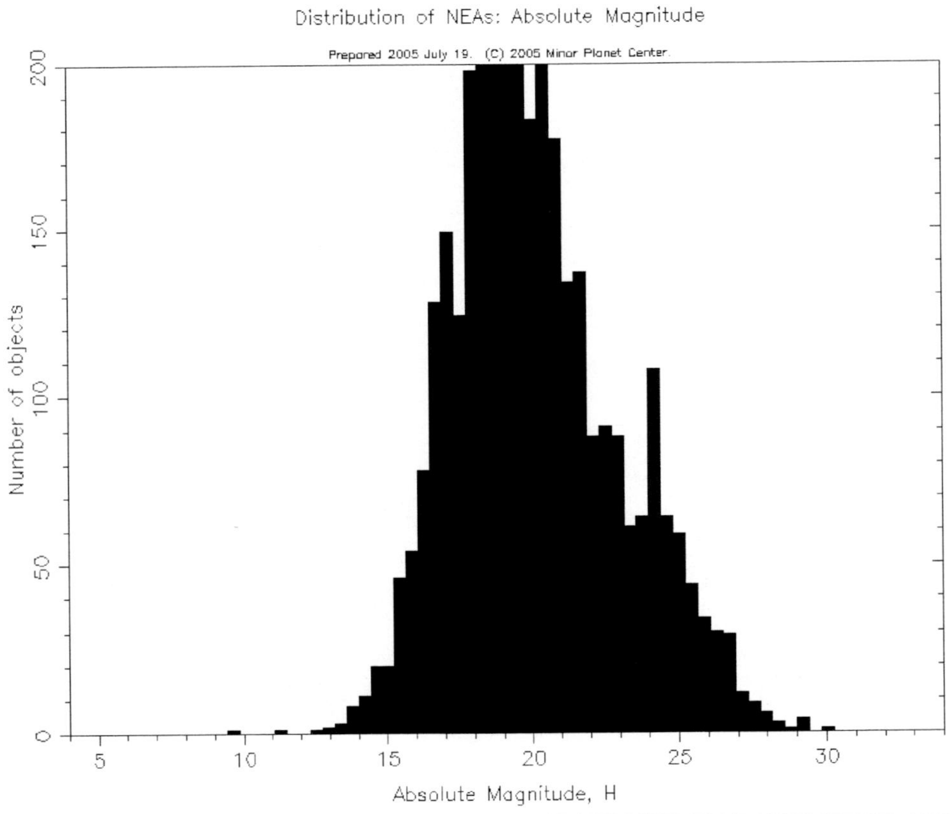

Figure 3. Total number of Near Earth Asteroids (July 2005).

4. Consequences of an Impact of a Celestial Body with Earth

The impact of a cosmic body with Earth essentially causes an explosion: an instantaneous conversion of kinetic energy of the impactor into fragmentation, comminution and cratering of the substrate; heating, melting and vaporisation of the projectile and target materials; kinetic energy of cratering ejecta and seismic shock waves penetrating the planet. A significant fraction of the energy is dissipated in the ecosphere, the thin shell of air, water and surface rocks and soils, whose constancy sustains and nurtures life. While the greatest damage is obviously at ground level, the stratosphere is badly polluted with dust on a global scale from impacts of objects exceeding 1 km diameter. Smaller impacts (e.g. by a 200m impactor), if into the ocean, can cause devastation thousands of kilometers away due to the efficient transmission of energy to great distances by *tsunamis* (Chapman, 2002, Ward and Aspaug, 2002). Examination of the cratering record on the Earth and terrestrial planets and especially on the moon, has demonstrated the continuity through the past 3.5 billion years of impact processes.

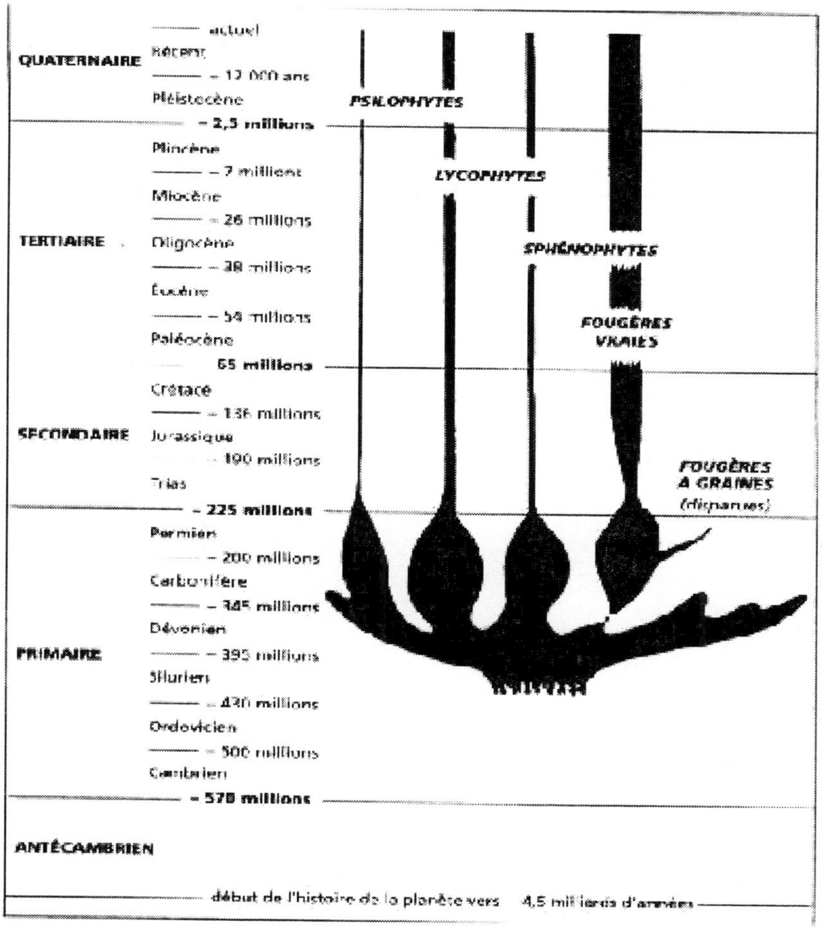

Figure 4. Almost extinction of the ferns.

The disaster of an impact at the level of fauna and flora is impressive. Below are two figures which were taken from the book *La botanique rédécouverte* (1999) from Aline Raynal, professor at the *Jardin des Plantes* in Paris, France. Figure4 shows the evolution of the ferns slowing down abruptly, 225 million years ago, epoch of the Yucatán impact. Figure5 is still more impressive and shows the almost complete extinction of the gymnospermia (conifers) and at the same time the overdeveloping of the angiospermia (fruit-bearing trees). The extinction of the dinosaurs falls at the same epoch (65 millions year ago).

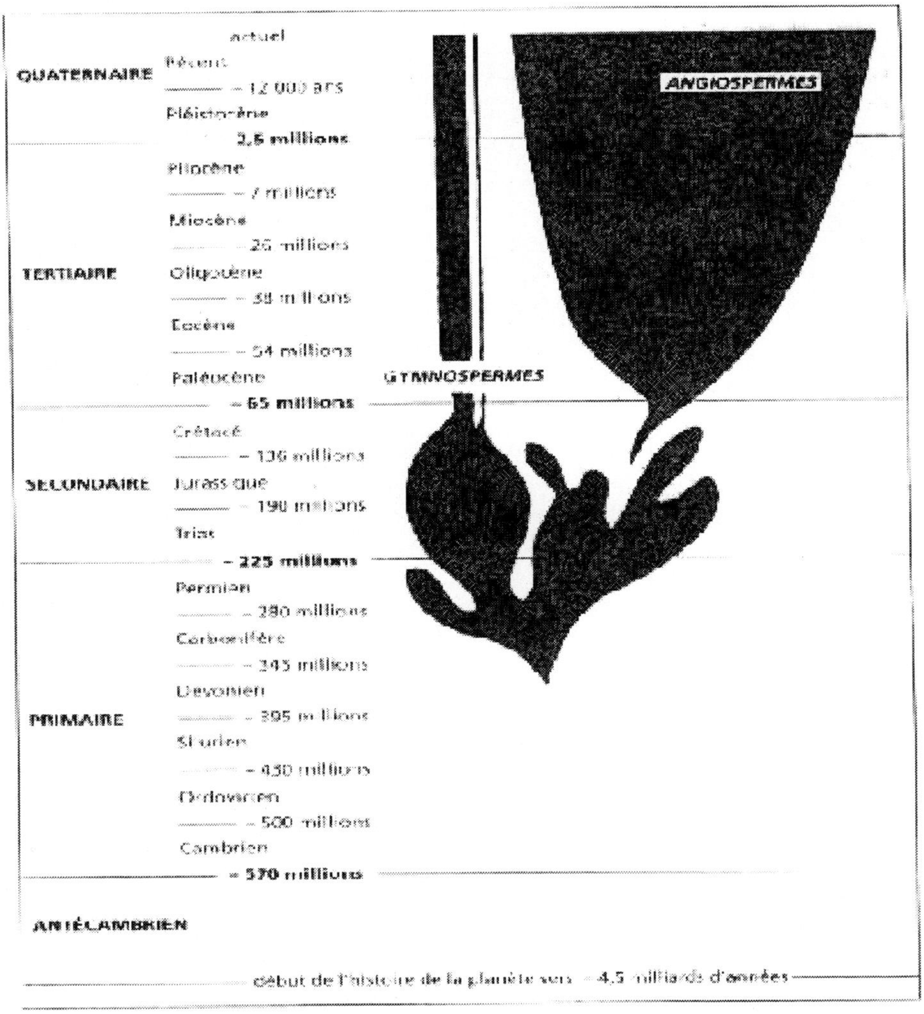

Figure 5: Overdeveloping of the angiospermia.

5. Sources of the PHAs

NEAs originate from five sources: asteroids clustering near the resonances v6 and 1:3 (Alindas), asteroids from the outer zones of the belt, Mars-crossers -asteroids that may cross the orbit of Mars- and comets perturbed by Jupiter and originating from the Kuiper Belt.

NEAs _and with them the PHAs_ are sightet at the entire sky. From inspection of the orbital elements of the NEAs (Table 5) we may get the impression, that some preferable directions are *preferred* by the possible impactors. Let us therefore look at the inclinations. Figure6a,b shows us a swarm of NEAs (the first 50 items from table 5). There seems hardly to exist a relation between semimajor axis and inclination for the involved objects (Fig6a, rectangles). However, the excentricity declines slowly with increasing inclination (Fig6b, stars)).

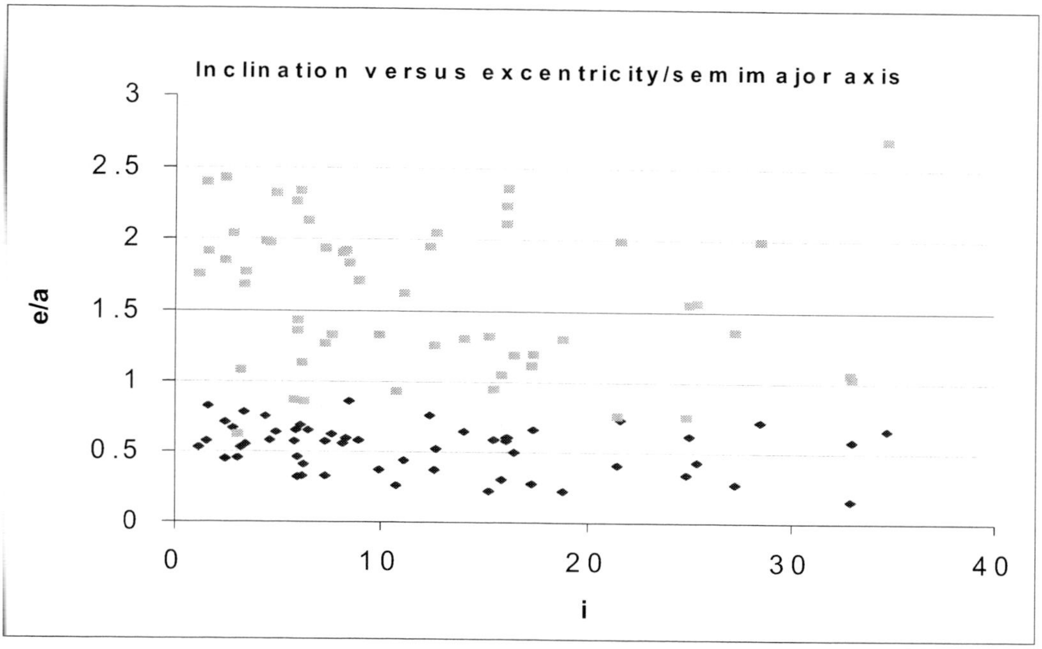

Figure 6: Semimajor axis/excentricity versus inclination.

Let us now plot semimajor axis versus inclination (Figure7) for the objects of Figure6. Here we notice clustering of the objects with a tendency to move upwards with increasing eccentricity. This is a consequence of the fact that earth-crossing asteroids are distinguished on the basis of semimajor axis and eccentricity. The Jacobi constant has the following form (Klose, 1928):

$$C= 1/a+2*Sqr(a(1-e2))*cosi. \quad (3)$$

Let us give an example. In September 1987 an Apollo-asteroid (1987 SB) has been discovered at the Bulgarian National Observatory (Elst et all, 1989). It is characterized by an eccentricity of 0.65, an inclination of 3° and a semimajor axis of 2.16. Apollo-Amor objects have C-values in the range of 2.7 and 3.0. With a value of C= 2.7 the object 1987 SB fits nicely into the Apollo-region. This means that if we are not considering the inclination, by putting semimajor axis versus eccentricity, then the eccentricity (semimajor axis) in equation (3) becomes dependent of the semimajor axis (eccentricity), explaining the particular behavior of the objects in Figure7.

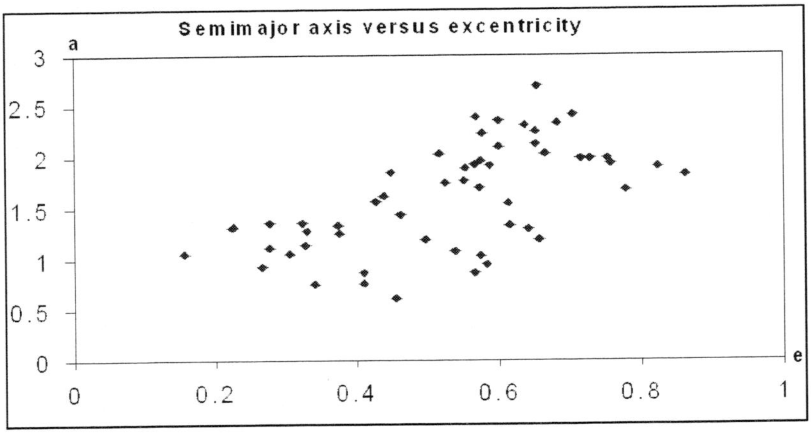

Figure 7. Clustering of a swarm of NEAs.

7: A strange object: comet/asteroid Elst-Pizarro.

Comets may also be considered as *potentially hazardous* if their orbits allow them to approach Earth. A special case is the object Elst-Pizarro that has been classified as an asteroid and as a comet. The object was discovered by E. W. Elst and G. Pizarro, during a routine search for new asteroids with the 1m ESO-Schmidt at the European Southern Observatory. Preliminary orbit calculation by Marsden (1996) showed that the orbit could be linked up with the *Themis* asteroids. However, a narrow dust tail remained for more than four months. Therefore an explanation had to be found for the persistence of the tail. Figure8 shows 133P/Elst-Pizarro, imaged on 2002 September 07 by the University of Hawaii 2.2m telescope on Mauna Kea. Nucleus and tail are indicated by white arrows. The tail extends about 3.5 arcminutes, corresponding to about 400.000 km.

Figure 8. Comet/asteroid Elst-Pizarro, 2.2m telescope on Mauna Kea (Hawaii).

The identification of 133P/Elst-Pizarro as a comet is not in doubt, but the origin of the mass loss is unclear. In 1996 the tail had been previously explained as debris released by an impact, but the discovery in 2002 of recurrent activity renders this interpretation implausible. The ejection of particles may be consistenly explained if the object is a barely active Jupiter-family comet, that has evolved into an asteroid-like orbit. A poor understanding of the precise effects of nongravitational forces on the long-term dynamical evolution of such bodies, results in uncertain knowledge of the fraction of inactive Jupiter-family comets among bona fide asteroids. 133P/Elst-Pizarro could be a true Themis family member upon which subsurface ice has been recently exposed, but would then still blur the asteroid-comet boundary by showing that small main-belt asteroids can preserve ice over astronomically long timescales. *Such a scenario would greatly complicate the determination of source regions for near-Earth objects, since the presence of significant ice in a body could no longer be considered a diagnostic of origin in the outer solar system and not in the main asteroid belt* (Hsieh, 2002).

6. Resonance Objects (Alinda Asteroids)

Alinda asteroids have semimajor axes of about 2.5 AU and eccentricities between 0.4 and 0.65. The objects are in a 1:3 orbital resonance with Jupiter and therefor in a 1:4 resonance with Earth. An object in this resonance has its orbital eccentricity steadily increased by gravitational interactions with Jupiter, until it approaches an inner planet that disrupts the resonance. Deorbited, the asteroids eventually come in collision with Earth or some other planet of the solar system. Several Alindas have their perihelia very close to Earth, resulting in a series of close encounters at almost exactly four-year intervals, due to the 1:4 resonance. One consequence of this is that if an Alinda asteroid happens to be in an unfavorable position for observing it, at the time of its close approach to Earth, this situation can persist for decades. We should not wonder that we find back a few PHAs among the Alinda-asteroids (Table 7). Figure9 shows a plot with the semimajor axis versus eccentricity for these objects. There is a tendency of declining semimajor axis with increasing eccentricity, but this appearance is inherent with the classification-scheme used for the Alinda-asteroids (Jacobi constant, Klose 1928). The apparent difference between Figure7 and Figure9 can be explained by the fact that in Figure7 the NEAs have been intermixed with each other, while in Figure9 (less scattering) we are looking at a single family of asteroids.

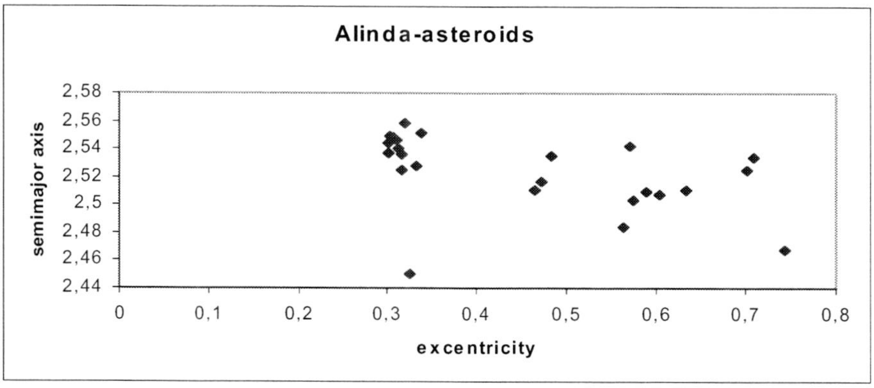

Figure 9. Alinda-asteroids: semimajor axis versus exentricity.

7. Hungaria- and Phocaea Asteroids

Fig 10 shows the distributions of the minor planets, semimajor axis versus inclination. The center of the cluster of Hungaria asteroids is at <a>=1.85 (semimajor axis between 1.78 and 2 AU) and <i>=23° (inclination between 16° and 34°). The Hungarias have eccentricities less than 0.18 and move just outside the orbit of Mars. The cluster of Phocaeas lies at <a>= 2.35 (semimajor axis between 2.25 and 2.5 AU), <i>= 24°(inclination between 18° and 32°) and eccentricities greater than 0.1. Hungaria asteroids dwell near the 8:9 resonance with Jupiter, reason why we suspect that they are regularly removed from this location in the main belt. The eccentricity increases steadily until the asteroid escapes from its cluster-position whereafter it may eventually be injected into the orbit of Earth. The problem remains why the Hungaria asteroids were originally residing at this particular place. The value of the inclination of the earth axis may (23°.5) may point to a *liaison* of these asteroids with Earth itself. But the intriguing aspect of the *empty place* below the clusters of the Hungaria- and Phocaea asteroids should imperatively be investigated. A substantial part of the NEAs may have found there their origin, flattening at the same time the lower part of the Hungaria asteroids. Let us look upon the NEA, 2003 NO4, that has been discovered some time ago and that could be probably linked with the cluster of Hungaria asteroids (Table 11). The orbital elements (i,a) of 2003 NO4 correspond nicely with the elements of a normal Hungaria

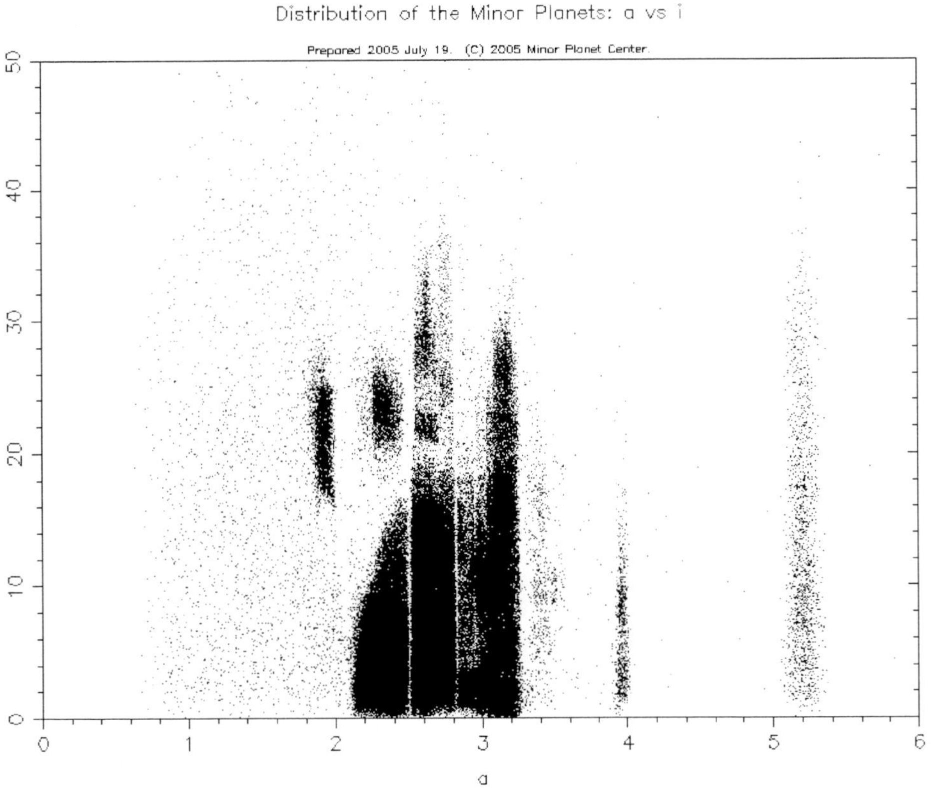

Figure 10. Minor Planets: Inclination versus semimajor axis.

asteroid, with exception of its eccentricity, which is in this case larger than 0.18. If we assume that these NEAs have their origin in the Hungaria cluster, than of course the asteroids have to find a way to leave the cluster, which is done by means of Jupiters gravitational field. 2005 PY16 has been discovered more recently. We notice here an elevated eccentricity and a lower inclination.

Table 11.

Orbital elements: 2003 NO4	Orbital elements: 2005 PY16
n 0.43884482 Peri. 171.04428	n 0.35485324 Peri. 193.40426
a 1.7149919 Node 135.44363	a 1.9759235 Node 159.45300
e 0.3127116 Incl. 22.6940	e 0.5247886 Incl. 6.41602
P 2.25 H 18.2 G 0.15	P 2.78 H 19.3 G 0.15

Let us look now at a plot which shows the Hungaria- and Phocaea asteroids, inclination versus semimajor axis (Figure11).

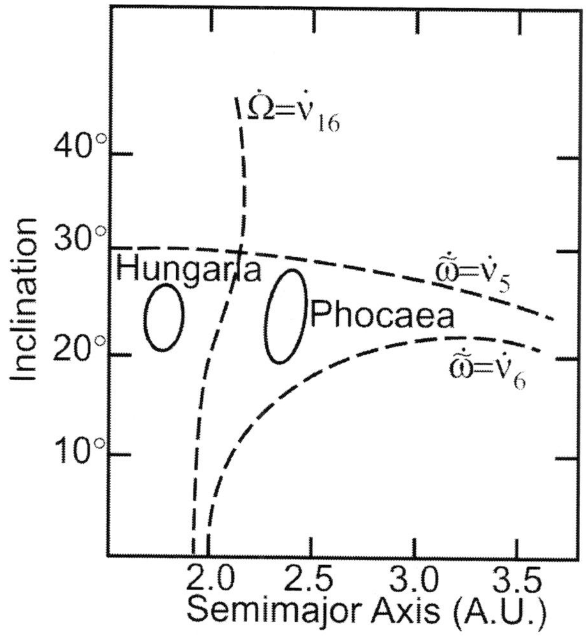

Figure 11 Hungaria and Phocaea asteroids.

The clusters are isolated from each other -and from other asteroids- by resonances. v5 and v6 are the eccentricity terms and v16 the inclination term of the perturbation by Jupiter (and Saturn). It is certainly no coincidence that below the cluster of the Hungaria asteroids space is almost empty (to be compared with Figure10). All asteroids once residing there have been removed by resonances during the past. A larger part of them hasmost probably found its way to Earth. The same applies to the Phocaea-asteroids. There seems indeed to exist a continuous transition from Phocaeas to Mars-crossers, which differ from the Phocaeas only by larger eccentricities which enable them to cross the orbit of Mars. We note additionaly that the part that has been removed by resonances follows neatly the shape of the resonances. However, we must consider the possibility that at the indicated spaces (E1 and E2) no asteroids have ever been present, actually due to the combined action of the resonances v6 and v16.

8. Conclusions

Collisions between asteroids and the *Yarkovsky Effect* -a very weak thermal impulse when the asteroid surface is heated by the Sun, affecting asteroids less than 10 km in diameter- may feed the resonances capable of injecting new NEAs over several million years into the inner part of the Solar System. Bottke et al. (19xx) estimate that 61% of NEAs of H<22 originate from the inner zone of the main belt, 24% from the central region, 8% from the outer zone and 6% from the Jupiter family of comets,a result that coincides with our investigation, that a effective source may be situated very close to Earth, with all consequences and possibilities of early detection. However much theoretical work as to be done to improve our knowledge of the mechanism that is responsible for the transport of the Hungaria- (and Phocaea asteroids from their cluster positions to Earth.

Annex 1

The symbols used in the list of Apollo-asteroids have the following meanings:

Number (and name):For numbered minor planet, this column contains the object's number and name (if named).

Designation: This is the object's provisional designation. Note that some of the low-numbered objects have old-style provisional designations, where the letters were recycled without regard to the calendar.

q: Perihelion distance (in AU).

Q: Aphelion distance (in AU).

H: Absolute visual magnitude.

M+, M-: The perpendicular distance to the earth's orbit when an object is at the earth's distance from the sun (in AU, omitted if greater than 0.3 AU). These quantities are given only for objects with perihelion distances less than 1.05 AU, although they are physically meaningful only when the orbit of an object intersects that of the earth. For Amor objects with perihelion distances less than 1.05 AU, a single value spans the two columns. For more information on these quantities, see Marsden, B. G. (1993) To Hit Or Not To Hit. In *Proceedings of the Near-Earth-Object Interception Workshop*, pp. 67-71. Eds. G. H. Canavan, J. C. Solem and J. D. G. Rather. Los Alamos National Laboratory.

N+, N-: The distances of an object's nodal points from the earth's orbit (in AU, omitted if greater than 0.3 AU). N+ refers to the distance from the ascending node, N- to the distance from the descending node. These quantities are only given for objects with perihelion distances less than 1.3 AU. For more information on these quantities see Marsden (1993, *ibid.*).

Min.: This quantity, in AU and given only in the list of the larger potentially dangerous minor planets, is the minimum value of N+, N-, M+ and M-. If the minimum value corresponds to one of the value of N, it is followed by a `D' (to denote that the minimum value occurs at the descending node) or by an `A' (to denote that the minimum value occurs at the ascending node). Note that quantity does not give any information on actual close approches (Information drawed out from the file "Explanation of Symbols" in "Various lists and plots" of the "Minor Planet Center", Massachusetts, USA (http:/cfa-www.harvard.edu/iau/lists/Lists.html).

Annex 2

Provisional Designations

The total number of new-style designations assigned by the Minor Planet Center (and its predecessors) is 544566. The table lists the number of new-style minor-planet provisional designations assigned in each year, as well as the number of omitted and deleted designations and the number of designations identified with numbered minor planets, comets and natural satellites. Also given are the number of designations involved in unnumbered identifications (Unn.) and the number of designations that are principal designations (Prin.).

Rank	Discoveries	Co	Between	Name(s)
1	50478		1997-2004	LINEAR
2	5123		1998-2004	LONEOS
3	5081		1985-2004	Spacewatch
4	4431		1995-2004	NEAT
5	3428		1960-1977	C. J. van Houten*
6	3161	79	1986-1999	E. W. Elst
7	2227	2	1991-2002	T. Kobayashi
8	1835		1998-2005	CSS
9	1196	306	1975-1989	S. J. Bus
10	947	110	1996-2001	K. Korlevic
11	931		1992-1993	UESAC
12	697		1987-2000	S. Ueda, H. Kaneda
13	664	6	1989-1993	H. E. Holt
14	605		1997-1999	ODAS
15	604	103	1965-1994	H. Debehogne
16	566		2000-2004	W. K. Y. Yeung
17	563	8	1977-1994	E. Bowell
18	563		1987-1997	K. Endate, K. Watanabe
19	552		1994-2000	Beijing Schmidt CCD Asteroid Program
20	519	18	1966-1986	N. S. Chernykh
21	496	351	1973-1995	E. F. Helin
22	436		1994-2000	BAO Schmidt CCD Asteroid Program
23	427		1995-2001	P. G. Comba
24	395		1914-1957	K. Reinmuth
25	368	14	1975-2002	R. H. McNaught

Table 1. Minor Planet Discoverers.

Rank	Discoveries	Between	Site name
1	50501	1980-2004	Lincoln Laboratory ETS, New Mexico
2	5539	1949-2004	Palomar Mountain
3	5123	1998-2004	Lowell Observatory-LONEOS
4	4963	1981-2004	Steward Observatory, Kitt Peak-Spacewatch
5	4763	1976-2003	European Southern Observatory, La Silla
6	2297	1999-2004	Palomar Mountain/NEAT
7	2227	1991-2002	Oizumi
8	1835	1998-2005	Catalina Sky Survey
9	1740	1995-2004	Haleakala-AMOS
10	1559	1975-2002	Siding Spring Observatory
11	1285	1966-1992	Crimea-Nauchnij
12	1162	1995-2001	Visnjan
13	992	1982-2000	Peking Observatory, Xinglong Station
14	821	1891-1959	Heidelberg-Konigstuhl
15	775	1977-2002	Klet Observatory, Ceske Budejovice
16	716	1987-2000	Kushiro
17	691	1973-2000	Lowell Observatory, Anderson Mesa Station
18	639	1987-1997	Kitami
19	606	1997-1999	Caussols-ODAS
20	508	1961-1995	Karl Schwarzschild Observatory, Tautenburg
21	478	1995-1999	Haleakala-NEAT/GEODSS
22	428	1995-2001	Prescott
23	406	1995-2003	Ondrejov
24	395	2001-2004	Desert Eagle Observatory
25	369	1999-2003	Fountain Hills

Table 2. Discovery sites.

Date	G-total	Total	Num.	Unnum.
2005 JULY 21	31148231	30863157	21609020	9254137
2005 JUNE 22	31144958	30863157	21609020	9254137
2005 MAY 23	29076150	28798017	20412550	8385467
2005 APR. 7	29070393	28798017	20412550	8385467
2005 FEB. 24	29065968	28798017	20412550	8385467
2005 JAN. 25	27193909	26929248	18923021	8006227
2004 DEC. 26	27188448	26929248	18923021	8006227
2004 NOV. 26	27183950	26929364	18923021	8006343
2004 OCT. 28	27180858	26929364	18923021	8006343
2004 SEPT. 2	25777637	25529678	17667099	7862579
2004 AUG. 30	25774634	25529679	17667099	7862580
2004 JULY 13	24467106	24226721	16316844	7909877
2004 JUNE 14	24462701	24226721	16316844	7909877
2004 MAY 4	24457805	24227286	16317409	7909877
2004 APR. 15	22740717	22512796	14701393	7811403
2004 MAR. 6	22737221	22512798	14701395	7811403
2004 FEB. 6	22709339	22487196	14676153	7811403
2004 JAN. 7	21002723	20783437	13260328	7523109
2003 DEC. 8	20999811	20783437	13260328	7523109

Table 2. Continued.

Date	G-total	Total	Num.	Unnum.
2003 NOV. 9	20996655	20783437	13260328	7523109
2003 OCT. 10	19738267	19528183	12238244	7289939
2003 SEPT.10	19735799	19529362	12238244	7291118
2003 AUG. 6	19073332	18870283	11583202	7287081
2003 JUNE 14	19068163	18870516	11583356	7287160
2003 MAY 1	17835056	17643161	10135467	7507694
2003 MAR. 18	18061511	17884877	10135012	7749865
2003 FEB. 16	17709465	17536021	9944176	7591845
2003 JAN. 6	16095934	15927481	8637644	7289837
2002 NOV. 20	16091657	15927481	8637644	7289837
2002 OCT. 21	15047801	14885649	7876425	7009224
2002 SEPT.21	15032031	14872166	7862942	7009224
2002 AUG. 22	14663187	14506528	7577328	6929200
2002 JULY 24	14229728	14076357	7144616	6931741
2002 JUNE 24	14228352	14076366	7144616	6931750
2002 MAY 26	13821888	13672511	6873821	6798690
2002 APR. 27	12871438	12724025	6282382	6441643
2002 MAR. 28	12869859	12724025	6282382	6441643
2002 FEB. 27	12244472	12100591	5813323	6287268
2002 JAN. 28	11626118	11484818	5408688	6076130
2001 DEC. 30	10434260	10294501	4858465	5436036
2001 NOV. 30	10432623	10294501	4858465	5436036
2001 NOV. 2	9295084	9158826	4340461	4818365
2001 OCT. 2	9293550	9158826	4340461	4818365
2001 SEPT. 2	8569563	8436698	4044338	4392360
2001 AUG. 4	8307627	8176594	3857318	4319276
2001 JULY 5	8173455	8043611	3750179	4293432
2001 JUNE 6	8055380	7926636	3647751	4278885
2001 MAY 9	7614471	7486552	3361104	4125448
2001 APR. 8	7613279	7486552	3361104	4125448
2001 MAR 9	7152219	7026881	3067420	3959461
2001 FEB. 8	6900401	6776125	2898820	3877305
2001 JAN. 9	6565826	6442557	2685858	3756699
2000 DEC. 11	6293060	6171238	2513767	3657471
2000 NOV. 11	5994340	5873652	2384537	3489115
2000 OCT. 13	5691892	5572441	2270108	3302333
2000 SEPT.13	5315726	5197079	2145798	3051281
2000 JULY 26	4988151	4870733	2008481	2862252
2000 JUNE 21	4771151	4655438	1941688	2713750
2000 MAY 23	4530870	4415853	1874284	2541569
2000 APR. 18	4194849	4080954	1799778	2281176
2000 MAR. 20	3828808	3715705	1654324	2061381
2000 FEB. 22	3609758	3497777	1584702	1913075
2000 JAN. 24	3414994	3304880	1509331	1795549

Table 3. Statistics of asteroid discovery.

Number	Name	Des.	q	Q	M+	M-	N+	N-	H
(99935)		2002 AV4	0.588	2.721	0.183	0.169			15.9
(99248)		2001 KY66	0.919	2.814	0.181	0.065	0.113		16.2
(98943)		2001 CC21	0.806	1.259	0.084	0.083	0.259	0.194	18.5
(96744)		1999 OW3	0.459	3.719					14.7
(96536)		1998 SO10	0.511	4.086					16.0
(96315)		1997 AP10	0.515	2.371	0.114	0.110			17.0
(90416)		2003 YK118	0.860	2.528	0.130	0.005		0.010	18.7
(90403)		2003 YE45	0.889	1.579	0.042	0.298	0.026		17.8
(90367)		2003 LC5	0.663	1.648	0.226	0.172	0.270		17.7
(90147)		2002 YK14	0.985	1.963			0.209		18.5
(90075)		2002 VU94	0.911	3.361	0.147	0.029	0.042		15.7
(89959)		2002 NT7	0.817	2.654	0.029		0.016		16.5
(89958)		2002 LY45	0.187	3.098	0.002	0.173	0.022		16.9
(88959)		2001 TZ44	0.751	2.696	0.098		0.052		17.6
(88710)		2001 SL9	0.775	1.348		0.201	0.281		17.6
(88254)		2001 FM129	0.438	1.926	0.026	0.010			17.3
(87311)		2000 QJ1	0.775	2.404	0.108	0.134	0.214		16.5
(87025)		2000 JT66	0.634	1.820					18.0
(87024)		2000 JS66	0.970	1.424	0.214	0.188	0.134	0.173	18.7
(86878)		2000 HD24	0.512	2.171	0.104	0.156			17.7
(86829)		2000 GR146	0.622	2.303	0.208	0.217			16.2
(86819)		2000 GK137	0.986	3.006	0.039	0.127		0.032	17.4
(86666)		2000 FL10	0.839	2.087		0.135	0.105		16.8
(86039)		1999 NC43	0.740	2.779	0.029	0.091		0.096	16.0
(85990)		1999 JV6	0.694	1.321	0.028	0.072	0.104	0.226	20.0
(85938)		1999 DJ4	0.957	2.748	0.117	0.034		0.028	18.6
(85818)		1998 XM4	0.966	2.348		0.125			15.5
(85774)		1998 UT18	0.942	1.867	0.235	0.040	0.033		19.2
(85713)		1998 SS49	0.694	3.155	0.000	0.078		0.001	15.8
(85640)		1998 OX4	0.813	2.352	0.002	0.062		0.009	21.2
(85585)	Mjolnir	1998 FG2	0.835	1.760	0.020	0.033	0.171	0.097	21.6
(85236)		1993 KH	0.850	1.618	0.008	0.157	0.009	0.274	18.9
(85182)		1991 AQ	0.486	3.942	0.018	0.035			17.1
(69230)	Hermes	1937 UB	0.622	2.688	0.003	0.006	0.038	0.015	17.5
(68950)		2002 QF15	0.693	1.421	0.023	0.186	0.020	0.142	16.4
(68548)		2001 XR31	0.962	2.451		0.029	0.011		17.0
(68372)		2001 PM9	0.945	2.291	0.006	0.135	0.008		18.9
(68348)		2001 LO7	0.337	3.965					14.7
(68346)		2001 KZ66	0.879	2.136	0.070	0.287		0.056	17.0
(68267)		2001 EA16	0.864	2.155	0.155		0.062		17.0
(68216)		2001 CV26	0.889	1.752	0.298	0.044	0.030		16.4
(67399)		2000 PJ6	0.851	1.752	0.238	0.087		0.079	18.4
(67381)		2000 OL8	0.603	2.037	0.013	0.034	0.038	0.098	19.9
(66253)		1999 GT3	0.216	2.452	0.199	0.275			18.0
(66008)		1998 QH2	0.909	1.943		0.084			16.0
(65909)		1998 FH12	0.502	1.682	0.039	0.011		0.106	19.2
(65733)		1993 PC	0.607	1.702	0.072	0.066			18.3
(65690)		1991 DG	0.909	1.945	0.179	0.048	0.065		19.1
(55532)		2001 WG2	0.545	3.044					16.3
(55408)		2001 TC2	0.853	1.347		0.146			18.8
(54509)		2000 PH5	0.773	1.236	0.010	0.002	0.080	0.016	22.7

Table 3. Continued

Number Name	Des.	q	Q	M+	M-	N+	N-	H
(53550)	2000 BF19	0.868	2.123	0.045	0.125	0.082		19.1
(53429)	1999 TF5	0.732	3.316		0.063	0.064		19.0
(53426)	1999 SL5	0.887	2.958		0.029	0.021		17.4
(53409)	1999 LU7	0.781	3.421	0.098	0.186		0.168	18.7
(53319)	1999 JM8	0.950	4.465	0.064	0.163		0.040	15.3
(52762)	1998 MT24	0.846	3.993		0.205		0.183	14.8
(52760)	1998 ML14	0.917	3.908	0.036	0.013	0.061		17.5
(52750)	1998 KK17	0.678	2.173	0.168	0.179	0.298		16.5
(42286)	2001 TN41	0.863	1.976	0.203			0.105	16.4
(41429)	2000 GE2	0.708	2.475	0.011	0.024	0.124		20.7
(40267)	1999 GJ4	0.257	2.421	0.162				15.3
(38239)	1999 OR3	0.859	3.207	0.162	0.068	0.103		17.9
(38086)	1999 JB	0.616	2.225					17.4
(37655) Illapa	1994 PM	0.366	2.591	0.272	0.027		0.097	17.7

Table 3a. Numbered (and named) Apollo-asteroids.

Number Name	Des.	q	Q	M+	M-	N+	N-	H
(37638)	1993 VB	0.918	2.901	0.005	0.086	0.013		19.4
(36284)	2000 DM8	0.661	2.306			0.272		15.0
(36236)	1999 VV	0.768	1.897			0.218		16.3
(35670)	1998 SU27	0.865	3.386	0.079	0.108		0.129	19.5
(35396)	1997 XF11	0.745	2.141	0.000	0.031	0.237	0.002	16.9
(35107)	1991 VH	0.973	1.300	0.219	0.047	0.276	0.013	16.9
(31669)	1999 JT6	0.899	3.372	0.165	0.013	0.021		16.0
(31662)	1999 HP11	0.722	2.472	0.158	0.262	0.181		19.5
(30997)	1995 UO5	0.556	2.564					17.2
(30825)	1990 TG1	0.781	4.098	0.151	0.074	0.162		14.7
(29075)	1950 DA	0.836	2.561	0.205	0.053		0.074	17.0
(27002)	1998 DV9	0.988	2.499	0.043	0.040	0.012		18.2
(26663)	2000 XK47	0.817	2.275	0.211	0.055		0.073	18.2
(26379)	1999 HZ1	0.681	2.531	0.147	0.134			18.4
(25330)	1999 KV4	0.969	2.112	0.227	0.211	0.294		16.8
(25143) Itokawa	1998 SF36	0.953	1.695	0.010	0.023		0.038	19.2
(24761) Ahau	1993 BW2	0.927	1.744	0.163		0.109		17.4
(24443)	2000 OG	0.416	4.212	0.165				16.2
(23187)	2000 PN9	0.758	2.932	0.045		0.025		16.1
(22771)	1999 CU3	0.750	2.404	0.067	0.155	0.124		17.0
(22753)	1998 WT	0.524	1.913	0.053	0.033			17.7
(22099)	2000 EX106	0.799	1.409	0.171	0.168		0.200	18.0
(20826)	2000 UV13	0.909	3.957		0.200		0.074	13.5
(20429)	1998 YN1	0.833	2.278	0.053	0.109		0.123	18.0
(20425)	1998 VD35	0.819	2.310	0.000	0.096	0.001		20.4
(20236)	1998 BZ7	0.903	3.173	0.062	0.098		0.243	17.6
(17511)	1992 QN	0.763	1.618	0.160	0.147		0.222	17.1
(17188)	1999 WC2	0.801	3.631	0.105		0.107		16.8
(17182)	1999 VU	0.619	2.155	0.144	0.152			17.2
(17181)	1999 UM3	0.790	3.973	0.183	0.079	0.143		16.3

Table 3a. Continued

Number	Name	Des.	q	Q	M+	M-	N+	N-	H
(16960)		1998 QS52	0.310	4.088	0.009	0.256	0.039		14.3
(16834)		1997 WU22	0.819	2.117	0.177	0.275	0.155		15.7
(16816)		1997 UF9	0.571	2.314					16.2
(14827)	Hypnos	1986 JK	0.955	4.738	0.034	0.018		0.175	18.3
(13651)		1997 BR	0.927	1.744	0.001	0.296		0.000	17.6
(12923)	Zephyr	1999 GK4	0.998	2.926	0.041	0.059		0.054	16.1
(12711)		1991 BB	0.863	1.509		0.097			16.0
(12538)		1998 OH	0.915	2.168	0.048		0.024		16.1
(11885)		1990 SS	0.894	2.511	0.093		0.092		18.6
(11500)		1989 UR	0.696	1.465	0.085	0.032	0.156	0.069	18.4
(11405)		1999 CV3	0.885	2.034	0.194	0.263	0.288	0.182	15.0
(11311)	Peleus	1993 XN2	0.983	3.252	0.206		0.105		16.5
(11066)	Sigurd	1992 CC1	0.870	1.914		0.113			15.2
(10636)		1998 QK56	0.918	2.850	0.129	0.212	0.218		17.5
(10563)	Izhdubar	1993 WD	0.738	1.275		0.139	0.207		16.9
(10165)		1995 BL2	0.613	1.857					17.1
(10145)		1994 CK1	0.700	3.104	0.077	0.061	0.269		17.3
(10115)		1992 SK	0.843	1.655	0.223	0.070		0.064	17.0
(9856)		1991 EE	0.848	3.647	0.035	0.150		0.088	17.4
(9202)		1993 PB	0.559	2.288					16.0
(9162)		1987 OA	0.605	2.388	0.081	0.096		0.278	18.3
(9058)		1992 JB	0.997	2.116	0.192	0.245	0.115		17.8
(8566)		1996 EN	0.858	2.156	0.035		0.016		16.5
(8507)		1991 CB1	0.684	2.690	0.232	0.251			17.8
(8201)		1994 AH2	0.738	4.330	0.166	0.114	0.232		16.3
(8176)		1991 WA	0.563	2.588	0.233		0.291		17.1
(8035)		1992 TB	0.721	1.962		0.277			17.3
(8014)		1990 MF	0.950	2.543	0.018	0.032		0.166	18.7
(7889)		1994 LX	0.825	1.698		0.172			15.3
(7888)		1993 UC	0.819	4.052	0.150		0.110		15.3
(7822)		1991 CS	0.938	1.307		0.141	0.159	0.033	17.4
(7753)		1988 XB	0.760	2.174	0.005	0.023	0.040	0.229	18.6
(7482)		1994 PC1	0.904	1.788		0.044	0.017		16.8
(7350)		1993 VA	0.826	1.886	0.088	0.126	0.155		17.3
(7341)		1991 VK	0.910	2.775	0.056	0.072		0.088	16.7
(7335)		1989 JA	0.913	2.628	0.262	0.043		0.045	17.0
(7092)	Cadmus	1992 LC	0.752	4.298	0.112	0.132		0.248	15.4
(6611)		1993 VW	0.874	2.518	0.068	0.114	0.186		16.5

Table 3b. Numbered(and named) Apollo-asteroids.

Number	Name	Des.	q	Q	M+	M-	N+	N-	H
(6489)	Golevka	1991 JX	0.986	4.009	0.039	0.031	0.278		19.2
(6455)		1992 HE	0.958	3.523					13.8
(6239)	Minos	1989 QF	0.676	1.628	0.028	0.041	0.208	0.210	17.9
(6063)	Jason	1984 KB	0.526	3.912	0.081	0.074			15.3
(6047)		1991 TB1	0.942	1.966	0.242		0.176		17.0

Table 3b. Continued

Number Name	Des.	q	Q	M+	M-	N+	N-	H
(6037)	1988 EG	0.636	1.905	0.024	0.034	0.249	0.229	18.7
(5828)	1991 AM	0.517	2.878					16.3
(5786) Talos	1991 RC	0.187	1.976	0.191	0.282			17.0
(5731) Zeus	1988 VP4	0.785	3.742	0.195	0.086		0.147	15.8
(5693)	1993 EA	0.527	2.016	0.008	0.041	0.057	0.250	17.0
(5660)	1974 MA	0.424	3.147		0.185			15.7
(5645)	1990 SP	0.830	1.880	0.212	0.076	0.085		17.0
(5496)	1973 NA	0.883	3.986	0.259			0.111	15.3
(5189)	1990 UQ	0.810	2.292	0.044	0.062		0.173	17.3
(5143) Heracles	1991 VL	0.416	3.249	0.063	0.141			14.0
(5131)	1990 BG	0.640	2.333				0.287	14.1
(5011) Ptah	6743 P-L	0.818	2.452	0.025	0.087		0.082	17.1
(4953)	1990 MU	0.556	2.687	0.040	0.137	0.193	0.073	14.1
(4769) Castalia	1989 PB	0.549	1.577	0.125	0.024	0.088		16.9
(4660) Nereus	1982 DB	0.953	2.024	0.005	0.021		0.028	18.2
(4581) Asclepius	1989 FC	0.657	1.387	0.004	0.046	0.018	0.183	20.4
(4544) Xanthus	1989 FB	0.781	1.302	0.228	0.208	0.202	0.259	17.1
(4486) Mithra	1987 SB	0.748	3.660	0.045	0.052		0.247	15.6
(4450) Pan	1987 SY	0.596	2.287	0.043	0.027	0.222	0.205	17.2
(4341) Poseidon	1987 KF	0.590	3.081	0.197	0.199			15.5
(4257) Ubasti	1987 QA	0.876	2.419		0.199			16.2
(4197)	1982 TA	0.526	4.074	0.120	0.087			14.6
(4183) Cuno	1959 LM	0.721	3.243	0.091	0.037		0.133	14.4
(4179) Toutatis	1989 AC	0.921	4.124	0.005	0.007			15.30
(4034)	1986 PA	0.589	1.530	0.140	0.023	0.290	0.061	18.1
(4015) Wilson-Harrington	1979 VA	0.993	4.285	0.047	0.048			15.99
(3838) Epona	1986 WA	0.449	2.561	0.173				15.5
(3752) Camillo	1985 PA	0.987	1.841		0.068			15.5
(3361) Orpheus	1982 HR	0.819	1.599	0.013	0.034	0.073		19.03
(3360)	1981 VA	0.633	4.302	0.215	0.124	0.181		16.3
(3200) Phaethon	1983 TB	0.140	2.403		0.025		0.115	14.6
(3103) Eger	1982 BB	0.907	1.903	0.297	0.146		0.119	15.38
(2329) Orthos	1976 WA	0.824	3.985	0.155			0.115	14.9
(2212) Hephaistos	1978 SB	0.359	3.974	0.124	0.203			13.87
(2201) Oljato	1947 XC	0.623	3.720	0.001	0.009	0.157	0.010	15.25
(2135) Aristaeus	1977 HA	0.795	2.404	0.014	0.270	0.014		17.94
(2102) Tantalus	1975 YA	0.904	1.675		0.116	0.029		16.2
(2101) Adonis	1936 CA	0.441	3.307	0.012	0.021			18.7
(2063) Bacchus	1977 HB	0.701	1.455	0.072	0.113	0.211	0.182	17.1
(1981) Midas	1973 EA	0.621	2.931	0.050	0.000	0.052	0.000	15.5
(1866) Sisyphus	1972 XA	0.874	2.914	0.189		0.110		13.0
(1865) Cerberus	1971 UA	0.576	1.584	0.267	0.162			16.84
(1864) Daedalus	1971 FA	0.563	2.359		0.276			14.85
(1863) Antinous	1948 EA	0.888	3.630	0.230	0.214			15.54
(1862) Apollo	1932 HA	0.648	2.295	0.028	0.032	0.124	0.192	16.25
(1685) Toro	1948 OA	0.771	1.963	0.063	0.127		0.123	14.23
(1620) Geographos	1951 RA	0.828	1.663	0.046	0.098	0.063	0.151	15.60
(1566) Icarus	1949 MA	0.187	1.969	0.040			0.162	16.9

Table 3c. Numbered (and named) Apollo-asteroids.

Year	Total	Removed	Identified with Omit Del	# mps	cmts	sats	Uun.	Prin.
2005	24995	75	3	1	5	0	4751	4116
2004	56689	289	3	428	13	0	13137	11842
2003	48085	323	3	1722	11	0	12989	14257
2002	64766	348	57	9614	9	0	16900	20672
2001	73193	194	19	21287	11	0	14658	21926
2000	72370	96	24	36978	10	1	15392	17147
1999	48871	90	20	27909	11	5	9143	8372
1998	32286	70	32	22022	6	5	5616	3262
1997	12231	32	0	7659	1	0	2378	963
1996	10250	31	2	5640	0	0	2038	633
1995	10833	20	1	4249	0	0	2407	953
1994	6336	7	2	3870	0	2	1177	375
1992	4886	21	6	3875	1	0	456	102
1991	5475	41	4	4519	2	0	296	76
1990	4943	23	16	4072	1	0	201	48
1989	3413	28	0	2922	1	2	110	28
1988	2518	20	6	2124	1	2	55	16
1987	2774	20	1	2231	0	1	50	3
1986	2770	10	2	2129	4	3	32	3
1985	1819	13	1	1464	0	0	23	2
1984	1516	1	0	1284	1	1	35	2
1983	1622	30	0	1189	2	0	29	9
1982	2703	24	2	2217	1	0	34	6
1981	3798	10	6	3290	2	0	140	118
1980	2116	2	2	1914	0	0	37	5
1979	2664	5	3	2291	0	0	81	22
1978	3220	5	3	2745	1	0	125	16
1977	2376	7	2	1996	2	0	58	2
1975	1329	13	2	1203	1	1	10	3
1974	914	2	1	825	0	0	6	1
1973	868	21	0	756	1	0	5	1
1972	895	3	0	835	0	0	9	1
1971	814	14	0	712	1	0	5	0
1970	554	0	3	482	0	1	0	0
1969	728	1	1	643	0	0	6	0
1968	385	0	1	318	0	0	0	0
1967	238	0	0	165	1	0	0	0
1966	319	0	2	223	0	0	1	0
1965	255	0	0	223	0	0	0	0
1964	351	1	1	315	0	1	0	0
1963	194	0	1	177	0	1	0	0
1962	329	4	1	312	0	0	0	0
1961	257	4	0	206	0	0	0	0
1960	137	0	1	104	1	0	0	0
1959	291	0	2	261	0	0	0	0

Table 3c. Continued

Year	Total	Removed	Identified with Omit Del	# mps	cmts	sats	Uun.	Prin.
1958	274	0	3	257	0	0	0	0
1957	418	0	4	289	0	0	0	0
1956	302	0	1	173	0	0	0	0
1955	599	0	8	408	0	0	1	0
1954	561	1	1	372	1	0	1	0
1953	739	0	3	528	0	0	2	0
1952	653	0	4	522	0	0	3	1
1951	736	0	0	619	0	0	1	1
1950	678	2	7	523	1	2	0	0
1949	563	1	26	454	0	0	1	0
1948	412	1	21	289	0	0	0	0
1947	170	0	4	111	0	0	0	0
1946	104	2	9	61	0	0	0	0
1945	74	0	24	39	0	0	0	0
1944	81	0	9	60	0	0	0	0
1943	180	0	11	129	0	0	0	0
1942	305	1	16	227	1	0	0	0
1941	291	0	27	190	1	0	0	0
1940	257	0	13	155	1	0	0	0
1939	291	0	11	192	2	1	0	0
1938	376	0	8	278	0	0	0	0
1937	279	0	2	196	0	0	0	0
1936	331	1	2	250	0	0	0	0
1935	350	0	2	277	0	1	0	0
1934	313	0	1	249	0	0	0	0
1933	340	0	1	260	0	0	0	0
1932	243	0	2	163	0	0	0	0
1931	441	1	2	348	1	0	0	0
1930	348	0	5	251	0	1	0	0
1929	328	0	17	253	0	3	0	0
1928	190	0	6	135	0	0	0	0
1927	145	0	9	94	0	0	0	0
1926	152	0	4	90	0	0	0	0
1925	106	0	1	66	1	0	0	0
1924	105	0	7	66	0	1	0	0
1923	50	0	1	37	0	0	0	0
1922	37	0	1	20	0	0	0	0
1921	53	0	1	27	0	0	0	0
1920	32	0	1	21	0	0	0	0
1919	38	0	4	23	0	0	0	0
1918	41	0	2	31	0	0	0	0
1917	80	0	4	64	0	0	0	0
1916	94	0	2	74	0	0	0	0
1915	82	0	5	61	0	0	0	0
1914	55	0	2	29	0	0	0	0
1913	70	0	3	39	0	0	0	0

Table 3c. Continued

Year	Total	Removed	Identified with Omit Del	# mps	cmts	sats	Uun.	Prin.
1912	73	0	3	40	0	0	0	0
1911	59	0	0	47	0	0	0	0
1910	39	0	0	36	0	0	0	0
1909	64	0	2	57	0	0	0	0
1908	91	0	5	71	0	0	0	0
1907	92	0	0	77	0	0	0	0
1906	102	0	5	86	0	0	0	0
1905	61	0	1	56	0	0	0	0
1904	45	0	3	36	0	0	0	0
1903	38	0	0	33	0	0	0	0
1902	50	0	3	42	0	0	0	0
1901	29	0	2	22	0	0	0	0
1900	26	0	5	15	0	0	0	0
1899	45	0	0	30	0	0	0	0
1898	9	0	1	5	0	0	0	0
1897	1	0	0	1	0	0	0	0
1896	7	0	0	6	0	0	0	0
1895	9	0	0	7	0	0	0	0
1894	4	0	0	4	0	0	0	0
1893	13	0	1	10	0	0	0	0
1892	12	0	0	3	0	0	0	0
1891	6	0	0	2	0	0	0	0
1890	1	0	0	0	0	0	0	0
1887	2	0	0	2	0	0	0	0
1881	1	0	1	0	0	0	0	0
1880	1	0	0	1	0	0	0	0
1878	2	0	0	1	0	0	0	0
1874	1	0	0	0	0	0	0	0
1873	1	0	0	0	0	0	0	0
1869	1	0	0	1	0	0	0	0
P-L	2369	0	0	1707	0	0	327	343
T-1	746	0	0	671	0	0	37	37
T-2	1396	0	0	1007	0	0	243	245
T-3	1430	0	0	899	0	0	299	303

Table 4. Provisional designations.

Designation (and name)	Prov. Des.	Min.	q	Q	H	Epoch	M	Peri.	Node	Incl.	e	a
	2005 PA5	0.050D	1.022	1.617	21.9	20050729	15.9	148.9	133.1	15.3	0.225	1.319
	2005 PJ2	0.049	0.409	1.984	20.0	20050818	308.2	128.7	326.6	17.4	0.658	1.196
	2005 PO	0.034	0.783	1.724	20.2	20050818	102.8	249.2	300.7	12.6	0.375	1.253
	2005 OE3	0.044A	0.896	2.235	20.2	20050818	338.3	58.3	321.2	25.4	0.428	1.565
	2005 OD3	0.025A	0.942	3.780	21.0	20050729	348.4	38.3	323.9	16.2	0.601	2.361
	2005 OX	0.015	0.509	2.152	20.4	20050818	328.9	100.1	323.9	7.6	0.617	1.331
	2005 NB7	0.005A	0.988	3.098	18.7	20050818	36.3	348.1	199.3	12.7	0.517	2.043
	2005 NZ6	0.010	0.250	3.419	17.4	20050818	340.3	47.8	39.8	8.5	0.864	1.834
	2005 NJ1	0.024	0.742	3.518	18.7	20050818	350.9	82.0	316.8	6.5	0.652	2.130
	2005 MO13	0.044	0.508	1.218	20.7	20050818	293.6	250.1	176.8	6.3	0.411	0.863
	2005 LO40	0.007	0.395	1.507	21.2	20050818	169.4	51.7	51.6	15.5	0.584	0.951
	2005 LW39	0.022	0.566	3.417	20.0	20050818	39.6	275.0	251.8	28.5	0.716	1.991
	2005 LX36	0.017	0.438	1.622	20.8	20050818	125.1	239.1	252.8	33.0	0.575	1.030
	2005 LW3	0.004	0.769	2.093	21.7	20050818	357.7	288.0	59.8	6.1	0.462	1.431
	2005 KJ10	0.026	0.791	3.047	21.3	20050530	342.2	233.7	81.4	8.3	0.588	1.919
	2005 JS108	0.028	0.919	1.794	19.1	20050818	339.7	243.7	111.0	6.0	0.323	1.356
	2005 JU81	0.049A	0.806	1.427	21.9	20050818	151.1	72.5	56.4	17.3	0.278	1.116
	2005 JE46	0.028	0.852	2.958	18.2	20050818	343.6	114.5	238.6	8.2	0.553	1.905
	2005 JU1	0.032	0.741	3.934	21.8	20050818	14.2	266.5	35.7	6.1	0.683	2.338
	2005 GC120	0.000	0.600	1.787	19.7	20050818	21.8	258.1	68.1	16.5	0.497	1.193
	2005 GD60	0.038	0.832	1.825	21.9	20050818	28.9	100.0	199.2	9.9	0.374	1.329
	2005 GE59	0.041	0.844	3.379	18.1	20050818	105.0	242.7	195.1	16.1	0.600	2.111
	2005 GR33	0.016D	0.480	1.078	22.0	20050818	54.4	334.6	22.7	27.9	0.384	0.779
	2005 GO22	0.019	0.338	3.490	18.7	20050818	68.8	18.7	62.1	1.6	0.824	1.914
	2005 GP21	0.007	1.015	1.602	19.6	20050818	265.3	1.3	10.1	18.8	0.224	1.308
	2005 GO21	0.032A	0.497	1.009	16.6	20050818	50.8	156.5	272.7	24.9	0.340	0.753
	2005 GY8	0.001	0.678	3.389	21.2	20050818	20.2	102.5	180.2	2.8	0.666	2.034
	2005 GJ8	0.031	0.793	2.740	20.3	20050818	84.9	214.3	252.9	3.4	0.551	1.767
	2005 GU	0.018	0.598	2.504	20.0	20050818	33.2	275.6	27.1	25.0	0.614	1.551
	2005 GL	0.016	0.733	1.376	21.1	20050818	41.0	265.1	43.7	15.9	0.305	1.054

Table 4. Continued

Designation (and name)	Prov. Des.	Min.	q	Q	H	Epoch	M	Peri.	Node	Incl.	e	a
	2005 FH	0.037A	0.930	4.471	17.5	20050818	55.2	317.8	144.3	34.7	0.655	2.701
	2005 ED318	0.011	1.019	2.676	20.8	20050818	32.7	164.0	82.1	2.4	0.448	1.848
	2005 EJ225	0.014	0.729	2.686	18.6	20050818	16.7	290.2	342.4	8.9	0.573	1.708
	2005 EW169	0.018	1.032	3.759	19.1	20050818	76.4	221.8	223.1	1.5	0.569	2.396
	2005 EY95	0.040	0.500	1.667	20.3	20050818	200.5	341.9	73.1	3.2	0.538	1.083
	2005 EO33	0.045	0.841	3.101	21.3	20050818	79.1	158.5	289.7	4.6	0.574	1.971
	2005 EE	0.025	0.759	1.502	20.9	20050818	228.0	284.7	111.0	6.2	0.328	1.130
	2005 EA	0.032	0.468	2.139	21.1	20050818	58.7	298.7	347.0	14.1	0.641	1.303
	2005 DD	0.033	0.837	3.032	20.1	20050818	105.3	251.8	157.6	7.3	0.567	1.934
	2005 CF41	0.034	0.909	2.331	18.9	20050130	131.7	215.4	127.9	11.1	0.439	1.620
	2005 CZ36	0.036D	0.944	3.525	17.5	20050818	9.4	139.1	117.0	16.1	0.578	2.234
	2005 CL	0.023	0.493	3.486	20.4	20050818	37.7	297.6	319.3	4.4	0.752	1.989
	2005 CJ	0.007	0.829	2.669	20.4	20050818	102.0	81.4	358.4	1.1	0.526	1.749
	2005 BG14	0.033	0.540	3.446	18.8	20050818	93.4	280.3	95.6	21.6	0.729	1.993
	2005 BY2	0.037	0.850	1.688	20.3	20050818	93.2	130.1	74.6	7.3	0.330	1.269
	2005 AY28	0.034	0.378	1.367	21.8	20050818	141.8	155.7	117.7	5.8	0.567	0.872
	2005 AV27	0.037	0.787	3.733	20.5	20050818	42.7	138.7	68.1	5.9	0.652	2.260
	2005 AN26	0.038	0.841	3.800	18.3	20050818	89.9	207.7	176.2	4.9	0.638	2.320
	2005 AD13	0.020	0.475	3.421	17.9	20050818	100.7	262.7	103.6	12.4	0.756	1.948
	2004 XZ130	0.004	0.337	0.898	20.2	20050818	319.4	4.7	211.8	3.0	0.455	0.618
	2004 XN50	0.009	0.373	2.991	18.8	20050818	132.0	252.9	62.8	3.3	0.778	1.682
	2004 XL35	0.021A	0.978	1.733	19.5	20050818	158.8	356.1	73.8	27.3	0.278	1.355
	2004 XN29	0.024	0.713	4.137	22.0	20050818	77.9	36.9	320.1	2.4	0.706	2.425
	2004 XP14	0.015A	0.891	1.224	19.3	20050818	137.2	275.9	281.1	32.9	0.157	1.057
	2004 XN14	0.015	0.683	1.180	20.1	20050818	150.3	115.7	121.0	10.7	0.266	0.931
	2004 XL14	0.018A	0.449	1.071	21.6	20050818	224.1	157.5	85.7	21.5	0.410	0.760
	2004 XO	0.005A	0.935	1.135	21.6	20050818	334.6	71.4	250.0	26.0	0.097	1.035
	2004 WS2	0.027	0.530	2.143	18.0	20050818	122.4	115.2	87.9	8.3	0.603	1.337
	2004 WK1	0.044	0.293	1.876	21.1	20050818	289.3	223.1	51.9	34.5	0.730	1.084
	2004 VG64	0.032	0.334	1.603	18.6	20050818	348.6	43.9	208.9	36.3	0.655	0.968
	2004 VD17	0.001	0.620	2.396	18.8	20050818	180.5	90.7	224.2	4.2	0.589	1.508

Table 4. Continued

Designation (and name)	Prov. Des.	Min.	q	Q	H	Epoch	M	Peri.	Node	Incl.	e	a
	2004 VC17	0.000	0.361	3.431	18.4	20050818	124.8	64.5	229.8	20.5	0.810	1.896
	2004 VW14	0.015	0.845	3.391	19.4	20050818	64.5	41.4	101.7	3.9	0.601	2.118
	2004 VC	0.038D	0.840	1.427	18.8	20050818	299.5	94.2	194.2	39.1	0.259	1.133
	2004 VB	0.042	0.862	2.055	20.8	20050818	121.6	71.4	52.5	10.9	0.409	1.458
	2004 VA	0.027	0.769	3.034	17.3	20050818	68.7	43.1	109.7	3.7	0.595	1.901
	2004 UV1	0.032A	1.007	4.665	17.9	20050818	45.7	20.0	76.7	20.8	0.645	2.836
	2004 UU1	0.023D	0.891	1.562	21.2	20050818	253.9	113.5	217.8	30.0	0.274	1.226
	2004 UR1	0.020	0.830	3.516	20.5	20050818	107.1	72.4	251.8	4.1	0.618	2.173
	2004 UL	0.021	0.093	2.440	18.8	20050818	158.2	149.4	39.7	23.7	0.927	1.266
	2004 UE	0.011D	0.940	4.324	21.2	20050818	59.0	215.0	227.5	15.1	0.643	2.632
	2004 TB18	0.006A	0.998	2.633	17.7	20050818	69.5	12.9	121.1	13.2	0.450	1.816
	2004 TL10	0.005	0.927	4.398	21.4	20050818	79.1	322.7	12.2	9.1	0.652	2.663
	2004 TG10	0.012	0.315	4.169	19.4	20050818	64.1	310.0	212.3	3.7	0.860	2.242
	2004 TB10	0.012D	1.000	1.206	21.3	20050818	122.6	149.7	12.9	22.5	0.094	1.103
	2004 TP1	0.027	0.788	1.792	20.7	20050818	157.2	87.6	30.2	7.5	0.389	1.290
	2004 TN1	0.007	0.862	4.678	21.8	20050818	53.4	231.3	215.9	8.6	0.689	2.770
	2004 SB56	0.048A	0.660	1.072	19.3	20050818	234.1	233.5	302.1	18.7	0.238	0.866
	2004 SW55	0.047A	0.951	1.903	20.7	20050818	154.6	352.7	75.2	8.8	0.334	1.427
	2004 SS	0.014	1.036	3.352	22.0	20050818	102.3	352.9	1.6	6.5	0.528	2.194
	2004 RZ164	0.003A	0.983	4.125	19.0	20050818	64.1	344.3	75.3	13.8	0.615	2.554
	2004 RF84	0.038D	0.857	3.471	18.4	20050818	65.1	122.4	344.8	23.2	0.604	2.164
	2004 RY10	0.049	0.633	2.712	21.4	20050818	124.1	283.0	166.1	15.5	0.621	1.673
	2004 RW10	0.039	0.966	3.745	21.1	20050818	109.1	69.6	207.8	3.1	0.590	2.356
	2004 RQ10	0.018	1.041	2.688	20.9	20050818	140.1	349.4	335.3	5.7	0.442	1.865
	2004 QT24	0.002A	0.847	1.418	18.3	20050818	168.0	76.5	36.2	17.7	0.252	1.133
	2004 QD14	0.019	0.624	1.261	20.6	20050818	140.1	109.1	75.5	6.2	0.338	0.942
	2004 QY2	0.045A	0.567	1.601	14.7	20050818	241.4	104.9	295.4	37.0	0.477	1.084
	2004 QX2	0.035	0.125	2.447	21.7	20050818	279.0	218.6	320.4	19.1	0.903	1.286
	2004 QB	0.005	0.736	3.486	19.8	20050818	135.4	110.7	132.1	11.0	0.651	2.111
	2004 PJ2	0.026	0.933	1.903	21.4	20050818	258.4	281.6	317.3	2.6	0.342	1.418
	2004 OT11	0.000A	0.939	3.367	17.3	20050818	142.4	326.1	295.3	19.9	0.564	2.153

Table 4. Continued

Designation (and name)	Prov. Des.	Min.	q	Q	H	Epoch	M	Peri.	Node	Incl.	e	a
	2004 OB	0.013	0.958	2.405	18.9	20050818	129.7	223.1	174.7	3.4	0.430	1.681
	2004 NL8	0.027	0.739	4.431	17.1	20050818	72.6	266.1	162.0	4.9	0.714	2.585
	2004 MP7	0.043	0.778	4.701	21.7	20050818	100.5	109.5	98.0	17.2	0.716	2.740
	2004 MD6	0.001	0.416	1.486	20.5	20050818	165.0	231.4	263.9	29.3	0.563	0.951
(99942) Apophis	2004 MN4	0.001	0.746	1.099	19.2	20050818	359.7	126.4	204.5	3.3	0.191	0.922
	2004 MX2	0.014	0.797	4.749	19.3	20050818	77.9	241.8	118.2	18.3	0.713	2.773
	2004 MD	0.026	0.888	2.947	20.2	20050818	175.2	123.9	85.8	9.7	0.537	1.918
	2004 LY5	0.019	0.405	3.249	21.5	20050818	151.5	295.0	87.4	16.0	0.778	1.827
	2004 LV3	0.005D	0.892	1.570	18.7	20050818	227.5	121.3	273.5	35.3	0.276	1.231
	2004 LC2	0.029	0.496	3.139	18.8	20050818	151.2	290.4	84.7	10.8	0.727	1.818
	2004 LJ1	0.008D	0.918	3.609	15.5	20050818	94.8	139.5	235.9	23.0	0.594	2.263
	2004 LJ	0.023	0.585	1.589	20.2	20050818	84.2	256.0	246.8	18.3	0.462	1.087
	2004 LB	0.004D	0.995	1.107	21.8	20050818	258.4	145.0	254.1	37.3	0.053	1.051
	2004 KB	0.037D	0.845	1.529	21.1	20050818	74.4	243.2	231.0	20.7	0.288	1.187
	2004 JA27	0.034	0.961	2.371	19.4	20050818	160.6	270.1	67.8	2.3	0.423	1.666
	2004 JG6	0.036D	0.298	0.973	18.9	20050818	28.7	352.9	37.1	18.9	0.531	0.635
	2004 JQ1	0.028D	0.656	1.731	20.1	20050818	278.3	272.6	77.1	32.8	0.450	1.194
	2004 HC39	0.000	0.906	2.743	21.5	20050818	170.7	224.0	56.2	14.7	0.503	1.824
	2004 HK33	0.004	0.904	2.870	17.6	20050818	155.3	221.3	104.9	5.4	0.521	1.887
	2004 HF12	0.030	0.750	3.522	20.5	20050818	167.4	104.0	23.5	13.3	0.649	2.136
	2004 HE12	0.001D	0.984	2.706	18.0	20050818	197.1	162.1	338.1	42.7	0.467	1.845
	2004 HD2	0.035	0.385	4.267	21.1	20050818	148.2	61.8	26.9	15.5	0.835	2.326
	2004 HW	0.010	0.977	4.401	17.2	20050818	100.5	62.3	220.4	0.8	0.637	2.689
	2004 GU9	0.009D	0.864	1.137	21.3	20050818	13.8	281.3	38.9	13.6	0.137	1.001
	2004 GE2	0.005	0.600	3.476	21.5	20050818	146.8	260.0	45.1	2.2	0.706	
	2004 GB2	0.012	0.735	3.498	21.0	20050818	180.9	253.9	210.3	12.5	0.653	2.117
	2004 GA1	0.011	0.807	4.067	17.5	20050818	98.2	54.2	261.9	7.8	0.669	2.437
	2004 FY31	0.026D	0.959	3.374	21.9	20050818	147.0	196.8	29.0	11.3	0.557	2.166
	2004 FJ11	0.026	0.990	3.511	21.0	20050818	129.2	230.8	22.1	2.4	0.560	2.250
	2004 FG11	0.021	0.438	2.753	20.9	20050818	220.9	227.2	84.9	3.1	0.725	1.596
	2004 FE5	0.038	0.498	1.992	20.0	20050818	52.2	243.3	189.6	18.3	0.600	1.245

Table 4. Continued

Designation (and name)	Prov. Des.	Min.	q	Q	H	Epoch	M	Peri.	Node	Incl.	e	a
	2004 FU4	0.008A	0.928	1.593	18.5	20050818	99.4	46.2	31.6	23.3	0.264	1.260
	2004 EW	0.022	0.713	1.267	20.9	20050818	261.9	55.8	343.5	4.7	0.280	0.990
	2004 DM44	0.041	0.716	4.233	21.1	20050818	149.0	95.5	351.0	10.5	0.710	2.475
	2004 DV24	0.012D	1.011	1.834	16.8	20050818	100.9	186.3	171.2	55.9	0.290	1.423
	2004 DC	0.005D	0.981	2.291	18.3	20050818	232.7	156.2	75.0	19.4	0.400	1.636
	2004 CO49	0.026	0.841	1.910	21.7	20050818	12.8	315.7	115.3	4.3	0.389	1.376
	2004 CL	0.019	0.596	1.974	20.8	20050818	64.3	274.5	109.6	13.3	0.536	1.285
	2004 BB103	0.022A	0.722	3.092	17.3	20050818	232.8	71.5	271.2	55.9	0.621	1.907
	2004 BL86	0.006A	0.897	2.108	19.0	20050818	333.9	311.1	126.8	23.7	0.403	1.502
	2004 BZ74	0.031	0.330	5.767	18.7	20050818	114.0	121.2	233.9	16.6	0.892	3.048
	2004 BE68	0.017D	0.977	2.542	18.5	20050818	269.5	191.1	211.0	15.8	0.445	1.760
	2004 BW58	0.040A	0.904	2.006	18.7	20050818	357.1	301.6	132.8	15.2	0.379	1.455
	2004 BV1	0.001D	0.929	1.994	19.9	20050818	78.6	223.0	96.8	12.1	0.365	1.461
	2004 AS1	0.013D	0.884	1.257	20.5	20050818	65.5	261.8	322.6	17.2	0.175	1.070
	2004 AR1	0.030	0.836	2.317	19.7	20050818	321.2	50.0	336.8	9.1	0.470	1.576
	2004 AE	0.038D	0.956	1.908	20.5	20050818	324.0	193.0	307.7	18.7	0.333	1.432
	2003 YM137	0.002	0.820	4.391	18.7	20050818	118.2	126.8	77.4	2.7	0.685	2.605
	2003 YH136	0.011	0.247	4.445	19.4	20050818	150.4	304.2	306.7	16.0	0.895	2.346
	2003 YL118	0.019	0.581	1.677	21.8	20050818	183.6	65.5	278.9	7.6	0.486	1.129
(90416)	2003 YK118	0.005	0.860	2.528	18.7	20050818	230.0	232.9	327.1	7.9	0.492	1.694
	2003 YG118	0.016	0.812	3.755	17.1	20050818	133.4	232.2	348.6	8.1	0.644	2.283
(90403)	2003 YE45	0.026A	0.889	1.579	17.8	20050818	350.8	306.7	286.3	19.0	0.280	1.234
	2003 YD45	0.024	0.753	4.227	21.0	20050818	166.6	104.2	253.0	8.4	0.697	2.490
	2003 YS17	0.027	0.639	1.222	21.7	20050818	194.8	134.5	99.2	6.5	0.313	0.930
	2003 YP17	0.044	1.027	2.974	19.1	20050818	175.3	31.0	144.5	4.9	0.487	2.000
	2003 YX1	0.006	0.644	1.113	20.9	20050818	113.7	222.8	89.9	5.8	0.267	0.879
	2003 YT1	0.010D	0.786	1.434	16.2	20050818	94.0	91.0	38.4	44.1	0.292	1.110
	2003 YP1	0.044	0.944	4.370	21.8	20050818	145.9	128.3	279.3	7.2	0.645	2.657
	2003 XB22	0.001D	0.852	1.959	19.3	20050818	274.7	118.2	108.9	29.9	0.394	1.406
	2003 XM	0.040	1.028	5.371	19.4	20050818	96.1	156.1	355.5	5.6	0.679	3.199
	2003 WH166	0.004	0.856	3.011	22.0	20050818	200.4	328.4	186.5	0.7	0.557	1.933

Table 4. Continued

Designation (and name)	Prov. Des.	Min.	q	Q	H	Epoch	M	Peri.	Node	Incl.	e	a
	2003 WC158	0.036	0.967	4.309	20.5	20050818	148.5	142.5	252.6	8.7	0.633	2.638
	2003 WY25	0.002A	0.997	5.159	21.1	20050818	112.3	9.0	69.3	5.9	0.676	3.078
	2003 WR21	0.024	0.826	1.412	19.6	20050818	52.6	107.9	85.9	9.3	0.262	1.119
	2003 WP21	0.031	0.492	4.026	21.8	20031117	341.8	123.6	38.0	4.3	0.782	2.259
	2003 WG	0.032	0.529	4.345	19.1	20050818	178.5	261.7	47.0	27.1	0.783	2.437
	2003 UX34	0.016	0.421	1.769	20.2	20050818	341.6	218.3	4.5	2.6	0.616	1.095
	2003 UW29	0.018	0.189	2.148	20.8	20050818	201.6	54.9	196.8	3.8	0.838	1.169
	2003 UC20	0.038	0.518	1.044	18.3	20050818	355.1	59.2	188.9	3.8	0.337	0.781
	2003 UV11	0.006	0.344	2.557	19.3	20050818	343.1	124.6	32.2	5.9	0.763	1.451
	2003 UX5	0.046A	0.950	1.932	20.4	20050818	344.1	50.2	50.6	21.5	0.340	1.441
	2003 TR9	0.023	0.491	2.792	21.2	20050818	345.0	70.7	198.7	13.5	0.701	1.642
	2003 TO9	0.019	0.966	3.893	21.9	20050818	165.5	225.1	198.1	3.1	0.602	2.430
	2003 TL4	0.037	0.480	1.073	19.4	20050818	113.6	321.8	220.2	12.1	0.382	0.776
	2003 TK2	0.023	0.819	3.865	19.9	20050818	198.7	320.9	1.1	4.3	0.650	2.342
	2003 SD220	0.041D	0.655	1.002	16.7	20050818	327.9	326.3	274.2	8.5	0.210	0.828
	2003 SS84	0.007	0.827	3.034	21.8	20050818	234.4	234.3	198.7	5.5	0.572	1.930
	2003 SK84	0.048	0.648	4.407	21.1	20050818	155.0	118.1	344.3	8.9	0.743	2.528
	2003 RN10	0.025A	1.024	3.437	15.8	20050818	218.0	4.6	267.3	39.6	0.541	2.231
	2003 RM10	0.012	0.755	2.938	20.3	20050818	298.6	287.0	341.6	13.7	0.591	1.846
	2003 RX7	0.038	0.794	1.665	18.5	20050818	46.9	245.7	241.1	10.4	0.354	1.229
	2003 RB5	0.033A	0.995	2.737	21.2	20050818	266.7	27.8	341.9	44.4	0.467	1.866
	2003 RS1	0.013	1.041	4.565	21.8	20050818	150.9	161.7	174.1	2.4	0.629	2.803
	2003 QO104	0.012	1.014	3.252	16.0	20050818	283.5	183.5	58.3	11.6	0.524	2.133
	2003 QQ47	0.001D	0.882	1.288	17.4	20050818	192.3	105.0	1.0	62.1	0.187	1.085
	2003 QC10	0.001	0.369	2.386	18.0	20050818	48.6	120.4	0.4	5.0	0.732	1.377
	2003 QH5	0.023A	0.982	1.538	20.3	20050818	190.2	321.1	299.3	17.6	0.220	1.260
	2003 NW1	0.031	0.878	3.986	18.6	20050818	183.2	116.1	267.3	6.0	0.639	2.432
	2003 NC	0.045	0.274	2.538	19.2	20050818	136.0	224.0	288.7	21.7	0.805	1.406
	2003 MK4	0.015D	0.885	1.276	20.9	20050818	234.9	109.5	282.7	22.3	0.181	1.080
	2003 MH4	0.007	0.952	2.974	20.0	20050818	300.3	322.8	260.0	3.9	0.515	1.963
	2003 MA3	0.013	0.661	1.551	21.8	20050818	240.9	228.8	152.7	1.4	0.402	1.106

Table 4. Continued

Designation (and name)	Prov. Des.	Min.	q	Q	H	Epoch	M	Peri.	Node	Incl.	e	a
	2003 MS2	0.008D	0.910	1.173	21.1	20050818	298.8	254.6	100.9	20.0	0.126	1.042
	2003 MU	0.008	1.038	3.070	20.4	20050818	245.0	175.5	143.8	5.9	0.495	2.054
	2003 KU2	0.032	0.872	4.514	17.7	20050818	149.4	246.5	105.5	5.4	0.676	2.693
	2003 KO2	0.008A	0.356	1.099	20.1	20050818	33.0	203.9	215.5	23.5	0.511	0.727
	2003 HG2	0.047D	0.919	1.205	21.8	20050818	163.8	225.3	214.3	19.8	0.135	1.062
	2003 HF2	0.003	0.361	1.865	19.4	20050818	84.6	230.7	190.2	3.1	0.675	1.113
	2003 HB	0.050	0.526	1.173	19.7	20050818	179.1	306.7	70.5	18.1	0.381	0.850
	2003 GP51	0.038	0.857	3.455	22.0	20050818	281.9	357.6	141.8	2.9	0.603	2.156
	2003 GQ22	0.027A	0.714	1.031	21.5	20050818	151.7	168.7	199.6	17.0	0.182	0.872
	2003 GG21	0.001	0.624	3.664	22.0	20050818	286.8	95.0	13.2	10.1	0.709	2.144
	2003 GY	0.020	0.942	1.818	20.1	20050818	99.4	333.9	322.1	4.7	0.317	1.380
	2003 FC5	0.034	0.750	3.084	18.3	20050818	345.4	270.5	189.5	5.8	0.609	1.917
	2003 FH	0.018	0.638	2.140	18.9	20050818	110.6	274.2	24.3	25.8	0.541	1.389
	2003 FG	0.044	0.354	2.129	19.7	20050818	219.3	312.8	7.1	15.6	0.715	1.241
	2003 EF54	0.040	0.848	2.368	20.1	20050818	35.8	353.3	268.8	3.0	0.472	1.608
	2003 ED50	0.030	0.642	2.190	20.8	20050818	124.1	93.9	173.6	33.7	0.547	1.416
	2003 EG16	0.020	0.753	4.046	19.2	20050818	219.7	109.8	152.2	20.3	0.686	2.399
	2003 EE16	0.000	0.547	2.289	19.9	20050818	237.2	259.9	126.8	0.6	0.614	1.418
	2003 DX10	0.045	0.811	1.940	20.4	20050818	139.1	193.8	62.0	3.1	0.411	1.376
	2003 CR20	0.001	0.569	3.676	18.9	20050818	260.9	88.1	177.8	5.0	0.732	2.122
	2003 CQ20	0.036	1.046	3.134	20.5	20050818	295.2	172.4	345.6	15.9	0.499	2.090
	2003 CG11	0.017	0.679	4.494	20.7	20050818	201.7	106.8	130.8	21.3	0.737	2.587
	2003 CR1	0.023	0.780	2.127	19.9	20050818	201.9	101.8	311.1	12.7	0.463	1.454
	2003 CC	0.039	1.010	1.991	20.2	20050818	71.7	103.1	136.4	2.3	0.327	1.501
	2003 CA	0.048	0.387	2.372	19.0	20050818	246.4	234.6	126.4	21.2	0.720	1.379
	2003 BR47	0.005	0.814	2.442	21.4	20050818	101.1	112.3	314.7	4.4	0.500	1.628
	2003 BK47	0.023	0.804	4.691	18.5	20050818	241.4	234.6	139.5	20.8	0.707	2.747
	2003 BD44	0.018	0.777	3.157	16.8	20050818	262.7	88.5	181.9	2.7	0.605	1.967
	2003 BB21	0.023	0.987	3.470	21.0	20050818	259.3	211.7	344.9	4.8	0.557	2.229
	2003 BH	0.022A	0.937	1.973	20.7	20050818	192.1	313.2	119.9	13.1	0.356	1.455
	2003 AF23	0.029	0.502	1.248	20.7	20050818	161.5	43.9	286.8	23.2	0.426	0.875

Table 4. Continued

Designation (and name)	Prov. Des.	Min.	q	Q	H	Epoch	M	Peri.	Node	Incl.	e	a
	2003 AD23	0.031	0.390	2.900	18.7	20050818	123.4	241.4	101.8	23.4	0.763	1.645
	2003 AC23	0.006	0.900	3.388	21.9	20050818	284.0	126.4	43.7	2.0	0.580	2.144
	2003 AA3	0.016A	1.010	1.832	20.2	20050818	199.6	346.9	106.5	13.8	0.289	1.421
	2003 AY2	0.030	0.792	2.851	19.2	20050818	355.2	285.5	275.4	10.3	0.565	1.822
	2002 YP2	0.019	0.492	2.663	19.1	20050818	148.4	281.2	52.7	20.6	0.688	1.577
	2002 XR14	0.010	0.713	3.096	18.1	20050818	326.5	120.1	75.7	2.1	0.626	1.905
	2002 WQ4	0.016	0.871	3.047	19.8	20050818	324.5	243.6	267.8	4.0	0.555	1.959
	2002 VX94	0.044	0.873	2.079	18.1	20050818	107.9	214.9	320.3	7.2	0.409	1.476
(90075)	2002 VU94	0.029	0.911	3.361	15.7	20050818	77.2	30.2	227.0	8.9	0.574	2.136
	2002 VR85	0.024	0.721	2.913	20.4	20050818	18.9	298.4	205.3	6.0	0.603	1.817
	2002 VP69	0.048D	0.949	3.085	18.2	20050818	316.8	188.6	307.1	10.2	0.530	2.017
	2002 VE68	0.019D	0.427	1.021	20.3	20050818	359.7	355.5	231.7	9.0	0.411	0.724
	2002 UK11	0.010	0.562	2.084	21.7	20050818	255.1	84.3	69.4	5.4	0.575	1.323
	2002 TR190	0.008D	0.905	1.249	19.0	20050818	118.1	104.8	24.3	26.9	0.160	1.077
	2002 TB70	0.015A	0.984	1.283	21.7	20050818	328.6	350.8	195.1	16.7	0.132	1.133
	2002 TP69	0.024	1.033	2.859	21.9	20050818	5.9	316.5	94.7	2.0	0.469	1.946
	2002 TD66	0.003	0.865	2.845	20.2	20050818	15.3	125.4	335.9	4.9	0.534	1.855
	2002 TW55	0.025D	0.712	3.523	18.0	20050818	5.7	251.1	33.7	59.4	0.664	2.117
	2002 SY50	0.001	0.530	2.882	17.6	20050818	69.4	99.2	34.6	8.7	0.689	1.706
	2002 SR41	0.032	0.551	1.614	20.3	20050818	118.3	258.0	248.0	11.6	0.491	1.083
	2002 SQ41	0.004	0.515	4.692	20.1	20050818	231.7	94.5	23.4	25.1	0.802	2.604
	2002 SZ	0.005	0.735	1.860	20.4	20050818	34.4	276.4	347.0	22.9	0.433	1.297
	2002 SM	0.031D	0.963	2.782	17.5	20050818	126.2	217.1	11.0	14.4	0.486	1.872
	2002 RW25	0.015	0.589	1.062	18.9	20050818	189.7	71.7	92.0	1.3	0.286	0.825
	2002 QW47	0.011	0.952	2.992	20.2	20050818	50.9	140.7	116.4	4.2	0.517	1.972
	2002 QQ40	0.028	0.529	1.901	21.1	20050818	20.6	356.4	104.2	1.7	0.564	1.215
(68950)	2002 QF15	0.020A	0.693	1.421	16.4	20050818	184.2	255.5	236.3	25.2	0.344	1.057
	2002 QH10	0.034	1.039	3.685	20.3	20050818	282.4	23.7	0.5	4.8	0.560	2.362
	2002 PE130	0.041A	0.976	4.145	18.3	20050818	252.0	33.4	357.6	15.6	0.619	2.560
	2002 PF43	0.047D	1.042	2.246	20.8	20050818	152.2	191.5	142.8	13.9	0.366	1.644
	2002 PD43	0.034	0.111	4.911	19.1	20050818	287.2	210.6	315.3	26.2	0.956	2.511

Table 4. Continued

Designation (and name)	Prov. Des.	Min.	q	Q	H	Epoch	M	Peri.	Node	Incl.	e	a
	2002 PZ39	0.004	0.666	2.273	19.0	20050818	282.1	259.2	329.8	1.7	0.547	1.470
	2002 PM6	0.034	0.180	2.216	17.7	20050818	149.4	224.2	304.3	19.2	0.850	1.198
	2002 OA22	0.039	0.709	1.163	19.4	20050818	272.5	318.3	174.4	6.9	0.243	0.936
	2002 OD20	0.030	0.861	1.867	18.8	20050818	77.8	274.9	260.2	4.2	0.369	1.364
	2002 NY40	0.002	0.592	3.503	19.0	20050818	352.7	268.1	146.8	5.9	0.711	2.048
	2002 NV16	0.035D	0.965	1.510	21.4	20050818	39.9	179.4	183.6	3.5	0.220	1.238
(89959)	2002 NT7	0.016A	0.817	2.654	16.5	20050818	68.5	300.6	132.2	42.3	0.529	1.735
	2002 NN4	0.010	0.496	1.257	20.1	20050818	77.7	222.2	259.7	5.4	0.434	0.877
	2002 MT3	0.042	0.871	4.747	19.9	20050818	253.1	25.3	168.4	6.5	0.690	2.809
	2002 MR3	0.047	0.572	2.445	21.4	20050818	277.8	67.0	110.5	9.8	0.621	1.508
	2002 MX	0.015	0.511	4.502	21.7	20050818	300.0	237.7	284.2	2.0	0.796	2.506
(89958)	2002 LY45	0.002	0.187	3.098	16.9	20050818	146.5	222.3	188.7	10.0	0.886	1.642
	2002 LT38	0.043	0.580	1.110	20.3	20050818	237.3	162.7	259.6	6.2	0.314	0.845
	2002 LY1	0.013	0.593	1.317	21.9	20050818	64.9	133.9	248.2	2.9	0.379	0.955
	2002 LX	0.037	0.823	4.202	21.1	20050818	301.7	345.3	197.8	3.2	0.673	2.513
	2002 LV	0.028D	0.919	3.713	16.5	20050818	301.9	224.2	132.3	29.5	0.603	2.316
	2002 KK8	0.043	1.045	2.862	21.0	20050818	62.2	174.0	77.1	24.5	0.465	1.953
	2002 KJ4	0.033D	0.999	3.534	19.5	20050818	344.5	155.2	75.7	27.8	0.559	2.266
	2002 KG4	0.048D	0.991	4.889	20.9	20050818	235.1	149.8	64.5	27.6	0.663	2.940
	2002 JV15	0.043	0.753	2.495	19.3	20050818	174.2	127.2	200.8	7.2	0.536	1.624
	2002 JE9	0.005	0.623	1.513	21.4	20050818	67.0	255.3	200.2	8.8	0.417	1.068
	2002 JZ8	0.001A	0.974	4.517	21.1	20050818	267.7	338.7	200.7	9.8	0.645	2.745
	2002 JX8	0.017	0.535	1.005	20.4	20050818	114.1	338.2	68.6	4.3	0.305	0.770
	2002 HK12	0.022	0.941	3.062	18.2	20050818	25.2	2.0	307.2	2.4	0.530	2.002
	2002 HP11	0.017	0.485	3.650	20.4	20050818	62.8	90.0	9.6	5.1	0.765	2.068
	2002 GZ8	0.005	0.968	4.611	18.2	20050818	272.4	159.9	310.9	5.3	0.653	2.790
	2002 GO5	0.036	0.441	3.350	17.9	20050818	122.6	64.5	23.4	13.8	0.768	1.895
	2002 GM5	0.009	0.647	3.588	21.9	20020327	337.6	274.0	14.0	7.3	0.694	2.118
	2002 GM2	0.007	0.425	3.979	18.5	20050818	41.7	80.0	343.5	3.4	0.807	2.202
	2002 GT	0.020	0.894	1.794	18.5	20050818	314.2	134.9	201.9	7.0	0.335	1.344
	2002 FG7	0.042	0.565	2.462	18.8	20050818	338.3	247.3	187.8	9.2	0.627	1.514

Table 4. Continued

Designation (and name)	Prov. Des.	Min.	q	Q	H	Epoch	M	Peri.	Node	Incl.	e	a
	2002 FV5	0.038	0.299	1.875	17.9	20050818	282.5	308.0	38.9	34.0	0.725	1.087
	2002 FU5	0.049	0.758	4.269	21.0	20050818	291.9	156.3	111.8	4.2	0.698	2.514
	2002 FQ5	0.035	0.787	3.044	20.7	20050818	72.4	80.3	196.6	7.6	0.589	1.915
	2002 FB3	0.003	0.303	1.220	16.4	20050818	167.7	148.2	203.7	20.3	0.602	0.761
	2002 FC	0.023	0.959	4.702	19.0	20050818	243.5	194.9	50.4	6.8	0.661	2.831
	2002 EZ11	0.001	0.220	2.008	18.1	20050818	88.1	317.5	52.1	2.4	0.802	1.114
	2002 EV11	0.050	0.226	3.963	20.1	20050818	66.3	216.3	185.3	12.0	0.892	2.095
	2002 EU11	0.015	0.747	3.937	21.9	20050818	328.6	274.6	346.2	2.9	0.681	2.342
	2002 EL6	0.031D	0.967	3.632	18.0	20050818	329.1	186.4	85.1	9.5	0.579	2.300
	2002 EY2	0.008A	0.886	2.566	19.0	20050818	263.8	50.9	343.6	21.8	0.487	1.726
	2002 DJ5	0.040	0.605	2.196	20.3	20050818	318.2	295.9	348.1	6.4	0.568	1.401
	2002 DU3	0.013	0.873	1.418	20.8	20050818	234.0	245.5	0.8	8.7	0.238	1.145
	2002 DO3	0.025	0.933	2.786	22.0	20050818	111.4	165.4	56.2	3.8	0.498	1.860
	2002 CD14	0.027	0.749	2.806	20.6	20050818	119.7	140.9	127.8	2.9	0.579	1.778
	2002 CU11	0.009D	0.860	1.581	18.6	20050818	148.2	110.6	157.8	48.8	0.295	1.220
	2002 CQ11	0.018	0.559	1.398	19.9	20050818	310.3	272.8	81.4	2.5	0.428	0.979
	2002 CZ9	0.024	0.848	1.802	22.0	20050818	63.8	80.6	143.0	5.0	0.360	1.325
	2002 CY9	0.009D	0.810	2.483	19.5	20050818	266.1	117.3	305.4	41.9	0.508	1.647
(89830)	2002 CE	0.024A	1.022	3.132	14.7	20050818	109.4	5.5	20.0	43.7	0.508	2.077
	2002 BM26	0.002	1.018	2.648	20.1	20050818	149.4	180.4	319.7	16.2	0.445	1.833
	2002 BK25	0.046	0.576	4.017	18.1	20050818	342.1	103.6	156.5	11.9	0.749	2.297
	2002 BG25	0.040D	0.944	1.520	20.9	20050818	171.4	239.3	316.6	48.8	0.234	1.232
	2002 AJ129	0.005	0.117	2.625	18.5	20050818	106.9	210.8	138.2	15.5	0.915	1.371
	2002 AM31	0.028	0.934	2.476	18.2	20050818	304.7	197.7	144.6	4.6	0.452	1.705
	2002 AC9	0.028	0.748	2.658	21.0	20050818	251.2	28.2	2.9	2.3	0.561	1.703
	2002 AT4	0.010	1.032	2.701	21.3	20050818	128.4	202.8	323.7	1.5	0.447	1.866
	2002 AS4	0.035	0.252	2.272	22.0	20020106	33.4	42.5	285.6	26.9	0.800	1.262
	2002 AP3	0.015	0.838	3.245	19.8	20050818	51.6	117.5	87.6	7.6	0.590	2.041
	2002 AZ1	0.021	0.707	3.523	22.0	20050818	36.9	96.9	113.9	8.1	0.666	2.115
	2002 AY1	0.026D	0.438	1.120	20.8	20050818	338.9	323.8	287.9	29.9	0.438	0.779
	2002 AW	0.005	0.796	1.343	20.8	20050818	292.7	118.0	163.0	0.6	0.256	1.069

Table 4. Continued

Designation (and name)	Prov. Des.	Min.	q	Q	H	Epoch	M	Peri.	Node	Incl.	e	a
	2002 AV	0.016	0.828	4.093	20.9	20050818	346.6	285.3	124.3	2.8	0.663	2.460
	2001 YB5	0.006	0.323	4.387	20.9	20050818	348.4	114.2	109.4	5.5	0.863	2.355
	2001 YE4	0.037	0.310	1.043	20.9	20050818	55.2	318.5	305.9	4.8	0.541	0.676
	2001 YV3	0.033	0.544	3.346	20.2	20050818	145.4	243.8	107.8	5.2	0.720	1.945
	2001 YP3	0.018	0.952	4.049	22.0	20050818	322.0	36.4	102.0	7.9	0.619	2.501
	2001 XN254	0.034	1.015	3.622	17.7	20050818	345.2	274.7	267.6	1.9	0.562	2.318
(68548)	2001 XR31	0.011A	0.962	2.451	17.0	20050818	332.2	24.6	302.1	22.7	0.436	1.707
	2001 XP31	0.009	0.704	1.593	21.9	20011127	55.4	81.6	241.4	6.1	0.387	1.148
	2001 XU30	0.013	0.738	3.656	19.8	20050818	26.1	105.4	73.5	8.9	0.664	2.197
	2001 XR30	0.016	0.823	1.771	21.0	20050818	115.9	294.3	248.0	11.9	0.365	1.297
	2001 XU10	0.014A	0.984	2.524	15.3	20050818	331.4	6.8	310.2	42.0	0.439	1.754
	2001 XT1	0.040	0.643	2.413	18.8	20050818	15.2	31.1	316.5	2.7	0.579	1.528
	2001 XP1	0.030D	0.722	5.068	17.8	20050818	252.1	246.2	268.5	39.3	0.751	2.895
	2001 XU	0.005	0.415	4.720	19.1	20050818	308.5	285.6	261.5	19.1	0.838	2.568
	2001 WL15	0.029	1.046	2.933	18.2	20050818	101.7	14.1	112.3	6.9	0.474	1.989
	2001 WN5	0.000	0.912	2.510	18.3	20050818	269.6	44.2	277.8	1.9	0.467	1.711
	2001 WS1	0.011A	1.010	4.190	17.0	20050818	327.3	4.4	1.6	13.2	0.611	2.600
	2001 WH1	0.042	0.491	4.435	20.4	20050818	331.3	107.3	68.3	15.6	0.800	2.463
	2001 VB76	0.039	0.951	1.966	20.4	20050818	351.3	248.3	259.6	4.2	0.348	1.458
	2001 VK5	0.007	0.617	1.922	17.8	20050818	279.6	263.8	54.3	19.4	0.514	1.270
	2001 VJ5	0.023	0.849	3.022	21.3	20050818	127.7	67.0	45.2	6.1	0.561	1.935
	2001 VC2	0.009D	0.903	1.178	21.2	20050818	83.3	103.8	67.2	12.5	0.132	1.041
	2001 VB	0.022	0.240	4.420	18.4	20050818	11.5	224.9	305.5	9.5	0.897	2.330
	2001 UZ16	0.047A	1.010	2.507	19.4	20050818	242.4	28.4	323.3	12.7	0.426	1.758
(89136)	2001 US16	0.031	1.013	1.698	20.2	20050818	282.7	67.0	176.0	1.9	0.253	1.356
	2001 UA5	0.026A	0.991	2.583	17.4	20050818	200.3	27.4	58.8	10.0	0.446	1.787
	2001 UY4	0.028	0.309	2.597	18.3	20050818	95.5	107.3	161.1	5.4	0.787	1.453
	2001 TO48	0.050	0.860	2.884	19.5	20050818	208.0	151.7	140.6	6.8	0.541	1.872
	2001 TD45	0.031	0.177	1.416	20.0	20050818	218.6	212.4	30.3	25.4	0.777	0.797
	2001 TC45	0.022	0.797	3.643	19.3	20050818	39.2	65.9	36.2	15.6	0.641	2.220
	2001 TX44	0.004	0.397	1.352	18.7	20050818	109.4	136.0	57.9	15.2	0.546	0.875

Table 4. Continued

Designation (and name)	Prov. Des.	Min.	q	Q	H	Epoch	M	Peri.	Node	Incl.	e	a
	2001 TA2	0.037	0.624	2.876	21.4	20010928	21.2	226.7	51.0	3.2	0.643	1.750
	2001 TX1	0.042	0.542	1.552	21.1	20050818	129.7	354.0	159.4	2.8	0.483	1.047
	2001 SN289	0.022D	0.879	2.688	16.4	20050818	329.1	225.6	357.2	53.3	0.507	1.784
	2001 SG286	0.000	0.887	1.835	21.1	20050818	201.2	55.9	241.3	7.8	0.348	1.361
	2001 SZ269	0.033	0.794	3.929	19.6	20050818	43.2	190.3	99.2	2.5	0.664	2.361
	2001 SY269	0.020	0.677	2.683	21.5	20050818	257.4	223.3	232.6	1.9	0.597	1.680
	2001 SN263	0.015	1.037	2.934	16.5	20050818	40.3	172.5	325.9	6.7	0.478	1.985
	2001 SG262	0.030	0.818	3.106	19.2	20050818	126.3	99.2	359.8	4.8	0.583	1.962
	2001 SX169	0.028	0.726	1.968	18.3	20050818	2.2	42.6	126.9	2.5	0.461	1.347
	2001 SK162	0.003	1.013	2.838	18.0	20050818	120.1	186.0	285.6	1.7	0.474	1.925
	2001 SO73	0.049	0.784	2.855	18.4	20050818	294.0	30.6	197.4	4.9	0.569	1.820
	2001 SG10	0.019	0.834	2.064	20.5	20050818	118.0	101.5	185.1	4.3	0.424	1.449
	2001 SQ3	0.029D	0.827	1.392	21.7	20050818	189.6	268.0	356.2	23.9	0.255	1.110
	2001 RA12	0.015A	0.926	3.146	18.1	20050818	145.5	325.9	311.7	17.1	0.545	2.036
	2001 QQ142	0.002A	0.980	1.864	18.5	20050818	85.0	337.7	83.3	9.3	0.311	1.422
	2001 QJ96	0.001	0.320	2.894	22.0	20010819	330.7	121.4	339.1	5.9	0.801	1.607
	2001 QC34	0.049	0.916	1.338	20.2	20050818	334.2	215.0	271.9	6.2	0.187	1.127
	2001 PT9	0.027	0.801	2.137	20.2	20050818	135.5	261.2	330.3	7.2	0.455	1.469
(68372)	2001 PM9	0.006	0.945	2.291	18.9	20050818	46.4	322.1	253.1	8.1	0.416	1.618
	2001 PJ9	0.048	0.647	2.951	18.5	20050818	278.7	290.7	284.3	10.5	0.640	1.799
	2001 OY13	0.006	0.815	1.821	20.8	20050818	295.8	291.1	284.3	10.3	0.382	1.318
	2001 MG1	0.023D	0.895	4.115	17.2	20050818	1.8	218.4	142.5	28.4	0.643	2.505
	2001 ME1	0.012	0.356	4.946	16.9	20050818	334.9	300.0	86.8	5.8	0.866	2.651
(68347)	2001 KB67	0.007	0.597	1.329	19.9	20050818	246.7	243.8	246.0	17.1	0.380	0.963
	2001 KF54	0.021	0.818	3.861	20.3	20050818	35.0	170.8	172.6	1.6	0.650	2.340
	2001 JV1	0.016	0.963	2.447	21.3	20050818	305.9	200.9	92.4	6.6	0.435	1.705
	2001 HZ7	0.048	0.737	2.199	19.7	20050818	197.6	312.6	156.5	5.4	0.498	1.468
	2001 HY7	0.036	0.537	1.291	20.5	20050818	112.5	211.0	205.4	5.2	0.412	0.914
	2001 HA4	0.005	0.552	4.821	17.6	20050818	6.5	91.6	358.4	17.1	0.794	2.687
	2001 HB	0.014	0.402	2.225	20.5	20050818	359.2	237.7	196.1	9.3	0.694	1.314
	2001 GT2	0.046	0.870	3.914	19.9	20050818	74.7	358.3	146.4	3.7	0.636	2.392

Table 4. Continued

Designation (and name)	Prov. Des.	Min.	q	Q	H	Epoch	M	Peri.	Node	Incl.	e	a
	2001 GR2	0.015	0.705	3.008	21.0	20050818	237.1	99.1	190.5	11.8	0.620	1.857
	2001 GQ2	0.003	0.603	1.824	20.1	20050818	36.9	280.2	37.2	21.8	0.503	1.214
	2001 GN2	0.020A	1.020	2.700	17.9	20050818	266.3	3.3	173.6	26.0	0.452	1.860
(88254)	2001 FM129	0.010	0.438	1.926	17.3	20050818	198.7	139.7	272.6	1.5	0.630	1.182
	2001 FE90	0.034	0.975	2.883	19.8	20050818	211.1	144.8	98.6	8.8	0.494	1.929
	2001 FB90	0.031	0.543	4.216	19.9	20010312	338.7	15.0	265.3	1.9	0.772	2.380
	2001 FD58	0.032	0.464	1.721	18.6	20050818	64.1	45.8	341.4	6.5	0.575	1.092
	2001 FC58	0.016	0.670	1.369	20.7	20050818	172.4	261.0	174.8	6.8	0.343	1.020
	2001 FA58	0.034	0.642	3.884	21.8	20050818	90.0	112.1	158.1	8.1	0.716	2.263
	2001 EB18	0.020D	0.860	1.247	19.3	20050818	326.7	101.9	155.5	50.1	0.184	1.053
	2001 EC	0.003	0.586	4.576	18.6	20050818	38.9	72.5	358.5	0.6	0.773	2.581
	2001 DF47	0.013	0.765	1.665	20.3	20050818	182.9	260.1	146.7	18.5	0.371	1.215
(68216)	2001 CV26	0.030A	0.889	1.752	16.4	20050818	70.7	48.6	18.3	18.0	0.327	1.320
	2001 CB21	0.030	0.689	1.379	18.5	20050818	38.5	271.6	353.9	7.9	0.334	1.034
	2001 BO61	0.012	0.457	3.093	17.9	20050818	300.5	78.3	160.3	9.1	0.742	1.775
	2001 BE10	0.042	0.520	1.127	19.1	20050818	164.5	30.6	297.9	17.5	0.369	0.823
	2001 AD2	0.009	0.354	1.725	19.6	20050818	205.6	111.1	211.4	1.7	0.660	1.039
	2000 YV137	0.033D	0.998	1.897	18.6	20050818	329.7	211.3	137.3	28.0	0.311	1.448
	2000 YN29	0.012	0.828	4.234	17.5	20050818	19.2	132.4	72.8	5.4	0.673	2.531
	2000 YG29	0.035A	0.965	5.369	18.8	20050818	298.3	358.3	92.5	18.9	0.695	3.167
	2000 YF29	0.022	0.938	2.045	20.4	20050818	169.0	27.9	124.9	6.3	0.371	1.492
	2000 YJ11	0.021A	1.008	1.615	20.7	20050818	60.8	338.9	65.1	7.3	0.231	1.311
	2000 WO148	0.046	1.024	2.260	20.7	20050818	60.7	323.6	150.3	4.4	0.376	1.642
	2000 WO107	0.002	0.200	1.623	19.3	20050818	209.5	213.6	69.4	7.8	0.781	0.911
	2000 WK10	0.022	0.441	2.516	18.9	20050818	273.9	244.3	58.2	14.8	0.702	1.479
	2000 UL11	0.022	0.777	3.474	20.1	20050818	177.5	151.0	333.8	2.2	0.635	2.126
	2000 UG11	0.005	0.826	3.032	20.4	20050818	266.8	240.4	224.4	8.9	0.572	1.929
	2000 TO64	0.025	1.045	4.441	16.8	20050818	25.3	177.6	223.8	37.5	0.619	2.743
	2000 TU28	0.004A	0.877	1.270	21.0	20050818	49.8	280.6	203.1	15.6	0.183	1.074
	2000 SU180	0.047	0.823	3.379	19.2	20050818	189.2	135.4	328.5	11.8	0.608	2.101
	2000 SP43	0.018	0.433	1.190	18.6	20050818	350.7	224.3	350.7	10.4	0.467	0.811

Table 4. Continued

Designation (and name)	Prov. Des.	Min.	q	Q	H	Epoch	M	Peri.	Node	Incl.	e	a
(87684)	2000 SY2	0.047	0.307	1.411	16.4	20050818	152.0	47.7	162.1	19.2	0.643	0.859
	2000 RD53	0.029D	1.022	2.552	20.1	20050818	16.0	192.6	176.1	9.3	0.428	1.787
	2000 RW37	0.001D	0.936	1.561	20.1	20050818	102.2	133.2	333.4	13.7	0.251	1.249
	2000 RS11	0.016A	0.870	1.693	19.5	20050818	66.1	301.7	172.3	17.1	0.321	1.281
	2000 QK130	0.002	0.872	1.490	20.9	20050818	19.8	66.2	174.0	4.7	0.262	1.181
	2000 QW69	0.018D	0.961	1.792	17.0	20050818	137.5	220.6	344.4	38.2	0.302	1.376
	2000 QW7	0.013	1.036	2.857	19.8	20050818	294.4	190.6	158.7	4.2	0.468	1.947
	2000 QV7	0.023	0.672	2.147	21.2	20050818	30.5	79.3	154.9	9.1	0.523	1.409
	2000 QS7	0.002	0.905	4.459	20.0	20050818	39.9	218.8	153.5	3.2	0.663	2.682
	2000 PP9	0.034	1.039	3.613	19.4	20050818	146.6	159.5	171.8	5.6	0.553	2.326
(23187)	2000 PN9	0.025A	0.758	2.932	16.1	20050818	298.0	293.5	164.4	51.3	0.589	1.845
	2000 PD3	0.023	0.811	3.184	18.4	20050818	248.8	109.7	299.1	7.7	0.594	1.998
(67381)	2000 OL8	0.013	0.603	2.037	19.9	20050818	171.0	266.6	295.1	10.7	0.543	1.320
	2000 OJ8	0.001	1.030	3.701	16.7	20050818	105.9	180.6	226.8	6.2	0.565	2.366
	2000 OH	0.005A	0.994	3.854	17.6	20050818	129.5	354.5	284.1	18.6	0.590	2.424
	2000 MU1	0.001	0.848	1.898	19.9	20050818	116.1	63.5	130.2	13.1	0.383	1.373
(67367)	2000 LY27	0.013	1.030	1.587	17.2	20050818	319.9	184.7	264.6	9.0	0.213	1.309
	2000 LB16	0.011A	0.797	1.685	18.5	20050818	211.8	285.3	80.9	50.7	0.358	1.241
	2000 LF3	0.015	0.879	4.281	21.6	20050818	78.7	222.6	83.2	14.9	0.659	2.580
	2000 KA	0.002	0.717	1.947	21.6	20050818	188.6	83.0	62.7	6.7	0.462	1.332
	2000 JG5	0.026	0.274	2.407	18.1	20050818	182.8	233.2	213.3	31.5	0.795	1.341
	2000 JF5	0.018	0.911	3.205	21.9	20000426	344.4	216.9	58.1	13.7	0.557	2.058
	2000 HA24	0.028	0.776	1.502	19.2	20050818	227.7	141.7	309.9	2.2	0.319	1.139
	2000 GV147	0.010	0.950	2.543	19.4	20050818	85.0	215.9	68.8	10.6	0.456	1.747
	2000 GJ147	0.016D	0.887	1.437	19.6	20050818	26.8	240.8	57.9	25.0	0.237	1.162
(86819)	2000 GK137	0.032D	0.986	3.006	17.4	20050818	285.3	150.1	165.0	10.1	0.506	1.996
	2000 GF2	0.016	0.835	1.847	20.5	20050818	120.1	108.0	176.1	9.6	0.377	1.341
(41429)	2000 GE2	0.011	0.708	2.475	20.7	20050818	210.0	297.6	351.7	2.2	0.555	1.592
	2000 EE104	0.008	0.710	1.299	20.3	20050818	39.2	281.0	26.0	5.2	0.293	1.005
(53789)	2000 ED104	0.050D	1.002	1.739	17.3	20050818	257.1	218.3	190.1	40.8	0.269	1.370
	2000 EW70	0.009	0.638	1.238	21.3	20050818	257.1	125.3	178.3	5.4	0.320	0.938

Table 4. Continued

Designation (and name)	Prov. Des.	Min.	q	Q	H	Epoch	M	Peri.	Node	Incl.	e	a
	2000 EU70	0.042	0.521	3.889	18.6	20000226	19.8	253.1	166.4	13.0	0.764	2.205
	2000 ET70	0.019D	0.830	1.064	18.4	20050818	101.7	46.4	331.2	22.3	0.124	0.947
	2000 EK26	0.011	0.819	3.991	18.0	20050818	192.7	305.4	126.6	15.6	0.659	2.405
	2000 EJ26	0.007	0.533	2.232	19.4	20050818	64.1	285.7	10.7	9.2	0.615	1.383
	2000 EH26	0.001	0.969	2.739	21.7	20050818	31.3	18.9	215.4	0.4	0.478	1.854
	2000 EE14	0.038A	0.309	1.015	17.1	20050818	222.4	197.8	155.8	26.5	0.533	0.662
	2000 ED14	0.041	0.362	1.308	20.8	20050818	320.2	310.0	4.0	13.8	0.567	0.835
	2000 EA14	0.043	0.891	1.343	20.9	20050818	314.6	206.0	204.0	3.6	0.203	1.117
	2000 DP107	0.020	0.851	1.881	18.2	20050818	58.3	289.7	358.8	8.7	0.377	1.366
	2000 DK79	0.037	1.041	2.512	15.7	20050818	186.4	2.4	43.5	60.7	0.414	1.777
	2000 DO1	0.013	0.456	2.398	20.4	20050818	40.0	302.5	336.1	3.4	0.681	1.427
	2000 DN1	0.012	0.951	4.812	19.7	20050818	36.5	146.2	42.3	7.8	0.670	2.882
	2000 CT101	0.001A	0.838	1.761	20.7	20050818	301.2	292.3	135.7	23.3	0.355	1.300
	2000 CP101	0.043	0.676	2.873	19.2	20050818	143.3	247.5	160.6	10.5	0.619	1.775
	2000 CO101	0.028A	0.979	1.172	19.3	20050818	64.9	65.1	353.2	15.3	0.090	1.075
	2000 CH59	0.024	0.498	1.228	19.6	20050818	121.5	109.0	214.3	3.3	0.423	0.863
	2000 CO33	0.006	1.015	3.643	21.1	20050818	200.1	178.9	319.7	18.2	0.564	2.329
	2000 CM33	0.043A	0.963	1.682	21.3	20050818	203.5	50.1	132.9	11.6	0.272	1.322
	2000 BO28	0.046	0.681	2.717	20.0	20050818	125.8	303.0	320.1	6.3	0.599	1.699
(53550)	2000 BF19	0.045	0.868	2.123	19.1	20050818	254.4	324.8	313.3	7.2	0.420	1.495
	2000 AF205	0.020	0.748	1.320	21.5	20050818	211.6	127.3	220.2	2.4	0.277	1.034
	2000 AZ93	0.010D	0.478	1.016	21.3	20050818	59.6	7.9	277.6	8.6	0.360	0.747
	2000 AF6	0.023	0.517	1.240	20.1	20050818	79.0	200.1	110.9	2.7	0.411	0.878
	2000 AC6	0.045	0.609	1.098	21.6	20050818	223.9	187.9	101.8	4.7	0.286	0.854
	2000 AA6	0.007	0.618	1.961	21.9	20050818	257.7	287.2	280.3	2.0	0.521	1.289
	1999 YR14	0.006A	0.991	2.317	18.9	20050818	278.5	9.4	3.2	3.7	0.401	1.654
	1999 YG3	0.030D	0.861	1.708	19.1	20050818	260.5	250.2	293.3	34.5	0.329	1.285
	1999 YD	0.021	1.005	3.922	21.1	20050818	170.4	61.7	9.7	1.4	0.592	2.463
	1999 XM141	0.028D	0.780	1.696	19.6	19991208	308.7	105.6	73.1	21.6	0.370	1.238
	1999 XL136	0.001	0.639	2.981	19.8	20050818	90.2	274.2	271.0	8.9	0.647	1.810
	1999 XK136	0.011	0.717	4.053	20.1	20050818	209.8	302.2	63.7	2.7	0.699	2.385

Table 4. Continued

Designation (and name)	Prov. Des.	Min.	q	Q	H	Epoch	M	Peri.	Node	Incl.	e	a
	1999 XS35	0.007A	0.953	34.769	17.2	20050818	27.8	333.0	49.2	19.4	0.947	17.861
	1999 VT25	0.047	0.554	1.769	21.4	20050818	144.8	319.0	222.2	5.1	0.523	1.161
	1999 VF22	0.036	0.344	2.281	20.5	20050818	339.3	271.3	3.8	3.9	0.738	1.312
	1999 VP11	0.004	0.452	1.717	18.6	20050818	125.4	61.5	208.2	17.3	0.583	1.084
	1999 VR6	0.048	0.528	3.963	21.2	19991029	340.8	293.8	213.3	8.6	0.765	2.246
	1999 UR	0.018	0.942	2.855	21.4	20050818	70.1	17.9	50.8	3.6	0.504	1.898
	1999 TF211	0.037D	0.943	3.950	15.0	20050818	133.1	161.1	348.4	39.0	0.615	2.447
	1999 TT16	0.022	0.730	3.604	19.6	20050818	269.3	148.3	328.5	2.0	0.663	2.167
	1999 TO13	0.041A	0.893	2.283	21.4	19991009	22.9	302.3	15.1	20.3	0.438	1.588
	1999 SG10	0.031	0.562	2.348	20.6	20050818	88.7	103.2	13.6	23.6	0.614	1.455
	1999 SO5	0.001	1.015	1.156	21.2	20050818	237.2	359.6	196.2	13.4	0.065	1.085
(53426)	1999 SL5	0.021A	0.887	2.958	17.4	20050818	100.3	43.1	239.3	22.8	0.539	1.922
	1999 RM45	0.000	0.598	2.764	19.5	20050818	291.7	88.5	163.1	10.9	0.644	1.681
	1999 RQ36	0.001	0.898	1.360	21.0	20050818	284.5	65.7	2.1	6.0	0.205	1.129
	1999 RD32	0.049	0.604	4.675	16.6	20050818	158.2	299.7	310.2	6.8	0.771	2.639
	1999 RB32	0.044	1.047	3.824	19.8	20050818	199.4	40.0	329.8	3.9	0.570	2.435
	1999 RR28	0.016	0.649	3.107	18.4	20050818	72.2	283.9	178.8	7.1	0.654	1.878
(86039)	1999 NC43	0.029	0.740	2.779	16.0	20050818	141.7	120.4	312.0	7.1	0.579	1.759
	1999 NB5	0.013	0.966	3.187	21.0	20050818	353.5	122.3	236.1	1.4	0.535	2.076
	1999 MN	0.006	0.226	1.122	21.4	20050818	243.6	9.9	80.8	2.0	0.665	0.674
	1999 MM	0.000	0.631	2.615	19.3	20050818	319.4	268.6	111.2	4.8	0.611	1.623
	1999 LX1	0.030	0.318	2.010	20.6	20050818	294.4	133.4	251.9	19.8	0.727	1.164
(66391)	1999 KW4	0.030A	0.200	1.084	16.5	20050818	214.4	192.6	244.9	38.9	0.688	0.642
(53319)	1999 JM8	0.040D	0.950	4.465	15.3	20050818	128.6	166.2	133.9	13.8	0.649	2.707
(85990)	1999 JV6	0.028	0.694	1.321	20.0	20050818	333.8	235.6	124.6	5.3	0.311	1.008
(31669)	1999 JT6	0.013	0.899	3.372	16.0	20050818	47.9	38.9	79.0	9.6	0.579	2.135
(85989)	1999 JD6	0.046	0.324	1.441	17.1	20050818	301.7	309.1	130.3	17.0	0.633	0.883
	1999 JU3	0.014D	0.963	1.415	19.4	20050818	51.2	211.3	251.7	5.9	0.190	1.189
	1999 JE1	0.029	0.394	2.252	20.0	20050818	1.8	302.6	53.2	20.9	0.702	1.323
	1999 GS6	0.019	0.599	1.783	19.6	20050818	15.7	134.7	314.7	2.0	0.497	1.191
	1999 GL4	0.025	0.840	3.397	19.5	20050818	47.0	293.7	178.7	7.2	0.603	2.119

Table 4. Continued

Designation (and name)	Prov. Des.	Min.	q	Q	H	Epoch	M	Peri.	Node	Incl.	e	a
(12923) Zephyr	1999 GK4	0.041	0.998	2.926	16.1	20050818	61.3	146.7	168.4	5.3	0.491	1.962
(38071)	1999 GU3	0.034A	1.029	3.148	19.6	20050818	37.8	8.8	195.6	12.7	0.507	2.089
	1999 FN53	0.049D	0.945	2.525	18.4	20050818	262.9	191.6	50.7	20.2	0.455	1.735
	1999 FA	0.000A	0.935	1.222	20.5	20050818	322.8	296.8	166.2	12.0	0.133	1.079
	1999 DB7	0.008A	0.971	1.441	20.2	20050818	300.2	29.8	157.7	10.8	0.195	1.206
(85938)	1999 DJ4	0.028D	0.957	2.748	18.6	20050818	186.8	197.5	20.0	9.2	0.483	1.852
	1999 CF9	0.022	0.710	2.835	17.9	20050818	238.3	89.8	157.4	5.5	0.600	1.772
	1999 BJ8	0.032	0.857	2.943	18.0	20050818	158.1	221.5	339.9	9.0	0.549	1.900
	1999 AQ10	0.001	0.717	1.157	20.3	20050818	318.9	299.5	327.4	6.6	0.234	0.937
	1999 AN10	0.015A	0.639	2.279	17.9	20050818	220.8	268.3	314.5	39.9	0.562	1.459
	1998 YM4	0.046	0.414	2.536	19.7	20050818	292.6	344.4	341.9	3.4	0.719	1.475
	1998 XD12	0.029	0.526	2.269	20.7	20050818	50.7	69.4	259.9	13.5	0.623	1.397
	1998 XX2	0.031	0.470	1.013	20.1	19981123	201.5	152.7	74.8	6.9	0.366	0.742
	1998 XN2	0.007	0.916	3.084	19.6	20050818	145.8	125.9	256.1	1.8	0.542	2.000
(33342)	1998 WT24	0.001	0.418	1.019	17.9	20050818	222.3	167.3	82.0	7.3	0.418	0.719
	1998 WB2	0.014	0.819	3.145	21.8	20050818	130.6	78.1	52.7	2.4	0.587	1.982
	1998 WZ1	0.019	0.959	3.366	19.9	20050818	27.3	137.2	345.1	4.3	0.557	2.162
(22753)	1998 WT	0.033	0.524	1.913	17.7	20050818	78.2	324.5	307.6	3.2	0.570	1.219
(20425)	1998 VD35	0.000	0.819	2.310	20.4	20050818	107.8	296.0	227.5	7.0	0.476	1.565
	1998 VF32	0.038	0.472	1.231	21.1	20050818	112.2	320.8	236.3	24.0	0.446	0.852
	1998 VO	0.029	0.830	1.317	20.4	20050818	107.8	75.9	228.3	10.1	0.227	1.074
(85774)	1998 UT18	0.033A	0.942	1.867	19.2	20050818	349.3	49.9	64.7	13.6	0.329	1.404
	1998 US18	0.034	0.838	4.410	20.4	19981014	7.8	115.2	207.6	9.7	0.681	2.624
	1998 SJ70	0.035	0.657	3.812	18.5	20050818	45.0	244.4	23.8	7.4	0.706	2.234
(85713)	1998 SS49	0.000	0.694	3.155	15.8	20050818	174.5	102.3	41.7	10.8	0.640	1.924
	1998 SL36	0.033D	0.809	1.980	20.5	19980924	291.8	116.5	353.3	19.2	0.420	1.395
	1998 SH36	0.012	0.467	1.709	20.6	20050818	319.5	278.5	218.1	2.1	0.571	1.088
(25143) Itokawa	1998 SF36	0.010	0.953	1.695	19.2	20050818	294.5	162.8	69.1	1.6	0.280	1.324
	1998 SZ27	0.015	0.448	1.358	20.5	19980924	91.2	47.5	166.8	23.4	0.504	0.903
	1998 ST27	0.010	0.385	1.254	19.5	20050818	335.1	322.4	197.6	21.0	0.530	0.819
	1998 SC15	0.001	0.746	1.802	19.5	20050818	214.8	277.3	198.8	16.1	0.415	1.274

Table 4. Continued

Designation (and name)	Prov. Des.	Min.	q	Q	H	Epoch	M	Peri.	Node	Incl.	e	a
	1998 SY14	0.008	0.950	4.744	20.6	20050818	143.7	21.7	40.2	3.5	0.666	2.847
	1998 SU4	0.028	0.480	1.830	21.0	19980904	41.9	241.8	350.8	23.4	0.584	1.155
	1998 SH2	0.011	0.760	4.627	21.0	19980904	10.7	259.9	14.3	2.5	0.718	2.694
(16960)	1998 QS52	0.009	0.310	4.088	14.3	20050818	356.4	242.2	261.3	17.7	0.859	2.199
	1998 QK28	0.011	0.978	3.683	19.4	20050818	334.8	204.6	174.8	7.9	0.580	2.330
	1998 QC1	0.013	0.811	3.156	19.6	20050818	151.4	115.1	308.7	9.6	0.591	1.983
	1998 QA1	0.024A	0.985	3.224	19.0	20050818	122.9	332.9	299.2	8.2	0.532	2.104
	1998 QP	0.016	0.743	2.844	21.5	20050818	307.0	77.9	326.9	9.4	0.586	1.793
(85640)	1998 OX4	0.002	0.813	2.352	21.2	20050818	144.0	116.8	299.9	4.5	0.486	1.583
(52768)	1998 OR2	0.011	1.035	3.746	16.2	20050818	8.1	173.7	27.3	5.9	0.567	2.391
(12538)	1998 OH	0.024A	0.915	2.168	16.1	20050818	312.5	321.6	220.8	24.5	0.406	1.541
(52760)	1998 ML14	0.013	0.917	3.908	17.5	20050818	303.7	20.1	338.9	2.4	0.620	2.412
	1998 MZ	0.002	0.575	2.117	19.3	20050818	254.8	40.5	121.0	0.1	0.573	1.346
	1998 KJ9	0.010	0.521	2.375	19.4	20050818	16.6	259.9	98.8	10.9	0.640	1.448
	1998 KN3	0.019	0.179	3.031	18.5	19980527	328.2	68.8	316.5	2.4	0.888	1.605
	1998 KM3	0.001	0.649	2.693	19.7	20050818	85.4	84.9	263.4	4.7	0.611	1.671
	1998 HH49	0.002	0.772	2.331	21.3	20050818	239.0	287.7	23.7	8.4	0.502	1.551
	1998 HT31	0.034	0.771	4.295	20.8	20050818	280.1	80.5	213.8	6.8	0.696	2.533
	1998 HD14	0.036	0.662	1.266	20.9	20050818	357.0	260.5	184.0	7.8	0.313	0.964
	1998 HL3	0.046	0.715	1.542	20.1	20050818	293.3	188.0	163.9	2.7	0.366	1.129
	1998 HJ3	0.011	0.506	3.464	18.9	20050818	194.9	93.6	224.1	6.5	0.745	1.985
	1998 HE3	0.003	0.491	1.266	21.9	20050818	208.2	309.0	53.9	3.4	0.441	0.879
	1998 HL1	0.037D	1.013	1.479	18.9	20050818	309.2	148.4	213.6	20.0	0.187	1.246
	1998 FF14	0.026A	0.862	1.645	20.4	20050818	14.8	301.2	359.9	38.7	0.312	1.254
(65909)	1998 FH12	0.011	0.502	1.682	19.2	20050818	264.9	284.4	108.6	3.5	0.540	1.092
	1998 FW4	0.001	0.678	4.314	19.6	20050818	337.2	75.4	3.1	3.6	0.728	2.496
	1998 FL3	0.048D	0.948	1.572	22.0	20050818	354.4	121.3	180.2	26.4	0.248	1.260
(85585) Mjolnir	1998 FG2	0.020	0.835	1.760	21.6	20050818	49.6	95.2	2.4	4.1	0.356	1.298
(27002)	1998 DV9	0.012A	0.988	2.499	18.2	20050818	87.7	0.7	130.4	8.7	0.433	1.744
	1998 CS1	0.016	0.629	2.357	17.8	20050818	17.9	97.4	111.0	7.8	0.579	1.493
	1998 BB10	0.018	0.732	1.813	20.4	20050818	150.8	259.1	124.4	11.5	0.425	1.272

Table 4. Continued

Designation (and name)	Prov. Des.	Min.	q	Q	H	Epoch	M	Peri.	Node	Incl.	e	a
	1998 BY7	0.027	0.800	3.248	21.5	20050818	201.2	90.0	122.4	3.3	0.605	2.024
(35396)	1997 XF11	0.000	0.745	2.141	16.9	20050818	250.3	102.8	213.9	4.1	0.484	1.443
	1997 XR2	0.009	0.860	1.293	20.8	20050818	35.8	84.7	250.9	7.2	0.201	1.077
	1997 WQ23	0.009	0.877	2.595	20.5	20050818	162.1	296.3	56.0	2.5	0.495	1.736
	1997 US2	0.005	0.568	2.781	19.8	20050818	171.3	99.9	66.2	3.2	0.661	1.674
	1997 QK1	0.002	1.006	4.589	20.2	20050818	257.0	2.5	307.1	2.9	0.640	2.798
	1997 NC1	0.035D	0.685	1.046	18.0	20050818	184.7	16.6	96.6	16.7	0.209	0.866
	1997 GD32	0.015	0.838	3.343	21.5	20050818	253.4	225.8	55.9	5.3	0.599	2.091
	1997 GL3	0.002	0.493	4.071	19.5	20050818	167.3	260.2	196.6	6.7	0.784	2.282
(13651)	1997 BR	0.000D	0.927	1.744	17.6	20050818	112.5	133.7	116.7	17.2	0.306	1.336
	1997 BQ	0.037	0.910	2.581	18.1	20050818	230.2	147.3	50.3	11.0	0.479	1.746
	1996 SK	0.003	0.495	4.362	17.0	20050818	97.8	283.7	198.1	2.0	0.796	2.429
	1996 RG3	0.004	0.791	3.209	18.5	20050818	18.3	299.9	158.4	3.6	0.604	2.000
	1996 JA1	0.014	0.763	4.352	21.0	20050818	82.2	247.0	57.3	21.8	0.702	2.558
	1996 JG	0.014	0.612	2.992	19.5	20050818	274.1	279.9	53.1	5.3	0.660	1.802
(65803) Didymos	1996 GT	0.039	1.013	2.276	18.4	20050818	307.8	319.2	73.3	3.4	0.384	1.644
	1996 FO3	0.030	1.024	1.861	20.6	20050818	182.9	162.6	333.7	5.8	0.290	1.443
	1996 FG3	0.028	0.685	1.423	18.2	20050818	138.3	23.9	299.9	2.0	0.350	1.054
	1996 EO	0.039A	0.804	1.879	19.2	20050818	93.1	64.5	356.6	21.6	0.401	1.341
(8566)	1996 EN	0.016D	0.858	2.156	16.5	20050818	318.6	125.1	164.2	38.0	0.431	1.507
	1996 AW1	0.039	0.737	2.316	19.5	19960118	49.4	228.6	117.9	4.7	0.517	1.527
	1996 AJ1	0.005	0.286	2.333	20.1	20050818	199.5	238.2	91.1	2.5	0.781	1.310
	1995 YR1	0.016	0.293	3.104	20.5	20050818	98.0	109.1	107.4	3.6	0.828	1.698
	1995 SA	0.022	0.871	4.026	17.4	20050818	265.0	51.5	171.5	20.2	0.644	2.449
	1995 CR	0.014	0.119	1.696	21.5	20050818	357.6	322.4	342.8	4.0	0.869	0.908
	1994 XL1	0.022D	0.318	1.024	20.9	20050818	4.6	356.5	252.7	28.2	0.526	0.671
	1994 XD	0.018	0.628	4.072	19.1	20050818	4.5	249.2	96.4	4.3	0.733	2.350
	1994 WR12	0.004	0.457	1.057	22.0	20050818	231.3	205.9	62.8	6.8	0.397	0.757
	1994 UG	0.025	0.882	1.592	21.1	20050818	84.6	225.4	13.6	5.1	0.287	1.237
	1994 RC	0.046	0.907	3.629	19.0	20050818	93.4	284.4	346.0	4.7	0.600	2.268
(7482)	1994 PC1	0.017A	0.904	1.788	16.8	20050818	151.1	47.6	117.9	33.5	0.328	1.346

Table 4. Continued

Designation (and name)	Prov. Des.	Min.	q	Q	H	Epoch	M	Peri.	Node	Incl.	e	a
(37655) Illapa	1994 PM	0.027	0.366	2.591	17.7	20050818	15.4	303.5	139.9	18.0	0.752	1.478
	1994 NE	0.042D	0.806	3.267	19.8	20050818	281.9	246.5	104.8	27.5	0.604	2.037
	1994 EK	0.031	0.772	3.515	20.0	19940309	24.4	97.3	334.7	6.0	0.640	2.144
	1994 CN2	0.014	0.951	2.195	16.7	20050818	83.4	248.2	99.4	1.4	0.395	1.573
	1994 CJ1	0.036	1.005	1.972	21.4	20050818	63.0	64.9	172.3	2.3	0.325	1.489
	1994 CC	0.011	0.955	2.319	17.9	20050818	51.6	24.7	268.8	4.6	0.417	1.637
	1994 AW1	0.036A	1.021	1.188	17.7	20050818	135.0	37.1	290.4	24.1	0.075	1.105
	1993 VD	0.017	0.393	1.359	21.6	20050818	232.2	253.6	2.7	2.1	0.551	0.876
(37638)	1993 VB	0.005	0.918	2.901	19.4	20050818	143.5	322.7	145.8	5.1	0.519	1.910
(52387)	1993 OM7	0.045D	1.039	1.525	17.8	20050818	272.2	195.4	297.6	24.2	0.190	1.282
(85236)	1993 KH	0.008	0.850	1.618	18.9	20050818	259.3	293.7	54.5	12.8	0.311	1.234
(5693)	1993 EA	0.008	0.527	2.016	17.0	20050818	68.7	258.8	97.2	5.1	0.585	1.271
(39572)	1993 DQ1	0.046	1.035	3.039	16.6	20050818	352.6	344.7	313.7	10.0	0.492	2.037
(65717)	1993 BX3	0.046	1.003	1.786	21.0	20050818	236.7	289.9	175.6	2.8	0.281	1.395
	1992 UY4	0.021	0.990	4.293	17.8	20050818	356.8	38.2	308.5	2.8	0.625	2.642
(5604)	1992 FE	0.037	0.551	1.303	16.4	20050818	110.5	82.4	312.0	4.8	0.405	0.927
(35107)	1991 VH	0.013D	0.973	1.300	16.9	20050818	171.1	207.1	139.5	13.9	0.144	1.136
	1991 RB	0.047	0.749	2.151	19.0	19910901	320.1	68.7	359.6	19.5	0.484	1.450
(6489) Golevka	1991 JX	0.031	0.986	4.009	19.2	20050818	197.8	66.8	211.0	2.3	0.605	2.497
	1991 JW	0.036A	0.915	1.161	19.3	20050818	56.4	301.9	54.0	8.7	0.118	1.038
	1991 GO	0.023	0.668	3.185	19.8	20050818	154.1	88.9	24.7	9.6	0.653	1.926
(9856)	1991 EE	0.035	0.848	3.647	17.4	20050818	63.4	115.2	169.0	9.8	0.623	2.247
(65690)	1991 DG	0.048	0.909	1.945	19.1	20050818	126.6	63.2	180.3	11.2	0.363	1.427
(7822)	1991 CS	0.033D	0.938	1.307	17.4	20050818	143.2	249.4	156.9	37.1	0.165	1.123
	1991 BN	0.024	0.869	2.018	19.2	20050818	238.6	80.7	269.0	3.4	0.398	1.444
(85182)	1991 AQ	0.018	0.486	3.942	17.1	20050818	132.6	241.0	341.5	3.2	0.780	2.214
(5189)	1990 UQ	0.044	0.810	2.292	17.3	20050818	287.7	159.5	135.4	3.6	0.478	1.551
	1990 UA	0.012	0.770	2.672	19.5	19901016	22.7	204.2	102.5	1.0	0.553	1.721
	1990 SM	0.024	0.491	3.711	16.2	20050818	341.3	106.9	136.9	11.6	0.766	2.101
	1990 OS	0.008	0.902	2.455	19.3	20050818	306.7	21.3	346.7	1.1	0.463	1.678
(4953)	1990 MU	0.040	0.556	2.687	14.1	20050818	182.5	77.6	77.9	24.4	0.657	1.621

Table 4. Continued

Designation (and name)	Prov. Des.	Min.	q	Q	H	Epoch	M	Peri.	Node	Incl.	e	a
(8014)	1990 MF	0.018	0.950	2.543	18.7	20050818	182.8	114.1	210.5	1.9	0.456	1.746
	1990 HA	0.013	0.781	4.360	16.0	20050818	269.5	308.3	184.8	3.9	0.696	2.570
	1989 VB	0.019	1.004	2.726	19.9	20050818	81.0	329.8	38.9	2.1	0.462	1.865
(11500)	1989 UR	0.032	0.696	1.465	18.4	20050818	295.5	289.3	234.6	10.3	0.356	1.080
(65679)	1989 UQ	0.015	0.673	1.157	19.4	20050818	233.5	14.9	178.4	1.3	0.265	0.915
	1989 UP	0.001	0.984	2.745	20.5	20050818	64.1	17.5	53.3	3.9	0.472	1.865
(6239) Minos	1989 QF	0.028	0.676	1.628	17.9	20050818	41.4	239.6	344.7	3.9	0.413	1.152
(4769) Castalia	1989 PB	0.024	0.549	1.577	16.9	20050818	150.9	121.3	325.7	8.9	0.483	1.063
(7335)	1989 JA	0.043	0.913	2.628	17.0	20050818	301.6	232.0	61.5	15.2	0.484	1.771
	1989 DA	0.042	0.981	3.337	18.9	20050818	72.8	139.5	349.3	6.5	0.546	2.159
(4179) Toutatis	1989 AC	0.005	0.921	4.124	15.30	20050818	72.7	280.0	122.8	0.4	0.635	2.523
(7753)	1988 XB	0.005	0.760	2.174	18.6	20050818	175.0	280.0	73.5	3.1	0.482	1.467
	1988 TA	0.006	0.804	2.279	20.8	20050818	321.3	104.7	195.1	2.5	0.479	1.541
	1987 SF3	0.042	1.045	3.459	18.5	20050818	114.3	134.4	187.2	3.3	0.536	2.252
(4450) Pan	1987 SY	0.027	0.596	2.287	17.2	20050818	169.1	291.5	312.1	5.5	0.587	1.442
(4486) Mithra	1987 SB	0.045	0.748	3.660	15.6	20050818	204.6	168.8	82.3	3.0	0.661	2.204
(4034)	1986 PA	0.023	0.589	1.530	18.1	20050818	55.3	296.5	158.1	11.2	0.444	1.060
(14827) Hypnos	1986 JK	0.018	0.955	4.738	18.3	20050818	22.2	238.0	58.1	2.0	0.665	2.846
(3362) Khufu	1984 QA	0.018	0.526	1.453	18.3	20050818	190.4	55.0	152.5	9.9	0.469	0.989
(3671) Dionysus	1984 KD	0.039D	1.007	3.389	16.3	20050818	174.9	204.2	82.2	13.5	0.542	2.198
(3200) Phaethon	1983 TB	0.025	0.140	2.403	14.6	20050818	115.8	322.0	265.4	22.2	0.890	1.271
	1983 LC	0.023	0.762	4.465	19.3	20050818	73.9	184.9	159.6	1.5	0.708	2.614
(3757)	1982 XB	0.020	1.017	2.652	18.95	20050818	37.7	17.1	75.0	3.9	0.446	1.835
(3361) Orpheus	1982 HR	0.013	0.819	1.599	19.03	20050818	240.2	301.6	189.7	2.7	0.323	1.209
(4660) Nereus	1982 DB	0.005	0.953	2.024	18.2	20050818	351.9	157.9	314.5	1.4	0.360	1.489
(3908) Nyx	1980 PA	0.031	1.042	2.812	17.4	20050818	107.9	126.0	261.7	2.2	0.459	1.927
	1979 XB	0.035	0.647	3.789	18.6	20050818	262.7	76.6	85.0	24.6	0.708	2.218
(4015) Wilson-Harrington	1979 VA	0.047	0.993	4.285	15.99	20050818	8.8	91.3	270.6	2.8	0.624	2.639
	1978 CA	0.026D	0.883	1.363	17.1	20050818	301.8	102.1	161.3	26.1	0.214	1.123
(2135) Aristaeus	1977 HA	0.014	0.795	2.404	17.94	20050818	32.2	290.8	191.3	23.1	0.503	1.600
(2340) Hathor	1976 UA	0.006	0.464	1.224	19.2	20050818	155.1	39.9	211.5	5.9	0.450	0.844

Table 4. Continued

Designation (and name)	Prov. Des.	Min.	q	Q	H	Epoch	M	Peri.	Node	Incl.	e	a
(2102) Tantalus	1975 YA	0.029A	0.904	1.675	16.2	20050818	49.4	61.6	94.4	64.0	0.299	1.290
(1981) Midas	1973 EA	0.000D	0.621	2.931	15.5	20050818	224.2	267.7	357.0	39.8	0.650	1.776
(5011) Ptah	6743 P-L	0.025	0.818	2.452	17.1	20050818	115.8	105.6	10.9	7.4	0.500	1.635
	6344 P-L	0.042	0.966	4.350	21.6	20050818	127.1	235.9	182.0	4.5	0.637	2.658
(2061) Anza	1960 UA	0.026	1.046	3.481	16.56	20050818	59.7	156.5	207.7	3.8	0.538	2.263
(4183) Cuno	1959 LM	0.037	0.721	3.243	14.4	20050818	224.9	235.4	295.7	6.8	0.636	1.982
(1620) Geographos	1951 RA	0.046	0.828	1.663	15.60	20050818	6.0	276.8	337.3	13.3	0.335	1.245
(1566) Icarus	1949 MA	0.040	0.187	1.969	16.9	20050818	116.7	31.3	88.1	22.9	0.827	1.078
(2201) Oljato	1947 XC	0.001	0.623	3.720	15.25	20050818	343.5	96.2	76.6	2.5	0.713	2.172
(69230) Hermes	1937 UB	0.003	0.622	2.688	17.5	20050818	279.6	92.4	34.5	6.1	0.624	1.655
(2101) Adonis	1936 CA	0.012	0.441	3.307	18.7	20050818	62.7	42.4	350.6	1.3	0.765	1.874
(1862) Apollo	1932 HA	0.028	0.648	2.295	16.25	20050818	343.1	285.8	35.8	6.4	0.560	1.471

References

[1] Elst, E.W., 2001, Are we approaching the end of asteroid-discovery?, *Planetary and Space Science*, 49, p.781-786

[2] van Houten, C.J., van Houten-Groeneveld, I, Herget P., Gehrels, T., 1970, The Palomar-Leiden survey of faint minor planets, *Astronomy and Astrophysics Supplement series*, 2, p. 239-448

[3] Marsden, B. G., 2005, *Minor Planet Electronic Circulars*, January 05

[4] Kirkwood, D., 1866, *Proceedings of the American Association for the Advance of Science*, 8, p.14

[5] Knezevic, Z., Milani, A., 1993, Asteroid proper elements, the big picture. *Asteroids, Comets, Meteors* p.143-155

[6] Cunningham, C. J., 1988, *Introduction to asteroids*. Willmann-Bell, Inc. p.59

[7] Meeus J., 1991, personal communication

[8] Ipatov, S. I., 1995, *Migration of small bodies to the earth: Solar System Research*, V. 29, p.261-286

[9] Menichella M., Paolicchi P., Farinella, P., 1996, *The main belt as a source of near-Earth asteroids: Earth, Moon and Planets*, V. 72, p.133-149

[10] Farinella, P., Froeschlé, C., and Gonczi, R., 1993, *Meteorite delivery and transport: Asteroids, Comets, Meteors:* p.205-222

[11] Elst, E. W., 2002, On the completeness of the discovery rate of the potentially hazardous asteroids, *Geological Society of America, Special Paper* 356, p.645-650

[12] Klose, A, 1928, *Astronomische Gesellschaft*, 63, p.333

[13] Elst, E. W., Ivanova, V., and Shkodrov, V., 1989, On the discovery of the Apollo object 1987 SB, *Astronomische Nachrichten*, 310, 2, p.157-158

[14] Marsden, B. G., 1996, I.A.U. Circulars,

[15] Hsieh, H. H., Jewitt, D. C., and Fernández Y. R., 2004, The strange case of 133P/Elst-Pizarro: A comet among the asteroids, *The American Journal*, 127, p.2997-3017

[16] Chapman, C. R., 2002, Catastrophic Events and Mass Extinctions, Impacts and Beyond, *Geological Society of America, Special Paper* 356, p.7-19

[17] Bottke, W. F., Jedicke, R., Morbidelli, A., Petit, J.-M., and Gladman B., 2000, Understanding the Distribution of Near-Earth Asteroids, *Science*, V. 288, Issue 5474, p.2190-2194

In: Space Science: New Research
Editor: Nick S. Maravell, pp. 245-285

ISBN 1-60021-005-8

Chapter 8

EFFECT OF ELECTROMAGNETIC RADIATION ON DYNAMICS OF COSMIC DUST PARTICLES

J. Klačka[1]*and M. Kocifaj*[2]†
[1]Faculty of Mathematics, Physics and Informatics, Comenius University, Mlynská dolina, 842 48 Bratislava, Slovak Republic
[2]Institute for Experimental Physics, University of Vienna, Boltzmangasse 5, 1090 Wien, Austria

Abstract

Effect of electromagnetic radiation on dynamics of arbitrarily shaped cosmic dust particles is investigated. Process of extinction, scattering, absorption and thermal emission is taken into account. Equation of motion is derived and physical meaning of the important terms is emphasized and discussed. The equation of motion is written in a relativistically covariant form. The equation of motion yields, as a special cases, both the results presented by Einstein in his relativistic paper written 100 years ago, and, also the Poynting-Robertson effect, well-known in astrophysics.

Applications of the derived equation of motion are shown for various systems in the Universe. The significance of the general equation of motion is always stressed by presenting results both for spherical and nonspherical dust particles. While Poynting-Robertson effect holds for spherical particles, nonspherical particles may behave in a different way than it would correspond to the Poynting-Robertson effect. The first application deals with the observed flux of interstellar dust entering the Solar System. Spherical interstellar dust grains can be captured only in small distances from the Sun and the grains are thermally destructed. The flux of interstellar dust may produce capture and surviving of nonspherical particles. The second application concentrate on motion of dust grains near commensurability resonances with a planet - mean motion resonances (MMR; a particle is in resonance with a planet when the ratio of their mean motions is the ratio of two small integers) - and possible capture of the grains in the resonances. Capture times for spherical particles can be larger than for nonspherical grains. However, nonspherical grains may exhibit an increase of semimajor axis before capturing in the MMR. This is caused by the effect of electromagnetic radiation on

*E-mail address: klacka@fmph.uniba.sk
†E-mail address: kocifaj@ap.univie.ac.at; on leave from Astronomical Institute, Slovak Academy of Sciences

nonspherical dust grain and this is the first known case when a cosmic body can be captured in the MMR during the process of an increase of semimajor axis.

Another application, of the general equation of motion for interaction between incident electromagnetic radiation and dust grain, concerns evolution of meteoroids ejected from comets. While observations yield that small dust grains are observed not only in the orbital plane of comet Encke near its aphelion, Poynting-Robertson effect cannot explain such a behaviour. The effect of electromagnetic radiation on nonspherical grains produces also change of orbital inclination. Thus, nonspherical dust grains near cometary aphelion have to be present also in the plane normal to the orbital plane of the comet.

Keywords: electromagnetic radiation, cosmic dust, relativistic covariant equation of motion, celestial mechanics, orbital motion, mean-motion resonances, comets

1. Introduction

The existence of a light pressure was proved by J. C. Maxwell, for the first time. Maxwell used equations for electromagnetic field, which were obtained by him in 1864 (Maxwell 1873). Later on, Italian physicist A. Bartoli (1876) has shown existence of the light pressure on the basis of thermodynamic considerations, more accurately, on the basis of the second law of thermodynamics. The effect of radiation pressure was experimentally proved by P. N. Lebedev from Moscow University: for solid bodies in 1899, and, for gases in 1905. At that time, Poynting (1904) formulated a problem of finding equation of motion for a perfectly absorbing spherical particle under the action of electromagnetic radiation. The second case, closely connected with the relativistic equation of motion for a free particle under the action of electromagnetic radiation, was presented by Einstein (1905), who calculated the change of energy and light pressure at arbitrary angle of incidence on a plane mirror. Robertson (1937) subsequently derived a correct equation of motion for a perfectly absorbing spherical particle. This result has been applied to astronomical problems for several decades and is known as the Poynting-Robertson (P-R) effect. Robertson's case was relativistically generalized by Klačka (1992), who showed that the generalized P-R effect holds only for the special case where the total momentum of the outgoing radiation per unit time is colinear with the incident radiation (in the proper/rest frame of reference of the particle), which may include radiation normally incident on a plane mirror. Other theoretical papers on the basic properties of the P-R effect were written during the last decades: Wyatt and Whipple (1950), Harwit (1963), Burns *et al.* (1979), Mediavilla and Buitrago (1989), Mignard (1992), Srikanth (1999), Williams (2002). Since real particles interact with electromagnetic radiation in a complicated manner and particles of various optical properties exist (e. g., Mishchenko *et al.* 2002), it is essential to have an equation of motion sufficiently general to cover a wide range of optical parameters, not just the limited cases previously investigated. As an attempt of formulating such a general equation of motion, we can mention Lyttleton (1976). Finally, Klačka (2000) succeeded in deriving relativistically covariant equation of motion. This equation of motion was derived in other possible way by Klačka and Kocifaj (2001), within the accuracy to the first order in v/c. The equations of motion for a moving particle have been constructed under the assumption that thermal emission from a particle is isotropic and does not exert a radiation pressure force on the particle in the particle frame of reference. Recently, Mishchenko (2001) has formulated the radiation pressure on arbitrarily shaped particles arising from an anisotropy of thermal emission. Equation of motion for an arbitrarily shaped particle moving in a radiation field taking into account the radiation pressure caused by an anisotropy of thermal emission as well as scattering and absorption of light was presented by Klačka (2004), in a relativistically covariant form. Experimental verification that real nonspherical dust particles move in a different way than spherical dust grains, under action of radiation pressure forces, may be found in Krauss and Wurm (2004).

We begin by reviewing in Sec. 2. and 3. the basic physical processes in proper and stationary frames. The equation of motion for simultaneous action of gravity and electromagnetic radiation is presented in Sec. 4.. Sec. 5. then applies these results to the calculation of osculating orbital elements, including the ejection of a dust particle from a parent body such as a comet or asteroid. The calculation for the special case of the P-R effect is carried out in Sec. 6. to first order in $\vec{v}/c$. The case when the disturbing acceleration is given in terms of velocity, is considered, as it was done already by Robertson (1937). Next, we briefly discuss, in Sec. 7., the effect of the solar wind on the secular changes in the orbital elements to the first order in $\vec{v}/u$, where u is the speed of solar wind particles. The other sections present several applications of the theory presented in the preceding chapters. Each section compares results for spherical dust particles and nonspherical grains and the effect of electromagnetic radiation force is stressed. Sec. 8. deals with possible capture of interstellar dust particles in the Solar System. Sec. 9. then treats motion of dust in zones of mean-motion resonances with planets. Orbital evolution of spherical and nonspherical dust grains ejected from a comet is shortly discussed in Sec. 10., which is concentrated on behaviour of inclination and perihelion motion. Finally, Sec. 11. summarizes the results: it compares evolution of spherical and nonspherical dust particles.

2. Proper Reference Frame of the Particle – Stationary Particle

The term "stationary particle" will denote a particle which does not move in a given inertial frame of reference. Primed quantities will denote quantities measured in the proper reference frame of the particle – rest frame of the particle.

The flux density of photons scattered into an elementary solid angle $d\Omega' = \sin\theta'\, d\theta'\, d\phi'$ is proportional to $p'(\theta', \phi')\, d\Omega'$, where $p'(\theta', \phi')$ is the "phase function". The phase function depends on orientation of the particle with respect to the direction of the incident radiation and on the particle characteristics; angles θ', ϕ' correspond to the direction (and orientation) of travel of the scattered radiation, θ' is the polar angle which vanishes for propagation along the unit vector $\vec{e}_1'$ of the incident radiation. The phase function fulfills the normalisation condition

$$\int_{4\pi} p'(\theta', \phi')\, d\Omega' = 1\,. \tag{1}$$

The momentum of the incident beam of photons which is lost in the process of interaction with the particle is proportional to the cross-section C'_{ext} (extinction). The part proportional to C'_{abs} (absorption) is emitted in the form of thermal radiation and the part proportional to $C'_{ext} - C'_{abs} = C'_{sca}$ is scattered. The differential scattering cross section $dC'_{sca}/d\Omega' \equiv C'_{sca}\, p'(\theta', \phi')$ depends on the polarization state of the incident light as well as on the incidence and scattering directions (e. g., Mishchenko *et al.* 2002).

The momentum (per unit time) of the scattered photons into an elementary solid angle $d\Omega'$ is

$$d\vec{p'}_{sca} = \frac{1}{c} \, S' \, C'_{sca} \, p'(\theta', \phi') \, \vec{K}' \, d\Omega' \, , \tag{2}$$

where the unit vector in the direction of scattering is

$$\vec{K}' = \cos\theta' \, \vec{e}_1' \, + \, \sin\theta' \, \cos\phi' \, \vec{e}_2' \, + \, \sin\theta' \, \sin\phi' \, \vec{e}_3' \, . \tag{3}$$

S' is the flux density of radiation energy (energy flow through unit area perpendicular to the ray per unit time). The system of unit vectors used on the RHS of the last equation forms an orthogonal basis. The total momentum (per unit time) of the scattered photons is

$$\vec{p'}_{sca} = \frac{1}{c} \, S' \, C'_{sca} \int_{4\pi} p'(\theta', \phi') \, \vec{K}' \, d\Omega' \, . \tag{4}$$

The momentum (per unit time) obtained by the particle due to the interaction with radiation – radiation force acting on the particle – is

$$\frac{d\,\vec{p'}}{d\,t'} = \frac{1}{c} \, S' \left\{ C'_{ext} \, \vec{e}_1' \, - \, C'_{sca} \int_{4\pi} p'(\theta', \phi') \, \vec{K}' \, d\Omega' \right\} \, + \, \vec{F}_e'(T') \, , \tag{5}$$

where the emission component of the radiation force acting on the particle of absolute temperature T' is (Mishchenko *et al.* 2002, pp. 63-66)

$$\vec{F}_e'(T') = - \frac{1}{c} \int_0^{\infty} d\omega' \int_{4\pi} \hat{\vec{r}}' \, K_e' \left(\hat{\vec{r}}', T', \omega' \right) \, d\hat{\vec{r}}' \, . \tag{6}$$

The unit vector $\hat{\vec{r}}' = \vec{r}'/r'$ is given by position vector $\vec{r}'$ of the observation point with origin inside the particle (the emitted radiation in the far-field zone of the particle propagates in the radial direction, i. e., along the unit vector $\hat{\vec{r}}'$), ω' is (angular) frequency of radiation,

$$K_e' \left(\hat{\vec{r}}', T', \omega' \right) = I_b' \left(T', \omega' \right) \left[K_{11}' \left(\hat{\vec{r}}', \omega' \right) - \int_{4\pi} Z_{11}' \left(\hat{\vec{r}}', \hat{\vec{r}}'', \omega' \right) d\hat{\vec{r}}'' \right] \, , \tag{7}$$

where K_{11}' is the (1,1) element of the particle extinction matrix, Z_{11}' is the (1,1) element of the phase matrix and the Planck blackbody energy distribution is given by the well-known relation

$$I_b' \left(T', \omega' \right) = \frac{\hbar \, \omega'^3}{4 \, \pi^3 \, c^2} \left\{ \exp \left(\frac{\hbar \, \omega'}{k \, T'} \right) \, - \, 1 \right\}^{-1} \, . \tag{8}$$

Thermal emission has to be included in the total interaction of the particle with electromagnetic radiation: if the particle absolute temperature is above zero, it can emit as well as scatter and absorb electromagnetic radiation.

For the sake of brevity, we will use dimensionless efficiency factors Q_x' instead of cross sections C_x': $C_x' = Q_x' \, A'$, where A' is geometrical cross section of a sphere of volume equal to the volume of the particle. Equation (5) can be rewritten to the form

$$\begin{aligned} \frac{d\,\vec{p'}}{d\,\tau} \;=\; & \frac{1}{c} \, S' \, A' \, \{ [Q'_{ext} \, - \, < \cos\theta' > \, Q'_{sca}] \, \vec{e}_1' \, + \\ & [\, - \, < \sin\theta' \, \cos\phi' > \, Q'_{sca}] \, \vec{e}_2' \, + \\ & [\, - \, < \sin\theta' \, \sin\phi' > \, Q'_{sca}] \, \vec{e}_3' \} \, + \, \sum_{j=1}^{3} F_{ej}' \, \vec{e}_j' \, , \end{aligned} \tag{9}$$

where $< x' >\equiv \int_{4\pi} x'\, p'(\theta', \phi')\, d\Omega'$ and $F'_{ej} \equiv \vec{F}'_e(T') \cdot \vec{e}'_j$. As for the energy, we assume that it is conserved: the energy (per unit time) of the incoming radiation E'_i, equals to the energy (per unit time) of the outgoing radiation (after interaction with the particle) E'_o. We will use the fact that time $t' = \tau$, where τ is proper time.

Summarizing important equations, we can write them in a short form

$$\frac{d\,\vec{p}'}{d\,\tau} = \sum_{j=1}^{3} \left(\frac{S'\,A'}{c}\, Q'_{pr,j} + F'_{ej} \right) \vec{e}'_j \ ; \ \frac{d\,E'}{d\,\tau} = 0 \, , \qquad (10)$$

where $Q'_{pr,1} \equiv Q'_{ext} - < \cos\theta' > Q'_{sca}$, $Q'_{pr,2} \equiv - < \sin\theta' \ \cos\phi' > Q'_{sca}$, $Q'_{pr,3} \equiv - < \sin\theta' \ \sin\phi' > Q'_{sca}$; $Q'_{pr,j}$, $j = 1, 2, 3$, are dimensionless efficiency factors for radiation pressure. We have added an assumption of equilibrium state when the particle mass does not change.

2.1. Summary of the Important Equations

Using the text concerning energy below Eq. (9) and the last Eq. (10), we may describe the total process of interaction in the form of the following equations (energies and momenta per unit time):

$$\begin{aligned} E'_o &= E'_i = A'\,S' \, , \\ \vec{p}'_o &= (1 - Q'_{pr,1})\,\vec{p}'_i - (Q'_{pr,2}\,\vec{e}'_2 + Q'_{pr,3}\,\vec{e}'_3)\,E'_o/c - \sum_{j=1}^{3} F'_{ej}\,\vec{e}'_j \, , \\ \vec{p}'_i &= (E'_i/c)\,\vec{e}'_1 \, , \end{aligned} \qquad (11)$$

The index "i" represents the incoming radiation, beam of photons, the index "o" represents the outgoing radiation. The relation for $\vec{p}'_o$ represents a generalization of the following cases:

i) $Q'_{pr,1} = 1$, $Q'_{pr,2} = Q'_{pr,3} = F'_{e1} = F'_{e2} = F'_{e3} = 0$ – Robertson (1937), Robertson and Noonan (1968), Srikanth (1999);

ii) $Q'_{pr,1}$ arbitrary, $Q'_{pr,2} = Q'_{pr,3} = F'_{e1} = F'_{e2} = F'_{e3} = 0$ – Klačka (1992);

iii) $Q'_{pr,1}$, $Q'_{pr,2}$, $Q'_{pr,3}$ arbitrary, $F'_{e1} = F'_{e2} = F'_{e3} = 0$ – Klačka (2000).

The changes of energy and momentum of the particle due to the interaction with electromagnetic radiation are

$$\begin{aligned} \frac{d\,E'}{d\,\tau} &= E'_i - E'_o = 0 \, , \\ \frac{d\,\vec{p}'}{d\,\tau} &= \vec{p}'_i - \vec{p}'_o \, . \end{aligned} \qquad (12)$$

3. Stationary Frame of Reference

By the term "stationary frame of reference" (laboratory frame) we mean a frame of reference in which particle moves with a velocity vector $\vec{v} = \vec{v}(t)$. The physical quantities measured in the stationary frame of reference will be denoted by unprimed symbols.

Our aim is to derive equation of motion for the particle in the stationary frame of reference. We will use the fact that we know this equation in the proper frame of reference – see Eqs. (11) and (12).

If we have a four-vector $A^\mu = (A^0, \vec{A})$, where A^0 is its time component and $\vec{A}$ is its spatial component, generalized special Lorentz transformation yields

$$\begin{aligned} A^{0'} &= \gamma\,(A^0 - \vec{v}\cdot\vec{A}/c)\,, \\ \vec{A}' &= \vec{A} + [(\gamma - 1)\,\vec{v}\cdot\vec{A}/\vec{v}^2 - \gamma\,A^0/c]\,\vec{v}\,, \end{aligned} \tag{13}$$

with inverse

$$\begin{aligned} A^{0} &= \gamma\,(A^{0'} + \vec{v}\cdot\vec{A}'/c)\,, \\ \vec{A} &= \vec{A}' + [(\gamma - 1)\,\vec{v}\cdot\vec{A}'/\vec{v}^2 + \gamma\,A^{0'}/c]\,\vec{v}\,, \end{aligned} \tag{14}$$

where $\gamma = 1/\sqrt{1 - \vec{v}^2/c^2}$.

As for four-vectors we immediately introduce four-momentum:

$$p^\mu = (p^0, \vec{p}) \equiv (E/c, \vec{p})\,. \tag{15}$$

3.1. Incoming Radiation

Applying Eqs. (14) and (15) to quantity $(E_i'/c, \vec{p}_i')$ (four-momentum per unit time – proper time is a scalar quantity) and taking into account also Eqs. (11), we can write

$$\begin{aligned} E_i &= E_i'\,\gamma\,(1 + \vec{v}\cdot\vec{e}_1'/c)\,, \\ \vec{p}_i &= \frac{E_i'}{c}\,\{\vec{e}_1' + [(\gamma - 1)\,\vec{v}\cdot\vec{e}_1'/\vec{v}^2 + \gamma/c]\,\vec{v}\}\,. \end{aligned} \tag{16}$$

Using the fact that $p^\mu = (h\nu/c,\ h\nu\vec{e}_1/c)$ for photons, we have

$$\begin{aligned} \nu' &= \nu\,w_1\,, \\ \vec{e}_1' &= \frac{1}{w_1}\,\{\vec{e}_1 + [(\gamma - 1)\,\vec{v}\cdot\vec{e}_1/\vec{v}^2 - \gamma/c]\,\vec{v}\}\,, \end{aligned} \tag{17}$$

where

$$w_1 \equiv \gamma\,(1 - \vec{v}\cdot\vec{e}_1/c)\,. \tag{18}$$

Inserting the second of Eqs. (17) into Eqs. (18), one obtains

$$\begin{aligned} E_i &= (1/w_1)\,E_i'\,, \\ \vec{p}_i &= (1/w_1)\,(E_i'/c)\,\vec{e}_1\,. \end{aligned} \tag{19}$$

We have four-vector $p_i^\mu = (E_i/c, \vec{p}_i) = (1, \vec{e}_1)E_i/c = (1/w_1, \vec{e}_1/w_1)\,w_1\,E_i/c \equiv b_1^\mu\,w_1\,E_i/c$.

For monochromatic radiation the flux density of radiation energy becomes

$$S' = n'\, h\, \nu'\, c\,; \quad S = n\, h\, \nu\, c\,, \tag{20}$$

where n and n' are concentrations of photons (photon number densities) in the corresponding reference frames. We also have continuity equation

$$\partial_\mu\, j^\mu = 0\,, \quad j^\mu = (c\, n, c\, n\, \vec{e}_1)\,, \tag{21}$$

with current density j^μ. Application of Eq. (13) then yields

$$n' = w_1\, n\,. \tag{22}$$

Using Eqs. (17), (20) and (22) we finally obtain

$$S' = w_1^2\, S\,. \tag{23}$$

Eqs. (11), (19) and (23) then together give $E_i = w_1\, S\, A'$, $\vec{p_i} = w_1\, S\, A'\, \vec{e}_1/c$.

3.2. Outgoing Radiation

The situation is analogous to that of the preceding subsection. It is only a little more algebraically complicated, since radiation may also spread out in directions given by unit vectors $\vec{e}_2$, $\vec{e}_3$. On the basis of Eqs. (18) – (19) we have

$$p_i^\mu = \frac{w_1\, E_i}{c}\, b_1^\mu\,, \tag{24}$$

where

$$\begin{aligned} b_1^\mu &= \left(\frac{1}{w_1}, \frac{\vec{e}_1}{w_1}\right)\,, \\ w_1 &= \gamma\,(1\, -\, \vec{v}\cdot\vec{e}_1/c)\,. \end{aligned} \tag{25}$$

We rewrite Eq. (11) to the form

$$\begin{aligned} E'_o &= E'_i = w_1\, E_i = A'\, S'\,, \\ \vec{p}\,'_o &= \vec{p}\,'_i\, -\, \sum_{j=1}^{3} \left(\frac{w_1\, E_i}{c}\, Q'_{pr,j}\, +\, F'_{ej}\right) \vec{e}\,'_j\,. \end{aligned} \tag{26}$$

We have two four-vectors in Eq. (26): $(E'_i/c, \vec{p}\,'_i)$, $(E'_o/c, \vec{p}\,'_o)$. Thus, also $\sum_{j=1}^3 \left\{(w_1\, E_i/c)\, Q'_{pr,j}\, +\, F'_{ej}\right\} \vec{e}\,'_j$ forms a four-vector. On the basis of Eqs. (24) – (25) we can immediately make a generalization of Eq. (25):

$$\begin{aligned} b_j^\mu &= \left(\frac{1}{w_j}, \frac{\vec{e}_j}{w_j}\right) \\ w_j &= \gamma\,(1\, -\, \vec{v}\cdot\vec{e}_j/c)\,, \quad j = 1, 2, 3\,. \end{aligned} \tag{27}$$

In general case, $b'^{\mu}_j = (1, \vec{e}'_j)$, $b^{\mu}_j = \Lambda_{\nu}{}^{\mu}\; b'^{\nu}_j$, $j = 1, 2, 3$, where $\Lambda_{\nu}{}^{\mu}$ represents general Lorentz transformation (see Appendix A; we have to use general Lorentz transformation, since also Thomas precession exists). The quantities $\left\{(w_1\, E_i/c)\; Q'_{pr,j} + F'_{ej}\;,\; j = 1, 2, 3\right\}$ behave as scalars. The only problem is how to use all the discussed four-vectors in order to fulfil Eq. (26). Simple idea $p^{\mu}_o = p^{\mu}_i - \sum_{j=1}^{3}(scalar_j b^{\mu}_j)$ does not work. Really, $\mu = 0$ would yield $E'_o = E'_i - c\sum_{j=1}^{3} scalar_j$, but this is not $E'_o = E'_i$ as it is required by Eq. (26). We have to use $p^{\mu}_o = p^{\mu}_i - \sum_{j=1}^{3}(scalar_j q^{\mu}_j)$, where four-vector q^{μ}_j reduces to 0 for $\mu = 0$ in the particle frame of reference. Such a four-vector is $q^{\mu}_j = b^{\mu}_j - u^{\mu}/c$, where u^{μ} is particle four-velocity $u^{\mu} = (\gamma c, \gamma \vec{v})$. Finally, Eq. (26) can be rewritten, in any frame of reference, in the following form:

$$p^{\mu}_o = p^{\mu}_i - \sum_{j=1}^{3}\left(\frac{w_1\, E_i}{c}\; Q'_{pr,j} + F'_{ej}\right)\left(b^{\mu}_j - u^{\mu}/c\right) . \tag{28}$$

3.3. Equation of Motion

In analogy with Eqs. (12), we have for the changes of energy and momentum of the particle due to the interaction with electromagnetic radiation

$$\begin{aligned} \frac{d\,E}{d\,\tau} &= E_i - E_o\,, \\ \frac{d\,\vec{p}}{d\,\tau} &= \vec{p}_i - \vec{p}_o\,. \end{aligned} \tag{29}$$

Using the fact that $(E/c, \vec{p})$, $(E_i/c, \vec{p}_i)$ and $(E_o/c, \vec{p}_o)$ are four-vectors, Eqs. (28) and (29) immediately yield

$$\frac{d\,p^{\mu}}{d\,\tau} = \sum_{j=1}^{3}\left(Q'_{pr,j}\;\frac{w_1\, E_i}{c^2} + \frac{1}{c}\; F'_{ej}\right)\left(c\; b^{\mu}_j - u^{\mu}\right) , \tag{30}$$

where p^{μ} is four-vector of the particle of mass m

$$p^{\mu} = m\; u^{\mu}\;, \tag{31}$$

four-vector of the world-velocity of the particle is

$$u^{\mu} = (\gamma\; c, \gamma\; \vec{v})\;. \tag{32}$$

It can be easily verified that:

i) the quantity $w\; E_i$ is a scalar quantity – see first of Eqs. (19);

ii) Eq. (30) reduces to Eq. (10) for the case of proper inertial frame of reference of the particle;

iii) Eq. (30) yields $d\,m/d\,\tau = 0$.

We introduce

$$\begin{aligned} b_j^0 &\equiv 1/w_j = \gamma\,(1 + \vec{v}\cdot\vec{e}_j^{\,\prime}\,/\,c)\,, \\ \vec{b}_j &\equiv \vec{e}_j\,/\,w_j = \vec{e}_j^{\,\prime} + [(\gamma - 1)\,\vec{v}\cdot\vec{e}_j^{\,\prime}\,/\,\vec{v}^{\,2} + \gamma\,/\,c]\,\vec{v}\,, \\ j &= 1, 2, 3 \end{aligned} \tag{33}$$

for the purpose of practical calculations. Physics of these relations corresponds to aberration of light.

We have derived an equation of motion for real dust particle under the action of electromagnetic radiation (including thermal emission). It is supposed that the equation of motion is represented by Eqs. (11) and (12) in the proper frame of reference of the particle. The final covariant form is represented by Eq. (30), or using $E_i = w_1\ S\ A'$ (see Eqs. (11), (19) and (23)),

$$\frac{d\,p^\mu}{d\,\tau} = \sum_{j=1}^{3}\left(\frac{w_1^2\ S\ A'}{c^2}\ Q'_{pr,j} + \frac{1}{c}\ F'_{ej}\right)\left(c\,b_j^\mu - u^\mu\right)\,. \tag{34}$$

In the case of general Lorentz transformation (Appendix A), the right-hand sides of Eq. (33) hold only to the first order in $\vec{v}/c$ (Eq. 27 holds).

To first order in $\vec{v}/c$, Eqs. (33) – (34) yield

$$\begin{aligned} \frac{d\,\vec{v}}{d\,t} &= \frac{S\ A'}{m\ c}\sum_{j=1}^{3} Q'_{pr,j}\ [(1 - 2\,\vec{v}\cdot\vec{e}_1/c + \vec{v}\cdot\vec{e}_j/c)\ \vec{e}_j - \vec{v}/c] + \\ &\quad \frac{1}{m}\sum_{j=1}^{3} F'_{ej}\left[\left(1 + \frac{\vec{v}\cdot\vec{e}_j}{c}\right)\vec{e}_j - \frac{\vec{v}}{c}\right], \\ \vec{e}_j &= (1 - \vec{v}\cdot\vec{e}_j^{\,\prime}/c)\,\vec{e}_j^{\,\prime} + \vec{v}/c\ \ ,\ \ j = 1, 2, 3\,. \end{aligned} \tag{35}$$

It is worth mentioning to stress that the values of Q'−coefficients depend on particle orientation with respect to the incident radiation – their values are time dependent.

It can be verified that Eq. (34) (or Eqs. 35 within the accuracy to the first order in $\vec{v}/c$) yields as special cases the situations discussed in Einstein (1905) and Robertson (1937) – Robertson's case is obtained simply by substituting $Q'_{pr,1} = 1$, $Q'_{pr,2} = Q'_{pr,3} = 0$, $F'_{e1} = F'_{e2} = F'_{e3} = 0$, Einstein's results require a little more calculations (see Appendix B).

3.4. Heuristic Derivation

Since we completely understand the physics of Eq. (30), we are able to present a short simple derivation of Eq. (35). We have:

i) $d\vec{v}^{\,\prime}/dt = \sum_{i=1}^{3}\left\{[S'A'/(mc)]\ Q'_{pr,i} + F'_{ej}/m\right\}\ \vec{e}_i^{\,\prime}$ (see Eq. 10);
   $dm/dt = 0$ is supposed;

ii) $S' = S(1 - 2\vec{v}\cdot\vec{e}_1/c)$ (see Eq. 23), due to the change of concentration of photons $n' = n(1 - \vec{v}\cdot\vec{e}_1/c)$ – Eq. (22) – and Doppler effect $\nu' = \nu(1 - \vec{v}\cdot\vec{e}_1/c)$ – Eq. (17);

iii) $\vec{e}'_j = (1 + \vec{v} \cdot \vec{e}_j / c) \vec{e}_j - \vec{v}/c$, $j = 1, 2, 3$ (aberration of light – see Eq. (33)).

Taking into account these physical phenomena, we finally obtain Eq. (35). However, only relativistic covariant formulation proves that $\{Q'_{pr,j}, F'_{ej}; j = 1, 2, 3\}$ behave as relativistically invariant quantities.

A little different derivation of the relation between S' and S, based on: a) stress-energy tensor (energy-momentum tensor), b) transformations of electric and magnetic fields, may be found in Klačka (1992: sections 2.3 and 2.5).

3.5. Continuous Distribution of Density Flux of Energy

For a continuous frequency distribution of density flux of energy, we can write

$$\frac{d\,\vec{p}'}{d\,\tau} = \sum_{j=1}^{3} \left\{ \frac{A'}{c} \int_0^{\infty} c\,h\,\nu' \, \frac{\partial n'}{\partial \nu'} \, Q'_{pr,j}(\nu')\, d\nu' \, + \, F'_{ej} \right\} \vec{e}'_j \equiv$$
$$\equiv \sum_{j=1}^{3} \left(\frac{S'\,A'}{c} \, \bar{Q}'_{pr,j} \, + \, F'_{ej} \right) \vec{e}'_j \, . \qquad (36)$$

Taking into account that concentration of photons fulfills $n' = w_1\, n$ (Eq. 22) and that Doppler effect yields $\nu' = w_1\, \nu$ (Eq. 17), we have $\partial n'/\partial \nu' = \partial n / \partial \nu$. Lorentz transformation finally yields

$$\frac{d\,p^{\mu}}{d\,\tau} = \sum_{j=1}^{3} \left\{ \frac{w_1^2\, A'}{c^2} \int_0^{\infty} c\,h\, \frac{\partial n}{\partial \nu} \, \nu \, Q'_{pr,j}(w_1\, \nu)\, d\nu \, + \, \frac{1}{c} \, F'_{ej} \right\} \times$$
$$\times \left(c\, b_j^{\mu} \, - \, u^{\mu} \right)$$
$$\equiv \sum_{j=1}^{3} \left(\frac{w_1^2\, S\, A'}{c^2} \, \bar{Q}'_{pr,j} \, + \, \frac{1}{c} \, F'_{ej} \right) \left(c\, b_j^{\mu} \, - \, u^{\mu} \right) . \qquad (37)$$

As a consequence, $dm/d\tau = 0$ (this corresponds to the condition $dE'/d\tau = 0$).

If the particle is not irradiated, then one has to use

$$\left(\frac{d\,p^{\mu}}{d\,\tau} \right)_e = \sum_{j=1}^{3} \, F'_{ej} \, b_j^{\mu} \, , \qquad (38)$$

instead of Eq. (37). As a consequence, Eq. (38) yields $(dm/d\tau)_e = \sum_{j=1}^{3} \, F'_{ej} \, / \, c$. Mass of the particle decreases due to the thermal emission, alone.

4. Gravitation and Electromagnetic Radiation

The generally covariant equation of motion can be immediately written on the basis of Eq. (37) (see e. g., Landau and Lifshitz 1975):

$$\frac{D\,p^{\mu}}{d\,\tau} = \sum_{j=1}^{3} \left(\frac{w_1^2\, S\, A'}{c} \, \bar{Q}'_{pr,j} \, + \, F'_{ej} \right) \left(b_j^{\mu} \, - \, u^{\mu} \, / \, c \right) . \qquad (39)$$

where the operator $D \, / \, d\tau$ is the "total" covariant derivative in the general relativistic sense and includes gravitational effects.

4.1. Gravity and Radiation – Equation of Motion to the First Order in v/c

We can write, on the basis of Eqs. (39), (27) and (33)

$$\begin{aligned} \frac{d\,\vec{v}}{d\,t} &= -\,\frac{G\,M_{\odot}}{r^2}\,\vec{e}_1 + \\ &\frac{G\,M_{\odot}}{r^2}\sum_{j=1}^{3}\beta_j\left[\left(1 - 2\,\frac{\vec{v}\cdot\vec{e}_1}{c} + \frac{\vec{v}\cdot\vec{e}_j}{c}\right)\vec{e}_j - \frac{\vec{v}}{c}\right] + \\ &\frac{1}{m}\sum_{j=1}^{3}F'_{ej}\left[\left(1 + \frac{\vec{v}\cdot\vec{e}_j}{c}\right)\vec{e}_j - \frac{\vec{v}}{c}\right] , \end{aligned} \tag{40}$$

where

$$\begin{aligned} \vec{e}_j &= (1 - \vec{v}\cdot\vec{e}\,'_j/c)\,\vec{e}\,'_j + \vec{v}/c\,, \;\; j = 1,2,3\,, \\ \beta_1 &= \frac{\pi\,R_{\odot}^2}{G\,M_{\odot}\,m\,c}\int_0^{\infty} B_{\odot}(\lambda)\,\{C'_{ext}(\lambda) - C'_{sca}(\lambda)\,g'_1(\lambda)\}\,d\lambda\,, \\ \beta_2 &= \frac{\pi\,R_{\odot}^2}{G\,M_{\odot}\,m\,c}\int_0^{\infty} B_{\odot}(\lambda)\,\{\,-\,C'_{sca}(\lambda)\,g'_2(\lambda)\}\,d\lambda\,, \\ \beta_3 &= \frac{\pi\,R_{\odot}^2}{G\,M_{\odot}\,m\,c}\int_0^{\infty} B_{\odot}(\lambda)\,\{\,-\,C'_{sca}(\lambda)\,g'_3(\lambda)\}\,d\lambda\,, \end{aligned} \tag{41}$$

if we use Sun as a source of gravitation and radiation. $R_{\odot}$ denotes the radius of the Sun and $B_{\odot}(\lambda)$ is the solar radiance at a wavelength of λ; G, $M_{\odot}$, and r are the gravitational constant, the mass of the Sun, and the distance of the particle from the center of the Sun, respectively. The asymmetry parameter vector $\vec{g}\,'$ is defined by $\vec{g}\,' = (1/C'_{sca}) \int \vec{n}\,'(dC'_{sca}/d\Omega')d\Omega'$, where $\vec{n}\,'$ is a unit vector in the direction of scattering; $\vec{g}\,' = g'_1\,\vec{e}\,'_1 + g'_2\,\vec{e}\,'_2 + g'_3\,\vec{e}\,'_3$, $\vec{e}\,'_j = (1 + \vec{v}\cdot\vec{e}_j/c)\,\vec{e}_j - \vec{v}/c$, $\vec{e}\,'_i \cdot \vec{e}\,'_j = \delta_{ij}$. We may mention that $\beta_1 \equiv \beta$, $\beta_2 \equiv \beta\,\bar{Q}'_{pr,2}/\bar{Q}'_{pr,1}$, $\beta_3 \equiv \beta\,\bar{Q}'_{pr,3}/\bar{Q}'_{pr,1}$ correspond to quantities used in Klačka and Kocifaj (2001).

5. Motion and Orbital Elements

Equation of motion is given by Eq. (39), or, by Eq. (40) to the first order in $\vec{v}/c$. We have to take into account that the non-dimensional parameter

$$\begin{aligned} \beta_1 &\equiv \beta = \frac{r^2\,S\,A'}{G\,M_{\odot}\,m\,c}\,\bar{Q}'_{pr,1} \equiv \frac{L_{\odot}\,A'}{4\,\pi\,G\,M_{\odot}\,m\,c}\,\bar{Q}'_{pr,1} = \\ &= \frac{0.02868}{12\,\pi}\,\bar{Q}'_{pr,1}\,\frac{A'\,[m^2]}{m\,[kg]} \end{aligned} \tag{42}$$

may change during the motion. Here $L_{\odot}$ is the rate of energy outflow from the Sun, the solar luminosity. β may change and at each point of the orbit all three parameters β_1, β_2 and β_3

have to be numerically calculated (except for a very special cases, when the β parameters are constant during the motion). Since β does not exhibit large changes during particle motion in the applications discussed below, we will use (osculating) orbital elements for the case when central acceleration is given by $-GM_{\odot}(1-\beta)\vec{e}_1/r^2$ (see Eq. 40). This is a very good approximation to reality. The central acceleration $-GM_{\odot}(1-\beta)\vec{e}_1/r^2$ defines Keplerian orbits within this approximation. As for long-term orbital evolution, we may use mean values of orbital elements – they may be defined as time mean values of orbital elements when true anomaly changes in 2 π radians.

Let us consider that dust particle is ejected from a parent body. Orbital elements of the parent body, at the moment of ejection, are: a_P, e_P, i_P, Ω_P, ω_P and Θ_P, i. e., semi-major axis, eccentricity, inclination, longitude of the ascending node, longitude of pericenter – longitude of perihelion for the case of Solar System – and position angle of the body measured from the ascending node in the direction of the motion of the body. The velocity vector of the parent body, in a given coordinate system with the origin in the dominant central body (Sun in the case of the Solar System) is:

$$\begin{aligned}
\vec{v}_P &= v_{PR}\,\vec{e}_{PR} + v_{PT}\,\vec{e}_{PT}\,,\\
v_{PR} &= \sqrt{\frac{G\,M_{\odot}}{a_P\,(1-e_P^2)}}\;e_P\,\sin f_P\,,\\
v_{PT} &= \sqrt{\frac{G\,M_{\odot}}{a_P\,(1-e_P^2)}}\;(1+e_P\,\cos f_P)\,,\\
\vec{r}_P &= \frac{a_P\,(1-e_P^2)}{1+e_P\,\cos f_P}\,\vec{e}_{PR}
\end{aligned}$$

$$\begin{aligned}
\vec{e}_{PR} &= (\cos\Omega_P\,\cos\Theta_P - \sin\Omega_P\,\sin\Theta_P\,\cos i_P\,,\\
&\quad \sin\Omega_P\,\cos\Theta_P + \cos\Omega_P\,\sin\Theta_P\,\cos i_P\,,\ \sin\Theta_P\,\sin i_P)\,,\\
\vec{e}_{PT} &= (-\cos\Omega_P\,\sin\Theta_P - \sin\Omega_P\,\cos\Theta_P\,\cos i_P\,,\\
&\quad -\sin\Omega_P\,\sin\Theta_P + \cos\Omega_P\,\cos\Theta_P\,\cos i_P\,,\ \cos\Theta_P\,\sin i_P)\,,\\
f_P &= \Theta_P - \omega_P\,. \qquad (43)
\end{aligned}$$

Radial and transversal velocity components are v_{PR} and v_{PT}, unit vector $\vec{e}_{PT}$ is normal to the radial vector and oriented in orientation of motion. True anomaly f_P equals 0 for pericenter/perihelion and π for apocenter/aphelion.

The initial conditions for the particle ejected with velocity $\vec{\Delta}$ from the parent body are:

$$\begin{aligned}
\vec{r}_{in} &= \vec{r}_P \equiv \frac{p_P}{1+e_P\,\cos f_P}\,\vec{e}_{PR}\\
\vec{v}_{in} &= \vec{v}_P + \vec{\Delta}\,, \qquad (44)
\end{aligned}$$

where $\vec{\Delta}$ is velocity with respect to the nucleus of the parent body, and

$$
\begin{aligned}
\vec{r}_{in} &= \frac{p_{in}}{1 + e_{in} \cos f_{in}} \vec{e}_{R\,in} , \\
\vec{v}_{in} &= v_{R\,in} \vec{e}_{R\,in} + v_{T\,in} \vec{e}_{T\,in} , \\
v_{R\,in} &= \sqrt{\frac{G M_{\odot} (1-\beta)}{p_{in}}} e_{in} \sin f_{in} , \\
v_{T\,in} &= \sqrt{\frac{G M_{\odot} (1-\beta)}{p_{in}}} (1 + e_{in} \cos f_{in}) . \\
\vec{e}_{R\,in} &= (\cos \Omega_{in} \cos \Theta_{in} - \sin \Omega_{in} \sin \Theta_{in} \cos i_{in} , \\
& \quad \sin \Omega_{in} \cos \Theta_{in} + \cos \Omega_{in} \sin \Theta_{in} \cos i_{in} , \sin \Theta_{in} \sin i_{in}) , \\
\vec{e}_{T\,in} &= (- \cos \Omega_{in} \sin \Theta_{in} - \sin \Omega_{in} \cos \Theta_{in} \cos i_{in} , \\
& \quad - \sin \Omega_{in} \sin \Theta_{in} + \cos \Omega_{in} \cos \Theta_{in} \cos i_{in} , \cos \Theta_{in} \sin i_{in}) , \\
f_{in} &= \Theta_{in} - \omega_{in} .
\end{aligned} \tag{45}
$$

We write ($\vec{e}_{PN} = \vec{e}_{PR} \times \vec{e}_{PT}$)

$$
\vec{\Delta} = \Delta v_R \, \vec{e}_{PR} + \Delta v_T \, \vec{e}_{PT} + \Delta v_N \, \vec{e}_{PN} . \tag{46}
$$

Using Eqs. (43) – (46), we finally obtain

$$
\begin{aligned}
p_{in} &= \frac{p_P}{1 - \beta} \frac{(v_{PT} + \Delta v_T)^2 + (\Delta v_N)^2}{v_{PT}^2} , \\
e_{in}^2 &= \left(\frac{1 + e_P \cos f_P}{1 - \beta} \frac{v_{TS}^2}{v_{PT}^2} - 1 \right)^2 + \\
& \quad \left(\frac{1 + e_P \cos f_P}{1 - \beta} \frac{v_{TS}}{v_{PT}} \right)^2 \left(\frac{v_{PR} + \Delta v_R}{v_{PT}} \right)^2 , \\
\cos i_{in} &= \frac{v_{PT} + \Delta v_T}{v_{TS}} \cos i_P - \frac{\Delta v_N}{v_{TS}} \cos \Theta_P \sin i_P , \\
\sin i_{in} \cos \Omega_{in} &= \frac{v_{PT} + \Delta v_T}{v_{TS}} \sin i_P \cos \Omega_P + \\
& \quad \frac{\Delta v_N}{v_{TS}} (\cos \Theta_P \cos i_P \cos \Omega_P - \sin \Theta_P \sin \Omega_P) , \\
\sin i_{in} \sin \Omega_{in} &= \frac{v_{PT} + \Delta v_T}{v_{TS}} \sin i_P \sin \Omega_P + \\
& \quad \frac{\Delta v_N}{v_{TS}} (\cos \Theta_P \cos i_P \sin \Omega_P + \sin \Theta_P \cos \Omega_P) ,
\end{aligned}
$$

$$
\begin{aligned}
e_{in} \cos f_{in} &= \frac{p_{in}}{p_P} \left(1 + e_P \cos f_P\right) - 1 , \\
e_{in} \sin f_{in} &= \frac{1 + e_P \cos f_P}{1 - \beta} \frac{v_{TS} \left(v_{PR} + \Delta v_R\right)}{v_{PT}^2} , \\
\sin i_{in} \cos \Theta_{in} &= \frac{v_{PT} + \Delta v_T}{v_{TS}} \sin i_P \cos \Theta_P + \\
& \frac{\Delta v_N}{v_{TS}} \cos i_P , \\
\sin i_{in} \sin \Theta_{in} &= \sin i_P \sin \Theta_P , \\
v_{TS}^2 &\equiv \left(v_{PT} + \Delta v_T\right)^2 + \left(\Delta v_N\right)^2 ,
\end{aligned}
\qquad (47)
$$

where v_{PR} and v_{PT} are given by expressions presented in Eq. (43) – $p_P = a_P \left(1 - e_P^2\right)$.

For the special case $\vec{\Delta} = 0$ Eq. (47) reduces to

$$
a_{in} = a_P \left(1 - \beta\right) \left(1 - 2\beta \frac{1 + e_P \cos f_P}{1 - e_P^2}\right)^{-1} , \qquad (48)
$$

$$
e_{in} = \sqrt{1 - \frac{1 - e_P^2 - 2\beta \left(1 + e_P \cos f_P\right)}{\left(1 - \beta\right)^2}} , \qquad (49)
$$

where $f_P \equiv \Theta_P - \omega_P$, $\omega_{\beta\ in}$ has to be obtained from

$$
\begin{aligned}
e_{in} \cos \left(\Theta_P - \omega_{in}\right) &= \frac{\beta + e_P \cos f_P}{1 - \beta} , \\
e_{in} \sin \left(\Theta_P - \omega_{in}\right) &= \frac{e_P \sin f_P}{1 - \beta} ,
\end{aligned}
\qquad (50)
$$

$$
\Omega_{in} = \Omega_P , \quad i_{in} = i_P , \quad \Theta_{in} = \Theta_P . \qquad (51)
$$

Physics of Eqs. (48) – (49) is following: meteoroid escapes from the Solar System when the orbital energy becomes positive and this can happen when the energy due to the radial component of the radiation force is included, without it being necessary for this force to exceed the gravitational attraction (Harwit 1963).

Complete equation of motion for dust particle orbiting the central body of mass $M_{\odot}$ under electromagnetic radiation of the central body and gravity interaction is given by Eq. (40). We can find osculating orbital elements for dust particle according to the following equations:

$$
\begin{aligned}
E &= \frac{1}{2}\,\vec{v}^2 - \frac{G\,M_\odot\,(1-\beta)}{r}\,,\\
\vec{H} &= \vec{r}\times\,\vec{v}\,,\\
p &= \frac{|\vec{H}|^2}{G\,M_\odot\,(1-\beta)}\,,\\
e &= 1\,+\,\frac{2\,p\,E}{G\,M_\odot\,(1-\beta)}\,,\\
a &= \frac{p}{1\,-\,e^2}\,,\;\; q = a\,\left(1\,-\,e\right)\,,\\
i &= \arccos(H_z/|\vec{H}|)\,,\\
\sin\Omega\,\sin i &= H_x/|\vec{H}|\,,\quad -\;\cos\Omega\,\sin i = H_y/|\vec{H}|\,,\\
\vec{e}_R &= \vec{r}/|\vec{r}| = (x,y,z)/r\,,\\
\vec{e}_N &= \vec{H}/|\vec{H}|\,,\\
\vec{e}_T &= \vec{e}_N\times\vec{e}_R\,,\\
\sin\Theta\,\sin i &= z/r\,,\\
\cos\Theta\,\sin i &= (\vec{e}_T)_z \equiv (y\,H_x\,-\,x\,H_y)/\left(r|\vec{H}|\right)\,,\\
\sin\left(\Theta\,-\,\omega\right) &= \vec{v}\cdot\vec{e}_R/\left(e\sqrt{G\,M_\odot\,(1-\beta)\,/p}\right)\,,\\
\cos\left(\Theta\,-\,\omega\right) &= \left\{\vec{v}\cdot\vec{e}_T/\left(\sqrt{G\,M_\odot\,(1-\beta)\,/p}\right)\,-\,1\right\}\,\frac{1}{e}\,,
\end{aligned}
\tag{52}
$$

where q is perihelion distance. If $i = 0$, then Θ has to be obtained from $\cos(\Omega+\Theta) = x/r$, $\sin(\Omega+\Theta) = y/r$, and, we may take Ω in an arbitrary way (e. g., the requirement that Ω is a continuous function of time may define Ω for $i = 0$). As for the value of inclination, we repeat the well-known fact: prograde orbits exist for $i \in \langle 0, \pi/2)$, retrograde orbits exist for $i \in (\pi/2, \pi\rangle$. The case $e > 1$ describes the case when the particle is ejected from the Solar System.

6. Poynting-Robertson Effect

Putting $\bar{Q}'_{pr,2} = \bar{Q}'_{pr,3} = F'_{e1} = F'_{e2} = F'_{e3} \equiv 0$ in Eq. (39), we obtain Poynting-Robertson effect (Robertson 1937, Klačka 1992):

$$\frac{D\,p^\mu}{d\,\tau} = \frac{w_1^2\,S\,A'}{c}\,\bar{Q}'_{pr,1}\left(b_j^\mu\,-\,u^\mu\,/\,c\right)\,. \tag{53}$$

Eq. (53) enables us some analytical calculations for changes of osculating orbital elements: it is supposed that $\bar{Q}'_1$ is a constant. We will consider the first order in $\vec{v}/c$ of Eq. (53).

We can write, on the basis of Eqs. (53), (40) – (42)

$$\frac{d\,\vec{v}}{d\,t} = -\,\frac{\mu}{r^2}\,\vec{e}_R\,+\,\frac{\mu}{r^2}\,\beta\,\left\{\left(1\,-\,\frac{\vec{v}\cdot\vec{e}_R}{c}\right)\,\vec{e}_R\,-\,\frac{\vec{v}}{c}\right\}\,, \tag{54}$$

where $\mu \equiv G\, M_{\odot}$, $\vec{e}_R \equiv \vec{e}_1$ and $\beta \equiv \beta_1$is a non-dimensional parameter ("the ratio of radiation pressure force to the gravitational force"; see also Eq. 42). Eq. (42) reduces to $\beta = 5.7 \times 10^{-5} \bar{Q}'_{pr,1} / (\varrho [g/cm^3]\ s[cm])$, for homogeneous spherical particle: ϱ is mass density and s is radius of the sphere.

6.1. Secular Changes of Orbital Elements

Using $-\ \mu\,(1-\beta)\,\vec{e}_R\ /\ r^2$ as a central acceleration determining osculating orbital elements, we can make a time averaging (T is time interval between passages through two following pericenters) in an analytical way

$$\begin{aligned}
\langle g \rangle &\equiv \frac{1}{T}\int_0^T g(t)dt = \frac{\sqrt{\mu\,(1-\beta)}}{a^{3/2}}\,\frac{1}{2\pi}\int_0^{2\pi} g(f)\left(\frac{df}{dt}\right)^{-1} df \\
&= \frac{\sqrt{\mu\,(1-\beta)}}{a^{3/2}}\,\frac{1}{2\pi}\int_0^{2\pi} g(f)\,\frac{r^2}{\sqrt{\mu\,(1-\beta)\,p}}\,df \\
&= \frac{1}{a^2\,\sqrt{1-e^2}}\,\frac{1}{2\pi}\int_0^{2\pi} g(f)\,r^2\,df\ , \qquad (55)
\end{aligned}$$

assuming non-pseudo-circular orbits and the fact that orbital elements exhibit only small changes during the time interval T; a is semi-major axis, e is eccentricity, f is true anomaly, $p = a(1-e^2)$; the second and the third Kepler's laws were used: $r^2\ df/dt = \sqrt{\mu(1-\beta)p}$ – conservation of angular momentum, $a^3/T^2 = \mu(1-\beta)/(4\pi^2)$.

Rewriting Eq. (54) into the form

$$\frac{d\vec{v}}{dt} = -\frac{\mu\ (1-\beta)}{r^2}\vec{e}_R - \beta\frac{\mu}{r^2}\left\{\left(\frac{\vec{v}\cdot\vec{e}_R}{c}\right)\vec{e}_R + \frac{\vec{v}}{c}\right\}\ , \qquad (56)$$

we can immediately write for components of perturbation acceleration to Keplerian motion:

$$F_R = -\ 2\,\beta\,\frac{\mu}{r^2}\,\frac{v_R}{c}\ ,\quad F_T = -\ \beta\,\frac{\mu}{r^2}\,\frac{v_T}{c}\ ,\quad F_N = 0\ , \qquad (57)$$

where F_R, F_T and F_N are radial, transversal and normal components of perturbation acceleration, and two-body problem yields

$$\begin{aligned}
v_R &= \sqrt{\frac{\mu\,(1-\beta)}{p}}\ e\sin f\ , \\
v_T &= \sqrt{\frac{\mu\,(1-\beta)}{p}}\ (1+e\cos f)\ . \qquad (58)
\end{aligned}$$

The important fact that perturbation acceleration is proportional to v/c ($\ll 1$) ensures the above metioned small changes of orbital elements during the time interval T.

Perturbation equations of celestial mechanics yield for osculating orbital elements (a – semi-major axis; e – eccentricity; i – inclination (of the orbital plane to the reference frame); Ω – longitude of the ascending node; ω – longitude of pericenter; Θ is the position angle of the particle on the orbit, when measured from the ascending node in the direction of the particle motion, $\Theta = \omega + f$):

$$
\begin{aligned}
\frac{da}{dt} &= \frac{2\,a}{1 - e^2} \sqrt{\frac{p}{\mu\,(1 - \beta)}} \left\{ F_R\, e\, \sin f + F_T\,(1 + e\, \cos f) \right\} , \\
\frac{de}{dt} &= \sqrt{\frac{p}{\mu\,(1 - \beta)}} \left\{ F_R\, \sin f + F_T \left[\cos f + \frac{e + \cos f}{1 + e \cos f} \right] \right\} , \\
\frac{di}{dt} &= \frac{r}{\sqrt{\mu\,(1 - \beta)\, p}}\, F_N\, \cos \Theta , \\
\frac{d\Omega}{dt} &= \frac{r}{\sqrt{\mu\,(1 - \beta)\, p}}\, F_N\, \frac{\sin \Theta}{\sin i} , \\
\frac{d\omega}{dt} &= -\frac{1}{e} \sqrt{\frac{p}{\mu\,(1 - \beta)}} \left\{ F_R \cos f - F_T \frac{2 + e \cos f}{1 + e \cos f} \sin f \right\} - \\
& \quad \frac{r}{\sqrt{\mu\,(1 - \beta)\, p}}\, F_N\, \frac{\sin \Theta}{\sin i}\, \cos i , \\
\frac{d\Theta}{dt} &= \frac{\sqrt{\mu\,(1 - \beta)\, p}}{r^2} - \frac{r}{\sqrt{\mu\,(1 - \beta)\, p}}\, F_N\, \frac{\sin \Theta}{\sin i}\, \cos i , \qquad (59)
\end{aligned}
$$

where $r = p/(1 + e \cos f)$.

Inserting Eqs. (57) – (58) into Eq. (59), one easily obtains

$$
\begin{aligned}
\frac{da}{dt} &= -\beta \frac{\mu}{r^2} \frac{2a}{c} \frac{1 + e^2 + 2e \cos f + e^2 \sin^2 f}{1 - e^2} , \\
\frac{de}{dt} &= -\beta \frac{\mu}{r^2} \frac{1}{c} \left(2e + e \sin^2 f + 2 \cos f \right) , \\
\frac{di}{dt} &= 0 , \\
\frac{d\Omega}{dt} &= 0 , \\
\frac{d\omega}{dt} &= -\beta \frac{\mu}{r^2} \frac{1}{c} \frac{1}{e} \, (2 - e \cos f) \sin f , \\
\frac{d\Theta}{dt} &= \frac{\sqrt{\mu\,(1 - \beta)\, p}}{r^2} . \qquad (60)
\end{aligned}
$$

It is worth mentioning that $da/dt < 0$ for any time t.

The set of differential equations Eqs. (60) has to be complemented with initial conditions. They are given by Eq. (47), or Eqs. (48) – (51).

By inserting Eqs. (60) into Eq. (55), taking into account that $e_{in} < 1$ (see Eq. 55), one can easily obtain for the secular changes of orbital elements:

$$
\langle \frac{da}{dt} \rangle = -\,\beta\, \frac{\mu}{c}\, \frac{2 + 3e^2}{a\,(1 - e^2)^{3/2}} , \qquad (61)
$$

$$
\langle \frac{de}{dt} \rangle = -\,\frac{5}{2}\, \beta\, \frac{\mu}{c}\, \frac{e_\beta}{a^2\,(1 - e^2)^{1/2}} , \qquad (62)
$$

$$\langle \frac{d\Theta}{dt} \rangle = \frac{\sqrt{\mu \ (1-\beta)}}{a^{3/2}} \ , \tag{63}$$

$$\langle \frac{di}{dt} \rangle = \langle \frac{d\Omega}{dt} \rangle = \langle \frac{d\omega}{dt} \rangle = 0 \ . \tag{64}$$

As an remark we may mention that a little more simple set of differential equations than that represented by Eqs. (61) – (62) is obtained when semi-major axis a is replaced by the quantity $p = a \ (\ 1 - e^2 \ )$. It can be easily verified that a new set of differential equations for secular evolution is given by the following equations:

i) $dp/dt = -2\beta(\mu/c)(1-e^2)^{3/2}/p$,

ii) $de/dt = -(5/2)\beta(\mu/c)[(1-e^2)^{3/2}/p]e/p$.

These two equations immediately yield $p = p_{in} \ (e/e_{in})^{4/5}$.
Equations i) and ii) enable to write:

i) $dp/dt = -2\beta(\mu/c) \ [1 - e_{in}^2(p/p_{in})^{5/2}]^{3/2} \ / \ p$,

ii) $de/dt = -(5/2)\beta(\mu/c) \ (e_{in}^{8/5}/p_{in}^2) \ (1-e^2)^{3/2} \ / \ e^{3/5}$.

These equations point out that near-circular / pseudo-circular orbits ($e \approx 0$) and orbits with small values of p are not described by the discussed secular changes of orbital elements and by the considered v/c approximation for the P-R effect.

7. Solar Wind Effect

We have calculated secular changes of orbital elements for the P-R effect up to the second order in v/c. The effects of the second order in v/c seem to be small to play an important role in Solar System studies. In reality, similar effect coming from the Sun exists and this effect may play more important role. Although this effect is not connected with electromagnetic radiation, its secular changes of orbital elements to the first order in v/c (more correctly, v/u, where u is the speed of solar wind particles) correspond to the secular changes for P-R effect. This effect is caused by solar wind particles hitting an interplanetary dust particle. The aim of this section is to obtain secular changes of orbital elements of the interplanetary dust particle under the action of the solar wind, to the first order in v/u.

Let us consider equation of motion in the form (we neglect decrease of particle mass)

$$\frac{d\,\vec{v}_{sw}}{d\,t} = \eta \ \frac{\beta}{\bar{Q}'_{pr,1}} \ \frac{\mu}{r^2} \ \frac{u}{c} \ \left\{ \left(1 \ - \ \frac{\vec{v} \cdot \vec{e}_R}{u} \right) \ \vec{e}_R \ - \ \frac{\vec{v}}{u} \right\} \tag{65}$$

where, as standardly in this paper, $\mu \equiv G \ M_{\odot}$ and non-dimensional parameter β is defined in Eq. (42) ("the ratio of radiation pressure force to the gravitational force"); $\eta \approx 1/3$. Adding the right-hand side of Eq. (54) to the right-hand side of Eq. (65), one obtains

$$\begin{aligned} \frac{d\,\vec{v}}{d\,t} \ &= \ - \ \frac{\mu}{r^2} \ \vec{e}_R \ + \ \beta \ \frac{\mu}{r^2} \ \left\{ \left(1 \ - \ \frac{\vec{v} \cdot \vec{e}_R}{c} \right) \ \vec{e}_R \ - \ \frac{\vec{v}}{c} \right\} \ + \\ & \eta \ \frac{\beta}{\bar{Q}'_{pr,1}} \ \frac{\mu}{r^2} \ \frac{u}{c} \ \left\{ \left(1 \ - \ \frac{\vec{v} \cdot \vec{e}_R}{u} \right) \ \vec{e}_R \ - \ \frac{\vec{v}}{u} \right\} \ . \end{aligned} \tag{66}$$

The first term in Eq. (66) is Newtonian gravity for two-body problem, the second term is the Poynting-Robertson effect in flat spacetime, the third term corresponds to the action of the solar wind up to the second order in v/u.

Eq. (66) may be written in the following form:

$$\frac{d\vec{v}}{dt} = -\frac{\mu}{r^2}\vec{e}_R + \beta\left(1+\frac{\eta}{\bar{Q}'_{pr,1}}\frac{u}{c}\right)\frac{\mu}{r^2}\vec{e}_R - -\beta\left(1+\frac{\eta}{\bar{Q}'_{pr,1}}\right)\frac{\mu}{r^2}\left(\frac{\vec{v}\cdot\vec{e}_R}{c}\vec{e}_R+\frac{\vec{v}}{c}\right). \qquad (67)$$

7.1. Secular Changes of Orbital Elements

Neglecting solar wind pressure term $\eta(\beta/\bar{Q}'_{pr,1})(\mu/r^2)(u/c)$, we can rewrite Eq. (67) to the following form:

$$\frac{d\vec{v}}{dt} = -\frac{\mu(1-\beta)}{r^2}\vec{e}_R - \beta\left(1+\frac{\eta}{\bar{Q}'_{pr,1}}\right)\frac{\mu}{r^2}\left(\frac{\vec{v}\cdot\vec{e}_R}{c}\vec{e}_R+\frac{\vec{v}}{c}\right), \qquad (68)$$

where the second term causes deceleration of the particle motion. We use $-\mu(1-\beta)\vec{e}_R/r^2$ as a central acceleration determining osculating orbital elements; β is considered to be a constant during the motion. Consideration analogous to those presented in Sec. 6.1., see Eqs. (55) – (60), lead to results analogous to Eqs. (61) – (64):

$$\begin{aligned}\langle\frac{da}{dt}\rangle &= -\beta\frac{\mu}{c}\left(1+\frac{\eta}{\bar{Q}'_{pr,1}}\right)\frac{2+3e^2}{a(1-e^2)^{3/2}}, \\ \langle\frac{de}{dt}\rangle &= -\beta\frac{\mu}{c}\left(1+\frac{\eta}{\bar{Q}'_{pr,1}}\right)\frac{5e/2}{a^2\sqrt{1-e^2}}, \\ \langle\frac{d\omega}{dt}\rangle &= 0. \end{aligned} \qquad (69)$$

Initial conditions are given by Eq. (47), or, Eqs. (48) – (51).

The results presented in this section hold for the most simple approximation of the solar wind action – only radial component of the solar wind particles is considered. Moreover, solar wind causes decrease of particle mass (see e. g., Leinert and Grűn 1990) and the secular change of particle mass m (present in β, see Eq. 42) is given as $dm/dt = -KA'_{eff} / (a^2\sqrt{1-e^2})$, where K is a constant depending on the material properties of the particle and A'_{eff} is the proper effective cross sectional area of the particle. If A'_{eff} (area toward the Sun) is changing during the particle motion, one has to use $dm/dt = -KA'_{eff}(1 - v_R/u)/r^2$ and no averaging is possible – considerations made in section 7. hold for the case $A'_{eff} \equiv A'$, where A' was defined above Eq. (16), and, moreover, A'_{eff} does not change during the particle motion (e. g., spherical particle).

Finally, solar wind causes also charging of meteoroids and Lorentz force has to be taken into account in the case of submicron grains.

8. Electromagnetic Radiation and Capture of Intestellar Dust Particles

The last two decades have shown that interstellar dust grains enter the Solar System (Baguhl *et al.* 1995, Grűn *et al.* 1994, Landgraf *et al.* 1999, Landgraf *et al.* 2000). Interstellar dust particles approach the Sun from the interstellar space with velocity $v_{\infty} = 26\ km/s$, approximately. This value is known on the basis of measurements on boards of interplanetary probes. The probes enable also determination of direction from which the flux of interstellar dust enter the Solar System. Sizes of particles correspond mainly to several tenths of micrometers, or to values about one micrometer.

An interesting question is orbital evolution of the interstellar dust particles entering the Solar System. Dealing with this problem, one has to take into account various forces which can influence motion of the particles. Jackson (2001) has considered gravitational force of the Sun and the Poynting-Robertson effect. It is not problem to take into account solar wind, as it is done in Eq. (68). Moreover, other nongravitational forces may play an important role. One possibility is considering a simple fact that real particles are nonspherical and general electromagnetic radiation force may be significant (Kocifaj and Klačka 2003). Moroever, since submicron dust particles have to be considered, the effect of interplanetary magnetic field cannot be neglected. Physically relevant equation of motion is (Kocifaj and Klačka 2004a):

$$\begin{aligned}
\frac{d\vec{v}}{dt} &= -\frac{G\,M_{\odot}}{r^2}\,\vec{e}_1 + \\
&+ \frac{G\,M_{\odot}}{r^2}\sum_{j=1}^{3}\beta_j\left[\left(1 - 2\,\frac{\vec{v}\cdot\vec{e}_1}{c} + \frac{\vec{v}\cdot\vec{e}_j}{c}\right)\vec{e}_j - \frac{\vec{v}}{c}\right] + \\
&+ \frac{q}{m}\,(\vec{v} - \vec{v}_{sw}) \times \vec{B}\,, \\
\vec{B} &= B_R\vec{e}_R + B_T\vec{e}_T\,, \\
B_R &= B_{R0}\,(r/r_0)^2\ \cos(\pi\ t[years]/11 + \varphi_0)\,, \\
B_T &= B_{T0}\,(r/r_0)\ \cos\vartheta\ \cos(\pi\ t[years]/11 + \varphi_0)\,, \\
B_{R0} &= B_{T0} = 3nT\,,\ r_0 = 1AU\,,
\end{aligned} \tag{70}$$

where $\vec{v}_{sw} = v_{sw}\vec{e}_R$ is the solar wind speed, $v_{sw} = 400\ km/s$ (and also $800\ km/s$) is used, $\vec{e}_T = \vec{e}_A \times \vec{e}_R$, $\vec{e}_A$ defines magnetic axis of the Sun, ϑ is the heliographic latitude, and different polarity of the magnetic field for northern and southern hemispheres is also considered. As for the magnetic field, presented in Eq. (70), the idea of Parker's spiral (Parker 1958) is used. This is standard approach of modeling of the effect of Lorentz force on motion of dust grains in the Solar System (Grűn *et al.* 1994). The process of charging of interplanetary dust grains due to the effect of solar wind is taken from Kimura and Mann (1998); the charge of the grains is $q = 4\,\pi\,\varepsilon_0\,U\,a_{eff}$, where ε_0 is the permittivity of vacuum, U is the surface potential given by Kimura and Mann (1998) and a_{eff} effective radius of the particle. U is a function of distance from the Sun and its value is + 3 V in the central part of the Solar System.

Eq. (70) was numerically solved under assumption that particle rapidly rotates around an axis of rotation which is fixed in space. The averaged values of dimensionless efficiency

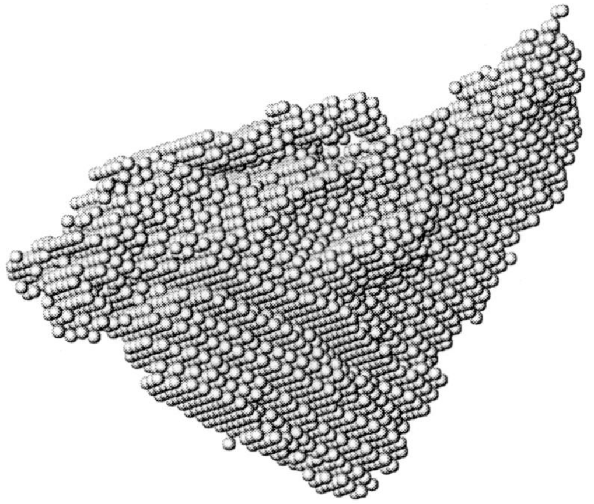

Figure 1. The digitized model of cosmic dust particle U2015 B10.

factors for radiation pressure $Q'_{pr,j}$, $j = 1, 2, 3$, are used in the β_j−quantities. Calculations were done for several sizes of particles and various orientations of rotational axes. Morphology of the considered particles was identical to grain U2015 B10 (archived in NASA collection: Clanton *et al.* 1984, Kocifaj *et al.* 1998), which appears to be representative for cosmic dust, see Fig. 1. Its aspect ratio (the ratio between the largest and smallest characteristic length of the particle) is equal to 1.4 so this value coincides well with the results of mid-infrared spectropolarimetry (Hildebrand and Dragovan 1995). These results indicate that aspect ratios of cosmic dust grains lie between 4/3 and 3/2. Magnesium-rich silicate particles were considered. Initial conditions were defined also by the value $v_\infty = 26\ km/s$ and various values of impact parameter, which is defined as the smallest distance between the particle and the Sun if no forces existed. An orientation of the particle rotational axis relatively to the incident radiation was recalculated in each new orbital point, and the process of particle rotation was taken into account. It means, the radiation scattering diagram was computed for each individual particle orientation (i. e. during rotation in fixed position, as well as in each new point of particle trajectory). The averaged components of the vector of mean propagation of scattered radiation for all directions, radial and both transversal, were calculated according to Draine and Flatau (2004). Numerical solution of the interaction of radiation with the particle was based on so-called Discrete Dipole Approximation (DDA, Purcell and Pennypacker 1973, Draine 1988, Draine and Flatau 1994).

What is the orbital evolution of interstellar dust particles when they approach to the Sun? Although small spherical (homogeneous spheres or spheres with spherically symmetric distributions of mass) submicron particles can reach the regions close to the Sun and may be captured in the Solar System (revolution of particles around the Sun), only small amount of them can survive. The majority of such particles hits the Sun (Fig. 2), the others are sublimated in the circumsolar zone $r < 3\ R_{Sun}$, where r is a distance from a center of the Sun and R_{Sun} is the solar radius.

The captured non-charged spherical dust grains would revolve the Sun (an example is shown in Fig. 3) in a zone where particles distances from the solar surface would be

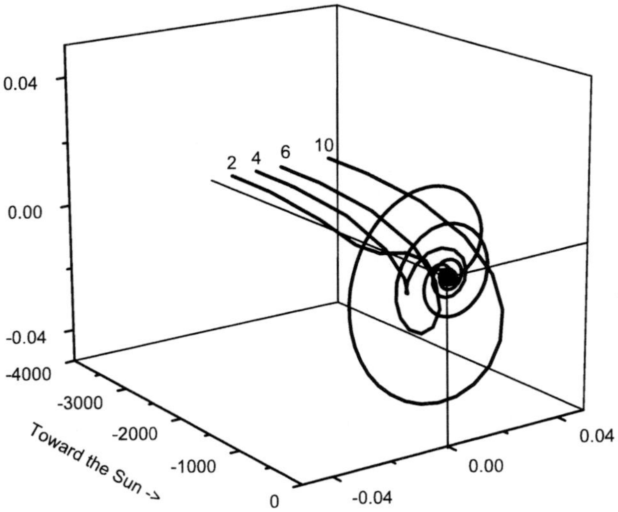

Figure 2. Trajectories of nonspherical interstellar dust particles in the Solar System (aspect ratio is close to 1.4). Position of the Sun is [0,0,0] (any position [x,y,z] in this cartesian coordinate system is measured in astronomical units). Particles move from infinity with impact parameters b = 2, 4, 6, 10 [solar radii].

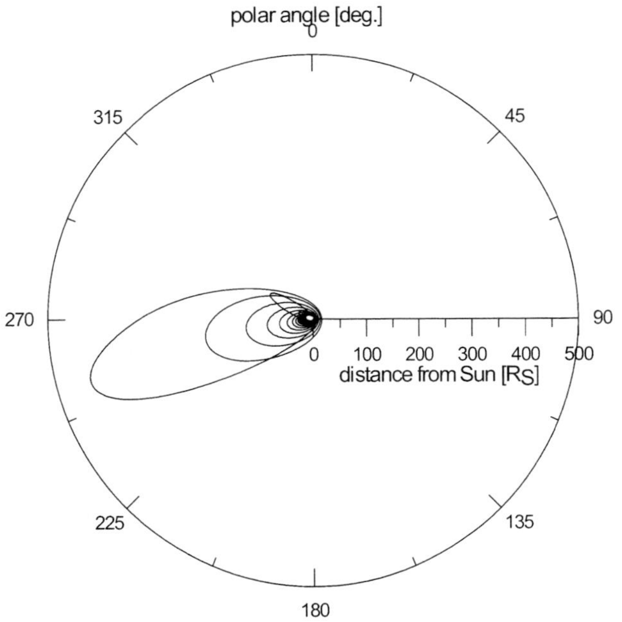

Figure 3. Trajectory evolution of non-spherical interstellar dust particle captured in the Solar System (projection to the ecliptic plane). The particle radius is $a_{eff} = 0.4\ \mu m$ and initial impact factor $b = 480\ R_{Sun}$.

less than 20% of the solar radius. If we consider non-charged nonspherical particles (or spherical, with structural/chemical inhomogeneity), the effect of non-radial components of radiation pressure may be strong and orbits of the particles may significantly differ from the orbits of spherical grains. The fundamental result is that nonspherical charged particles may be captured, and they can even survive in the Solar System.

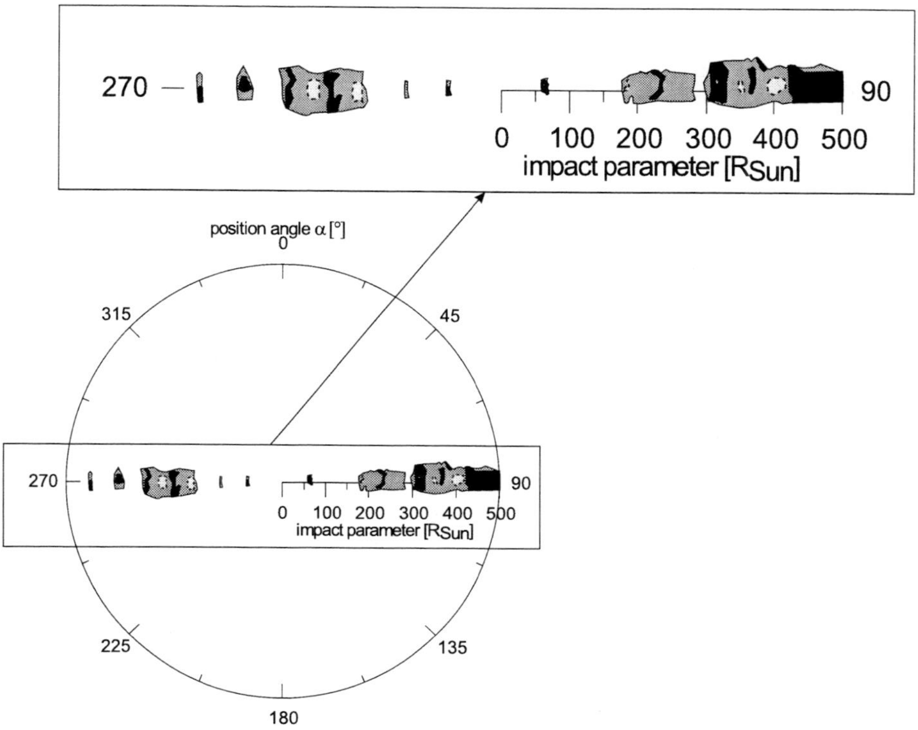

Figure 4. Capture area D for magnesium-rich silicate dust (radius $a_{eff} = 0.4\ \mu m$). The region of D for spherical dust is slightly shaded and bounded by dashed curve. The region of D for non-spherical dust corresponds to dark shaded planes bounded by solid curves. The zones, in which the particles will survive (i.e. $r_{min} > 3\ R_{Sun}$) are depicted by black color. The solar wind speed is assumed to be 400 $km.s^{-1}$. Here the angle α characterizes the position of the particle within dust stream in the plane perpendicular to direction of motion, which is parallel to flowline between particle and the Sun. The cases $\alpha = 90°$ or $270°$ hold for particles moving in the vicinity of ecliptic plane.

Charged nonspherical interstellar grains can be captured only in a very narrow belt near ecliptic plane and surviving charged grains (grains that are not sublimated, i. e. their perihelion distances from the Sun are large enough) exist mainly for impact parameters b $>$ 150 solar radii (Fig. 4). When assuming a solar wind speed of 400 km/s, then 20% of the captured particles of the effective radius 0.4 μm will survive. The corresponding percentage for particles of the radius 0.6 μm is 7%. As for the fast solar wind (wind speed of 800 km/s), the resulting values are quite small: 5% for particles of the effective radius 0.4 μm, 0.3% for effective radius 0.6 μm and 0.1% for 0.8 μm.

An interesting results was obtained for a quantity analogous to the third Kepler's law. In the third Kepler's law we have $a^3/T^2 = G\ M_{\odot}/(4\pi^2)$, where a is semi-major axis and T is period of revolution around the Sun, if the mass of the object is of negligible mass in

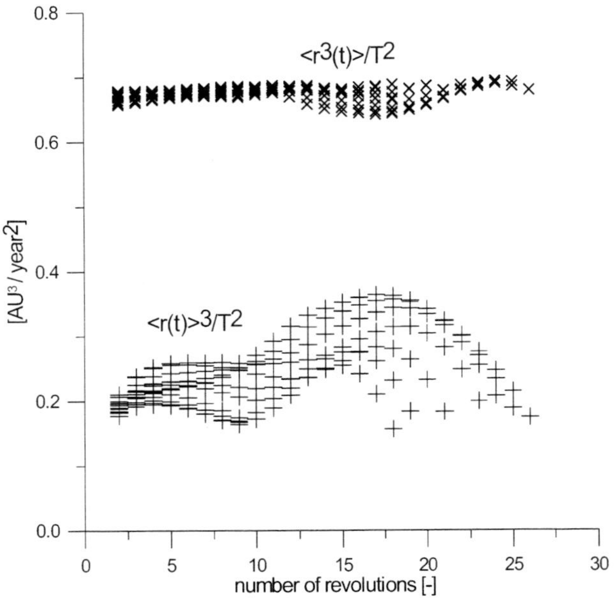

Figure 5. A dependency of physical quantities $\langle r^3(t)\rangle/T^2$ and $\langle r(t)\rangle^3/T^2$ on number of revolutions of dust grains around the Sun. The distribution contains averaged data over each revolution for particles moving in region $b \in (300, 500)$ solar radii. In this region the most important captures of submicron-sized particles were identified. The particles were initially moving in the vicinity of ecliptic plane.

comparison with the mass of the Sun $M_\odot$. Thus, fundamental result of the third Kepler's law is that the quantity a^3/T^2 is a constant for all objects moving in various orbits. If we consider charged interstellar dust particles captured in the Solar System, then similar quantity $\langle r(t)^3\rangle/T^2$ behaves almost as a constant, although each individual particle exhibit various distances from the Sun in the process of orbiting the Sun. The quantity then the quantity $\langle r(t)^3\rangle$ expresses time averaging (over time interval of length T) of the immediate distance from the Sun. The used interval T is defined as a time interval between two equal positions of the particle in a fixed plane with respect to the Sun. The situation is shown in Fig. 5, where also $\langle r(t)\rangle^3/T^2$ is presented (Kocifaj and Klačka 2004b).

As for small submicron particles, the Lorentz force is generally considered to be dominant. However, as for the surviving captured interstellar grains, the solar radiation force is the most significant factor: non-radial components of the radiation pressure are the most relevant components of the considered forces. Lorentz force makes the total mass of the surviving charged particles about two orderes of magnitude larger than the mass of the surviving non-charged particles.

9. Motion of Dust near Resonances with Planets

Commensurability resonances are well-known since the time of Kirkwood, who found gaps in distribution of asteroidal semi-major axes (histogram: number of known asteroids versus their semi-major axes) in the asteroid belt in 1857. A body is in resonance with a planet when the ratio of their mean motions (mean motion $n = 2\pi/T$, where T is orbital period) is the ratio of two small integers. Physics of these resonances – mean-motion resonances – with Solar System planets is successfully studied mainly during the last thirty years (Wisdom 1980). The orbital evolution of dust grains near resonances with planets is intensively studied since the end of 80-ties of the 20-th centurys. Besides gravitational forces of the Sun and planet(s), also the effect of solar electromagnetic radiation in the form of the Poynting-Robertson effect is usually taken into account, including effect of solar wind: e. g., Jackson and Zook (1989), Šidlichovský and Nesvorný (1994), Beaugé and Ferraz-Mello (1994), Marzari and Vanzani (1994), Liou and Zook (1995), Liou *et al.* (1995), Liou and Zook (1997). Qualitatively new results are obtained when we admit that real particles are nonspherical and instead of the Poynting-Robertson effect (based on the assumption that material within dust grains is distributed in a spherically symmetric way) we use Eqs. (40) – (41) in modelling mean-motion resonances. We can refer to Klačka *et al.* (2005a, 2005b).

According to the third Kepler's law we have

$$\begin{aligned} a^3\, n^2 &= G\, M_{\odot}\, (1\, -\, \beta)\ , \\ a_P^3\, n_P^2 &= G\, (M_{\odot}\, +\, m_P)\ , \end{aligned} \tag{71}$$

where a, a_P are semi-major axes of the particle characterized by parameter β (see Eq. 42) and a planet with mass m_P, n and n_P are mean motions characterizing revolutions around the Sun of the mass $M_{\odot}$. The first part of Eq. (71) uses the fact that central Keplerian acceleration is given by the sum of solar gravitational acceleration and radial component of radiation pressure acceleration, as it was discussed at the beginning of Sec. 5.. Eq. (71) yields

$$a = (1\, -\, \beta)^{1/3}\, \left(\frac{n_P}{n} \right)^{2/3}\, \left(1\, +\, \frac{m_P}{M_{\odot}} \right)^{-1/3}\, a_P\ . \tag{72}$$

If the dust particle is in a resonance with a planet, we can define the q–th order exterior resonance by the relation $n_P/n = (p+q)/p$, where p and q are integer numbers. Similarily, the q–th order interior resonance is defined by the relation $n_P/n = p/(p+q)$, where p and q are integer numbers. In terms of orbital periods: $T/T_P = (p+q)/p$ for exterior , $T/T_P = p/(p+q)$ for interior resonances. On the basis of these definitions and Eq. (72), we can immediately write

$$a = (1\, -\, \beta)^{1/3}\, \left(\frac{p+q}{p} \right)^{2/3}\, \left(1\, +\, \frac{m_P}{M_{\odot}} \right)^{-1/3}\, a_P\ , \tag{73}$$

for the semimajor-axis of the dust particle in the q–th order exterior resonance with the planet of mass m_P. Similar relation can be obtained for the interior resonance.

Section 9.1. will deal with mean-motion resonances for spherical particles and in Sec. 9.2. we will present some important results for nonspherical dust grains.

9.1. Mean-Motion Resonances and Spherical Dust Particles

Let us consider that spherical dust grain is under action of gravitational forces generated by the Sun and a planet moving around the Sun. Moreover, the grain is evolving also under action of solar electromagnetic radiation and solar wind, in reality. Equation of motion of the particle is

$$\frac{d\vec{v}}{dt} = -\frac{\mu(1-\beta)}{r^2}\vec{e}_R - \beta\left(1+\frac{\eta}{\bar{Q}'_{pr,1}}\right)\frac{\mu}{r^2}\left(\frac{\vec{v}\cdot\vec{e}_R}{c}\vec{e}_R+\frac{\vec{v}}{c}\right) - - \mu\frac{m_P}{M_\odot}\left\{\frac{\vec{r}-\vec{r}_P}{|\vec{r}-\vec{r}_P|^3}+\frac{\vec{r}_P}{|\vec{r}_P|^3}\right\}, \tag{74}$$

where $\vec{r}$ and $\vec{r}_P$ are position vectors of the particle and the planet with respect to the Sun, and, as standardly, $\mu = GM_\odot$ and G is the gravitational constant. Eq. (68) was used and the term generated by gravitational force of the planet was added.

If the particle is ejected from a parent body (comet, asteroid, planet, satellite of a planet), then its secular evolution is given by Eq. (69). The important property is that secular evolution of semi-major axis of the particle is a decreasing function of time. In this way, the particle approaches to the Sun and sometimes it may reach a suitable conditions for capture into a mean-motion resonance. Being in the resonance, its secular evolution is characterized by a constant value of semi-major axis defined by Eq. (72) or Eq. (73) (or, by analogous equation for the interior resonance). Particle can be captured for thousands or even billions years. It depends on the size of the particle (we remind that $\beta \propto size^{-1}$), mass of the planet, type of the resonance (the smaller values of p and q, the longer capture times), and, also on the initial orbital conditions at the instant of the capture. Finally, when the capture is ended, Eq. (69) correctly desribes behaviour of the particle: secular evolution of semi-major axis is, again, a decreasing function of time.

We have already mentioned that secular evolution of semi-major axis is characterized by its constant value when the particle is in a resonance with a planet. What can be said about secular evolution of the eccentricity of the particle in such a situation? In order to show what really happens, we will suppose that the planet is moving in a circular orbit around the Sun. In this case, we have a special gravitational problem of three bodies together with a presence of small nongravitational forces. The gravitational problem is called *the circular restricted problem of three bodies* in celestial mechanics. At the end of the 19-th century French astronomer F. F. Tisserand found a quantity which does not change during motion of the third body, whose mass is negligible in comparison to the masses of the planet and the Sun (see, e. g. Brouwer and Clemence 1961). For our case, defined by Eq. (74), we can write the Tisserand's quantity in the form

$$C_T = \frac{1-\beta}{a} + 2\sqrt{\frac{(1-\beta)\,a\,(1-e^2)}{a_P^3}}\cos I, \tag{75}$$

where e is eccentricity of the particle, characterized by parameter β, and I is the inclination of the particle orbital plane with reference to the plane of the planetary orbit. We have to stress that Tisserand's quantity C_T does not change only for the special case of the restricted problem of three bodies. However, Eq. (75) enables to find secular change of the eccentricity of the particle captured in a resonance. For this purpose, we will consider $I = 0$ in Eq. (75). We can write

$$\frac{dC_T}{dt} = \frac{\partial C_T}{\partial a}\left(\frac{da}{dt}\right)_{total} + \frac{\partial C_T}{\partial e}\left(\frac{de}{dt}\right)_{total}, \tag{76}$$

for the total time derivative of the Tisserand quantity C_T defined by Eq. (75). However, according to Eq. (74), time derivatives of semi-major axis and eccentricity of the particle are caused by gravitational perturbations of the planet (these terms will be denoted by the subscript G) and nongravitational perturbations caused by solar electromagnetic radiation and solar wind (these terms will be denoted by the subscript NG):

$$\begin{aligned} \left(\frac{da}{dt}\right)_{total} &= \left(\frac{da}{dt}\right)_G + \left(\frac{da}{dt}\right)_{NG} , \\ \left(\frac{de}{dt}\right)_{total} &= \left(\frac{de}{dt}\right)_G + \left(\frac{de}{dt}\right)_{NG} . \end{aligned} \tag{77}$$

On the basis of Eqs. (76) – (77) we can write

$$\begin{aligned} \frac{dC_T}{dt} &= \frac{\partial C_T}{\partial a} \left\{ \left(\frac{da}{dt}\right)_G + \left(\frac{da}{dt}\right)_{NG} \right\} + \\ &\quad \frac{\partial C_T}{\partial e} \left\{ \left(\frac{de}{dt}\right)_G + \left(\frac{de}{dt}\right)_{NG} \right\} . \end{aligned} \tag{78}$$

According to Tisserand, gravitational terms alone do not change the value of C_T:

$$\frac{\partial C_T}{\partial a} \left(\frac{da}{dt}\right)_G + \frac{\partial C_T}{\partial e} \left(\frac{de}{dt}\right)_G = 0 . \tag{79}$$

Putting Eq. (79) into Eq. (78):

$$\frac{dC_T}{dt} = \frac{\partial C_T}{\partial a} \left(\frac{da}{dt}\right)_{NG} + \frac{\partial C_T}{\partial e} \left(\frac{de}{dt}\right)_{NG} . \tag{80}$$

If we are interested in secular changes of orbital elements a and e, then the particle stay in the resonance is characterized by the relation $(da/dt)_{total} = 0$ and Eq. (76) reduces to

$$\frac{dC_T}{dt} = \frac{\partial C_T}{\partial e} \left(\frac{de}{dt}\right)_{total} . \tag{81}$$

Eqs. (80) – (81) finally give for the total secular change of the eccentricity of the particle

$$\left(\frac{de}{dt}\right)_{total} = \left(\frac{de}{dt}\right)_{NG} + \frac{\partial C_T/\partial a}{\partial C_T/\partial e} \left(\frac{da}{dt}\right)_{NG} . \tag{82}$$

Calculating partial derivatives of C_T defined by Eq. (75), and using also Eq. (69) for secular change of eccentricity of the particle caused by nongravitational forces, one finally obtains

$$\frac{de}{dt} = \beta \frac{\mu}{c} \left(1 + \frac{\eta}{\bar{Q}'_{pr,1}}\right) \frac{\sqrt{1-e^2}}{a^2 \, e} \left\{ 1 - \frac{n}{n_P} \frac{1 + 3e^2/2}{(1-e^2)^{3/2}} \right\} , \tag{83}$$

if also definition represented by Eq. (72) is used, and also the fact that $m_P \ll M_\odot$ was taken into account. Moreover, the symbol of angle brackets – denoting secular evolution – in Eq. (69) was also omitted.

Eq. (83) determines secular evolution of eccentricity of the particle characterized by the values β and $\bar{Q}'_{pr,1}$, if the particle is captured into a mean-motion resonance. This equation enables to find detail evolution. However, important qualitative behaviour can be found very easily. If we take some special mean-motion resonance, we already know the value n/n_P. If the initial (secular) eccentricity is small enough, then de/dt is always positive and the eccentricity of the particle is an increasing function of time, during the stay of the particle in the exterior mean-motion resonance. Eccentricity of the particle can only asympotically approach to a limiting value e_{lim} given by condition that the term in the combined brackets is zero: $n_P/n = (1+3e_{\text{lim}}^2/2) \; / \; (1-e_{\text{lim}}^2)^{3/2}$. If the initial eccentricity is large enough, then de/dt is always negative and the eccentricity of the particle is a decreasing function of time, during the stay of the particle in the mean-motion resonance. Eccentricity of the particle can only asympotically approach to the limiting value e_{lim}. Similar considerations can be done for interior resonances: secular eccentricity of the particle is always a decreasing function of time.

9.2. Mean-Motion Resonances and Nonspherical Dust Particles

Orbital evolution of spherical dust grain in the restricted problem of three bodies, with simultaneous action of solar electromagnetic radiation (and solar wind), is characterized by three important results. In the zone of mean-motion resonances with a planet the fundamental results are:

i) secular evolution of semi-major axis is a decreasing function of time near the resonances ($da/dt = 0$ in the resonance),

ii) secular evolution of eccentricity in the exterior resonance is either an increasing function of time or a decreasing function of time and the values of eccentricity cannot reach a limiting value e_{lim}, given by condition $n_P/n = (1+3e_{\text{lim}}^2/2) \; / \; (1-e_{\text{lim}}^2)^{3/2}$,

iii) secular evolution of eccentricity in the interior resonance is a decreasing function of time.

The crucial question emerges: Are the features, presented above, typical features for real dust particles?

Real dust particles are nonspherical in shape. Thus, the total electromagnetic force represented by Eqs. (40) – (42) have to be used, instead of the Poynting-Robertson effect. Equation of motion for the particle in the problem of three bodies, with the action of electromagnetic radiation, is

$$
\begin{aligned}
\frac{d\,\vec{v}}{d\,t} \;&=\; -\,\frac{G\,M_{\odot}}{r^2}\,\vec{e}_1 \,+ \\
&+\,\frac{G\,M_{\odot}}{r^2}\sum_{j=1}^{3}\beta_j\left[\left(1\,-\,2\,\frac{\vec{v}\cdot\vec{e}_1}{c}\,+\,\frac{\vec{v}\cdot\vec{e}_j}{c}\right)\vec{e}_j\,-\,\frac{\vec{v}}{c}\right]\,- \\
&-\,G\,m_P\left\{\frac{\vec{r}-\vec{r}_P}{|\vec{r}-\vec{r}_P|^3}\,+\,\frac{\vec{r}_P}{|\vec{r}_P|^3}\right\}\,, \qquad (84)
\end{aligned}
$$

if thermal emission is neglected. The following questions arise for the case of nonspherical particles: Can be the particle captured into a mean-motion resonance with a planet? How does look secular orbital evolution of the particle near the resonances?

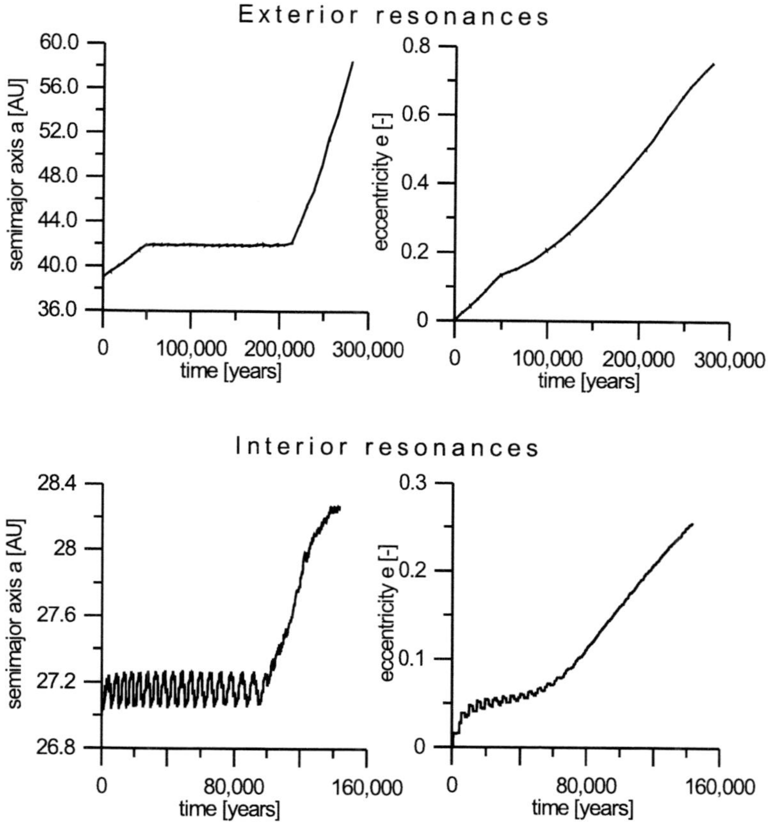

Figure 6. Time evolution of semimajor axis and eccentricity for nonspherical dust grains of effective radius 2 microns. The plots show exterior resonance 5:3 of iron-rich silicates with Neptune, and interior resonance 7:8 of magnesium-rich silicates with Neptune.

We made numerical calculations for the particle of a shape identical to the particle U2015 B10. It was supposed that particle rapidly rotates around a rotational axis fixed in space (various rotational axes were considered) and averaging over the corresponding azimuthal angle in orientation with respect to the Sun were done; values of β_j obtained in this way were used in Eq. (84). The simulations were done for the restricted problem of three bodies. This enables comparison between behaviour of spherical dust grains (the results can be obtained even in an analytical way) and nonspherical grains. As for the questions formulated above, the conclusion is that spherical dust particles do not yield representative results for real dust grains. Although not only spherical grains can be captured in mean-motion resonances, orbital evolution near the resonances cannot be characterized by a simple description as it was obtained for spherical grains (grains with spherically symmetric dustribution of mass). Figure 6 shows possible evolution for outer and inner resonances with planet Neptune. In both cases, dust grains of effective radius 2 microns are considered. As for the inner mean-motion resonance with Neptune, it is defined by the ratio $T : T_P = 7 : 8$, and our magnesium-rich silicate (mass density 4 g/cm^3) nonspherical particle is characterized by semi-major axis in the resonant zone 27.02 AU $\leq a \leq$ 27.28 AU, while theoretical value for $\beta = 0.036$ (its change is less than 5% for various orientations) is $a = 27.164$ AU, according to Eq. (73). The outer resonance with Neptune $T : T_P = 5 : 3$ is for iron-rich silicate (mass density 6 g/cm^3) nonspherical particle, and, semi-major axis in the resonant zone is 41.81 AU $\leq a \leq$ 42.02 AU, while theoretical value for $\beta = 0.024$ is $a = 41.908$ AU. The figures show that nonspherical dust grains can be captured even in the case when secular change of semi-major axis is an increasing function of time near the zone of the resonance, and, no limiting value od eccentricity exists in the resonance.

9.3. Mean-Motion Resonances and Arbitrarily Shaped Dust Particles: Fundamental Results

Resonant captures of arbitrarily shaped dust grain exist for exterior and interior mean-motion resonances with planets. Ejection of the grain from the exterior resonance is due to the grain close encounter with the planet. While capture of spherical grain in a resonance is characterized by decrease of semi-major axis and eccentricity outside the resonance (see Eq. 69 for semi-major axis and eccentricity), nonspherical grain may exhibit also an increase of semimajor axis (and eccentricity) outside the resonance. Resonant trapping is not normally expected for diverging orbits such as the trapping presented in Fig. 6: this is the first case when a physically justified force can generate such a behaviour.

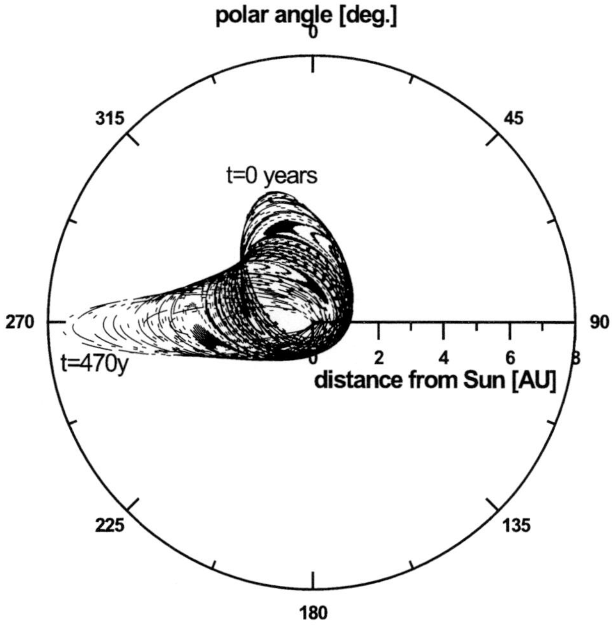

Figure 7. Projection of orbital evolution for non-spherical 2- μm sized particle to the plane of parent body. Dust particle is ejected at aphelion of comet Encke. Zero ejection velocity is considered.

10. Cometary Particles and Radiation Forces

Let us consider ejection of dust particle from a comet. The process of ejection is considered in Eqs. (43) – (45), or, in terms of orbital elements, in Eqs. (48) – (51) for zero ejection velocities, or, in Eq. (47) for non-zero ejection velocities: the results hold for arbitrarily shaped dust particles. If one considers only spherical dust grains, the effect of solar gravity and solar electromagnetic (and corpuscular) radiation yields that inclination of the particle orbit does not change in time (Eqs. 60) and secular orbital evolution is described by Eqs. (61) – (64), or, by Eq. (69). As a special case, there does not exist perihelion motion – secular change of longitude of perihelion (the third equation in Eq. 69). However, if we take ejection of 2-μm sized nonspherical dust particle U2015 B10 from aphelion of comet Encke (semi-major axis 2.2 AU, eccentricity 0.85, inclination 12.4°), we obtain also perihelion motion, as it is shown in Fig. 7. Another relevant result is that nonspherical particles may exhibit change of orbital plane. As a consequence, nonspherical dust particles have to be observed also perpendicular to the orbital plane of the comet. In contrary, the radiation forces acting over time to disperse spherical particles after ejection, will spread material along the orbit and in the orbital plane, but this does not account for dispersion of material normal to the orbital plane (Sykes *et al.* 1986). Fig. 8 shows evolution of inclination for nonspherical particle ejected from perihelion of comet Encke for several particle rotational axes fixed in space. It turns out that aphelion ejection velocities for the comet Encke, with values up to $|\vec{\Delta}| = 0.1\ km/s$, do not significantly influence orbital evolution of the particles. The values for ejection velocities are based on various models (Whipple 1951, Jones 1995, Crifo 1995).

In reality the dynamical behaviour of dust particles ejected from a comet will strongly depend on initial conditions (e. g., on the position of the starting point – perihelion or aphelion) and the orientation (including slope) of the particle rotation axis relative to the orbital plane. It can be also easily shown how the particle density and size predetermines the efficiency of the gravitational as well as the electromagnetic forces. The general trend is: the smaller nonspherical particle, the larger deviation of its orbital evolution from the evolution of volume identical spherical particle. However, orbital evolution even for micron sized real nonspherical particle is still significantly different from the evolution of spherical particle. The runs for 2 μm-sized particle showed that spherical particle rapidly spirals in toward the Sun and its inclination is conserved – this case corresponds to the P-R effect. Although the 2 μm-sized nonspherical particle may finally also hit the Sun, its orbital evolution is much more complicated and requires longer time for the final fall than it is for the case of spherical particle. The larger particle density, the greater time is required for reaching the Sun.

Numerical calculations showed that the perihelion distance of irregularly shaped particle may be even an increasing function of time as well as of an oscillating form at least for some time interval. Analogous results hold for inclination (between the instantaneous particle orbital plane and the ecliptic – orbital plane of the Earth around the Sun). Time evolution of eccentricity of the osculating ellipse also differs from that given by the P-R effect. Orbital motion of such interplanetary dust particle differs from the motion given by the Poynting-Robertson effect. It is caused by the fact that realistically shaped dust particles exhibit important nonforward and nonbackward scattering efficiences in comparison with

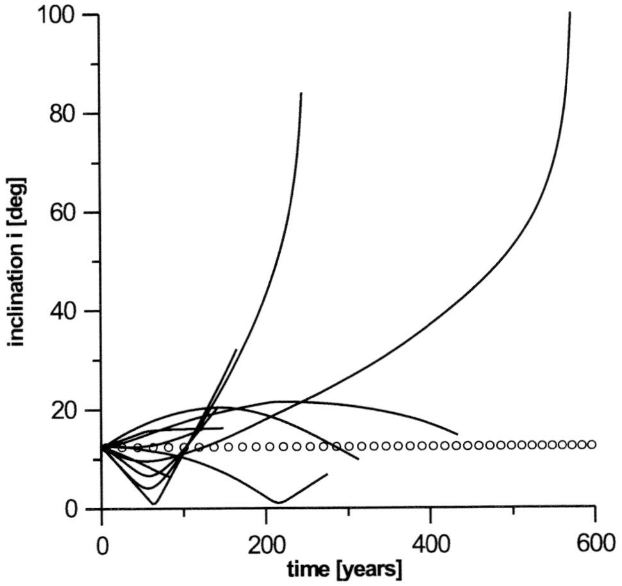

Figure 8. Secular change of inclination for non-spherical 2- μm sized dust particle ejected at aphelion of comet Encke. Computational results for various particle rotation axes are presented. Solid curves: $|\vec{\Delta}| = 0\ km.s^{-1}$, short dashed curves: $\Delta v_R = 0.1\ km.s^{-1}$ (outward the Sun), long dashed curves: $\Delta v_R = -0.1\ km.s^{-1}$ (i.e. sunward orientation). All the cases exhibit almost identical behaviour for a given rotation axis. Circles correspond to 2-μm sized spherical dust particle: $\Delta v_R = 0\ km.s^{-1}$.

the result when the P-R effect holds. Therefore, the P-R effect cannot be simply applied at least for some sizes of cosmic dust particles to simulate their motion in space.

11. Summary and Conclusions

The contribution deals with the effect of electromagnetic radiation (light) on dynamics of cosmic dust particles. The paper presents derivation of relativistically covariant equation of motion for dust particle under the action of electromagnetic radiation – see Eq. (39). The equation yields, as special cases, the results obtained already by Einstein in 1905 and by Robertson in 1937. Application of the equation of motion concentrates on various systems in the Universe: possible capture of interstellar dust grains in the Solar System, mean-motion resonances with planets, and, orbital evolution of dust ejected from a comet.

Differences between behaviour of spherical and nonspherical particles are pointed out. Although spherical particles often enable analytical calculations, orbital evolution of spheres cannot be considered as a representative evolution for real cosmic dust particles. The crucial effect turns out to be the effect of electromagnetic radiation on motion of dust grains.

A flux of interstellar dust particles entering the Solar System may lead to a result that some of the grains are captured in the Solar System. Spherical interstellar grains can be captured almost only in the vicinity of the Sun, where they are destroyed due to high temperatures. On the other hand, nonspherical grains can be captured and they can revolve the Sun in larger distances from the Sun.

Arbitrarily shaped dust particle may be captured in a mean-motion resonance with a planet oribiting a central star (e. g., Sun). The capture of spherical grain is characterized by a decrease of semimajor axis (and eccentricity) before the capture in the resonance. Nonspherical grains may be captured even in the case when secular change of semimajor axis is an increasing function of time, near the resonance. Resonant trapping is not normally expected for diverging orbits. Moreover, the result is the first case when a physically justified force can generate such a behaviour.

The effect of solar electromagnetic radiation and gravitational force of the Sun influence the motion of spherical dust particles in the way that orbital inclination of the particle does not change – conservation of the orbital plane – and perihelion motion of the particle does not exist. Change of orbital inclination and perihelion motion of a nonspherical particle may exist under the effect of solar electromagnetic radiation.

Acknowledgments

This paper was supported by Project M772-N02 by the Fonds FWF and by VEGA grant No. 1/3074/06. We specially thank Prof. Ch. Dellago for computer resources at the Institute of Experimental Physics.

Appendix A: General Lorentz Transformation and Some Comments on Equation of Motion

(Reference to equation of number (j) of this appendix is denoted as Eq. (A.j). Reference to equation of number (i) of the main text is denoted as Eq. (i).)

Let us have a four-vector $A^{\mu} = (A^0, \vec{A})$, where A^0 is its time component and $\vec{A}$ is its spatial component. Since generalized special Lorentz transformations do not form a group (in general, composition of two generalized special Lorentz transformations is not a generalized special Lorentz transformation), we will consider more general Lorentz transformation. This Lorentz transformation can be written as

$$A^{'\mu} = \Lambda^{\mu}{}_{\nu} \, A^{\nu} \, , \qquad (A.1)$$

where summation over repeated indices is supposed (and also in all the following equations) – 0, 1, 2, 3 for Greek letters and 1, 2, 3 for Latin letters – and its inverse

$$A^{\mu} = \Lambda_{\alpha}{}^{\mu} \, A^{'\alpha} \, . \qquad (A.2)$$

Important property of the Lorentz transformation

$$A^{''} = \Lambda \, A \qquad (A.3)$$

is, that it can be composed of the following two transformations:

$$A^{'} = L \, A \, ,$$

$$A'' = R\,A' , \tag{A.4}$$

where L corresponds to generalized special Lorentz transformation and R represents rotation in 3-dimensional space:

$$A'^{\mu} = L^{\mu}{}_{\nu}\, A^{\nu} , \tag{A.5}$$

and (see Eq. 13)

$$\begin{aligned} L^{0}{}_{0} &= \gamma , \\ L^{0}{}_{i} &= L^{i}{}_{0} = -\,\gamma\,(\vec{v}/c)_i , \quad i = 1,2,3 , \\ L^{i}{}_{j} &= \delta_{ij} + (\gamma - 1)\,(\vec{v})_i\,(\vec{v})_j \,/\, \vec{v}^2 , \quad i = 1,2,3 , \quad j = 1,2,3 , \end{aligned} \tag{A.6}$$

where $\delta_{ij} = 1$ if $i = j$ and $\delta_{ij} = 0$ if $i \neq j$,

$$A^{\mu} = L_{\alpha}{}^{\mu}\, A'^{\alpha} , \tag{A.7}$$

where (see Eq. 14)

$$\begin{aligned} L_{\alpha}{}^{\beta} &= \eta_{\alpha\,\rho}\,\eta^{\beta\,\gamma}\,L^{\rho}{}_{\gamma} , \\ \eta_{\alpha\,\beta} &= diag(+1, -1, -1, -1) \end{aligned} \tag{A.8}$$

and

$$\begin{aligned} L_{0}{}^{0} &= \gamma , \\ L_{0}{}^{i} &= L_{i}{}^{0} = \gamma\,(\vec{v}/c)_i , \quad i = 1,2,3 , \\ L_{i}{}^{j} &= \delta_{ij} + (\gamma - 1)\,(\vec{v})_i\,(\vec{v})_j \,/\, \vec{v}^2 , \quad i = 1,2,3 , \quad j = 1,2,3 , \end{aligned} \tag{A.9}$$

$$\begin{aligned} R^{0}{}_{0} &= 1 , \\ R^{0}{}_{i} &= R^{i}{}_{0} = 0 , \quad i = 1,2,3 , \\ R^{i}{}_{j} &= r_{ij} , \quad i = 1,2,3 , \quad j = 1,2,3 , \end{aligned} \tag{A.10}$$

where r_{ij} can be expressed in terms of Euler angles, and, moreover, orthogonality conditions are fulfilled:

$$\begin{aligned} &\sum_{i=1}^{3} r_{ij}\, r_{ik} = \delta_{jk} , \quad j = 1,2,3 , \quad k = 1,2,3 , \\ &\sum_{j=1}^{3} r_{ij}\, r_{kj} = \delta_{ik} , \quad i = 1,2,3 , \quad k = 1,2,3 . \end{aligned} \tag{A.11}$$

On the basis of Eqs. (A.3) and (A.4) we can write

$$\Lambda^{\mu}{}_{\nu} = R^{\mu}{}_{\varrho}\, L^{\varrho}{}_{\nu} \,. \tag{A.12}$$

Eqs. (A.6), (A.10) and (A.12) yield

$$\begin{aligned} \Lambda^{0}{}_{0} &= L^{0}{}_{0} = \gamma \,, \\ \Lambda^{0}{}_{i} &= L^{0}{}_{i} = -\,\gamma\,(\vec{v}/c)_i \,, \quad i = 1, 2, 3 \,. \end{aligned} \tag{A.13}$$

Finally, requirement $A^{\mu} A_{\mu} = A'^{\mu} A'_{\mu}$ yields,

$$\Lambda^{\mu}{}_{\nu}\, \Lambda_{\mu}{}^{\varrho} = \delta^{\varrho}_{\nu} \,, \tag{A.14}$$

where $\delta^{\varrho}_{\nu} = 1$ if $\varrho = \nu$ and $\delta^{\varrho}_{\nu} = 0$ if $\varrho \neq \nu$; Eq. (A.1) was also used: $A'^{\mu} = \Lambda^{\mu}{}_{\nu}\, A^{\nu}$, $A'_{\mu} = \Lambda_{\mu}{}^{\nu}\, A_{\nu}$.

Incoming Radiation

Applying Eqs. (A.1) and (A.13) to quantity $(E_i/c, \vec{p}_i)$ (four-momentum per unit time – proper time is a scalar quantity) and taking into account also the fact that $\vec{p}_i = E_i/c\,\vec{e}_1$, we can write

$$E'_i = E_i\, w_1 \,, \tag{A.15}$$

where

$$w_1 \equiv \gamma\,(1 \;-\; \vec{v}\cdot\vec{e}_1/c) \,. \tag{A.16}$$

Using the fact that $p^{\mu} = (h\,\nu, h\,\nu\,\vec{e}_1)$ for photons, we have

$$\nu' = \nu\, w_1 \,. \tag{A.17}$$

We have four-vector $p^{\mu}_i = (E_i/c, \vec{p}_i) = (1, \vec{e}_1) E_i/c = (1/w_1, \vec{e}_1/w_1)\; w_1\; E_i/c \equiv b^{\mu}_1\; w_1\; E_i/c$. We have found a new four-vector b^{μ}_1, which is given as $b'^{\mu}_1 = \Lambda^{\mu}{}_{\nu}\; b^{\nu}_1$ in the proper frame of reference of the particle: $b'^{\mu}_1 = (1, \vec{e}\,'_1)$. The transformation of space components between b^{μ}_1 and b'^{μ}_1 corresponds to aberration of light.

For monochromatic radiation the flux density of radiation energy becomes

$$S' = n'\, h\, \nu'\, c \,; \quad S = n\, h\, \nu\, c \,, \tag{A.18}$$

where n and n' are concentrations of photons (photon number densities) in the corresponding reference frames. We also have continuity equation

$$\partial_{\mu}\, j^{\mu} = 0 \,, \quad j^{\mu} = (c\, n, c\, n\, \vec{e}_1) \,, \tag{A.19}$$

with current density j^{μ}. Application of Eqs. (A.7) and (A.19) then yields

$$n' = w_1\, n \,. \tag{A.20}$$

Using Eqs. (A.17), (A.18) and (A.19) we finally obtain

$$S' = w_1^2\, S \,. \tag{A.21}$$

Eqs. (A.15) and (A.21) then together give $E_i = w_1\, S\, A'$, $\vec{p}_i = w_1\, S\, A'\, \vec{e}_1/c$.

Appendix B: Einstein's Example

(Reference to equation of number (j) of this appendix is denoted as Eq. (B j). Reference to equation of number (i) of the main text is denoted as Eq. (i).)

Let us consider a plane mirror moving (at a given moment) along x-axis (system S) with velocity $\vec{v} = (v, 0, 0)$, $v > 0$; the mirror is perpendicular to the x-axis (the plane of the mirror is parallel to the yz-plane). A beam of incident (hitting) photons is characterized by unit vector $\vec{S}' = (\cos\theta', \sin\theta', 0)$ in the proper frame (primed quantities) of the mirror. Reflected beam is described by the unit vector $\vec{e}' = (-\cos\theta', \sin\theta', 0)$ (in the proper frame S').

The problem is: Find equation of motion of the mirror in the frame of reference S.

Solution 1: Trivial Manner

Consider one photon (frequency f') in the proper frame of the mirror. Since the directions (and orientations) of the incident and outgoing photons are characterized by

$$\vec{S}' = \left(+\cos\theta', \sin\theta', 0\right) ,$$
$$\vec{e}' = \left(-\cos\theta', \sin\theta', 0\right) , \tag{B.1}$$

we can immediately write

$$p_i'^{\mu} = \frac{h\,f'}{c}\,\left(1, +\cos\theta', \sin\theta', 0\right) ,$$
$$p_o'^{\mu} = \frac{h\,f'}{c}\,\left(1, -\cos\theta', \sin\theta', 0\right) , \tag{B.2}$$

for the four-momentum of the photon before interaction with the mirror and after the interaction.

As a consequence, the mirror obtains four-momentum

$$p'^{\mu} = p_i'^{\mu} - p_o'^{\mu} = \frac{h\,f'}{c}\,\left(0, 2\cos\theta', 0, 0\right) . \tag{B.3}$$

Application of the special Lorentz transformation to Eq. (B.3) yields

$$p^{\mu} = \frac{h\,f'}{c}\,2\,\gamma\,\left(\cos\theta'\right)\,\left(\beta, 1, 0, 0\right) , \tag{B.4}$$

where, as standardly abbreviated,

$$\gamma = 1 / \sqrt{1 - \beta^2} \;;\quad \beta = v / c . \tag{B.5}$$

On the basis of Eq. (B.4), we can immediately write equation of motion of the mirror

$$\frac{dp^{\mu}}{d\tau} = \frac{E_i'}{c}\,2\,\gamma\,\left(\cos\theta'\right)\,\left(\beta, 1, 0, 0\right) , \tag{B.6}$$

where E_i' is the total energy (per unit time) of the incident radiation measured in the proper frame of reference.

Solution 2: Application of General Theory Presented in Sec. 3. of the Main Text

We have to choose orthonormal vectors in the systems S': we will use $\vec{e}_1' \equiv \vec{S}'$ and $\vec{e}_2'$ and one can easily find

$$\vec{e}_1' \equiv \vec{S}' = \left(+ \cos\theta', \sin\theta', 0\right) ,$$
$$\vec{e}_2' = \left(- \sin\theta', \cos\theta', 0\right) . \qquad (B.7)$$

We have to write ($Q'_{pr,3} = 0$)

$$\vec{p}' = \frac{h\, f'}{c} \left(Q_1' \, \vec{e}_1' + Q_2' \, \vec{e}_2'\right) . \qquad (B.8)$$

On the basis of Eqs. (B.3), (B.7) and (B.8) we have ($Q'_{pr,3} = 0$)

$$Q'_{pr,1} = 2 \left(\cos\theta'\right)^2 , \; Q'_{pr,2} = - 2 \left(\sin\theta'\right) \left(\cos\theta'\right) . \qquad (B.9)$$

Other prescription yields (see Eq. 33)

$$\begin{aligned} b_1^0 &= \gamma \left(1 + \vec{v} \cdot \vec{e}_1'/c\right) = \gamma(1 + \beta \cos\theta') , \\ \vec{b}_1 &= \vec{e}_1' + \left[(\gamma - 1)\vec{v} \cdot \vec{e}_1'/\vec{v}^2 + \gamma/c\right] \vec{v} = \left(\gamma \cos\theta' + \gamma\beta, \sin\theta', 0\right) , \end{aligned} \qquad (B.10)$$

$$\begin{aligned} b_2^0 &= \gamma \left(1 + \vec{v} \cdot \vec{e}_2'/c\right) = \gamma(1 - \beta \sin\theta') , \\ \vec{b}_2 &= \vec{e}_2' + \left[(\gamma - 1)\vec{v} \cdot \vec{e}_2'/\vec{v}^2 + \gamma/c\right] \vec{v} = \left(-\gamma \sin\theta' + \gamma\beta, \cos\theta', 0\right) . \end{aligned} \qquad (B.11)$$

Inserting Eqs. (B.9) – (B.11) (and $F'_{ej} = 0$ for $j = 1, 2, 3$) into Eq. (30), one obtains

$$\begin{aligned} \frac{dp^\mu}{d\tau} &= \frac{E_i'}{c} \left\{ \left[2(\cos\theta')^2\right] \left(b_1^\mu - \beta^\mu\right) + \left[-2(\sin\theta')(\cos\theta')\right] \; \left(b_2^\mu - \beta^\mu\right) \right\} \\ &= \frac{E_i'}{c} 2\gamma \left(\cos\theta'\right) \left(\beta, 1, 0, 0\right) . \end{aligned} \qquad (B.12)$$

Unit vectors $\vec{e}_1' \equiv \vec{S}'$ and $\vec{e}_2'$ are used. They are orthonormal in the system S'. However, corresponding vectors are not orthogonal in the system S.

$$\begin{aligned} \vec{e}_1 &= \frac{1}{w'} \left\{ \vec{e}_1' + \left[(\gamma - 1)\, \vec{v} \cdot \vec{e}_1'/\vec{v}^2 + \gamma/c\right] \vec{v} \right\} , \\ w' &= \gamma \left(1 + \vec{v} \cdot \vec{e}_1'/c\right) , \end{aligned} \qquad (B.13)$$

and analogous equation holds for vector $\vec{e}_2$. Inserting Eqs. (B.7), one obtains:

$$\vec{e}_1 = \left\{ \frac{\cos\theta' + \beta}{1 + \beta \cos\theta'}, \frac{\sin\theta'}{\gamma \left(1 + \beta \cos\theta'\right)}, 0 \right\} ,$$
$$\vec{e}_2 = \left\{ \frac{- \sin\theta' + \beta}{1 - \beta \sin\theta'}, \frac{\cos\theta'}{\gamma \left(1 - \beta \sin\theta'\right)}, 0 \right\} . \qquad (B.14)$$

It can be easily verified that scalar product of these two vectors is nonzero, in general.

Comparison with Einstein's Result

Inserting $E_i' = w^2 \; S \; A'_{mirror} \; \cos\theta'$ into Eq. (B.6) (or Eqs. (B.12)) and using Eq. (B.14) ($\vec{e}_1 \equiv (\cos\theta, \sin\theta, 0)$) for the purpose of obtaining $\cos\theta' = (\cos\theta - \beta)/(1 - \beta \; \cos\theta)$, one easily obtains:

i) $dE/d\tau \; = \; 2 \; \gamma^3 \; S \; A'_{mirror} \; (\cos\theta \; - \; \beta)^2 \; \beta$; using definition of radiation pressure $dE/dt \equiv P \; v \; A'_{mirror}$, we have $P = 2 \; (S \; / \; c) \; (\cos\theta - \beta)^2 \; / \; (1 - \beta^2)$, or,

ii) $dp/d\tau \; = \; 2 \; \gamma^3 \; (S \; A'_{mirror} \; / \; c) \; (\cos\theta \; - \; \beta)^2$; using definition of radiation pressure $P \equiv (dp/dt) \; / \; A'_{mirror}$, we have $P = 2 \; (S \; / \; c) \; (\cos\theta - \beta)^2 \; / \; (1 - \beta^2)$.

Result for P is consistent with the result presented in Einstein (1905).

References

[1] Baguhl M., Grűn E., Hamilton D., Linkert G., Riemann R., Staubach P., 1995. The flux of interstellar dust observed by Ulysses and Galileo. *Space Sci. Rev.* **72**, 471-476.

[2] Bartoli A., 1876. Sapora il movimenti prodotti dalla luce e dal calore, Le Mounier, Florence.

[3] Beaugé C., Ferraz-Mello S., 1994. Capture in exterior mean-motion resonances due to Poynting-Robertson drag. *Icarus* **110**, 239-260.

[4] Brouwer D., Clemence G. M., 1961. Celestial Mechanics. Academic Press, New York and London, 598 pp.

[5] Burns J. A., Lamy P. L., Soter S., 1979. Radiation forces on small particles in the Solar System. *Icarus* **40**, 1-48.

[6] Clanton V. S., Gooding J. L., Mckay D. S., Robinson G. A., Warren J. L., Watts L. A., 1984. *Cosmic dust catalog (particles from collection flag U2015)*, NASA, Johnson Space Center (Houston, Texas) 70/1 10.

[7] Crifo J. F., 1995. A general physicochemical model of the inner coma of active comets. I. Implications of spatially distributed gas and dust production. *Astrophys. J.* **445**, 470-488.

[8] Draine B. T., 1988. The discrete-dipole approximation and its application to interstellar graphite grains. *Astrophys. J.* **333**, 848-872.

[9] Draine B. T., Flatau P. J., 1994. Discrete-dipole approximation for scattering calculations. *J. Opt. Soc. Am.* **A 11**, 1491-1499.

[10] Draine B. T., Flatau P. J., 2004. User Guide for the Discrete Dipole Approximation Code DDSCAT.6.1," Freeware, http://arxiv.org/abs/astro-ph/0409262.

[11] Einstein A., 1905. Zur Elektrodynamik der bewegter Kőrper. *Annalen der Physik* **17**, 891-920.

[12] Grün E., Gustafson B., Mann I., Baguhl M., Morfill G. E., Staubach G., Taylor A., Zook H. A., 1994. Interstellar dust in the heliosphere. *Astron. Astrophys.* **286**, 915-925.

[13] Harwit M., 1963. Origins of the Zodiacal dust cloud. *J. Geophys. Res.* **86**, 2171-2180.

[14] Hildebrand R. H., Dragovan M., 1995. The shapes and alignment properties of interstellar dust grains. *Astrophys. J.* **450**, 663-666.

[15] Jackson A. A., 2001. The capture of interstellar dust: the pure Poynting-Robertson case. *Planet Space Sci.* **49**, 417-424.

[16] Jackson A. A., Zook H. A., 1989. A Solar System dust ring with the Earth as its shepherd. *Nature* **337**, 629-631.

[17] Jones J., 1995. The ejection of meteoroids from comets. *Mon. Not. R. Astron. Soc.* **275**, 773-780.

[18] Kimura H., Mann I., 1998. The electric charging of interstellar dust in the solar system and consequences for its dynamics. *Astrophys. J.* **499**, 454-462.

[19] Klačka J., 1992. Poynting-Robertson effect. I. Equation of motion. *Earth, Moon, and Planets* **59**, 41-59.

[20] Klačka J., 2000. Electromagnetic radiation and motion of real particle. http://xxx.lanl.gov/abs/astro-ph/0008510.

[21] Klačka J., 2004. Electromagnetic radiation and motion of a particle. *Cel. Mech. and Dynam. Astron.* **89**, 1-61.

[22] Klačka J., Kocifaj M., 2001. Motion of nonspherical dust particle under the action of electromagnetic radiation. *J. Quant. Spectrosc. Radiat. Transfer* **70**, 595-610.

[23] Klačka J., Kocifaj M., Pástor P., 2005a. Motion of dust near exterior resonances with planets. *Journal of Physics: Conference Series* **6**, 126-131.

[24] Klačka J., Kocifaj M., Pástor P., 2005b. Effect of radiation on nonspherical particles in resonances with large planets. In: *8th Conference on Electromagnetic and Light Scattering by Nonspherical Particles: Theory, Measurements and Applications* , F. Moreno, J. J. López-Moreno, O. Munoz and A. Molina (eds.), Instituto de Astrofisica de Granada, pp. 156-159.

[25] Kocifaj M., Kapišinský I., Kundracík F., 1998. Optical effects of irregular cosmic dust particle U2015 B10. *J. Quant. Spectrosc. Radiat. Transfer* **63**, 1-14.

[26] Kocifaj M., Klačka J., 2003. The capture of interstellar dust: the pure electromagnetic radiation case. *Planet. Space Sci.* **51**, 617-626.

[27] Kocifaj M., Klačka J., 2004a. The capture of interstellar dust: the Lorentz force case. *Planet. Space Sci.* **52**, 839-847.

[28] Kocifaj M., Klačka J., 2004b. Dynamical behaviour of interstellar dust particles in the solar system. *J. Quant. Spectrosc. Radiat. Transfer* **89**, 165-177.

[29] Krauss O., Wurm G., 2004. Radiation pressure forces on individual micron-size dust particles: a new experimental approach, *J. Quant. Spectrosc. Radiat. Transfer* **89**, 179-189.

[30] Landau L. D., Lifshitz E. M., 1975. The Classical Theory of Fields, 4th ed., Pergamon Press, New York (In Russian: Teoria polia. Nauka, Moscow, 1973).

[31] Landgraf M., Augustsson K., Grűn E., Gustafson B. S., 1999. Deflection of the local interstellar dust flow by solar radiation pressure. *Science* **286**, 2319-2322.

[32] Landgraf M., Baggaley W., Grűn E., Krűger H., Linkert G., 2000. Aspects of the mass distribution of interstellar dust grains from in-situ measurements. *J. Geophys. Res.* **105 (A5)**, 10343-10352.

[33] Leinert Ch., Grűn E., 1990. Interplanetary dust. In: *Physics of the Inner Heliosphere I*, R. Schwenn and E. Marsch (eds.), Springer-Verlag, Berlin Heidelberg, pp. 207-275.

[34] Liou J-Ch., Zook H. A., 1995. An asteroidal dust ring of micron-sized particles trapped in the 1:1 mean motion resonance with Jupiter. *Icarus* **113**, 403-414.

[35] Liou J-Ch., Zook H. A., 1997. Evolution of interplanetary dust particles in mean motion resonances with planets. *Icarus* **128**, 354-367.

[36] Liou J-Ch., Zook H. A., Jackson A. A. 1995. Radiation pressure, Poynting-Robertson drag, and solar wind drag in the restricted three body problem. *Icarus* **116**, 186-201.

[37] Lyttleton R. A., 1976. Effects of solar radiation on the orbits of small particles. *Astrophys. Space Sci.* **44**, 119-140.

[38] Marzari F., Vanzani V., 1994. Dynamical evolution of interplanetary dust particles. *Astron. Astrophys.* **283**, 275-286.

[39] Mediavilla E., Buitrago J., 1989. The Poynting-Robertson effect and its generalization to the case of an extended source in rotation. *Eur. J. Phys.* **10**, 127-132.

[40] Maxwell J. C., 1873. A Treatise on Electricity and Magnetism, Oxford.

[41] Mignard F., 1992. On the radiation forces. In: *Interrelations Between Physics and Dynamics for Minor Bodies in the Solar System*, D. Benest and C. Froeschle (eds.), Frontières, pp. 419-451.

[42] Mishchenko M., 2001. Radiation force caused by scattering, absorption, and emission of light by nonspherical particles. *J. Quant. Spectrosc. Radiat. Transfer* **70**, 811-816.

[43] Mishchenko M., Travis L. D., Lacis A. A., 2002. Scattering, Absorption and Emission of Light by Small Particles. Cambridge University Press, Cambridge (UK), 445 pp.

[44] Parker E. N., 1958. Dynamics of the interplanetary gas and magnetic fields. *Astrophys. J.* **128**, 664-676.

[45] Poynting J. M., 1904. Radiation in the Solar System: Its effect on temperature and its pressure on small bodies. *Philosophical Transactions of the Royal Society of London* **Series A 202**, 525-552.

[46] Purcell E. M., Pennypacker C. R., 1973. Scattering and absorption of light by non-spherical dielectric grains. *Astrophys. J.* **186**, 705-714.

[47] Robertson H. P., 1937. Dynamical effects of radiation in the Solar System. *Mon. Not. R. Astron. Soc.* **97**, 423-438.

[48] Robertson H. P., Noonan T. W., 1968. Relativity and Cosmology. Saunders, Philadelphia.

[49] Srikanth R., 1999. Physical interpretation of the Poynting-Robertson effect. *Icarus* **140**, 231-234.

[50] Sykes M. V., Lebofsky L. A., Hunten D. M., Low F., 1986. The discovery of dust trails in the orbits of periodic comets. *Science* **232**, 1115-1117.

[51] Šidlichovský M., Nesvorný D., 1994. Temporary capture of grains in exterior resonances with Earth: Planar circular restricted three-body problem with Poynting-Robertson drag. *Astron. Astrophys.* **289**, 972-982.

[52] Whipple F. L., 1951. A comet model II. Physical relations for comets and meteors. *Astrophys. J.* **113**, 464-474.

[53] Williams I. P., 2002. The effect of radiation on the motion of meteoroids. In: *Optics of Cosmic Dust*, G. Videen and M. Kocifaj (eds.), Kluwer Academic Publishers, Dordrecht, pp. 283-300.

[54] Wisdom J., 1980. The resonance overlap criterion and the onset of stochastic behavior in the restricted three-body problem. *Astron. J.* **85**, 1122-1133.

[55] Wyatt S. P., Whipple F. L., 1950. The Poynting-Robertson effect on meteor orbits. *Astrophys. J.* **111**, 558-565.

In: Space Science: New Research
Editor: Nick S. Maravell, pp. 287-310
ISBN 1-60021-005-8

Chapter 9

MAGNETIC RECONNECTION IN THE EARTH'S MAGNETOTAIL: RECONSTRUCTION METHOD AND DATA ANALYSIS

T. Penz[1,2*]***, V. V. Ivanova***[3]***, V. S. Semenov***[3]***, R. Nakamura***[1]***,***
I. B. Ivanov[4,5]***, H. K. Biernat***[1,2]***, M. F. Heyn***[4]***, V. A. Sergeev***[3]***,***
and I. V. Kubyshkin[3]

[1] Space Research Institute, Austrian Academy of Sciences,
Schmiedlstr. 6, A–8042 Graz, Austria
[2] Institute of Physics, University of Graz, Universitätsplatz 5,
A-8010 Graz, Austria
[3] Institute of Physics, State University St. Petersburg, 198504 Russia
[4] Institute for Theoretical Physics, Technical University Graz,
Petersgasse 16, A-8010 Graz, Austria
[5] Petersburg Nuclear Physics Institute, Gatchina, 188300 Russia

Abstract

Magnetic reconnection is a topological restructuring of magnetic field lines, leading to a conversion of magnetic energy into kinetic energy and a heating of the plasma. It takes place in regions with strong magnetic gradients: the Earth's magnetopause and the plasma sheet in the magnetotail, during solar and stellar flare events, and in laboratory plasmas on the Earth. It is also involved in astrophysical phenomena like the generation of relativistic jets and processes in accretion disks. The Earth's magnetosphere is the only region where signatures of magnetic reconnection can be measured in-situ by satellites. Based on the method developed by Semenov et al. [1], we solve the direct problem of magnetic reconnection and calculate the magnetic field and plasma flow disturbances in a compressible plasma caused by transient reconnection. In order to get the reconnection rate from measured data, we invert the problem. Since the inverse problem is ill–posed, it is treated with the method of regularization. Application of this method to Cluster measurements from September 8th, 2002, where a series of Earth-ward propagating 1–minute scale magnetic field and plasma flow variations were observed outside of the plasma sheet, showed good agreement for the z-component of

*E-mail address: thomas.penz@oeaw.ac.at

the reconstructed magnetic field. The reconnection rate is about 1 mV/m and the reconnection region is located at about 24–25 Earth radii in the magnetotail.

Keywords: Magnetic reconnection, Magnetotail, Cluster, Magnetohydrodynamics

1. Introduction

Magnetic reconnection is a fundamental plasma process allowing the conversion of magnetic energy into kinetic and thermal energy of the plasma. The occurrence of thin current layers associated with a plasma flow and the presence of magnetic fields inevitably leads to the process of magnetic reconnection. In fact, magnetic reconnection is caused by a change in the plasma resistivity leading to the release of stored magnetic energy.
On scales larger than the characteristic length of the plasma (e.g., the gyroradius), the motion of the plasma is coupled with the magnetic field. This is called the frozen–in condition, which implies that plasma and magnetic fields from distinct regions do not intermix and stay separated. In the upper panel of Fig. 1, this situation is displayed for the simple configuration of two antiparallel magnetic fields, which are kept apart by a thin current layer. If magnetic flux is added to the current sheet from both sides, free energy is accumulated in the current sheet. After some time, the gradients across the current sheet become too strong, resulting in a violation of the frozen–in condition. If such a violation takes place, reconnection is initiated.

Because of the highly non–uniform structure of the current sheet, it is very likely that a local breakdown of the frozen–in condition takes place, leading to the formation of the so–called diffusion region. Early models of magnetic field annihilation as proposed by Sweet [2] and Parker [3] assumed a diffusion region with a size comparable to the whole current sheet. It was quickly realized that this process is much too slow to account for rapid energy release expected in many astrophysical processes. A way to avoid this problem was proposed by Petschek in 1964 [4]. In contrast to the Sweet–Parker model, where homogeneous resistivity is assumed, Petschek proposed a localized enhancement of the resistivity in a highly conducting fluid. Thus, the diffusion region, where the assumption of an ideal conducting plasma does not hold, becomes much smaller than the scale size of the system (the black area in the middle panel of Fig. 1). Petschek showed that the diffusion region is the source for large amplitude MHD waves. In this model, most of the inflowing plasma crosses the shock waves and not the diffusion region anymore. From this point of view, the shock waves transport information into a wide region of ideal plasma, allowing for much higher reconnection rates than in the Sweet–Parker model. It was also shown by simulations using resistive MHD equations that the Sweet–Parker configuration appears for a spatially constant plasma resistivity, while Petschek–type structures arise for a strongly localized plasma resistivity [5, 6]. This insight is also supported by numerical simulations [7]. The newly formed region bounded by the shock waves is called outflow region, where the inflowing plasma is heated and accelerated, leading to the formation of bi–directional jets [8] and fast plasma flows, which are important features indicating the occurrence of magnetic reconnection. Also the connection of the magnetic field lines takes places over the outflow region (lower panel of Fig. 1).

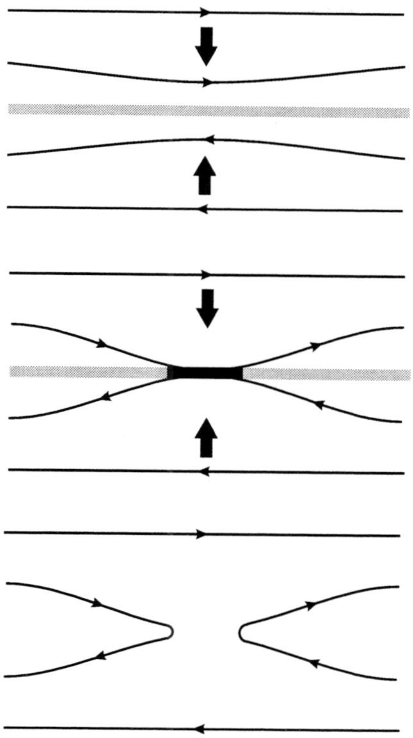

Figure 1. Different stages during the process of magnetic reconnection. The upper panel shows the initial configuration of antiparallel magnetic fields separated by a thin current sheet (dashed area). If the magnetic fields are convected towards each other, strong gradients occur, leading to the formation of a diffusion region (black area), where the frozen–in condition is not valid anymore (middle panel). In this region, the magnetic field lines can be broken and reconnected (lower panel). This leads to a connection between plasma regions which were previously separated.

In the last century, many sites were discovered where magnetic reconnection plays a crucial role [9]. It is observed in laboratory fusion machines like tokamaks [10, 11]. In order to investigate the fundamental physics involved in magnetic reconnection in laboratory settings, different devices like the Magnetic Reconnection Experiment (MRX) device were constructed [12]. It revealed the appearance of different shapes of the diffusion region as well as the importance of the shear angle between the magnetic field lines and the influence of boundary effects.
Moving away from the Earth, other plasma environments suitable for reconnection can be found. The first environment where magnetic reconnection was expected is the Sun. The earliest models [13, 2] have aimed at explaining the large amount of energy released in solar flares. Reconnection on the Sun is inferred from different observations, including high speed flows of X–ray and EUV features, and energetic particle events in hard X–ray images [14, 15], which have been observed by several spacecrafts like Yohkoh, SOHO, and TRACE. The application of Petschek–like reconnection to solar flares gives typical recon-

nection times consistent with observations [16]. Reconnection has also been proposed as a mechanism for the heating of solar and stellar coronae to extremely high temperatures of more than 10^6 K [17]. The importance of reconnection in plasma environments outside of our solar system is still under debate. The two astrophysical topics to which reconnection theory has been extensively applied are stellar flares [18] and accretion disks [19]. Reconnection theory also enters into models of protostellar collapse, extragalactic jets, galactic magnetic fields, and even galactic clusters [9].

However, the only place in space were magnetic reconnection can be observed in–situ is the terrestrial magnetosphere. At the dayside magnetopause, energy and plasma from the solar wind can enter the terrestrial magnetosphere because of the reconnection of the interplanetary magnetic field (IMF) and the Earth's magnetic field [20]. Of particular interest is the transient feature of magnetic reconnection, leading to the formation of so–called flux transfer events (FTEs), which are flux tubes connecting the solar wind plasma with the magnetospheric plasma. The first in-situ measurements of flux transfer events caused by transient reconnection were reported by Russell and Elphic in 1978 [21] for the dayside magnetopause. They defined FTEs as signatures with a bipolar variation of the magnetic field component perpendicular to the current sheet, a deflection of the tangential component, and an increase in the magnetic field strength. In the same year, Haerendel et al. [22] found indications for reconnection processes also at the high–latitude dayside magnetopause. In the following years, a collection of studies supported the interpretation that FTEs are burst of magnetic reconnection at the magnetopause appearing most frequently during intervals of southward directed IMF [23, 24]. However, newer observations with the Cluster satellites show that for a northward–orientated IMF, reconnection can appear tailward of the cusp [25]. There are also strong indications that reconnection at the dayside magnetopause can be continuous over an extended period of time [26].

Magnetic flux reconnected at the dayside is convected into the tail lobes, where reconnection is initiated in a region between 15 and 30 R_e in the magnetotail [27, 28]. It was proposed by Sergeev et al. [29] that FTE–like structures appear also in the Earth's magnetotail. They performed a qualitative comparison between measurements of ISEE–1 and model predictions from a transient reconnection model, which showed a general agreement. In later work, these events were referred to as nightside flux transfer events (NFTEs) [30]. These are short–term events in the substorm–time plasma sheet, which can be described by impulsive variations of the reconnection rate in models of transient reconnection. They are characterized, similar to FTEs, as a bipolar variation of the B_z–component and a deflection of the x–component of the magnetic field. Additionally, the velocity shows an upward flow of plasma in the beginning, followed by a strong flow directed downward to the current sheet. Such structures noticed in the tail plasma sheet are often referred to as individual bursts of BBF, as transient plasma sheet expansions, as plasmoids or flux ropes, as well as NFTEs [30, 31, 32]. The appearance of reconnection structures in the magnetotail was also confirmed by several other satellite missions, like Geotail [33] or Cluster [34, 35].

The aim of this paper is to show recent evidence for magnetic reconnection from Cluster measurements and introduce a model for the reconstruction of the reconnection rate from satellite measurements. In Section 2, multi–spacecraft measurements of the Cluster mission are discussed with emphasize to the implications for magnetic reconnection. The theoretical model used to analyze NFTEs in the terrestrial magnetotail is described in Section 3. Section 4 is devoted to the reconstruction of the reconnection rate from Cluster measurements. Also the site of reconnection with respect to the satellite is estimated.

2. Observational Features of Magnetotail Reconnection

Whereas the dayside magnetosphere is compressed by the solar wind, the nightside magnetosphere is stretched out into a long magnetotail. When the interplanetary magnetic field (IMF) is southward, the merged field lines at the dayside are transported towards the nightside, reconnected in the distant tail and then transported back towards the dayside, creating inward and Earthward flows in the central plasma sheet. The flux transport rates at the dayside and at the nightside are, however, balanced only in an average sense and the unbalanced transport is the ultimate cause of a substorm when magnetic reconnection takes place closer to the Earth, at radial distance of 15 - 30 R_E in the tail. The processes associated with this near-Earth reconnection determine different scales of magnetotail dynamics.
Cluster traversed the magnetotail covering regions Earthward of 19 R_E during the past summer seasons since July 2001. The tetrahedron scale was between 250 km and 4000 km, which was changed every year, so that characteristics at different scales could be identified. The four spacecraft observations enable us for the first time to differentiate spatial from temporal disturbances. In this section, we highlight several Cluster observations when Cluster was located near the X-line and more remote observations showing the effects of the processes in the ion diffusion region.

2.1. Observation near the X-line

The four spacecraft observations enable us to obtain spatial gradients of different parameters continuously. For example, the current density obtained from the gradient in the magnetic field is an essential parameter for magnetotail diagnostics. Using the gradient information obtained from the four Cluster spacecraft during current sheet crossings, current sheet structures in the vicinity of X-line were reconstructed by Runov et al. [34] as illustrated in Fig. 2. A thin current sheet with a half-thickness of about one ion gyro radius was found for the crossing closest to the X-line, whereas the outer crossings showed bifurcated current sheet profiles. Changes in the curvature of the field for the different current sheet crossings illustrated in the figure were consistent with a X-line motion from Earthward and tailward of the spacecraft. Furthermore, consistent features of the field disturbance associated with Hall-current at both side of X-line were identified, confirming that the spacecraft traversed the ion diffusion region.

The importance of multi-scale processes from heavy ions to electron kinetics in the reconnection processes have been identified by several Cluster observations. During thin current sheet intervals associated with crossings of the X-line during storm-time substorms, the O^+ pressure and density were observed to be dominated compared to those of H^+ [36]. In such current sheets, the O^+ were observed to execute Speiser-type serpentine orbits across the tail and were found to carry about 5-10% of the cross-tail current [36]. The O^+ in the reconnection region was suggested to experience a ballistic acceleration [37] based on the observation of a large amplitude bipolar electric field. These large-amplitude

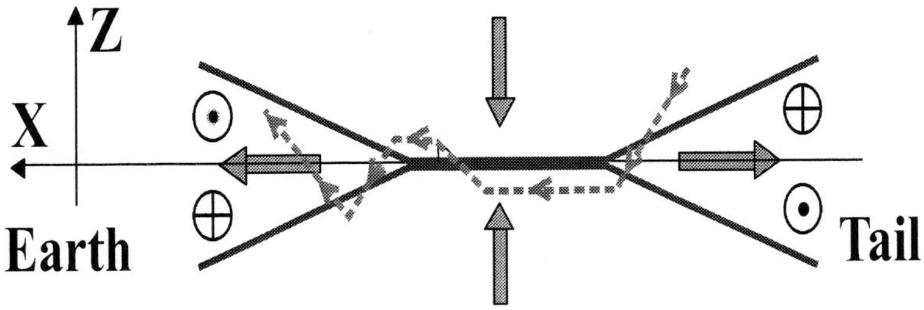

Figure 2. Sketch of Cluster observations of Hall magnetic fields and current sheet structure around the reconnection region; the red dashed line indicates the trajectory of Cluster (after Runov et al. [34]).

(up to 50 mV/m) solitary waves, identified as electron holes, have been observed during several plasma sheet encounters that have been identified as the passage of a magnetotail reconnection X-line [38]. The electron holes were seen near the outer edge of the plasma sheet, within and on the edge of a density cavity, at distances in the order of a few ion inertial lengths from the center of the current sheet. Based on a detailed comparison with simulations, Cattell et al.[38] suggested that the observed nonlinear wave modes, electron holes, may play an important role in reconnection by scattering and energizing electrons.

2.2. Observations Away from the Reconnection Site

Whereas in situ observations of the ion diffusion region depend on the rare chance that the spacecraft being located at the right place at the right time, different effects of reconnection can be detected remotely from the reconnection site and can still contain useful information on temporal and spatial characteristics of reconnection. For example, the limited area of the Hall currents flowing in the ion diffusion region indicates that the current has to be closed including regions outside of the ion diffusion region. At the lobe side, the closure of the Hall current takes place via cold electron flowing into the ion diffusion region. At the outflow region, on the other hand, the accelerated electrons can carry the current into the ion diffusion region. Cluster observed such field-aligned electrons related to the Hall current system consistent with previous observations, but also obtained fine structures indicating multiple temporal or spatial properties of reconnection [39, 40, 41]. Fig. 3 illustrates multi-point observations by Cluster in the northern hemisphere and by Geotail in the southern hemisphere detecting these field aligned currents possibly connected to the ion diffusion region from Nakamura et al. [39]. The four Cluster satellites also allowed to determine the scale size of these field aligned currents, suggesting that the scale size of the downward current was at maximum comparable to the ion inertia length so that it plausibly connects to the near-Earth X-line and is driven by Hall effects in the reconnection region. This interhemispheric observation supported the theoretical prediction [42] that the downward current region is thin because on the lobe side, the ions may travel a substantial fraction of the ion inertia length until their motion separates from that of the electrons.

Using multi-composition plasma observation by Cluster, slow-mode shocks connected

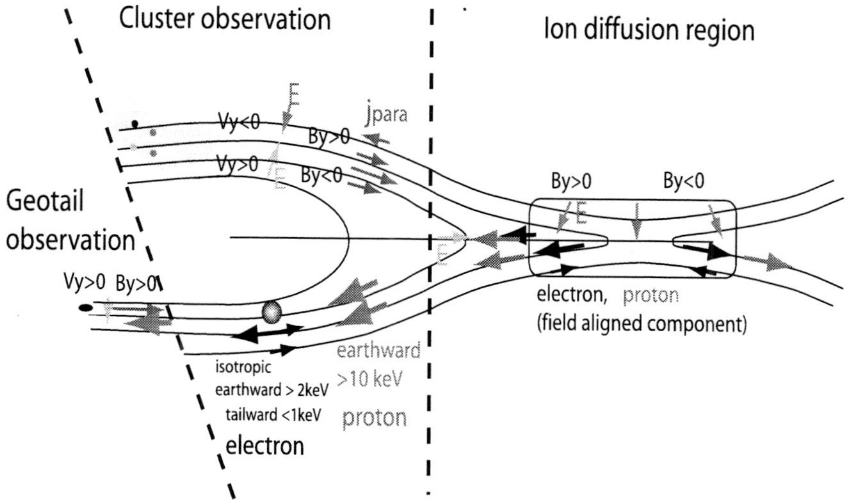

Figure 3. Summary of observations by Cluster and Geotail during a transient entry into the plasma sheetd during a substorm event and illustration of the possible relationship to the reconnection region. For Cluster observations field signatures are showing in the northern hemisphere, while particle signatures are illustrated in the southern hemisphere; adapted from [39].

to the ion diffusion region have been analyzed by Eriksson et al. [43] taking into account also the contribution from oxygen ions during a substorm X-line event when Cluster observed fast tailward and Earthward flows. The successful joint Walén and slow shock analyses on the tailward flows within the plasma sheet presented further evidence in favor of Petschek-type reconnection at distances $X_{GSM} >$ -19 R_E of the near-Earth magnetotail.

Cluster succeeded to obtain detailed characteristics of Earthward propagating southward then northward magnetic field disturbances related to plasmoids/flux ropes [44, 45], travelling compression regions [46, 47], and nightside flux transfer events [48, 49, 35]. Multi-point analysis by Cluster were used to measure the current density and check the force-free model [46] and energetic particle boundaries [45] to show the structures of the plasmoid/flux rope. Yet, since the plasma flows jetting towards the Earth are significantly influenced by the strong dipolar field and pressure gradient, it still remains unknown to what extent these structures can be treated as motions of a stable structure in the analysis. As described in the later sections, similar magnetic features were rather interpreted as transient profiles of a remote X-line due to its change in the reconnection rate and used to determine the location of the X-line [48, 49, 35]. Observationally, the determination of the field topology would be a key to differentiate whether these structure are coming from a single X-line or are signatures of multiple X-lines.

In the plasma sheet, bursty bulk flows (BBF) are the most clear signatures of the consequence of reconnection and have been intensively studied using Cluster spacecraft measurements based on multi-point data analysis. A statistical analysis was performed by Nakamura et al. [50] to estimate the typical scale size of BBF. Different types of magnetic field disturbances accompanied by the BBF were studied such as dipolarization [51], low frequency wave activity [52], and turbulence [53]. Using Cluster, together with the other spacecraft and ground-based observations, BBF-associated field-aligned and ionospheric current system have been obtained [54, 55].

3. The Petschek–Type Model of Magnetic Reconnection

In the following, we present a model to reconstruct the reconnection rate developed by Semenov et al. [67] for an incompressible plasma extended to a compressible one. First we show the solution of the direct problem, giving the behavior of NFTEs for a prescribed reconnection rate. To reconstruct the reconnection rate, we solve the inverse problem. This enables us to achieve the reconnection rate from magnetic field measurements.

3.1. Basic Configuration

In order to describe the temporal evolution of FTEs, time–dependent Petschek–type models of reconnection were developed [1, 56, 57, 58, 59]. For our purpose, we consider the structure of the Earth's magnetotail as oppositely directed magnetic fields, which are separated by a current sheet, the so–called plasma sheet. In terms of ideal MHD, this current sheet can be described as an infinitely thin tangential discontinuity. The normal components of the magnetic field and plasma velocity are zero ($B_n = 0$, $v_n = 0$) across a tangential discontinuity, while the density and the tangential components of the magnetic field and velocity can change arbitrarily, subject only to the requirement that the total pressure, which is the sum of the thermal and the magnetic pressure, $p + B^2/(8\pi)$, stays constant [60]. Due to these requirements, there is no mass flow and no magnetic connection across a tangential discontinuity. As a result, there will be no electric field component along the discontinuity ($E_t = 0$).
Due to the local enhancement of the plasma resistivity, a tangential electric field is generated, causing a normal component of the magnetic field. Thus, the conservation laws for a tangential discontinuity are violated, and the surface gets non–linearly unstable, decaying into a system of MHD wave modes according to the general Riemann problem [61, 60]. The most general scheme of a decay looks like $S^+(R^+)AS^-(R^-)C(T)S^-(R^-)AS^+(R^+)$, where S^+, S^- are fast and slow shocks, R^+, R^- are fast and slow rarefaction waves, and A, C, T are Alfvén, contact, and tangential discontinuities, respectively. If we assume that the total pressure is conserved in lowest order, no strong fast shocks are present in the system. For symmetric conditions, no contact discontinuity and no rarefaction waves will appear. Due to this process, reconnection–associated disturbances start to propagated along the current sheet in the form of pairs of shocks. Until the reconnection electric field ceases, the shocks are detached to the reconnection site forming a teardrop–shaped region of heated and accelerated plasma, the so–called outflow regions (shaded areas in Fig. 4).

After the electric field has dropped to zero, the shocks detach from the reconnection site and continue to propagate along the current sheet (Fig. 4). These propagating shocks are causing disturbances in the ambient plasma environment, which can be measured by a satellite. In the following, we will show how these measured disturbances can be calculated by using a time–dependent Petschek–type model of transient magnetic reconnection.

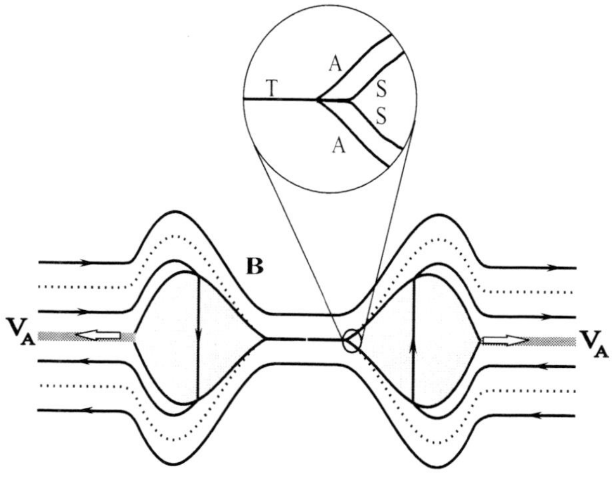

Figure 4. Shocks propagating away from the reconnection site with local Alfv én velocity. The small figure shows the internal structure of the shocks according to the general Riemann problem, where the tangential discontinuity decays into a slow shock S and an Alfvén wave A. The shaded areas are the so–called outflow regions.

3.2. MHD Treatment of the Problem for a Compressible Plasma

After the local reconnection of magnetic flux according to the Petschek-type model of magnetic reconnection, the plasma and magnetic field behavior in the outer regions can be described within the frame of ideal magnetohydrodynamics [1, 61],

$$\frac{\partial \rho}{\partial t} + \nabla \cdot (\rho \, \mathbf{v}) = 0 \, , \tag{1}$$

$$\frac{\partial \, (\rho \, \mathbf{v})}{\partial t} + \nabla \cdot \left[\rho \, \mathbf{v} \, \mathbf{v} + p \, \mathbf{I} - \frac{1}{4\,\pi} \left(\mathbf{B} \, \mathbf{B} - \frac{B^2}{2} \, \mathbf{I} \right) \right] = 0 \, , \tag{2}$$

$$\frac{\partial}{\partial t} \left(\frac{\rho \, v^2}{2} + \rho \, e + \frac{B^2}{8\,\pi} \right) + \nabla \cdot \left[\rho \, \mathbf{v} \left(\frac{v^2}{2} + e + \frac{p}{\rho} \right) + \frac{1}{4\,\pi} \mathbf{B} \times (\mathbf{v} \times \mathbf{B}) \right] = 0 \, , \tag{3}$$

$$\frac{\partial \mathbf{B}}{\partial t} + \nabla \cdot (\mathbf{B} \, \mathbf{v} - \mathbf{v} \, \mathbf{B}) = 0 \, . \tag{4}$$

These equations describe the conservation of mass (Eq. 1), momentum (Eq. 2), energy (Eq. 3), and magnetic flux (Eq. 4), respectively. ρ, $\mathbf{v}$, $\mathbf{B}$, p, and e are the mass density, the plasma velocity, the magnetic field, the isotropic pressure, and the internal energy, respectively, and $\mathbf{I}$ is the unit dyadic.

These ideal MHD equations can be linearized with respect to constant background quantities, in the following denoted by superscript (0). The linearization allows for the description of all first order quantities (indicates by superscript (1)) in terms of a displacement vector ξ,

$$\mathbf{v}^{(1)} = \left(\frac{\partial}{\partial t} + \mathbf{v}^{(0)} \cdot \nabla \right) \xi \, , \tag{5}$$

$$\mathbf{B}^{(1)} = \mathbf{B}^{(0)} \cdot \nabla \xi - \mathbf{B}^{(0)} \nabla \cdot \xi \,, \tag{6}$$

$$\rho^{(1)} = -\rho^{(0)} \nabla \cdot \xi \,, \tag{7}$$

$$p^{(1)} = c_S^2 \rho^{(1)} \,, \tag{8}$$

$$P^{(1)} = -\rho^{(0)} \left[u^2 \nabla \cdot \xi + (\mathbf{v}_A \cdot \nabla) \, \mathbf{v}_A \cdot \xi \right] \,. \tag{9}$$

Here, $P^{(1)}$ is the first order total pressure, while $\mathbf{v}_A$ and c_S are the Alfvén velocity and the sound speed, respectively, and $u^2 = v_A^2 + c_S^2$.
Substituting these expressions (Eq. 5–9) into the linearized MHD equations, applying a Laplace transformation with respect to time t, and a Fourier transformation with respect to the spatial coordinate x, the z–component of the displacement vector $\zeta(p, k, z)$ fulfills the ordinary differential equation

$$\frac{\partial^2}{\partial z^2} \zeta - q^2(p, k) \zeta = 0 \,, \tag{10}$$

with

$$q^2 = \frac{p^4 + u^2 k^2 p^2 + v_A^2 c_s^2 k^4}{p^2 u^2 + v_A^2 c_s^2 k^2} \,. \tag{11}$$

The solutions of Eq. 10 which satisfy the first order total pressure balance at $z = 0$ and vanish at infinity are

$$\zeta(p, k, z) = \frac{\tilde{L}}{L + \tilde{L}} Q(p, k) \mathrm{e}^{-qz} \qquad z > 0 \,, \tag{12}$$

$$\tilde{\zeta}(p, k, z) = -\frac{L}{L + \tilde{L}} Q(p, k) \mathrm{e}^{qz} \qquad z < 0 \,, \tag{13}$$

where

$$L(p, k) = -\rho^{(0)} \frac{p^2 + v_A^2 k^2}{q(p, k)} \,. \tag{14}$$

Tilde indicates the same function evaluated for the lower half space. Because of the finite boundary layer at $z = 0$, the source function $Q(p, k) = \zeta(p, k, 0) - \tilde{\zeta}(p, k, 0)$ is non–zero and can be found from the solution of the Riemann problem [1].
The expression for the displacement vector in the coordinate–time space can be found by using the so–called Cagniard–deHoop method [62, 63, 64, 59]. This method allows an analytical solution of the inverse Fourier transformation. The inverse Laplace transformation can be rewritten leading to an expression for the displacement vector as a convolution integral of the reconnection rate and an integration kernel in time, which can be solved numerically. After this procedure, the z–component of the displacement vector in coordinate–time space is found as

$$\zeta(x, z, t) = \frac{B}{\pi} \int_C \Re \frac{\tilde{L}}{L + \tilde{L}} Q(s) F(t - \tau(s)) d\tau \,, \tag{15}$$

where $F(t)$ is the reconnected flux, $s = k/p$ and the integration is performed in the complex plane along a path C where the integration variable τ is real. From Eqs. 5–9 it is possible to derive the coordinate–time expressions for all relevant MHD parameters, like the pressure,

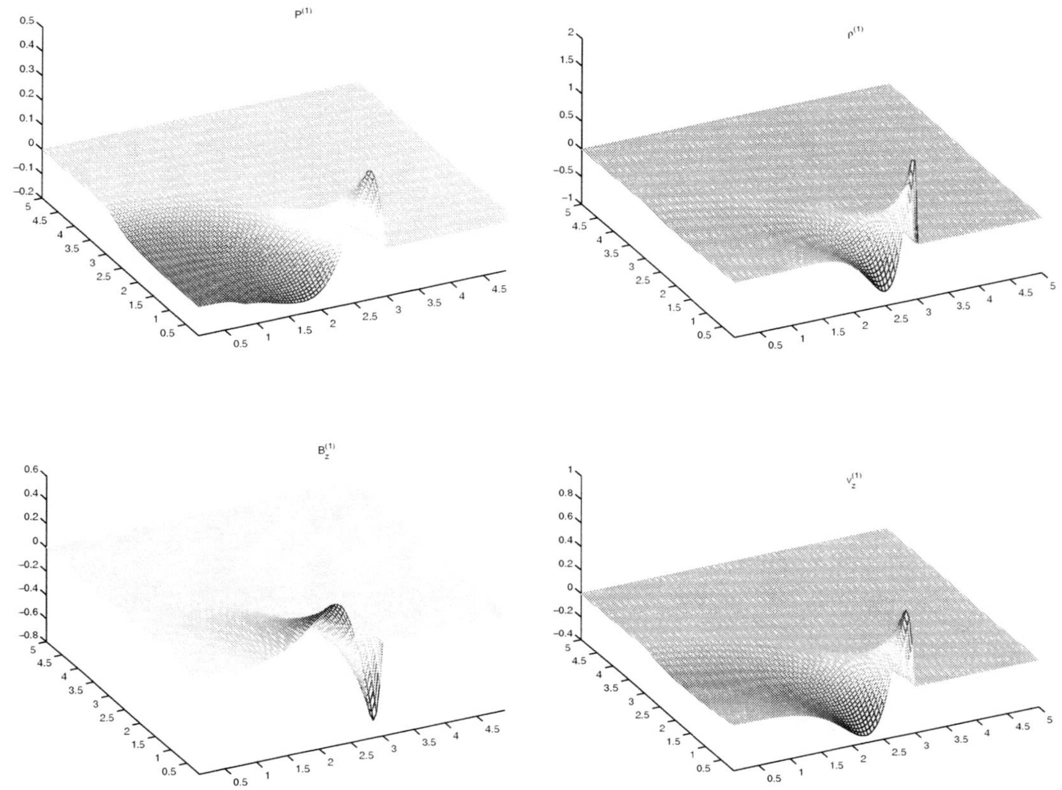

Figure 5. Variations of the total pressure (upper left panel), the plasma density (upper right panel), the z–component of the magnetic field (lower left panel), and the z–component of the flow velocity (lower right panel) for the upper half plane. The magnetic fields are $B = 1$ and $\tilde{B} = -1$, $t = 3$, and $\beta = 0.1$.

the plasma density, the magnetic field and the flow velocity for a given reconnection rate. The variation of the z–component of the magnetic field, which is used in the following for the reconstruction method, is given as

$$B_z^{(1)}(x, z, t) = \frac{B^2}{\pi} \int_C \Re \frac{\tilde{L}}{L + \tilde{L}} i\, s\, Q(s)\, E(t - \tau(s))\, d\tau\,. \qquad (16)$$

The expressions for the other quantities have a similar mathematical structure. In Fig. 5, the variations of the main MHD parameters are shown for fixed time ($t = 3$) in the x–z–plane. The reconnection rate is assumed to be

$$E(t) = \frac{b^2 e^2}{20} t^2 e^{-b\,t}\,,$$

where $b = 4$.

It is possible to simulate magnetic field measurements along a satellite trajectory by using Eq. 16. Fig. 6 shows a time series of $B_z^{(1)}$ measured by a satellite fixed in space. The characteristic asymmetric bipolar variation expected for the B_z–component is clearly visible,

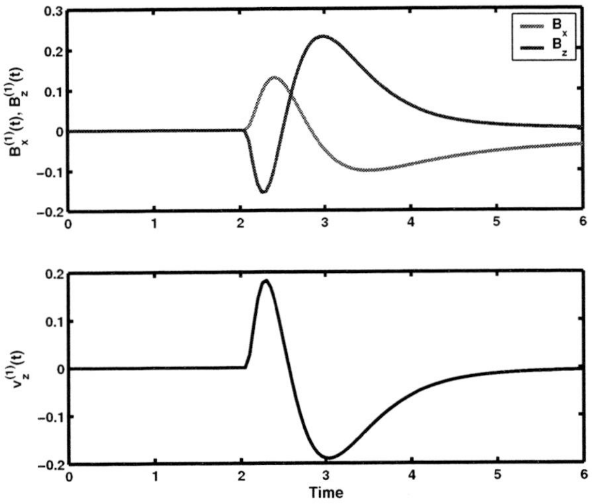

Figure 6. The normal (B_z) and the tangential (B_x) component of the magnetic field (upper panel) and the plasma flow velocity normal to the current sheet (v_z; lower panel). All characteristic features expected for NFTEs, like the bipolar variation of B_z and the anticorrelated behavior of v_z can be seen.

and also the deflection of the x–component of the magnetic field. Additionally, the velocity behaves as proposed for NFTEs, with a upward flow of plasma followed by a strong flow directed downward to the plasma sheet.

4. Solution of the Inverse Problem

Since we are interested to reconstruct the reconnection rate $E(t)$ from satellite measurements of the magnetic field components, e.g., $B_z(t)$, we have to solve an inverse problem. An inverse problem can be seen as a model of a phenomenon characterized by φ, which belongs to a certain space of models P. Furthermore, let u be the observed indirect attributes of the phenomenon with $u \in U$. An operator A relates the two values by

$$A\varphi = u \,. \tag{17}$$

As a rule, the φ attributes cannot be directly observed.
The main job when solving an inverse problem is to find out, whether the chosen model is compatible with the experimental data. The mathematical difficulty of solving such problems is that the inverse operator A^{-1} defined throughout its domain $AP \subset U$ is not continuous. Hence there is the conventional division into the classes of well–posed and ill–posed problems. A well–posed problem, as defined by Hadamard [65], must meet the requirements that:

1. for each data set u, Eq. 17 is solvable in the entire space U,

2. the solution is unique,

3. the solution is stable, meaning that small perturbations of u result in small perturbations of the solution φ.

If a problem does not meet one or more of these criteria, the problem is considered to be ill–posed. These kind of problems can be solved by using the theory of regularization, which we apply in the following.
To reconstruct the reconnection electric field from the magnetic field, we apply a Laplace transformation to the convolution integral (16) giving

$$B_z^{(1)}(p) = K(p)E(p)\,, \tag{18}$$

where $K(p)$ represents the kernel of the convolution integral (for convenience, the factor B^2/π in front of the integral in Eq. 16 is put into $K(p)$). Thus, the reconnection electric field is given as $E(p) = B_z^{(1)}(p)/K(p)$. From this expression it is obvious that we are faced with an ill–posed inverse problem. If $K(p) \to 0$, requirement 1 is not fulfilled. Additionally, if $K(p) \sim 0$, small perturbations in the measured data $B_z^{(1)}(p)$ will cause large perturbations in the result $E(p)$.
To avoid this problem, we introduce a regularization parameter $M(p)$ [66] so that

$$E(p) = \frac{B_z^{(1)}(p)}{K(p) + M(p)}\,. \tag{19}$$

The regularization operator $M(p)$ is defined as [67]

$$M(p) = \begin{cases} 0 & |p| < R_{max} \\ \infty & |p| > R_{max} \end{cases}\,. \tag{20}$$

This operator does not influence the electric field for small values of p, but when the functions tend to zero for large p ($p > R_{max}$), the denominator goes to infinity, so that the reconnection electric field is zero in Laplace space and large oscillations are suppressed. The value of R_{max} is found from internal parameters of the numerical Laplace transformation.

5. Reconstruction of the Reconnection Rate and Site

After the observation of FTE signatures, several attempts were made to reconstruct and analyze different features of the reconnection process involved in the generation of FTEs. Southwood [68] predicted that FTE signatures would be observed by a satellite regardless of whether or not it actually penetrates the FTE. This suggestion was verified by Farrugia et al. [69], who reproduced the magnetic field signatures outside the flux tube by considering the flow of an inviscid, incompressible plasma over a semi-circular cylinder. In a new approach, Walthour et al. [70, 71] used a method based on integral transforms for inferring the cross–sectional size, shape, and the speed of propagation of a thin, infinitely long obstacle corresponding to a flux tube.
Another approach to this topic was used by several authors [72, 73, 74] who developed a method based on the Grad–Shafranov equation to reconstruct two-dimensional space

plasma structures in magnetohydrostatic equilibrium. Hu and Sonnerup [74] applied this model to two magnetopause crossings by the spacecraft AMPTE/IRM, and reconstructed magnetic field structures. Sonnerup et al. [75] used this method to give a rough estimation of the reconnection rate. However, the propagation of a reconnected flux tube can rather be described by the inertia force of the plasma and the Maxwellian tension, which is essentially a time–dependent process, and therefore this process can be hardly understood in a magnetohydrostatic approach.
Recently, different models based on multi–spacecraft measurements were developed to determine the position of the reconnection site with respect to the satellite. Wild et al. [76] used data from Cluster and Geotail positioned at the high– and the low–latitude dayside magnetopause. They consider the motion of reconnected flux tubes away from a user–defined location and compare the flux tube motion with the observed FTE–like signatures. In another approach, Fuselier et al. [77] determine the reconnection inflow velocity into the magnetosphere, and use Cluster and IMAGE data to estimate the approximate reconnection site. A third approach to reconstruct the reconnection rate, which was discussed above, was developed by Semenov et al. [67] for an incompressible plasma. It was applied to Cluster measurements in the magnetotail by Semenov et al. [48] and Penz et al. [49]. In the following, three applications of this method are presented.

5.1. Cluster Observations on September 8th, 2002

On September 8th, 2002, a favorable constellation of multiple spacecraft and ground–based observations allowed to reconstruct the time sequence of a substorm and to model the near-Earth magnetic configuration [35]. Between 20 and 23 UT, an isolated substorm with a peak AE of about 400 nT occurred. A clear growth phase was observed after the arrival of a southward–orientated IMF after 20 UT. The auroral breakup, the intensification of a westward electrojet, and Pi2 pulsations consistently indicated the expansion phase onset at 21:18 UT in the 22–24 MLT sector. The Cluster tetrahedron was centered in the middle of the magnetotail at $[-16.7; 0.2; 4.5]$ R_e GSM. The satellites exited from the thinning plasma sheet shortly after 21:00 UT. Therefore, they were located outside of the plasma sheet at the time of interest. After 21:17 UT, a series of Earthward propagating 1 minute scale variations of the magnetic field and plasma flow components consistent with the picture of multiple NFTEs/flux ropes were observed (Fig. 7). The first NFTE appeared at about 21:17 UT and propagated Earthward at a speed v_x=625 km/s; v_y=-72 km/s (determined from timing of magnetic variations, [35]). After this NFTE, the plasma sheet continued to be thin for some 20 min until the transient plasma sheet expansions start to be observed. This is a favorable situation, because if the plasma sheet is thin, the approximation of a tangential discontinuity as an initial state is better justified.

Fig. 7 shows the measured B_x, B_z, and v_z from the four Cluster satellites during the time interval from 21:16 to 21:26 UT. The GSM magnetic field data are obtained from the fluxgate magnetometer (FGM) experiment [78] with 1 s time resolution. The O^+ moments with 4 s time resolution were measured by the Composition and Distribution Function Analyser (CODIF) of the Cluster Ion Spectrometry (CIS) experiment [79] observed at the Cluster spacecrafts. The O^+ data was only used if the O^+ density exceeded 0.005 cm^{-3}.

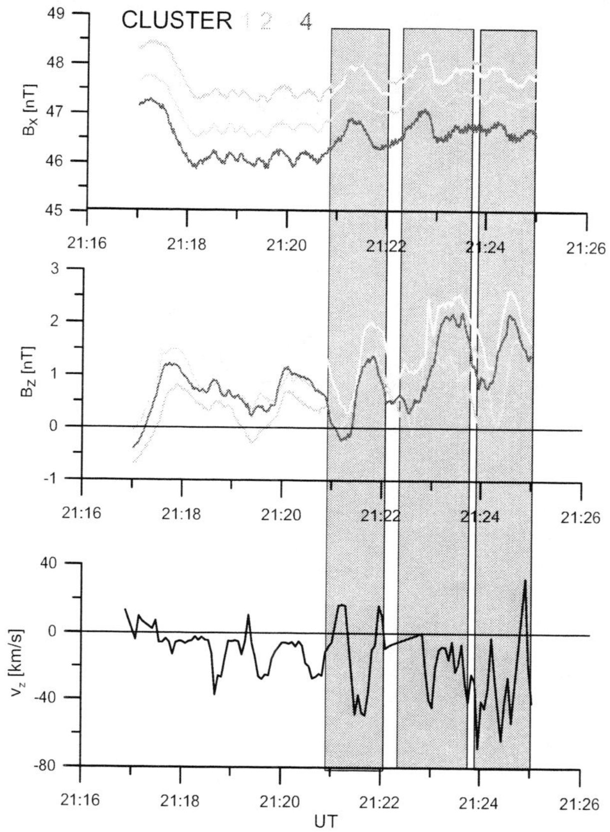

Figure 7. Cluster measurements of B_x (upper panel), B_z (middle panel), and v_z (lower panel). We analyze the NFTEs occurring at 21:21 UT, 21:22:30 UT, and 21:24 UT (shaded areas) [48].

We applied our model to the NFTEs starting at 21:21 UT, 21:22:30 UT, and 21:24 UT, ignoring the first NFTE at 21:17 UT (in which the interaction of reconnected flux tube with previously closed plasma sheet flux tubes should be more pronounced than in the following development). A comparison with the results of the theoretical model (Fig. 6) shows that the expected features for the perturbations are indeed found in the observations : the asymmetric bipolar variation of B_z, a compression of B_x, as well as a plasma flow v_z of cold O^+ ions directed to the plasma sheet are clearly visible.

5.1.1. Reconstruction of the Events on September 8th, 2002

To evaluate the integration kernel $K(p)$ we also need to know the spacecraft location with respect to the reconnection site (Fig. 8). Fortunately, the actual z–position of the neutral sheet (about +1 R_e in z–direction) is known from the modeling made in [35]. Therefore, the z–distance between the satellite and the reconnection site is approximately 3.5 R_e.

The determination of the x–distance is done by using a global minimization routine. In

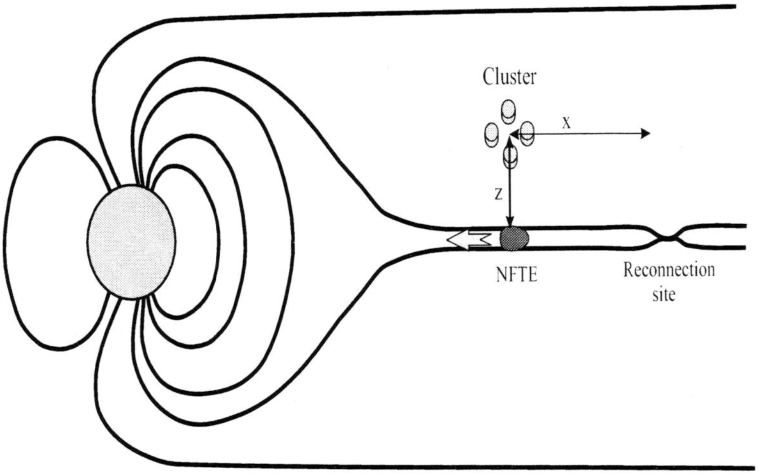

Figure 8. The position of the Cluster satellites with respect to the reconnection site. The z–distance is found from the modeling done by Sergeev et al. [35], while the x–distance is reconstructed by using a minimization routine.

our time–dependent model this is possible, because the shape of the shock is changing if the x–distance from the reconnection site is increasing. We use the measured B_z component as an input data, calculate the electric field in Laplace space, which should be strictly positive, but since we do not know the x–distance it can appear negative somewhere. Therefore, we take the module of the electric field and recalculate $\tilde{B}_z$ out of it. This procedure can be summarized as

$$B_z(t) \Rightarrow B_z(p) \Rightarrow E(p) \Rightarrow E(t) \Rightarrow |E(t)| \Rightarrow \tilde{B}_z(t)\,.$$

Then we minimize the difference between B_z(t) and $\tilde{B}_z$(t) with a least square approach in order to find the x–distance as the value of x where the difference between the initial and the reconstructed magnetic field has a minimum. We limit the search to distances less than 35 R_e, which corresponds to the region of the near Earth neutral line (NENL), where reconnection most likely takes place. The local velocity of the disturbances is determined by using multipoint timing analysis [80], giving about 700 km/s. We assume that this velocity is approximately the Alfvén velocity. Additionally, this analysis shows that the propagation velocity is directed mainly Earthward in x–direction with a small y–component, which is a preferable configuration for our 2-D model. Now we know all quantities used for the normalization. Thus, we are able to analyze the events by using the method presented in the previous sections.
The reconnection electric field reconstructed for the NFTE starting at 21:21 UT in Fig. 7 is 1.1 mV/m over a time period of about 50 s from C1. C2 and C3 give a reconnection rate of 1.0 mV/m, while C4 gives 1.1 mV/m (Fig. 9).

The distance x between C1 and the reconnection site is 8.6 R_e, so that the reconnection site is located at 24.7 R_e in the magnetotail (Fig. 10). For C2, $x = 8.6$, giving 24.9 R_e as the location of the reconnection site. From C3 x is found to be 8.4, giving the reconnection

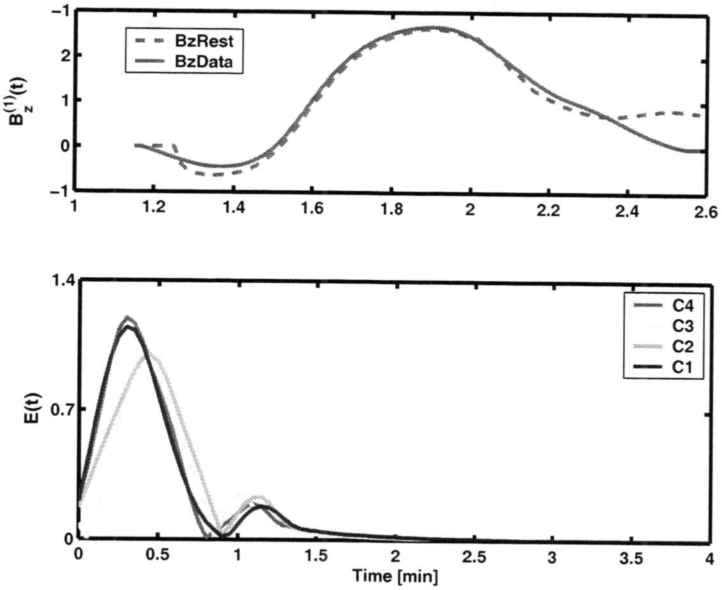

Figure 9. The lower panel shows the reconstructed reconnection rate from the four Cluster satellites, which has an amplitude of 1–1.1 mV/m and a duration of about 50 s. The upper panel shows a comparison between the measured $B_z^{(1)}$ and the restored $B_z^{(1)}$ from the reconstructed reconnection rate for C4.

site at 24.6 R$_e$. C4 gives $x = 7.6$, corresponding to a reconnection site at 24.3 R$_e$. One can see that the reconstructed amplitude of the reconnection electric field and the reconnection site are consistent among the four satellites. For example, C3 is located about 0.5 Re lower than C2, therefore the observed amplitude in the magnetic field perturbation at C3 is 1.7 nT compared with 1.3 nT at C2. Yet, the reconstructed amplitude of the electric field is quite similar, i.e., 1.0 mV/m from C3 and from C2. The same can be found for the x–distance, where C4 is located closest to the reconnection site, and therefore the model gives the smallest distance between the satellite and the reconnection site. Therefore our model shows that the observed features at all four satellites are associated with the same reconnection event.

Analysis of the event starting at 21:22:30 UT was done only with C2 and C4, since the other two satellites did not observe pronounced NFTE–like signatures. The reconnection rates are found to be 0.95 and 1.0 mV/m for C2 and C4, respectively. C2 gives $x = 8.6$, leading to a distance of 24.9 R$_e$ for the reconnection site. From C4, x is found to be 8.25 R$_e$, giving a distance of 24.8 R$_e$ in the magnetotail. For the event starting at 21:24 UT, data from C2, C3, and C4 are used. The reconnection rate is found to be 0.95 mV/m (C2), 0.91 mV/m (C3), and 0.9 mV/m (C4). According to these data, the reconnection site is located at 24.9 R$_e$ (C2), 23.9 R$_e$ (C3), and 24.5 R$_e$ (C4) in the magnetotail.

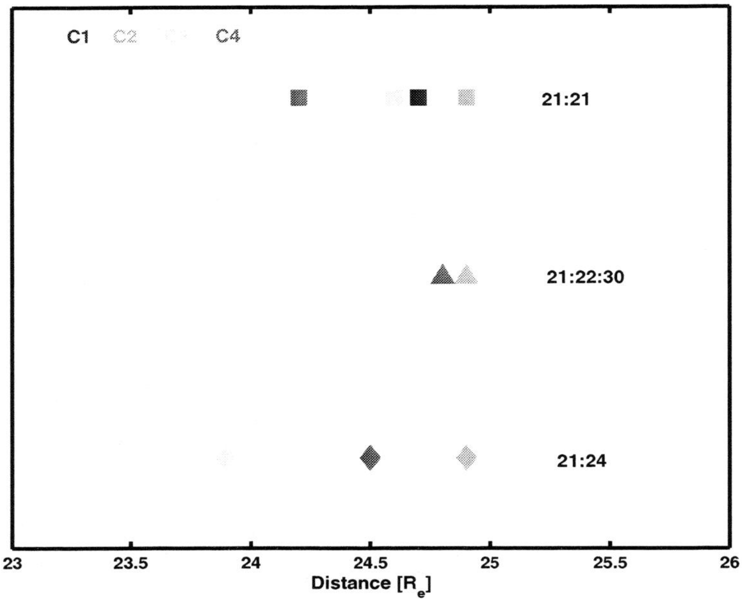

Figure 10. Location of the reconnection site reconstructed for the NFTE starting at 21:21 UT (squares), 21:22:30 UT (triangles), and 21:24 UT (diamonds) from the Cluster satellites.

6. Conclusion

We present a model to reconstruct the reconnection rate and the reconnection site from satellite measurements in a compressible plasma. As already predicted [48], the reconnection rate is smaller than inferred from the incompressible model. Also the distance between the reconnection site and the satellite decreased, while the duration of the pulses increase. All these features are consistent with the qualitative estimates done by Semenov et al. [48]. However, a significant reduction of the noise resulting from the solution of the inverse problem is achieved compared with the incompressible model [48]. Also the variations of the reconstructed reconnection rate and site between the satellites decreased.

Several simplifying assumptions were made to allow a treatment of the problem within an analytical model:

- x-distance: The x–distance between the satellite and the reconnection site is determined by using a minimization routine. There exists the possibility that more that one minimum occurs, meaning that there is no single solution for the problem considered. To avoid this problem, we applied our method only to the range of x–distances, where reconnection most likely takes place, namely to distances less than 35 R_e, and run it with different starting points.

- Homogeneous background density : We assume that there is a homogeneous background density in the magnetotail. If the density changes significantly between the point of observation and the starting point of the disturbances, this will influence the propagation speed of the perturbations and may modify the results.

- Alfvén velocity: We assume that the perturbations in the magnetic field are moving approximately with Alfvén velocity. If the Alfvén velocity is not constant between the satellite and the reconnection site, the results will change. Since the Alfvén velocity is used for the normalization of the length scales, a variation of the Alfvén velocity will also give a variation of the spatial distances. If the Alfvén velocity decreases, also the length scales will decrease.

However, we found that the variations observed by the Cluster spacecrafts outside of the plasma sheet boundary are consistent with the impulsive model of magnetic reconnection used in this work. The characteristic bipolar variation of the B_z–component and the anti-correlated behavior of the perpendicular plasma flow velocity are reproduced. Application to NFTEs occurring on September 8th, 2002, gives an amplitude of the reconnection electric field in the range of 0.9–1.2 mV/m. This amplitude of the reconnection electric field is consistent with estimations of the magnetotail reconnection rate obtained from ground–based measurements [81, 82]. The time duration of the reconnection pulse is in the order of 50 s, while the reconnection site is located at about 23.9–24.9 R_e tailwards.

Acknowledgements

The authors thank A. Runov for fruitful discussions. This work is supported by the RFBR grants No. 04–05–64935 and No. 03–05–20012 BNTS, by the Austrian "Fonds zur Förderung der wissenschaftlichen Forschung" under project P17099–N08, by project No. I.12/04 from the "Österreichischer Austauschdienst", and by INTAS Grant Nr. 03–51–3738. V.S.S. acknowledges financial support from the Technical University Graz and the hospitality of the Institut für Theoretische Physik during scientific visits to Graz. T.P. acknowledges support by the "Österreichische Forschungsgemeinschaft" and by a Paul-Urban scholarship from the Institute of Physics of the University of Graz during visits in St. Petersburg. V.V.I. acknowledges support by INTAS "Fellowship Grant for Young Scientists" Nr. 04–83–3816, and by the Nansen community. Also acknowledged is support by the Austrian Academy of Sciences, "Verwaltungsstelle für Auslandsbeziehungen".

References

[1] Semenov, V. S.; Heyn, M. F.; Ivanov, I. B. *Phys. Plasmas* 2004, 11, 62–70.

[2] Sweet, P. A. In *Electromagnetic Phenomena in Cosmic Physics*; Lehnert, B.; Ed.; Cambridge University Press: London, UK, 1958, pp 123–134.

[3] Parker, E. N. *Astrophys. J. Suppl.* 1963, 8, 177–211.

[4] Petschek, H. E. In *The Physics of Solar Flares*; Hess, W. N.; Ed.; NASA Spec. Publ.: Washington, DC, USA, 1964; SP–50, pp 425–440.

[5] Erkaev, N. V.; Semenov, V. S.; Jamitzky, F. *Phys. Rev. Lett.* 2000, 84, 1455–1458.

[6] Erkaev, N. V.; Semenov, V. S.; Alexeev, I. V.; Biernat, H. K. *Phys. Plasmas* 2001, 8, 4800–4809.

[7] Birn, J.; Drake, J. F.; Shay, M. A.; Rogers, B. N.; Denton, R. E.; Hesse, M.; Kuznetsova, M.; Ma, Z. W.; Bhattacharjee, A.; Otto, A.; Pritchett, P. L. *J. Geophys. Res.* 2001, 106, 3715–3720.

[8] Phan, T. D. et al. *Nature* 2000, 404, 848–850.

[9] Priest, E.; Forbes, T. *Magnetic Reconnection: MHD Theory and Applications* ; Cambridge University Press: Cambridge, UK, 2000.

[10] Paré, V. K. In *Magnetic reconnection in space and laboratory plasmas* ; Hones, E. W.; Ed., Geohysical Monograph: Washington, DC, USA, 1984; 30, pp 341–346.

[11] Yamada, M.; Levinton, F. M.; Pomphrey, N.; Budny, R.; Manickam, J.; Nagayama, Y. *Phys. Plasmas* 1994, 1, 3269–3276.

[12] Yamada, M.; Ji, H.; Hsu, S.; Carter, T.; Kulsrud, R.; Bretz, N.; Jobes, F.; Ono, Y.; Perkins, F. *Phys. Plasmas* 1997, 4, 1936–1944.

[13] Parker, E. N. *J. Geophys. Res.* 1957, 62, 509–520.

[14] Scholer, M. *Lecture Notes in Physics* 2003, 612, 9–27.

[15] Biernat, H. K.; Semenov, V. S.; Penz, T.; Miklenic, C.; Veronig, A.; Hanslmeier, A.; Vrsnak, B.; Heyn, M. F.; Ivanov, I. B.; Ivanova, V. V.; Kiehas, S.; Langmayr, D. *Hvar Obs. Bullet.* 2006, in press.

[16] Uzdensky, D. A. *Astrophys. J.* 2003, 587, 450–457.

[17] Axford, W. I.; McKenzie, J. F. *Astrophys. Space Sci.* 1996, 243, 1–4.

[18] Mullan, D. J. In *The M–type stars*; Johnson, J. R.; Querci, F. R.; Ed.; NASA Spec. Publ: Washington, DC, USA 1986; SP–492, pp 455–479.

[19] Verbunt, F. *Space Sci. Rev.* 1982, 32, 379–404.

[20] Cowley, S. W. H. In *Magnetic Reconnection in Space and Laboratory Plasmas* ; Hones, E. W.; Ed.; Geohysical Monograph: Washington, DC, USA, 1984; 30, pp 375–378.

[21] Russell, C. T.; Elphic, R. C. *Space Sci. Rev.* 1978, 22, 681–715.

[22] Haerendel, G.; Paschmann, G.; Sckopke, N.; Rosenbauer, H.; Hedgecock, P. C. *J. Geophys. Res.* 1978, 83, 3195–3216.

[23] Rijnbeek, R. P.; Cowley, S. W. H.; Southwood, D. J.; Russell, C. T. *J. Geophys. Res.* 1984, 89, 786–800.

[24] Kawano, H.; Russell, C. T. *J. Geophys. Res.* 1997, 102, 11307–11314.

[25] Twitty, C.; Phan, T. D.; Paschmann, G.; Lavraud, B.; Rème, H.; Dunlop, M. *Geophys. Res. Lett.* 2004, 31, L19808, doi:10.1029/2004GL020646.

[26] Frey, H. U.; Phan, T. D.; Fuselier, S. A.; Mende, S. B. *Nature* 2003, 426, 533–537.

[27] Walker, R. J.; Ogino, T.; Raeder, J.; Ashour–Abdalla, M. J. *Geophys. Res.* 1993, 98, 17235–17250.

[28] Lopez, R. E.; Benitez–Marquez, E.; Wiltberger, M. J.; Lyon, J. G.; Figueroa, R. *Adv. Space Res.* 2003, 31, 1167–1176.

[29] Sergeev, V. A.; Semenov, V. S.; Sidneva, M. V. *Planet. Space Sci.* 1987, 35, 1199–1212.

[30] Sergeev, V. A.; Elphic, R. C.; Mozer, F. S.; Saint–Marc, A.; Sauvaud, J. A. *Planet. Space Sci.* 1992, 40, 1551–1572.

[31] Ieda, A.; Machida, S.; Mukai, T.; Saito, Y.; Yamamoto, T.; Nishida, A.; Terasawa, T.; Kokubun, S. J. *Geophys. Res.* 1998, 103, 4453–4465.

[32] Slavin, J. A.; Lepping, R. P.; Gjerloev, J.; Fairfield, D. H.; Hesse, M.; Owen, C. J.; Moldwin, M. B.; Nagai, T.; Ieda, A.; Mukai, T. J. *Geophys. Res.* 2003, 108, 1015, doi:10.1029/2002JA009557.

[33] Nagai, T.; Shinohara, I.; Fujimoto, M.; Hoshino, M.; Saito, Y.; Machida, S.; Mukai, T. J. *Geophys. Res.* 2001, 106, 25929–25950.

[34] Runov, A.; Nakamura, R.; Baumjohann, W.; Treumann, R. A.; Zhang, T. L.; Volwerk, M.; Vörös, Z.; Balogh, A.; Glassmeier, K.–H.; Klecker, B.; Rème, H.; Kistler, L. *Geophys. Res. Lett.* 2003, 30, 1579, doi:10.1029/2002GL016730.

[35] Sergeev, V. A.; Kubyshkina, M. V.; Baumjohann, W.; Nakamura, R.; Amm, O.; Pulkkinen, T.; Angelopoulos, V.; Mende, S. B.; Klecker, B.; Nagai, T.; Sauvaud, J.–A.; Slavin, J. A.; Thomsen, M. F. *Ann. Geophys.* 2005, in press.

[36] Kistler, L.; Mouikis, C.; Moebius, E.; Klecker, B.; Sauvaud, J.–A.; Rème, H.; Korth, A.; Marocci, M.; Lundin, R.; Parks, G.; Balogh, A. J. *Geophys. Res.* 2005, 110, A06213, doi:10.1029/2004JA010653.

[37] Wygant, J. R.; Cattell, C. A.; Lysak, R.; Song, Y.; Dombeck, J.; McFadden, J.; Mozer, F. S.; Carlson, C.; Parks, G.; Lucek, E. A.; Balogh, A.; Andre, M.; Rème, H.; Hesse, M.; Mouikis, C. J. *Geophys. Res.* 2005, submitted.

[38] Cattell, C.; Dombeck, J.; Wygant, J.; Drake, J. F.; Swisdak, M.; Goldstein, M. L.; Keith, W.; Fazakerley, A.; Andre, M.; Lucek, E.; Balogh, A. J. *Geophys. Res.* 2005, 110, A01211, 10.1029/2004JA010519.

[39] Nakamura, R.; Baumjohann, W.; Nagai, T.; Fujimoto, M.; Mukai, T.; Klecker, B.; Treumann, R. A.; Balogh, A.; Rème, H.; Sauvaud, J.–A.; Kistler, L.; Mouikis, C.; Owen, C. J.; Fazakerley, A.; Dewhurst, J.; Bogdanova, Y. J. *Geophys. Res* 2004, 109, A05204, doi:10.1029/2003JA010174.

[40] Alexeev, I. V.; Owen, C. J.; Fazakerley, A. N.; Runov, A.; Dewhurst, J.; Balogh, A.; Rème, H.; Klecker, B.; Kistler, L. *Geophys. Res. Lett.* 2005, 32, L03101, doi:10.1029/2004GL021420.

[41] Asano, Y.; Nakamura, R.; Baumjohann, W.; Runov, A.; Vörös, Z.; Volwerk, M.; Zhang, T. L. *Geophys. Res. Lett.* 2005, 32, L03108, doi:10.1029/2004GL021834.

[42] Treumann, R. A.; Jaroschek, C. H.; Nakamura, R.; Runov, A.; Scholer, M. *COSPAR Colloquia Series*, 2005, in press.

[43] Eriksson, S.; Oieroset, M.; Baker, D.; Mouikis, C.; Vaivads, A.; Dunlop, M.; Rème, H.; Ergun, R.; Balogh, A. J. *Geophys. Res.* 2004, 109, A05212, doi:10.1029/2003JA010534.

[44] Slavin, J. A.; Lepping, R.; Gjerloev, J.; Fairfield, D.; Acuna, M.; Goldstein, M.; Balogh, A.; Dunlop, M.; Kivelson, M.; Khurana, K.; Fazakerley, A.; Owen, C. J.; Rème, H.; Bosqued J. *Geophys. Res. Lett.* 2003, 30, 1362, doi:10.1029/2003GL016411.

[45] Zong, Q.–G.; Fritz, T. A.; Pu, Z. Y.; Fu, S. Y.; Baker, D. N.; Zhang, H.; Lui, A. T.; Vogiatzis, I.; Glassmeier, K.–H.; Korth, A.; Daly, P. W.; Balogh, A.; Rème, H. *Geophys. Res. Lett.* 2004, 31, L18803, doi:10.1029/2004GL020692.

[46] Slavin, J. A.; Owen, C. J.; Dunlop, M. W.; Boralev, E.; Moldwin, M. B.; Sibeck, D. G.; Tanskanen, E.; Goldstein, M. L.; Fazaderley, A.; Balogh, A.; Lucek, E.; Richter, I.; Rème, H.; Bosqued, J. *Geophys. Res. Lett.* 2003, 30, 2208, doi:10.1029/2003GL018438.

[47] Owen, C. J.; Slavin, J. A.; Fazakerley, A. N.; Dunlop, M. W.; Balogh, A. *Geophys. Res. Lett.* 2005, 32, L03104, 10.1029/2004GL021767.

[48] Semenov, V. S.; Penz, T.; Ivanova, V. V.; Sergeev, V. A.; Biernat, H. K.; Nakamura, R.; Heyn, M. F.; Kubyshkin, I. V.; Ivanov, I. B. *J. Geophys. Res.* 2005, 110, A11217, doi:10.1029/2005JA011181.

[49] Penz, T.; Semenov, V. S.; Ivanova, V. V.; Biernat, H. K.; Sergeev, V. A.; Nakamura, R.; Kubyshkin, I. V.; Ivanov, I. B.; Heyn, M. F. *Adv. Space Res.* 2006, in press.

[50] Nakamura, R.; Baumjohann, W.; Mouikis, C.; Kistler, L. M.; Runov, A.; Volwerk, M.; Asano, Y.; Vörös, Z.; Zhang, T. L.; Klecker, B.; Rème, H.; Balogh, A. *Geophys. Res. Lett.* 2004, 31, L09804, doi:10.1029/2004GL019558.

[51] Nakamura, R.; Baumjohann, W.; Klecker, B.; Bogdanova, Y.; Balogh, A.; Rème, H.; Bosqued, J.-M.; Dandouras, I.; Sauvaud, J.–A.; Glassmeier, K.–H.; Kistler, L.; Mouikis, C.; Zhang, T. L.; Eichelberger, H.; Runov, A. *Geophys. Res. Lett.* 2002, 29, 1942, doi:10.1029/2002GL015763.

[52] Volwerk, M.; Nakamura, R.; Baumjohann, W.; Treumann, R. A.; Runov, A.; Vörös, Z.; Zhang, T. L.; Asano, Y.; Klecker, B.; Richter, I.; Balogh, A.; Rème, H. *J. Geophys. Res.* 2003, 108, 1429, doi:10.1029/2003JA010155.

[53] Vörös, Z.; Baumjohann, W.; Nakamura, R.; Runov, A.; Zhang, T. L.; Volwerk, M.; Eichelberger, H.; Balogh, A.; Horbury, T. S.; Glassmeier, K.–H.; Klecker, B.; Rème, H. *Ann. Geophys.* 2003, 21, 1955–1964.

[54] Grocott, A.; Yeoman, T.; Nakamura, R.; Cowley, S. W. H.; Frey, H.; Rème, H.; Klecker, B. *Ann. Geophys.* 2004, 22, 1061–1075.

[55] Nakamura, R.; Amm, O.; Laakso, H.; Draper, N.; Lester, M.; Grocott, A.; Klecker, B.; McCrea, I.; Balogh, A.; Rème, H.; Andre, M. *Ann. Geophys.* 2005, 99, 553–566.

[56] Biernat, H. K.; Heyn, M. F.; Semenov, V. S. *J. Geophys. Res.* 1987, 92, 3392–3396.

[57] Semenov, V. S.; Kubyshkin, I. V.; Lebedeva, V. V.; Rijnbeek, R. P.; Heyn, M. F.; Biernat, H. K.; Farrugia, C. J. *Planet. Space Sci.* 1992, 40, 63–87.

[58] Semenov, V. S.; Kubyshkin, I. V.; Rijnbeek, R. P.; Biernat, H. K. In *Physics of magnetic reconnection in high–temperature plasmas*; Ugai, M.; Ed.; Research Signpost: Trivandrum, India, 2004, pp 35–68.

[59] Heyn, M. F.; Semenov, V. S. *Phys. Plasmas* 1996, 3, 2725–2741.

[60] Heyn, M. F.; Biernat, H. K.; Rijnbeek, R. P.; Semenov, V. S. *J. Plasma Physics* 1988, 40, 235–252.

[61] Akhiezer, A. I.; Akhiezer, I. A.; Polovin, R. V.; Sitenko, A. G.; Stepanov, K. N. *Plasma Electrodynamics; Pergamon Press*: Oxford, UK, 1975.

[62] Lamb, H. *Philos. Trans. R. Soc. London Ser. A* 1904, 203, 1–42.

[63] Cagniard, L. *Reflexion et Refraction des Ondes Seismiques Progressive*; Gauthier–Villard: Paris, France, 1939; English translation by Flinn, E. A.; Dix, C. H. *Reflection and refraction of progressive seismic waves*; McGraw–Hill: New York, USA, 1962.

[64] deHoop, A. T. *Appl. Sci. Res. B* 1960, 8, 349–356.

[65] Hadamard, J. *Le problème de Cauchy et les équations aux dérivées partielles Lineaires Hyperboliques*; Hermann: Paris, France, 1932.

[66] Tikhonov, A. N.; Arsenin, V. Y. *Solutions of ill-posed problems*; John Wiley: New York, USA, 1977.

[67] Semenov, V. S.; Penz, T.; Heyn, M. F.; Ivanov, I. B.; Kubyshkin, I. V.; Biernat, H. K.; Ivanova, V. V. In *Solar–planetary relations*; Biernat, H. K.; Lammer, H.; Vogl, D. F.; Mühlbachler, S.; Eds.; *Research Signpost*: Trivandrum, India, 2005, pp 261–302.

[68] Southwood, D. J. *Adv. Space Res.* 1985, 5, 7–14.

[69] Farrugia, C. J.; Elphic, R. C.; Southwood, D. J.; Cowley, S. W. H. *Planet. Space Sci.* 1987, 35, 227–240.

[70] Walthour, D. W.; Sonnerup, B. U. Ö.; Paschmann, G.; Lühr, H.; Klumpar, D.; Potemra, T. J. *Geophys. Res.* 1993, 98, 1489–1504.

[71] Walthour, D. W.; Sonnerup, B. U. Ö.; Elphic, R. C.; Russell, C. T. *J. Geophys. Res.* 1994, 99, 8555–8563.

[72] Hau, L.–N.; Sonnerup, B. U. Ö. *J. Geophys. Res.* 1999, 104, 6899–6918.

[73] Hu, Q.; Sonnerup, B. U. Ö. *Geophys. Res. Lett.* 2001, 28, 467–470.

[74] Hu, Q.; Sonnerup, B. U. Ö. *J. Geophys. Res.* 2003, 108, 1011, doi:10.1029/2002JA009323.

[75] Sonnerup, B. U. Ö.; Hasegawa, H.; Paschmann, G. *Geophys. Res. Lett.* 2004, 31, L11803, doi:10.1029/2004GL020134.

[76] Wild, J. A.; Milan, S. E.; Cowley, S. W. H.; Bosqued, J. M.; Rème, H.; Nagai, T.; Kokubun, S.; Saito, Y.; Mukai, T.; Davies, J. A.; Cooling, B. M. A.; Balogh, A.; Daly, P. W. *Ann. Geophys.* 2005, 23, 445–460.

[77] Fuselier, S. A.; Trattner, K. J.; Petrinec, S. M.; Owen, C. J.; Rème, H. *J. Geophys. Res.* 2005, 110, A06212, doi:10.1029/2004JA010805.

[78] Balogh, A. et al. *Ann. Geophys.* 2001, 19, 1207–1217.

[79] Rème, H. et al. *Ann. Geophys.* 2001, 19, 1303–1354.

[80] Harvey, C. C. In *Analysis methods for multi–spacecraft data* ; Paschmann, G.; Daly, P. W.; Eds.; ISSI SR–001, 1998.

[81] Blanchard, G. T.; Lyons, L. R.; de la Beaujardière, O.; Doe, R. A.; Mendillo, M. J. *Geophys. Res.* 1996, 101, 15265–15276.

[82] Østgaard, N.; Moen, J.; Mende, S. B.; Frey, H. U.; Immel, T. J.; Gallop, P.; Oksavik, K.; Fujimoto, M. *Ann. Geophys.* 2005, 23, 123–134.

In: Space Science: New Research
Editor: Nick S. Maravell, pp. 311-327
ISBN: 1-60021-005-8

Chapter 10

RESEARCH ON AERODYNAMICS OF LARGE BOLIDES

V.P. Stulov
Institute of Mechanics Moscow Lomonosov State University, Moscow

Abstract

Crushing and evaporation are basic physical processes at the entrance of natural space bodies of size 0.01-100 m into atmospheres of planets. The relative role of evaporation is characterized by value of mass loss parameter, which is proportional to the relation of kinetic energy of a body mass unit to effective enthalpy of evaporation. Examples of real large bolides for various values of this parameter are given. The offered general approach helps to understand a plenty of observant data of a different level of reliability, and also allows to separate preliminary real hypotheses from fantastic.

Introduction

Modern researches of meteors and bolides reveal two limiting cases of these phenomena. In the first case the question is the physics of meteors generated by particles of the small sizes, less than 1 mm. This case is investigated most in details. The entrance of such particles is accompanied by their full combustion in the high layers of atmosphere, at altitudes of 80-100 and more kilometers. These events occur often enough. There is an extensive observant material and numerous analytical researches. The last are reduced, basically, to attempts to reproduce a curve of luminosity of a meteor along its trajectory to determine observable parameters: mass of a meteoric particle, structure and other physical and chemical parameters. There is a significant progress in this area. However some authors mark{celebrate} essential divergences of the theory and observations (see, for example, Lebedinets, 1980).

In other limiting case of very much meteoric body sizes, presence of atmosphere is not considered as essential. There is a collision of a body with a planet, giving numerous consequences, elementary of which is formation of craters on the solid surface. Recently the similar phenomena are investigated in connection with their probable influence on formation of planets at their early stages.

In the given work the intermediate situation, i.e. an entrance into atmosphere of planets of natural space bodies in the size in a range of 0.01-100 m, is considered. These phenomena occur essentially less often, than micrometeors. However the significant observant material recently has collected. There is a hope for its updating in connection with development of observant technical equipment. As stimulus to studying this material serves not only cleanly scientific interest to natural phenomena, in particular, to properties of material bodies in a space, but also improvement of estimations of so-called asteroid hazard which reality increases in process of development of the civilization.

We study bolides using methods and results of hypersonic aerodynamics as a meteoroid moves in atmospheric gas in the conditions corresponding{meeting} the fluid dynamics. It has appeared that not only the body size, but also speed of entrance and thermo-physical properties of the body can determine substantially the type of interaction with an atmosphere and the basic consequences.

The elementary theory of meteors shows that the solution for trajectories depends on two dimensionless parameters α and β (Stulov, Mirskii, Vislyi, 1995)

$$\alpha = \frac{1}{2} c_d \frac{\rho_0 h_0 S_e}{M_e \sin\gamma}, \beta = \frac{c_h V_e^2}{2 c_d H^*}. \tag{1}$$

In the solution of the trajectory equations a corner of trajectory γ, coefficients of drag c_d and heat exchange c_h, and also effective enthalpy of destructions H^* are constant. In formulas (1), speed V_e, mass of body M_e and the area of middle section S_e at the entrance into atmosphere (index *«e»*), and also height of a homogeneous atmosphere h_0 and density of gas on sea level ρ_0 are present. Parameter α characterizes intensity of braking as it is equal to the relation of mass of the atmosphere column with cross-section section S_e and heght h_0 to the body mass. Parameter β is proportional to the relation of kinetic energy of a body mass unit to effective enthalpy of evaporation.

The obvious formula for a trajectory containing α and β allows to estimate easily a relative role of evaporation and braking. Also it is easy to determine at what stage of body movement in atmosphere its crushing takes place. It has been made earlier in work of the author (Stulov, 1998a). It has been shown that for large bodies ($\alpha \leq 1$) in a wide range of parameter β values ($\beta \geq 1$) process of the body evaporation precedes its braking. For stone bodies crushing accompanies evaporation; more fragile ice bodies at enough big speed of entrance can start to be splitted up even prior to the beginning of heating and evaporation.

The relative role of crushing and evaporation can be characterized in value of parameter β. Small β correspond to entrance enough heat-resistant objects with rather small speeds. If solidity of the body is not large there is its crushing on set of fragments, at rather small role of mass loss due to blowing awat a liquid film and evaporation of the body material on its frontal surface. As a result this case corresponds to fall of numerous fragments on the planet surface with formation of meteoric and crater fields.

At moderate β, roles of crushing and evaporation during meteoroid interaction with atmosphere are comparable. Depending on morphological properties of a body, crushing can occur or on set of fragments in one or several stages, or gradually by consecutive separation of small fragments from the parental body. Fine fragments are braked more quickly and lag behind from large. Because of it a role of evaporation on fine fragments fast becomes secondary.

At last, great values β correspond to primary gasification of the body, or as single object, or in the form of a cloud of fragments. Thus evaporation (and crushing) takes place at rather early stage of entrance into atmosphere, so braking of meteoroid as a solid body is insignificant. Asymptotic solution (Stulov, 1994) of the trajectory equations for $\beta >> 1$ shows that evaporation comes to the end practically at the speed equal to speed of entrance.

Below examples of widely known real falling, corresponding three described above ranges of parameter β, are given. They are the widely known Sikhote-Aline meteoric rain (S.-A.m., 12.02.1947, Fesenkov, 1951), the Beneshov bolide, recorded by the stations of Czech Republic which are the part of the European bolide network (B., 07.05.1991, Spurný, 1994) and, at last, falling the Tunguska space body (TSB, 30.06.1908, Korobeinikov, Tshushkin, Shurshalov, 1991), investigated up and down, however, till now causing serious disagreements and discussions, including messages on detection of fragments of an other planet spacecraft (the program of telechannel TVC, August, 2004) . In brackets after the name of each event, there are specified: an abbreviation of the name used further, date of event and the reference. Under the reference we shall give here a following explanation. Scientific (and pseudo-scientific) literature on each event is rather extensive. Here there are those papers from which the author took the numerical data used below.

In table 1, values of β for each of three events are given. Below for some of initial numbers, corresponding sources are specified. Calculated values of c_h and β are received by means of the approximate method, in detail described in the monography (Stulov, Mirskii, Vislyi, 1995). As already it was specified above (after formulas (1)), the coefficient of heat exchange c_h in the table is considered to be constant. It was estimated on parameters of entrance into atmosphere and characteristic height of real event (destruction, flash, etc.). Calculated values of c_h depend on speed of entrance V_e, the initial characteristic size of body R and density of air.

Table 1.

Event	V_e, km/s	H^*, 10^3J/g	c_h, estimated	c_h, calculated[5]	β, estimated	β, calculated[5]
S.-A.m.	14.5[1]	8[4]	0.010	0.018	0.13	0.23
B.	21.18[2]	2[5]	0.012[2]	0.018	1.35	2.03
TSB	35[3]	2[5]	0.100	0.090	30.63	27.57

In the table there are used the following references: 1 (Fesenkov, 1951), 2 (Spurný, 1994), 3 (Korobeinikov, Tshushkin, Shurshalov, 1991), 4 (Chyba Ch. F., Thomas P.J., Zahnle K.J., 1993), 5 (Stulov, Mirskii, Vislyi, 1995). In all cases it is accepted $c_d = 1$. From the table follows, that calculated and estimated values of β are close, anyway, both calculated, and estimated values of β for three real events differ approximately on the order of value.

Thus, if parameter α is responsible for intensity of braking at passage of atmosphere, parameter β characterizes a role of thermochemical destruction and considerably influences the basic consequences of the meteoroid entrance into atmosphere. Mechanical crushing takes place almost always. Therefore stated above judgement about parameters α and β concern both to initial meteoroid, and to its fragments.

Crushing of Meteoric Bodies

There are various approaches to the solution of the problem of meteoric bodies crushing under action of aerodynamic loading. In the elementary cases semi-empirical rules of separation of an initial parental body on fragments are offered. More complex approach assumes the solution of the problem of the theory of solidity at variable external loading. We shall discuss more in detail these two points of view.

Earlier three models for calculation of destruction and movement of fragments in an atmosphere have been offered: slow consecutive crushing (Baldwin, Sheaffer, 1971), instant destruction (Grigoryan, 1979), fast consecutive crushing (Ivanov, Рыжанский, 1997). We had been constructed analytical solutions for calculation of trajectories according to these models (Stulov, 1998b; Stulov, Stulov, 1999; Stulov, Titova, 2001a.). Then comparison of calculated trajectories on these models for specific variants of falling (Stulov, Titova, 2001b) has been given.

The ideas incorporated in the models, only partially reflect the real phenomena. So, analysis of Sikhote-Aline falling has shown (Krinov, 1975; Krinov, Tsvetkov, 1979) that destruction of the space body occured in three discrete stages. Besides, small flat fragments were separated from large fragments during movement. Apparently, this small fragments were observed by eyewitnesses in the form of sparks. Further, the description of in detail documentary Beneshov bolide (Borovicka, Spurný, 1996) testifies that in the beginning on the certain site of its trajectory ($24 < h < 42$ km) from a parental body were consistently separated rather fine fragments; in the end of this site has occured the explosion-like destruction, accompanied almost instant going out the bolide, i.e. fast braking of products of destruction.

Apparently from this brief description, analytical models can be used for the description of separate stages of real falling.

Attempts to calculate destruction of a meteoric body in atmosphere by statement and solution of solidity problems have been undertaken. The equations of separation dynamics for meteoric body have been constructed and solved in the work of Grigoryan, 1979. It was shown that the braked and collapsing object transfers its energy to atmospheric gas volume with cross-section sizes considerably exceeding the initial size of the meteoric body.

Korobeinikov and co-authors (1994) have calculated the intense condition of a meteoric body in process of introduction into atmosphere. It has appeared that the point of the maximal pressure appears inside of the body. Gradually the area of breaking points extends down to achievement of a surface of the body, then its disintegration begins. Authors emphasize necessity of the account of initial heterogeneity and casual defects for a meteoric body.

The rests of meteoroid after destruction move in the form of a swarm of fragments. Below some results of numerical modeling the initial stage of this movement (Zhdan, Stulov, Stulov, 2004a; Zhdan, Stulov, Stulov, 2004b) are given. The stage is characterized by intensive aerodynamic interaction of elements in the swarm.

To study this question the model of a supersonic flow around a system of small number of spheres has been used. The purpose of calculations is definition of aerodynamic coefficients of spheres in the swarm and studying of changes of shock wave configurations depending on distance between bodies. These data promote understanding mechanics of fragments scattering after destruction of a body in a supersonic stream.

Here as an example the supersonic flow of three configurations is considered: 1 - two spheres with a line of the centers across a stream, 2 - three spheres in apexes of an equilateral triangle, 3 - six spheres in apexes and the middle of the sides of an equilateral triangle; in cases 2 и 3 the side of thr triangle is perpendicular to the stream. The centers of spheres lay in one plane containing a vector of initial flow velocity. As the size of configurations the size h, equal to half of distance between the nearest points of two next spheres, related to radius of sphere R is used. Comparison of aerodynamic properties of three simple configurations probably will allow to draw the general conclusions about systems of bodies in more complex cases, including both for an irregular arrangement of bodies, and in case of some changes of their form. All the calculations presented here are lead for systems of bodies in a stream of the perfect gas with parameters of M = 6, $\gamma = 1.4$.

Aerodynamic coefficient of drag c_x and cross-section force c_y were calculated under following formulas:

$$c_x = \frac{p_\infty}{0.5\rho_\infty V_\infty^2 S}\int_\Sigma p\cos\theta d\sigma,$$

$$c_y = \frac{p_\infty}{0.5\rho_\infty V_\infty^2 S}\int_\Sigma p\sin\theta d\sigma. \qquad (2)$$

Here $S = \pi R^2$ is the area of middle sections of the sphere, $\rho_\infty V_\infty^2/p_\infty = \gamma M^2$, pressure p in the integral is related to pressure in the initial stream p_∞, corner θ is counted in the meridional plane from a forward point of sphere

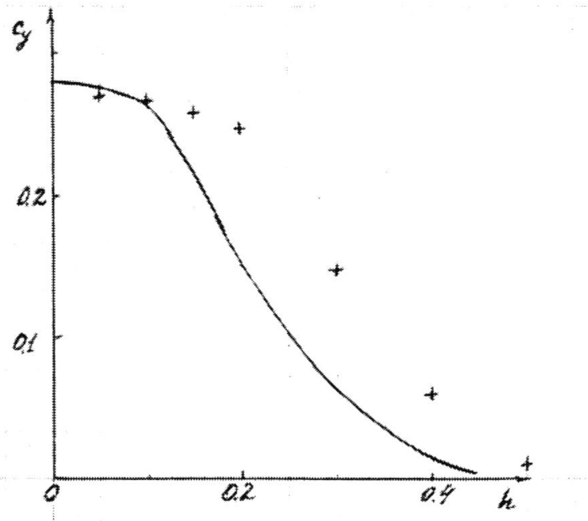

Figure 1. Cross-force coefficient for two spheres with the line of centers across the stream (Mach number M = 6) and two semi-cylinders (Shuvalov, Artemieva,Trubetskaja, 2000).

The supersonic flow of system 1 was studied earlier as the elementary model of fragments scattering a meteoric body after its destruction. In work (Passey, Melosh, 1980) it was offered a rather rough model of a divergence of two fragments in cross direction to a stream. According to this model a constant force acts on two fragments. The force is equal to product of stagnation pressure and middle section, i.e. one has $c_y = 2$. From formulas of uniformly accelerated movements, value of cross-section speed on distance of the order of the characteristic size of a fragment is received. Unfortunately, this model has affirmed in the literature, in the further over and over was used, and even has entered into the review (Nemchinov, Popova, Teterev, 1999).

The early experimental and approximated researches (Dubrovina, Stulov, 1989) have shown that, really, aerodynamic interaction of spheres in a configuration 1 stops on distances of the order of the fragment characteristic size when the head shock wave reflected in Mach configuration is tangent the surface of sphere. However numerical solutions for system 1 have found out significant reduction of cross-section force in comparison with the value accepted in work (Passey, Melosh, 1980). It has appeared that the value of c_y not too strongly depends on the fragment form. On Figure 1 dependence $c_y(h)$ for system 1 (a continuous line) and for two half-cylinders, received by cutting of the cylinder with height R and radius of basis R along a plane passing through its axis and containing the velocity of initial stream is shown. Data for half-cylinders are published earlier (Artemieva, Shuvalov, 1996) and taken by us from later publication (Shuvalov, Artemyeva, Trubetskaja, 2000). The satisfactory consent as on value of max $c_y \approx 0.27$, and on the greatest distance of interaction $h \approx 0.5$ is visible.

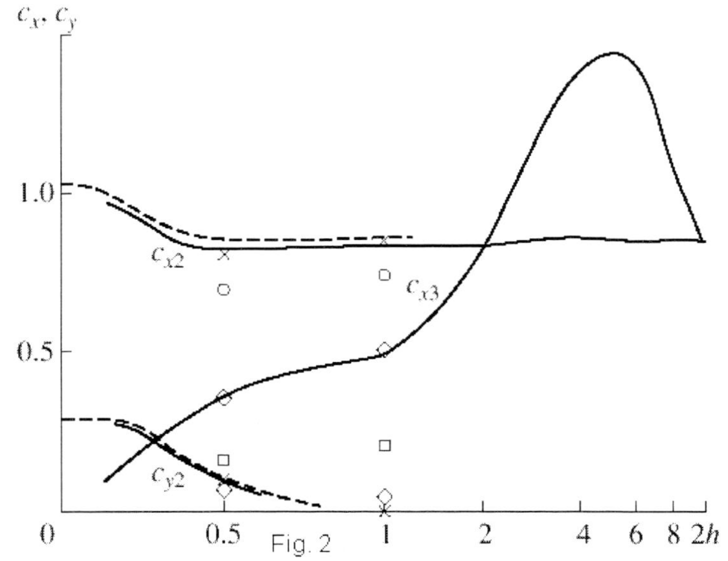

Fig. 2

Figure 2. Coefficients of drag and cross force in systems 1 (shaped lines), 2 (continuous lines) and 3 (badges:o - c_{xfm}, x - c_{xfa}, c_{yfa}, ⌢ - c_{xsm}, c_{ysm}, ⌣ - c_{xb}).

Results of calculations of the flow around system 1 are given also on Figure 2 by shaped lines. Certainly, during the initial moments after destruction, replacement of fragments with spheres is not quite correct and cross-section force will be defined by stagnation pressure, however development of flow between fragments at once will lead to pressure drop between them.

Aerodynamics of system 2 was investigated in comparison to results of a flow around system 1. We shall designate parameters of forward sphere in system 2 by index *f*, and parameters of back sphere by index *b*. Coefficient of drag c_{xf} and cross-section force c_{yf} of the forward sphere, and also coefficient of drag c_{xb} of the back sphere are shown on Figure 2 by continuous lines. It is visible that presence of the third sphere only slightly changes conditions of the flow around the forward spheres. This little change is expressed in some reduction both c_x and c_y of the forward sphere due to increase in ground pressure because of presence of the third sphere. The form of the head shock wave before two forward spheres almost does not change.

Let's discuss the flow about the back sphere. At small distances h before it the shock wave limited on scope because of non-uniformity of the initial stream is formed. With growth of h scope of a shock wave increases and drag c_{xb} monotonously increases. The situation does not change and at distances when interaction of forward spheres stops, i.e. at $h > 0.5$. With increase of h on a frontal part of back sphere the reflected shock wave falls, so the value of c_{xb} continues to increase. At the further increase of h, the point of reflection, and then a lateral part of head shock waves of forward spheres fall on a head shock wave of back sphere. During all period of interaction of shock waves of forward spheres with back sphere 3, the value of c_{xb} exceeds the coefficient of drag of a single sphere in an infinite stream. It is shown on Figure 2 where at $2h \geq 1.0$ the scale on an absciss axis is changed on logarithmic one. Here a new factor of interaction of bodies in system in a supersonic stream is noted: interaction of the reflected shock waves with bodies results not only in occurrence of cross-section force, but also in substantial growth of drag.

Properties of the flow around system 3 in the certain degree repeat the tendency noted above. In system 3, the following system of designations is accepted: parameters of spheres in the front on the stream row accept an index *f*, the second row accept an index *s*, and the back sphere accept an index *b*. Further, angular spheres are supplied with an additional index *a*, the spheres located in the middle of the triangle sides are supplied with an index *m*. Coefficients of drag and cross-section force of spheres *fa* and *sm* and coefficient of drag of spheres *fm* and *b* are given on Figure 2. These values are shown by badges. Decoding of badges is given in the signature to Figure 2. We shall note here more essential decrease in drag of sphere *fm* in comparison with sphere *fa* for which this decrease practically repeats reduction c_{xf} in system 2. It occurs owing to greater increase in ground pressure at sphere *fm* because of presence of the third layer made by sphere *b* in a configuration 3. At the same time, drag of sphere *b* increases with growth h more slowly, than resistance of sphere *f* in system 2 or spheres *sm* in system 3.

The form of a head shock wave with increase h changes practically the same as it takes place in configurations 1 and 2. There is a transition from the common shock wave before system to individual shock waves before each sphere in a head part of system. Obviously factor of interaction of the reflected shock waves noted above with head shock waves before spheres of back lines also will take place and lead to appreciable increase in drag of these spheres. The picture of a flow around system 3 is shown on Figure 3 for case at $h = 0.5$.

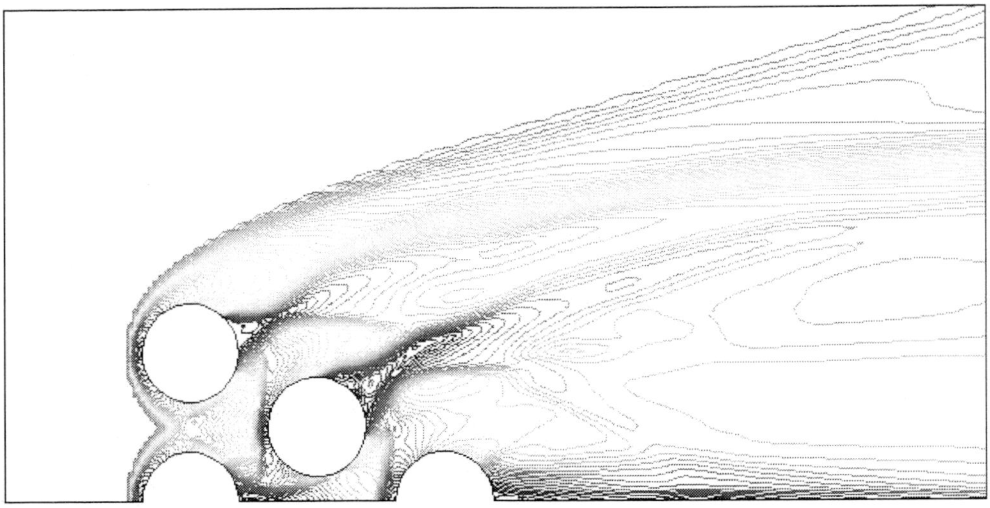

Figure 3. The flow around system 3, $h = 0.5$; there are shown lines of Mach constant.

The basic conclusion of research consists in the fact that at a supersonic flow around system of bodies, distortion of the stream due to «collectivity» of obstacles as though is carried downwards on a stream. Small reduction of drag of head spheres arises due to increase in ground pressure. The last, however, practically does not affect reorganization of the head shock wave which with growth of distance between spheres monotonously breaks up to individual shock waves. The characteristic distance between bodies for this disintegration also does not change and quite corresponds to a simple case of a flow of pair spheres with a line of the centers across the stream.

Bolides with the Moderate Values of Mass Loss Parameter

The majority of bolides of Prairie and European bolide networks correspond to this range of value of parameter β. Trajectories of the largest bolides of Prairie networks, USA (MacCrosky, Shao, Posen, 1978; MacCrosky, Shao, Posen, 1979) have been processed by the least squares method in which as trial functions solution for a single body with ablation were used. Values of β for eighteen bolides Tab. 7.1 (Stulov, Mirskii, Vislyi, 1995) (excepting bolide 39470*) lay in a range $0.48 < \beta < 2.76$. As it is known, in all cases, except for bolide Lost City, there was not revealed, any traces of falling meteoric substance, despite of careful searches in the areas predicted by extrapolation to the Earth of observable sites of trajectories.

Among bolides of the European network the significant attention of researchers was given to the bolide Beneshov, recorded in Czech Republic on May 7, 1991. According to opinion of one of observers (Spurný, 1994), it was one of the brightest and in details documentary bolide.

The big observant material on results of the Canadian camera network during 1974-85 was published in the work Halliday, Griffin, Blackwell, 1996.

One of fundamental problems of meteoric physics is the definition of preatmospheric mass of bodies which form bolides. Intensity of the meteoric phenomenon serves is defined by kinetic energy of a body at the approach to a planet atmosphere. As it is known, speed of bodies at the entrance into the Earth's atmosphere lays in rather narrow range $11 < V_e < 72$ km/s, so the range of values of the speed contribution to kinetic energy does not exceed 50 times. At the same time value of a meteoric body mass can change in essential wider range, from shares of gram (micrometeors) up to hundreds thousand tons (the Tunguska space body), i. e. on 12-14 orders. Besides the entrance speed is rather simply determined in observation of the initial site of the atmospheric trajectory. On the contrary, reliable ways of definition of the entrance mass, containing an estimation of accuracy of result, now are absent.

Approaches known from the literature to an estimation of the entrance mass can be divided conditionally on two groups. The so-called photometric methods using luminosity of bolides enter into the first. More often a method of definition of photometric mass M_{ph} on the basis of the formula

$$M_{ph} = -2\int_{t_t}^{t_b} \frac{I}{\tau V^2} dt. \tag{3}$$

Here I is intensity of a luminescence along an observable site of a trajectory, t_b, t_t are initial and final times for this site, τ is a factor of proportionality. The formula (3) is based on the assumption that defining contribution to luminosity is given by vapors of the body material. Detailed calculations show that it is impossible to consider it is proved, especially for large bodies.

In work of Halliday, Griffin, Blackwell, 1996 attempt is made to modernize a photometric method of definition of preatmospheric meteoric masses. Firstly as the luminous efficiency τ it is taken the variable, proved (calibrated) in early works of authors. It is considered that factor τ depends on speed of bolide movement: $\tau = 0$, $V \le 10$ km/s; $\tau = 0.04$, $10 < V < 36$ km/s; $\tau = 0.069(36/V)^2$, 36 km/s $\le V$. Secondly, for large bolides the value of the possible mass which remained after going out the bolide, and fell as possible meteorites, is calculated. This mass is determined with cleanly dynamic way by means of the local value of deceleration received by numerical differentiation of the bolide speed distribution on time in the end of the light site of the trajectory. If this dynamic mass appeared to be significant, the conclusion about possible falling meteorites was made. The full preatmospheric mass of the meteoroid was estimated as the sum of photometric mass and final dynamic mass. The hypothesis about falling, however, has not proved to be true in actual searches of meteorites.

We have made calculations of preatmospheric mass for bolides of the Canadian camera network by means of the least squares method with use of observant data about the light site of the trajectory. As trial function the form of a trajectory for the limited values of mass loss parameter β (the formula (6), see below) was used.

Earlier this method was applied to bolides of the Prairie network, USA; results are given in monography Stulov, Mirskii, Vislyi, 1995. In case of the Canadian camera network, the elementary variant of a method consisting in calculation of parameters α and β under obvious formulas (7.2) of monography Stulov, Mirskii, Vislyi, 1995 was used. Results of calculations for four cases are given in table 2. According to discussion of the elementary variant of the least squares method, given in the monography, two bases for approximation of an observable trajectory (the second column of the table) here were used. As it is specified in the monography, the basis of the seven points, not containing an initial site of the trajectory with small braking where $v \geq 0.9$ is preferable.

Table 2.

No. of bolide (Tables 4 and 6)	Basis for least squares method	α	β	M_e, g	MI, g (Table 4)	MT, g (Table 4)
018	7 points (p.) t, s: 3.00 and 6 next points (n.p.)	30.60	0.9391	6160	6000	450
	10 p. t, s: 1.00; 2.00 and 8 n.p.	39.55	0.3119	2850		
223	7 p. t, s: 4.51 and 6 n.p.	19.78	1.6461	80250	240000	8000
	10 p. t, s: 2.07; 3.51 and 8 n.p.	11.67	2.9458	390780		
567	7 p. t, s: 12.08 and 6 n.p.	111.62	1.3430	6210	42000	210
	10 p. t, s: 7.08; 9.08; 11.08 and 7 n.p.	98.66	1.6415	8990		
925	7 p. t, s: 17.50 and 6 n.p.	44.49	1.3885	724290	1300000	40000
	10 p. t, s: 10.00; 14.00 and 8 n.p.	48.36	1.6415	454650		

As one would expect, comparison to calculations of work Halliday, Griffin, Blackwell, 1996 shows various tendencies. We shall note first of all that here there is no so obvious divergence between photometric and dynamic masses as it was at processing of the Prairie network data. It is consequence of new calibration of the parameter τ. At the same time use of the photometric formula still does not guarantee a correct estimation of true preatmospheric meteoroid mass. Addition of the dynamic mass which have remained after going out of the bolide, does not change the situation essentially. This value practically in all cases examined by us is essentially less than photometric mass. Besides, this value is determined in the work of Halliday, Griffin, Blackwell, 1996 by means of local values of deceleration which have small accuracy of definition.

Data processing supervision of the Canadian network will be continued by us.

Other approach based on value of luminosity is developed by I.V.Nemchinovym with employees and named by a method of radiating radius (references to original works are available in the work of Borovicka, Popova *et al.*, 1998). The authors have calculated the table of luminosity depending on radius of a body, speed and altitude of flight using the method of non-stationary analogy. The radius of an observable body is determined then under this table on value of observable luminosity at the same values of speed and altitude of flight.

The second group of methods should be considered dynamic as the body mass is determined on the basis of braking in the atmosphere. Parameter α defines an area of the maximal braking. As it contains the body size twice (S_e и M_e), it is convenient to rewrite it so

$$\alpha = \frac{1}{2} c_d \frac{\rho_0 h_0}{\sin\gamma} \frac{A_e}{\rho_m^{2/3} M_e^{1/3}}, \tag{4}$$

Here A_e is factor of the body form.

In dynamic methods, parameter α (4) is defined by comparison of a calculated trajectory with observable one. After that at given A_e and density of body ρ_m it is determined M_e under the formula (4). The lack of such approach is necessity of the a priori task of values A_e and ρ_m which are not determined in supervision.

The dynamic approach based on the solution for a trajectory in variables v, y (dimensionless speed and height) and the least squares method, was used by us earlier (Stulov, Mirskii, Vislyi, 1995). Other approach considering crushing, was offered by authors of work (Ceplecha *et al.*, 1993). It was named by them the method of gross-fragmentation.

Below the detailed analysis of results of the Beneshov bolide observation is given. certain The values of parameters α and β are determined by the least squares method. Losses of the bolide mass both due to observable separation of fragments, and by means of evaporation are calculated.

Observation and registration of the Beneshov bolide are made by stations of the Czech part of European bolide network on May 7, 1991.. Speed of almost vertical entrance into atmosphere has made over 21 km/s. The shone site of the trajectory reached from 97 up to 16 km on height. Time of movement has made 5.2 s. The estimation of photometric mass of the has given value of 15 t. Since height of 42 km, the separation of fragments from the basic body was observed. In total before achievement of height of 24 km, 4 fragments were separated. The first fragment, in turn, was divided in two pieces during movement. The last observable crushing the basic body on three large and set of fine splinters have occured at h = 24 km. Final fading the bolide took place at height of 19 km at speed of still compact plenty V = 5.2 km/s (Borovicka, Spurný, 1996).

The inverse problem for the trajectory of the Beneshov bolide was solved by the least squares method. As trial functions, it was chosen solutions for a trajectory of a uniform body without ablation

$$y = \ln \alpha - \ln(-\ln v), \tag{5}$$

and for a trajectory of a uniform body in view of ablation

$$y = \ln \alpha - \ln(-\ln v) + 0.83\beta(1 - v), \tag{6}$$

and also a trajectory for model of slow consecutive destruction. Last is described by following system of formulas

$$\frac{\alpha_d}{2} e^{-\frac{4}{3}y} - \frac{\alpha_d}{2} e^{-\frac{4}{3}y_0} = \frac{e^{-\beta/3}}{(3/\beta)^{1/3}} \frac{\widetilde{\Delta}_0}{3}, \alpha_d = \alpha \left(\frac{\rho_0 V_e^2}{\sigma_t} \right)^{1/3}, \widetilde{\Delta}_0 = \int_{\beta v^2/3}^{\beta v_0^2/3} \frac{e^t dt}{t^{4/3}}. \tag{7}$$

Here σ_t is meteoroid solidity on a stretching. The height and speed in a point of crushing y_0 and v_0 are determined from a condition

$$\rho_0 V_e^2 \exp(-y_0) v_0^2(y_0) = \sigma_t, v_0(y_0) = \exp(-\alpha e^{-y_0}). \tag{8}$$

For simplification of calculations following approximation of integral $\widetilde{\Delta}_0$ is used:

$$\frac{e^{-\beta/3}}{(3/\beta)^{1/3}}\frac{\widetilde{\Delta}_0}{3} = \left(v^{-2/3} - v_0^{-2/3}\right)\exp\left[-\beta^*(v_0 - v)\right] \tag{9}$$

Parameter β* differs from β in the numerical factor, giving the best approximation of integral $\widetilde{\Delta}_0$.

The formula for a trajectory (7) is applied to the analysis of an observable trajectory of the Beneshov bolide under condition of $y < y_0$, i.e. below height of the first crushing. At calculations it was accepted $\sigma_t = 1.6\times10^7$ dyn/cm^2. If to accept that appreciable braking begins below height of the first crushing, i.e. $v_0 = 1$, so chosen value of σ_t according to (8) provides valueof y_0, close to observable, i.e. $h = 42$ km. The assumption of the beginning braking after the beginning crushing also proves to be true in supervision: $v = 0.98$ at $h = 42$ km. Thus, formulas (7) - (9) describe the trial function containing two free parameters α and β*.

Application of the least squares method to the observable trajectory of the Beneshov bolide has shown, that the best approximation is provided with the formula (6) at following values of free parameters: α = 7.8, β = 1.5. We note that the received value of mass loss parameter β quite corresponds to the calculations and the estimations given in Table 1. These values allow to estimate the entry mass of the bolide and the average ablation factor. Accepting the numerical values offered by observers: $c_dA_e = 1.21$, $\rho_m = 3.7$ g/cm^3, $\gamma = 80.6^\circ$, $V_e = 21.12$ km/s, we shall receive

$$M_e = 28 \text{ kg}, \sigma = 0.0067 \text{ s}^2/\text{km}^2. \tag{10}$$

The second value (10) corresponds to data of Spurný (1994). As to initial mass M_e, this value depends on the assumption of the spherical form of the body. For example, suppose that the body has the form of a plate 1x1x0.25. Then $A_e = 2.52$, $c_d = 2$, and $M_e = 2023$ kg.

Thus, the best approximation of an observable trajectory turns out by means of model of a uniform body with ablation. This conclusion received by the least squares method, at first sight seems paradoxical. It is possible to explain it only by the assumption that the lump sum of the observable fragments which have separated from the basic body on a greater part of its trajectory (for example, at $42 > h > 24$ km), does not make the main share of full meteoroid mass. Observant data about parameters of the fragments, given by Borovicka and Spurný (1996), allow to lead corresponding estimations.

According to observation, the final mass of four observable fragments was $M_t = 1.82$ kg. On the basis of the data adduced by observers about ablation factors for fragments, calculations of their total initial mass, i.e. mass in the moment of their separation from the parental body, have been lead (Barry, Stulov, 2003)). It was received the following value of total initial mass of these fragments: $M_b = 7.19$ kg.

As the bolide velocity in each point of the trajectory is known from observation, it is possible to determine the total mass loss on any site only due to ablation on the basis of known integral of the equations of movement and mass loss. Calculations show that before final destruction at $h = 24$ km on all trajectory at $\beta = 1.5$ the mass equal to 15.81 kg is carried away, i.e. it is more than twice exceeding total mass of the individual fragments. This circumstance explains that the trajectory in variables "speed-height" is described by model of a single uniform body with ablation.

Results of estimation of entry mass of the Beneshov bolide by photometric and dynamic methods are given in Table 3.

Table 3.

Method	**Observations, M_{ph}**	**Observations, M_{ph}**	**Method of radiating radius**	**Method of a gross-fragmentation**	**The least squares method**
References	1	2	3	4	This work
M_e, kg	15000	13000	3000-4000	82	Sphere: 28; a plate 1x1x0.5: 523

Sources of the results shown in the table: 1 (Spurný, 1994), 2 (Borovicka, Spurný, 1996), 3 (Borovicka, Popova *et al.*, 1998), 4 (Ceplecha, etc., 1993).

From data of the table it follows that the divergence of estimations of the bolide preatmospheric mass by various methods is still great. Rapproachement is possible as on a way of specification of physicomechanical properties of a meteoric body, in particular, its form, and by means of perfection of methods of the calculation based on an estimation of bolide luminosity.

Entrance into Atmosphere of the Tunguska Space Body

Significant excess of the specific (on a mass unit) kinetic energy brought into atmosphere by the Tunguska object, above the quantity of energy necessary for evaporation of a mass unit of its substance, was a reason of rather unusual consequences of Tunguska falling on June 30, 1908. We mean full absence of craters, and also the fall and burn of the taiga in huge territory which characteristic size approximately on four orders surpasses the probable geometrical size of the Tunguska space body. These features show that influence of the Tunguska space body rests on the Earth's surface, most likely had purely gas dynamical character.

In work (Stulov, 1994) for the first time it has been shown that asymptotic soluyion of the equations of meteoric physics at great values of parameter β has a following appearance

$$v = 1, \quad m = 1 - 2\alpha\beta e^{-y} \tag{11}$$

In other words, at big β fast evaporation of a meteoric body takes place for the lack of braking. After full evaporation there should be a fast braking of products in a mix with air. This idea has been developed in the subsequent works of the author (Stulov, 1997; Stulov, 1998a; Stulov, 1998b).

In second half of the 90th experts discussed in details a new grandiose bolide phenomenon: falling of fragments of comet Shoemaker-Levy 9 on the Jupiter in the summer of 1994. It has appeared that some natural observations of this phenomenon quite keep within the offered hypothesis of bolides behaviour болидов at big values of β. In publications of the American authors (see, for example, Hammel, Beebe *et al.*, 1995), photos of huge gas emissions out of the Jovian atmosphere, received by means of the Hubble telescope, after the entrance of large fragments of the comet into atmosphere are given. Apparently, it were clots of hot evaporation products of fragments in a mix with the atmospheric gases, thrown out of atmosphere under action of buoyancy forces.

In June, 2003 at the conference «95 years to the Tunguska problem» in Moscow, V.V.Shuvalov has made the report on the basis of work published before (Shuvalov, Artemieva, 2002) on numerical modeling an large bolide entrance into atmosphere of planets.

The numerical solution has completely confirmed the basic features of the asymptotic soluyion (11).

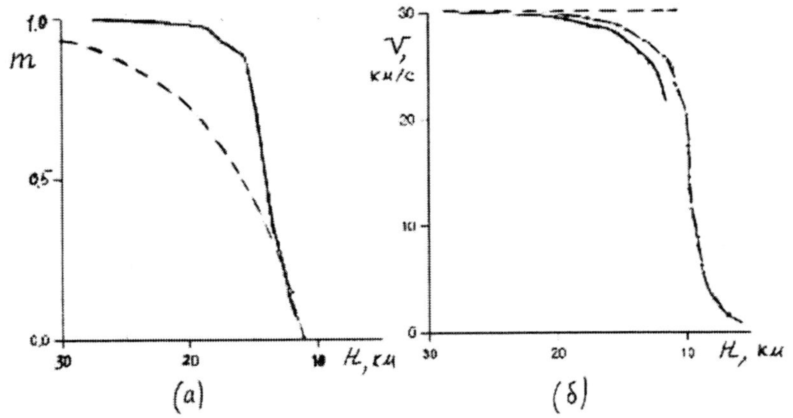

Figure 4. Relative mass of the solid meteoroid (a) and its velocity (b) as functions of flight altitude; continuous lines - numerical results (Shuvalov, Artemieva, 2002), shaped lines - asymptotic solution (11) at $\alpha = 0.077$ ($\rho_m = 1.5$ g/cm^3), $\beta = 30$.

By the numerical solution of the nonviscous equations of gas dynamics, calculation of movement of a spherical body in radius of 30 m in the Earth's atmosphere perpendicular to its surface is made. Entry conditions: $V_e = 30$ km/s at $h = 30$ km. Full evaporation of a firm phase comes to the end at height of $h = 11$ km, thus «visible speed of a meteor», i.e. the meteoroid rests and products of evaporation surrounding it, makes 25-26 km/s. At the subsequent stage of calculation «the air-vapor jet» (« air-vapor jet ») is braked till zero speed at height of 4 km then there is a rise of hot gas volume in atmosphere due to buoyancy forces. The characteristic cross-section radius of a jet during the moment of braking exceeds 2 km. From this brief abstract follows that in work (Shuvalov, Artemieva, 2002) for the first time, the basic properties of the asymptotic form of a bolide trajectory, received on the basis of the meteoric physics equations (Stulov, 1994), are shown by calculatios. It is a question about practically full ablation of meteoroid in conditions of insignificant braking. These properties are shown on Figure 4, where change of relative mass $m = M/M_e$ (a) and speed of body V (b) depending on height of flight H are given. Continuous lines show the numerical solution (Shuvalov, Artemieva, 2002), shaped lines show the solution (11) constructed on bolide parameters at meteoroid density $\rho_m = 1.5$ g/cm^3. The stroke-dashed line on Figure 4 (b) shows change of gas volume speed which is not defined in the asymptotic solution.

Let's note that this calculation does not reproduce all major details of entrance of the Tunguska space body into the Earth's atmosphere. However, it is possible to assume that change of entry conditions, for example, up to values of V_e = 35 km/s, R_e= 38 m (Korobeinikov, Tshushkin, Shurshalov, 1991) will allow to reproduce falling "an air-vapor jet" on the taiga and it{her} растекание, accompanied by burn and fall of the wood.

The Conclusion

The certain dependence of a bolide character from value of mass loss parameter β is noted. This dependence allows to interpret more confidently large falling, excepting, in particular, fantastic or semi- fantastic scripts.

As a result of numerical experiment, coefficients of drag and cross-section force of fragments at the initial stage after crushing a meteoric body are received. It has appeared that for two fragments with a line of the centers across a stream the coefficient of cross-section force in some times less of the value used earlier in the literature.

In case of big values of parameter β prevailing process is evaporation of fragments of the destroyed meteoric body. It occurs for the lack of significant braking. Therefore interaction of the bolide with the planet surface can have purely gas dynamical character.

The author thanks Dr. M.A.Smirnov for discussion of work.

References

[1] Artemieva N.A., Shuvalov V.V. Interaction of shock waves during the passage of disrupted meteoroid through atmosphere // *Shock waves*. 1996. V. 5. No.6, pp. 359-367.

[2] Baldwin B., Sheaffer Y. Ablation and breakup of large meteoroids during atmospheric entry // *J. Geophys. Res.* 1971. V. 76. No. 19, pp. 4653-4668.

[3] Barry N.G., Stulov V.P. Features of crushing the Beneshov bolide // *Astron. vestn.* 2003. V. 37. No. 4, pp. 332-335 (in Russian).

[4] Borovicka J., Spurný P. Radiation study of two very bright terrestrial bolides and an application to the comet S-L 9 collision with Jupiter // *Icarus*. 1996. V. 121, pp. 484-510.

[5] Borovicka J., Popova O.P., Nemtchinov I.V., Spurný P., Ceplecha Z. Bolides produced by impacts of large meteoroids into the Earth's atmosphere: comparison of theory with observations I. Benešov bolide dynamics and fragmentation // *Astron. and Astrophys.* 1998. V. 334, pp. 713-728.

[6] Ceplecha Z., Spurný P., Borovicka J., Kecliková J. Atmospheric fragmentation of meteoroids // *Astron. and Astrophys.* 1993. V. 279, pp. 615-626.

[7] Chyba Ch. F., Thomas P.J., Zahnle K.J. The 1908 Tunguska explosion: atmospheric disruption of a stony asteroid // *Nature*. 1993. V. 361, pp. 40-44.

[8] Dubrovina I.A., Stulov V.P. Determination of the greatest distance of aerodynamic interaction for pair spheres and cylinders in a supersonic stream // *Vestn. of the Moscow State University. Ser. 1. Mathematics, mechanics*. 1989. No. 5, pp. 46-49 (in Russian).

[9] Fesenkov V.G. Orbit of the Sikhote-Aline meteorite // *Meteoritics*. Publishing house AS of the USSR. 1951. Iss. IX, pp. 27-31 (in Russian).

[10] Grigoryan S.S. On movement and destruction of meteorites in atmospheres of planets // *Kosmich. issled.* 1979. V. 17. No. 6, pp. 875-893 (in Russian)..

[11] Halliday I., Griffin A.A., Blackwell A.T. Detailed data for 259 fireballs from the Canada camera network and inferences concerning the influx of large meteoroids // *Meteoritics and Planetary Science*. 1996. V. 31, pp. 185-217.

[12] Hammel H.B., Beebe R.F., Ingersoll A.P., Orton G.S., Mills J.R., Simon A.A., Chodas P., Clarke J.T., De Jong E., Dowling T.E., Harrington J., Huber L.F., Karkoschka E., Santon C.M., Toigo A., Yeomans D., West R.A. HST Imaging of Atmospheric Phenomena Created by the Impact of Comet Shoemaker-Levy 9 // *Science*. 1995. V. 267, pp. 1288-1296.

[13] Ivanov A.G., Ryzhanskii V.A. Fragmentation of a small celestial body at its interaction with atmosphere of a planet // *DAN*. 1997. V. 353. No.3, pp. 334-337 (in Russian).

[14] Korobejnikov V.P., Chushkin P.I., Shurshalov L.V. Complex modeling flight and explosion of a meteoric body in atmosphere // *Astron. vestn.* 1991. V. 25. No.3, pp 327-343 (in Russian).

[15] Korobejnikov V.P., Vlasov V.I., Volkov D.B. Modeling destruction of space bodies at movement in atmospheres of planets // *Math. modeling*. 1994. V. 6. No.9, pp. 61-75 (in Russian).

[16] Krinov E.L. Crushing Sikhote-Aline meteoric body // *Meteoritics*. Publishing house AS of the USSR. 1975. Iss. 34, pp. 3-14 (in Russian).

[17] Krinov E.L., Tsvetkov V.I. Sikhote-Aline meteoric rain as a classical meteoric falling // *Meteoritics*. Publishing house AS of the USSR. 1979. Iss. 38, pp. 19-26 (in Russian).

[18] Lebedinets V.N. *Dust in the top atmosphere and space*. Meteors. L.: Hydrometeopub. 1980. 250 p. (in Russian).

[19] Mak-Kroski R.E., Shao T.I., Pozen A. Bolides of the Prairie network. 1. The general data and orbits // *Meteoritics*. Publishing house AS of the USSR. 1978. Iss. 37, pp. 44-59 (in Russian).

[20] Mak-Kroski R.E., Shao T.I., Pozen A. Bolides of the Prairie network. 2. Trajectories and curves of shine // *Meteoritics*. Publishing house AS of the USSR. 1979. Iss. 38, pp. 106-156 (in Russian).

[21] Nemchinov I.V., Popova O.P., Teterev A.V. Entrance of large meteoroids into atmosphere: the theory and observations // *Eng.-phys. J.* 1999. V. 72. No.6, pp. 1233-1266 (in Russian).

[22] Passey Q.R., Melosh H.J. Effects of Atmospheric Breakup on Crater Field Formation // *Icarus*. 1980. V. 42, pp. 211-233.

[23] Shuvalov V.V., Artemeva N.A., Trubetskaja I.A. Modelling movement of a destroyed meteoroid with evaporation // *Astron. vestn.* 2000. V. 34. No.1, pp. 55-67 (in Russian).

[24] Shuvalov V.V., Artemieva N.A. Numerical modeling of Tunguska-like impacts // *Planet. and Space Sci.* 2002. V. 50, pp. 181–192.

[25] Spurný P. Recent fireballs photographed in central Europe // *Planet. and Space Sci.* 1994. V. 42. No. 2, pp. 157–162.

[26] Stulov V.P. Interaction of a comet nuclei with atmosphere of planet // *Int. Conf. "Modern problems of the theoretical astronomy". Abstracts*. St-Petersburg. 1994. V. 3, p. 62 (in Russian).

[27] Stulov V.P., Mirskij V.N., Vislyi A.I. *Fairball aerodynamics*. M.: Nauka, 1995, 236 p. (in Russian).

[28] Stulov V.P. Movement of a comet nuclei in the top atmosphere of a planet // *Vestn. of the Moscow State University. Ser. 1. Mathematics, mechanics*. 1997. No. 1, pp. 63-65 (in Russian).

[29] Stulov V.P. Gasdynamical model of the Tunguska fall // *Planet. and Space Sci.* 1998a. V. 46. No. 2/3, pp. 253–260.

[30] Stulov V.P. Gasdynamical model of large bolides // *Vestn. of the Moscow State University. Ser. 1. Mathematics, mechanics*. 1998b. No. 5, pp. 39-45 (in Russian).

[31] Stulov V.P. An analytical model of consecutive crushing and ablation of a meteoric body in atmosphere // *Astron. vestn.* 1998c. V. 32. No.5, pp. 455-458 (in Russian).

[32] Stulov V.P., Stulov P.V. Braking and ablation of a meteoroid after its destruction in atmosphere // *Astron. vestn.* 1999. V. 33. No.1, pp. 45-49 (in Russian).

[33] Stulov V.P., Titova L.Ju. A model of meteoroid fragmentation in atmosphere // *DAN*. 2001a. V. 376. No.1, pp. 53, 54 (in Russian).

[34] Stulov V.P., Titova L.Ju. The comparative analysis of crushing models for meteoric bodies // *Astron. vestn.* 2001b. V. 35. No.4, pp. 345-349 (in Russian).

[35] Zhdan I.A., Stulov V.P., Stulov P.V. Aerodynamic interaction of two bodies in a supersonic stream // *DAN*. 2004a. V. 396. No.2, pp. 191-193 (in Russian).

[36] Zhdan I.A., Stulov V.P., Stulov P.V. Characteristic elements of a destroyed body in a supersonic stream // *DAN*. 2004b. V. 399. No.2, pp. 199-201 (in Russian).

In: Space Science: New Research
Editor: Nick S. Maravell, pp. 329-351
ISBN: 1-60021-005-8

Chapter 11

SPACE WEATHER

Juhani Huovelin
Observatory, University of Helsinki, Finland

Abstract

A qualitative discussion of the background, the present status, and the future prospects of space weather research, monitoring, and services is presented. The elements of the interconnection between the Sun and the Earth are described, and their relations with space weather phenomena are explained. The effects of space weather from the near Earth space to Earth's surface are discussed, and a short summary of present instruments, space facilities, and space weather services is given. Aspects of future needs and directions are also discussed.

Introduction

The term *space weather* refers to time-variable conditions in the space environment that may damage space-borne and ground-based technological systems. The term was introduced in the early 1990's among researchers in the field of solar-terrestrial physics and atmospheric science. Space weather includes phenomena, which can seriously disturb or damage the basic infrastructure that supports our society, including communication, navigation, transportation, electric power transmission, and gas and oil pipelines, and its potential effects become even more significant as our dependence on technology increases with more sophisticated and versatile services and applications based on information technology and automation. In the worst case, a space weather storm may even be lethal for the unlucky crew in a manned spaceflight, if sufficient shielding against the energetic radiation is not available and used.

Understanding different physical phenomena involved in the chain from the Sun to the Earth's surface (i.e. the space weather chain) requires scientific research, which is important and interesting for the scientists themselves. It does, however, also provide essential information for activities related to the community, like space weather risk management, forecasting, and measures to be taken to minimise damages by space weather effects. Observations and theoretical studies of the solar wind, for example, are important for solar-terrestrial research, but they also are essential to understand space weather processes in the magnetosphere and ionosphere, at lower levels of the atmosphere, and on the ground level.

The consequences of space weather contain a multitude of both direct and indirect influences. The overall costs of the disturbances are difficult to determine, but they have an increasing trend. As a rough estimate, we can speak of at least hundreds of millions of US dollars, when we consider the deleterious consequences of a major space weather storm. This is comparable to damages caused by a hurricane or an earthquake. Luckily enough, big space weather storms with serious ground level consequences are less frequent than the more familiar atmospheric and geological phenomena, and also human beings on the ground level are rarely the direct "targets" of space weather effects. On the other hand, the coverage of the effects of a space weather storm is often in a bigger scale than that of an earth-bound natural catastrophe, because the scale and power of the source phenomena are much bigger.

An important further motivation for advanced studies of space weather effects, forecasting of it, and finally shielding from the effects of space weather is offered by the urge of human kind to reach away from the Earth to other planets, manned space stations and bases on the Moon, and missions to other planets. All these carry with them the need to survive over long periods in a hostile space environment, with unavoidable need for effective and reliable precautions against space weather. In addition to effects of microgravity and need for social adaptation required by the very limited spatial volume available in a spacecraft or a space station, we have to prepare for a totally different radiation environment when we enter the era of living in space for up to several years required by the above activities. Thus we should forget dreams and future visions of luxurious cabins and large halls in gigantic spacecraft, and prepare the growing generation that urges to become astronauts to prepare for extremely modest living conditions, where it will be lethal to go for a walk outside during a stormy space weather.

Various science aspects of space weather and their influence on the human civilization are currently investigated by a large number of research groups and individual scientists around the world. There are also several national and multinational organizations, which provide space weather services on a regular basis. What is the way forward? A coherent approach is needed which includes, 1) developing various strategies and methods in parallel, 2) improving in the use of the monitored data, 3) further development of facilities, detection techniques, and technology, and 4) active communication between the developers and potential users of space weather information. All this is necessary to guarantee a successful future of scientific and community-related activities that are in the realm of space weather effects. A global endeavour to meet this global challenge is definitely our guideline for the future.

Solar Activity

Among all stars in our galaxy, the Milky Way, the Sun belongs to the category of very calm and inactive stars, and it is a relatively weak source of e.g. X-rays. The normal levels of solar X-ray radiation (quiescent Sun) are about one millionth of the total electromagnetic radiation (luminosity) of the Sun. Even in the limited category of single 'solar-type' stars, most of the other stars even in the neighbourhood of the Sun exhibit large observable variations in brightness and spectrum [e.g. Huovelin, Saar, Tuominen 1988]. These indicate presence of activity phenomena, including dark and cool spot regions that are much larger than any of those ever seen on the Sun. The surface coverage of sunspots on solar disk can be up to about 2 %, while most of other observed spotted stars have cool spots, which are generally more than 5 % of the visible disk, and in some cases cover nearly half of it [e.g. O'Neill, Saar, Neff 1996]. Due to the proximity of 'our star', however, we can observe solar activity phenomena with much better accuracy than other stars, and are physically connected to and affected by them.

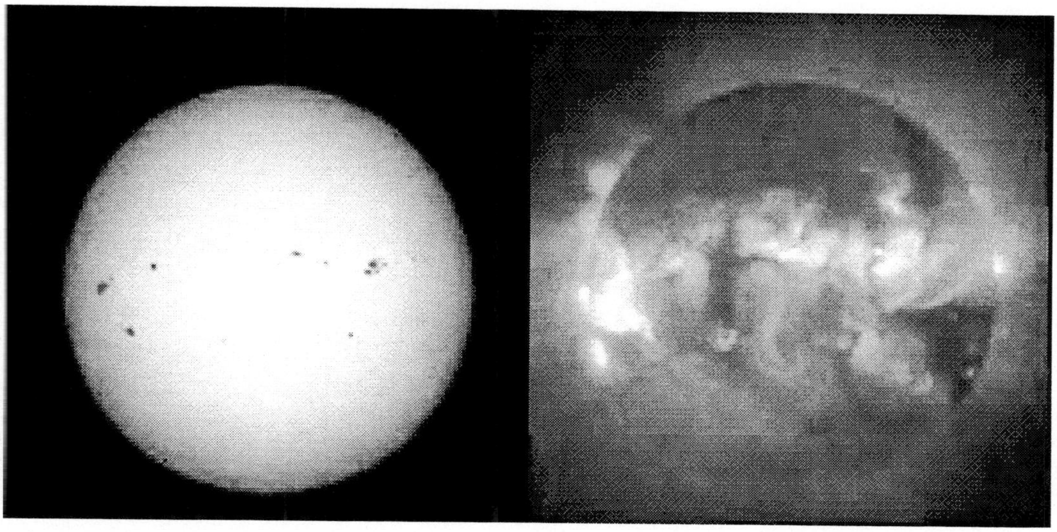

Figure 1. The Sun in visible light (left) and in X-rays (right). Sunspots (leftmost figure) are surrounded by hot active regions (rightmost figure), where solar flares are triggered. Open magnetic field lines guide the solar wind and CMEs to the interplanetary space on the dark areas seen in X-rays (http://www.sunblock99.org.uk/sb99/home/home_f.html).

The presence of a corona on the Sun and other stars is related with the presence of an outer convective zone just below the surface (photosphere) and rotation of the star, the period of which varies from the equator to the poles (differential rotation). The gas in the outer convective zone is turbulent, and combining the effects of differential rotation and turbulence within the convective zone gives rise to a stellar dynamo, where a seed magnetic field in a star is amplified by the complex motions of electrically conductive gas in the convective zone. Hence an outer convective zone is a prerequisite for the presence of stellar coronae. A similar basic mechanism, i.e., a hydromechanical dynamo, is believed to be the source of the terrestrial magnetic field, and magnetic fields in also other planets.

The measured temperature of the solar atmosphere rises from the photosphere to a certain level in the corona, which is believed to relate with the magnetic field above solar surface in the following way. The gas pressure in the convective zone exceeds the magnetic pressure within the magnetic flux tubes, and they are dragged around and twisted by the differential rotation and turbulence. Energy of these motions is transported along the flux tubes up to solar corona. The gas pressure and density decrease with increasing height in the solar atmosphere, and the magnetic pressure in the flux tubes, which rise from under the solar surface and reach to the heights of corona, exceeds the pressure of the surrounding gas in the corona. Due to the change in the pressure balance, the roles between the flux tubes and surrounding gas are exchanged, and the extra energy loaded to the flux tubes under the Sun's surface is dissipated to the surrounding gas in the corona. This concept is quite reasonable, and may well be the correct qualitative explanation. The energy transport mechanism along the flux tubes and the dissipation mechanism in the corona are, however, poorly understood. There are at least two theories to explain them. A theory presented by Parker [Parker 1988], suggest that a nearly uniform and dense distribution of very small flares, called nanoflares, are the source of energy, which would explain the persistent solar corona and its high temperature. Nanoflares are flares in very small magnetic structures that cannot be resolved with present solar telescopes. Direct observational proof of the theory would require an X-ray or far ultraviolet telescope with spatial resolution of the order 100 km or better on the Sun's surface. Another theory suggests that the energy is transported to corona by hydromagnetic waves, which are then damped out providing the energy sufficient to heat the corona. Probably both of these mechanisms play a role.

The interaction between the Sun and the Earth which gives rise to space weather phenomena, originates from the Sun. In the origin there are a number of time variable phenomena related to the solar magnetic field, which we call by the common term 'solar activity'. Solar activity reveals itself in active regions, which are connected to local strong enhancements of magnetic field penetrating the surface of the Sun. Typical features in an active region are dark/cool sunspots, also called *maculae,* and bright/hot *plage* areas surrounding them, also called *faculae*. These are long lasting structures in the photosphere and chromosphere of the Sun, and they are also observed in optical wavelengths.

Prominences are often arc shaped features that extend up to the corona. *Quiescent* prominences are very stable, and can sometimes be seen slowly evolving over several months. *Eruptive* prominences are less stable, often extending very high from the solar surface, and carry particles with speeds well over 1000 km/s. Prominences are seen at the edges of solar active regions, and are probably associated with the local magnetic field. They exhibit the structure of the field in the surroundings of an active region, and the difference between quiescent and eruptive prominences is probably associated with the stability of the local magnetic field. Prominences, as well as most of the other features in solar active regions can be best observed in UV and X-rays, and also radio waves, where the relative enhancement compared to the emission of surrounding atmosphere is much bigger, and the features are well recognized. There is also a continuous outflow of ionized gas (electrons, protons and a smaller fraction of heavier ions) from outer solar corona via open solar magnetic field lines, which is called the *solar wind.*

There are significant transient enhancements of solar radiation and mass flow from the Sun during eruptions, which occur in unpredictable intervals of time. The most energetic types of these events are related with solar corona, and are called *flares* and *Coronal Mass Ejections* (CME). The solar magnetic field undergoes cyclic variation with a long term mean period of approximately 22 years, which is seen as an 11 year cycle in the number of sunspots, average frequency and intensity of solar flares and CMEs, as well as emission in X-rays, ultraviolet, and optical lines such as CaII H+K and H-alpha. During the solar cycle maximum, also coronal activity is at significantly high level, and strong flares and CMEs are frequent. During solar cycle minimum, there are very few, if any, big flares or large CMEs. There are large variations in the cyclic behaviour of the Sun. The presently ongoing cycle, for example, still shows high activity and large flares, although we should be arriving to the cycle minimum soon. Also the long record of sunspot cycles over roughly four hundred years shows a variation from 7 to14 years in the length of cycles.

During solar eruptions, a very large amount of magnetic energy is released in a huge explosion extending to the solar corona up to hundreds of thousands kilometres from the visible photosphere. The burst can include a flash of light throughout the whole electromagnetic spectrum with the clearest intensity change in X-rays and Gamma-rays, whence it is called *flare*. The reason of strong enhancement in X-ray and Gamma-ray emission is the powerful heating of gas associated with flares. While the X-ray emission of the solar corona normally corresponds with a temperature of a few (2-3) million Kelvin (K), the temperature increases typically to 10 – 20 million K, and can be as high as 100 million K in a flare. The flare heating mechanism is an intensively studied but still an unresolved theoretical problem in solar physics. It is known that the heating is associated with release of magnetic energy, but the exact mechanism and energy transport in the solar corona is not fully understood. Solar flares are always spatially associated with active regions. The first note on a scientific observation of a solar flare is from 1st of September 1859, when R.C. Carrington and R. Hodgson independently observed a flare in white light when they were watching sunspots. During a flare explosion, also large amounts of electrons, protons, and ionized atomic nuclei are accelerated to high energies and ejected away from the Sun. The duration of a flare event depends on the peak brightness, and typically varies from a few minutes to several hours. The biggest flares, e.g. that of October 28, 2003, can last nearly 10 hours, as measured from the rise and decay of X-rays. The peak power of X-ray emission (1.5 - 12.4 keV) during the same flare was more than 10^{19} W derived from data by the GOES environmental satellites. The total energy release during the full extent of a big flare is approximately 10^{25} J, which is comparable to a 100 million megaton hydrogen bomb, or 10 million simultaneous volcanic eruptions! On the other hand, put in solar context, the total flux of electromagnetic radiation emitted by the Sun every second is 10 times more than this. The electromagnetic radiation is dominated by photospheric emission, which helps us understand why the Sun is so bright!

It was long believed that all or most of the solar-based magnetic disturbances and geomagnetic storms experienced on Earth are due to solar flares. This is easy to understand, since only large flares could be seen with the available ground-based telescopes and sometimes even visually, and large flares often are associated with significant amounts of particles with high energies being ejected to interplanetary space. Theories explaining the phenomenon were actively studied and developed by e.g. the famous American astronomer George E. Hale [Hale 1931], and by 1960's, flare-based explanation for geomagnetic storms had become one of the cornerstones of solar-terrestrial physics.

It was not until the early 1970's when the phenomena called Coronal Mass Ejections (CME) were clearly distinguished from solar flares, using the first special telescopes called *coronagraphs* flown to space, e.g. on the manned US Skylab space station in 1973. Skylab carried onboard also X-ray and UV cameras for observations of solar activity and flares with wavelengths unobservable at Earth, which greatly helped in seeing the difference and spatial separation between those two types of solar events. This also explains why the coronagraphs already in use at Earth-based telescopes from 1930's, e.g. at Pic du Midi Observatory, had not been able to resolve the difference. With the new observations, however, scientists became aware of something fundamental being misunderstood in the old concepts of solar-terrestrial physics. During the following two decades, the observational knowledge of CMEs increased dramatically, and the scientific community grew finally mature enough to accept the fact that most CMEs are not associated with solar flares [e.g. Kahler 1992, Gosling 1993]. A new era in solar-terrestrial physics and space weather research had begun.

According to present understanding, flares are usually not the source of CMEs [e.g. Gosling 1993]. The events can occur nearly simultaneously, but the origin and the location are not the same. Solar flares seem to be connected with disturbances in small-scale magnetic features, and their origin is relatively close to the photosphere in association with active regions. CMEs, on the other hand, are triggered at higher levels of solar atmosphere away from active regions by perturbations and rapid changes in large-scale solar magnetic field.

Solar Wind and Phenomena in Interplanetary Space

The solar wind plays an important role in solar–terrestrial physics since it carries the effects from the Sun to the Earth and interacts with the Earth's magnetic field. Observations of solar wind, and theories on its origin existed already during the first half of the 19th century. However, a real breakthrough was to come shortly after that era, and should be mentioned as being one, but certainly not the only, cornerstone of solar wind research. In 1958, American physicist Eugene N. Parker published his derivation of a hydrodynamic model of solar corona, which led to the surprising result that the gas pressure at infinity cannot reach zero value, and, in fact, the expansion occurs at a supersonic (500 km/s) velocity. Parker's courageous interpretation was that the solar corona must expand continuously, leading to the existence of solar wind [Parker 1958]. Due to the opposition and criticism against his solution, Parker had to write another article on the same subject, which he published two years later [Parker 1960]. A further conclusion by Parker was that the out-flow of plasma turn the magnetic field lines away from the Sun in the outer parts of the corona, thus enabling matter being carried away to interplanetary space. An observational proof for Parker's theoretical work was to come a few years later. Also studies of solar wind interactions were boosted by Parker's fundamental work on solar and interplanetary plasma physics. An ADS (Smithsonian/NASA Astrophysics Data System) search shows that he published as the only author a total of 75 scientific refereed papers on subjects related with solar and solar-terrestrial physics between 1952 and 2001. Parker's more extensive work began with the first serious approach to the modern solar dynamo theory [Parker 1955], continuing with the aforementioned solution of the problem of solar wind [Parker 1958, 1960)], and further theoretical work touching virtually every important aspect of solar-terrestrial physics.

Long term data from 1982-2003 available in the OMNIWeb database (http://nssdc.gsfc.nasa.gov/omniweb/) gives an average speed of solar wind outflow of about 450 km/s at Earth distance. There are theories on the dynamics of solar wind, e.g. velocity along the radial path outwards from the Sun, but no systematic in situ measurements exist in the interplanetary space along the path to Earth orbit. There are studies involving extensive observational data on the interactions of the solar wind with Venus (e.g. Russell, Vaisberg 1983), and mainly theoretical studies on Mercury [e.g. Ip, Kopp 2002], and comets in their passage within Earth orbit [e.g. Luhmann 1995]. Extensive data in the Sun-Earth axis is still available only from the first Lagrange (L1) point to the Earth and nearer, which is just one hundredth of the distance to the Sun.

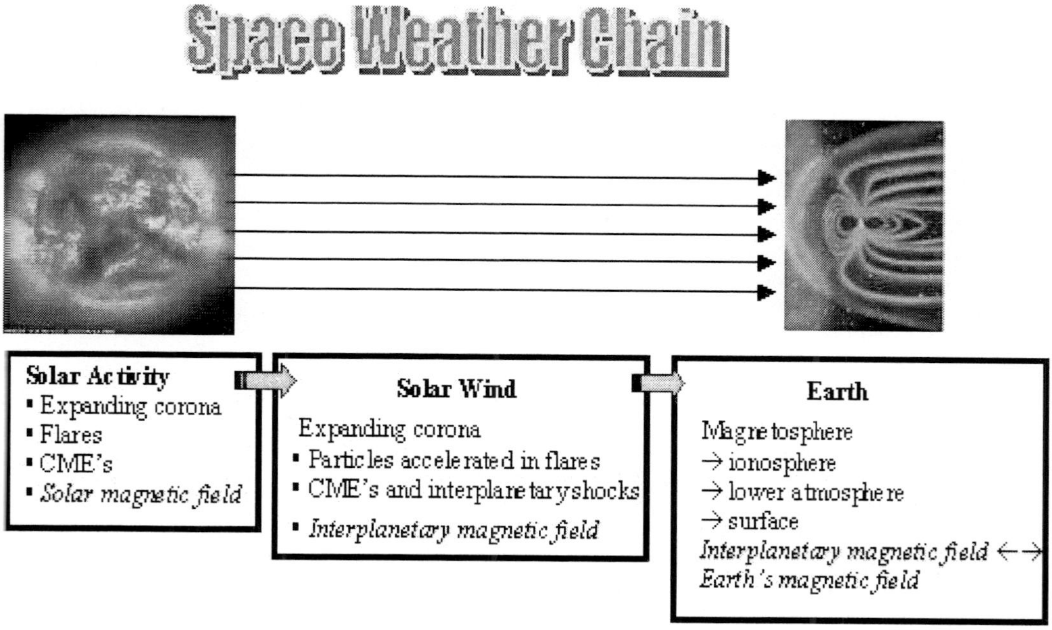

Figure 2. A Schematic summary of Solar-Terrestrial connection including the key elements of the space weather chain.

A CME can consist of slow particles with average speed of as low as 50 km/s (slow CME), or fast particles with speed over 1000 km/s (fast CME). The mass associated with a large CME can be as high as 10^{16} g. Slow CMEs do not cause significant perturbations in the solar wind, Earth's magnetosphere or ionosphere. The fast CMEs, on the other hand, induce massive chains of disturbance in the solar wind, driving shocks in front of them when the CME collides with the much slower solar wind ions. The compression of matter in the collision with solar wind causes an enhancement of magnetic field behind the shock front. Large geomagnetic storms occur when the magnetic field interacts with the geomagnetic field. The shock also accelerates a fraction of solar wind ions to energies of several MeV. Studies indicate that these particular ions may be the source of almost all large particle flux enhancements in the neighbourhood of the Earth instead of those accelerated in flares [e.g. Reames 1992].

While the bulk of energetic MeV-class particles from solar events arrive at Earth distance in about a day, there is also a very highly energetic component of particle flux from solar events with particles of energies up to 1 GeV, which arrive at Earth distance in a few tens of minutes to one hour. These particles are produced by acceleration near a flare site or interplanetary shocks that precede a large CME. Particles with very high energies can also be transported from flaring region directly to Earth's magnetic poles via interplanetary magnetic field lines, if the Earth's location is magnetically connected to the flare.

In some rare circumstances, it is possible to see the other side of the coin, as demonstrated by the period of very low intensity of solar wind in 1999. During the near absence of the breeze of spatially flat distribution of solar wind, very narrow streams of electrons were observed near the Earth. The geometrical properties and energy distribution of the coronal electron flow were presumably propagated unchanged with the electrons from the Sun to the Earth distance. During normal conditions such phenomena can never be observed, since the solar wind dominates the particle flow from the Sun, and its pressure hides the dynamical details of the mass flow in solar corona. Another special aspect connected to the solar wind is its role in shielding the Earth from the galactic cosmic rays. Especially during strong solar events the contribution of cosmic radiation in the Earth's ionosphere is strongly depressed, and the solar wind has similar albeit weaker effect at all times. The relative importance of the solar wind and cosmic radiation in some ionosphere processes are, however, still under discussion.

Knowing and understanding the solar wind has a major role in future progress of space weather forecasting, as demonstrated by the following general examples. The occurrence of space weather effects and magnetic storms after a CME depends on the location of the CME on the Sun, the exact geometry of the interplanetary magnetic field that originates from this location, and the Earth's location with respect to the field. Sometimes the particles even from a big solar event can be channelled away from the Earth and no magnetic storms follow, although the event is observed on the Sun. There are also cases when a magnetic storm can occur without any preceding observation of CMEs or flares on the Sun. This can occur if the interplanetary magnetic field and the connected solar wind are suitably curved on the way to Earth. Thus, to be able to predict the relation of each solar event to what is going to happen in the neighbourhood of Earth, it is vital to know the structure and time variations of the solar wind and the interplanetary magnetic field within the whole space inside the Earth's orbit. Such a model does not yet exist, and reliable predictions cannot be made at the moment.

The electromagnetic radiation from solar flares traverses the passage from the Sun to the distance of Earth in 8 minutes relatively undisturbed. Major effects occur only when the short wavelength UV, X-ray, and Gamma-ray photons interact with the upper layers of terrestrial atmosphere, or penetrate the material of a satellite or its instruments. The flux density of the radiation decreases as the inverse square of distance to the source, i.e. the flare region on the Sun. The density of solar wind is so low that absorption of the electromagnetic radiation by solar wind particles is negligible, and flux estimates need no modification due to this. Considering the electromagnetic radiation from a flare on the Sun, making predictions and modelling is much simpler and more straightforward than with the particle events and magnetic field, since there is no coupling with the solar wind, and the heated optically thin plasma within the flare emits radiation to all directions at equal intensity (i.e., isotropic radiation). When occurring on the Earth's side of the Sun, the radiation enhancement is always observed and all its consequences follow, while the flares obscured by the Sun never cause effects due to the electromagnetic radiation here on Earth.

Phenomena from Magnetosphere to Surface

In the solar–terrestrial chain (the space weather chain) the last links are the most remarkable and important for space weather business, and some of them are familiar also to the general public. In physical terms they begin when the solar wind reaches the solar side boundary of the Earth's Magnetosphere. Magnetospheric processes follow this and propagate the effects to the ionosphere. Ionospheric phenomena, like the aurora, are then seen during periods of enhanced activity. In the ionosphere the effects are then propagated via the coupling with the geoelectromagnetic field down to the Earth's surface, causing disturbances in electrical power lines, and several other effects. The overall process chain appears to be fairly straightforward. However, knowing this is not very useful as such, but requires subsequent scientific research and modelling for real world estimates of quantitative parameters that can then be used for practical purposes.

Due to the pressure of solar wind the Earth's magnetic field becomes asymmetric. The magnetosphere has a comet-like shape with the head towards the Sun, and tail extending far away from Earth at the night side. Phenomena that occur in the magnetosphere and ionosphere are complex [e.g. Vassiliadis et al. 2000] and not yet fully understood. The solar wind is the energy source of the magnetosphere, and observations and monitoring of the solar wind are thus of utmost importance to understand the processes in the magnetosphere and ionosphere.

Considering the effectiveness of the solar wind and consequential interactions in near Earth space, a crucial parameter is the direction of the interplanetary magnetic field (IMF) carried by the solar wind. If the direction of the field is southward, i.e. opposite to the geomagnetic field, there will be large effects within the magnetosphere, since the energy transfer from the IMF to the geomagnetic field is very efficient. Other important factors are the velocity and dynamic pressure of solar wind. A lot of theoretical work and semi-empirical modelling has been made to work out the problems and issues involved. There are several alternatives for the theoretical explanation of the energy input from the solar wind into the magnetosphere [e.g., Vasyliunas et al. 1982]. This leads to the problem on the distribution of the energy consumption inside the magnetosphere [e.g., Østgaard and Tanskanen, 2003]. Solar wind observations can also be used in conjunction with neural networks or empirical models for a parametric description of magnetosphere and ionosphere [Klimas et al. 1999, Liou et al. 2004]. The same approach can be used to determine the geoelectromagnetic field at the Earth's surface [Kappenman 2000, Lundstedt 1999]. Magnetohydrodynamic (MHD) simulations of the solar wind–magnetosphere–ionosphere system are directly based on physical laws and, within their applicability limits they are the most precise tool to predict the conditions of the magnetosphere using solar wind measurements as input [e.g. Fedder et al. 1995, Janhunen 1996, Palmroth et al. 2003, Powell et al. 1999, Raeder et al. 1998]. With present computational power, however, fully realistic simulations are not yet possible.

Space Weather Effects

Space weather influences life and technological systems on Earth and in space. The effects on the ground level have existed ever since there has been such technological systems and devices as electrical power production, and long lines of electrically conducting material, e.g. gas and oil pipelines, that can sense the ambient electromagnetic field of the Earth and its variations. From the beginning of the "space age", the era when human civilization extended its territory to space outside the Earth's atmosphere, effects of space radiation and the electromagnetic field in space have also hampered our activities out there.

Even during quiescent periods of the Sun and "normal space weather", the space around the Earth is not an easy environment. There are regions around the Earth, the Van Allen radiation belts, where the space radiation level is high at all times. Also the anomalous depression of the Earth's magnetosphere above the south Atlantic (SAA) is a region with high radiation levels. Highly energetic protons and electrons within these regions disturb very severely the operation of electronic devices such as detectors for electromagnetic radiation and certain types of electronic components, and most satellites cannot operate safely inside these radiation zones. Extra Vehicle Activities (EVA) of spacemen are also not recommended due to enhanced radiation dose within the radiation belts and the SAA. During long operations, the electronic components, solar cells, and materials degrade from the accumulated space radiation. In low Earth orbits, satellite surfaces wear out by bombardment of the atoms in the upper atmosphere, and the atoms also increase drag, which may eventually cause spiralling down and destruction of a satellite. Also very highly energetic particles from outside the solar system (cosmic radiation) penetrate the electronics of spacecraft and can cause e.g. changes in the state of memory chips, which are called with a common term 'single event effects' (SEE). Furthermore, the unblocked ultraviolet radiation from the Sun causes accelerated ageing of surface materials, and it is dangerous for unshielded human eyes. All these are more or less known and fairly well predictable drawbacks of our space activities. Lack of proper atmosphere and air pressure makes the outer space even more challenging environment for humans.

As the effects of enhanced solar wind and magnetospheric storms reach the satellites and space probes and human beings that operate in the outer space, the previously well-known space radiation environment turns into an unknown and dangerous territory. Effects from the solar activity and the consequent increase of space radiation fills up temporarily, even for days, a large volume of space outside the radiation belts with equally or even much more harmful energetic particles. Disturbances due to shock effects and increased solar wind pressure also change the shape of the Earth's magnetosphere and the radiation belts, and load up more destructive power within them. All these combined make the passage of each satellite along its orbit an adventure, which includes unpredictable temporary effects and, in the worst case, may lead to irrecoverable damages to electric devices, and may endanger the lives of spacemen and space women.

During a magnetic storm the effects of the strongly enhanced electron radiation can occur anywhere, in the radiation belts, in the SAA, in the auroral zone, or in the interplanetary space. The effects include charge-up of satellites. Especially large potential differences that accumulate between dielectric and normal metal surfaces can cause eventual discharges that damage materials and propagate as electrical transients to spacecraft electronics. Highly energetic electrons can also penetrate deep inside the electronic components, and cause local discharging. The electric pulses from both above cases can resemble real signal, and cause "phantom commands" or "phantom data", leading to very unpredictable consequences. In the worst case the consequence can be loss of satellite power. These electrical transients often follow the injection of electrons towards the Earth from the magnetotail during magnetic storms.

The highly energetic protons and heavier ions penetrate deeply into satellite and space vehicle structures, and some of them can even pass through satellites. During such passages, the particles interact with the materials, causing secondary radiation, ionisation, and also permanent structural changes e.g. in detector crystals. The recoverable damages can be avoided by powering off the devices before exposure to the radiation, while the permanent lattice changes within the crystals are, to some extent, unavoidable. The non-ionising energy loss (NIEL) of protons is the most destructive process alternative in the interactions with materials, since it may cause permanent and irreversible material damages. The energy loss associated with ionisation is often compensated by recovery of the atomic electron configuration. This, however, causes fluorescence and scattering, which may add poorly distinguishable background to X-ray and Gamma-ray detectors.

The highly energetic proton radiation after flares and CME's is the most dangerous element of space weather for humans in space. The radiation dose during such event can be so high, that an unshielded astronaut can receive several times the recommended upper limit of yearly dose within a few hours. The auroral zones also increase in size and become regions of avoidance for EVA for spacecraft. In addition to this, also high altitude intercontinental flights via polar caps are not recommended during magnetic storms and increased solar wind, since the radiation doses of airline passengers may be significantly enhanced. Fortunately not all flares and CMEs cause the most dangerous proton radiation.

Definitely the best known of space weather effects in the uppermost atmosphere of the Earth are the auroras, which are caused by electrons precipitated ("rained") to the upper ionosphere from the magnetosphere above it. The actual height of the aurora electrons is 100-400 km. The auroras are concentrated on northern and southern polar areas, which is due to the structure of the Earth's dipolar magnetic field. The enhancement of this rain of electrons during magnetic storms is a consequence of the interaction between the amplified solar wind and the Earths' magnetosphere, which accelerates electrons to high energies and conducts them above the poles. The electrons emit bremsstrahlung up to X-ray energies (tens of keV) when they interact with the atoms at the higher atmosphere. The X-rays are not observed from ground. The atmospheric atoms are excited by interactions with the electrons, and emit visible light as they de-excite. The auroras in their colourfulness and beauty manifest the, perhaps only, positive effect related with magnetic storms.

Although the enhanced electromagnetic radiation during flares has a secondary role considering space weather, there are several measurable and significant effects caused by it. The time frame of these starts from 8 minutes after the start of the flare on the Sun and can continue a few hours. These effects occur on the sunward side of the Earth. Basically all effects from UV and X-ray emission are follow-up of ionisation caused by the enhanced energetic irradiance. This also heats the ionosphere, which expands and reaches higher altitudes. The density of the upper atmosphere near the lowest orbits of satellites can increase hundredfold or more in the worst case. This increases significantly the atmospheric drag of the satellites, causes need for relocation and orbit manoeuvring, and can decrease the orbit lifetime. The key parameter considering the radio frequency effects is the total electron content (TEC) of the ionosphere, which is increased significantly by the UV and X-ray photons. As a consequence, long distance radio signals are perturbed and disrupted, and e.g. satellite telemetry can be corrupted and there can be scintillation effects. Satellite signals can even be completely blocked by the increased absorption in the ionosphere. Radio bursts, which are sometimes associated with a strong flare may also interfere with satellite downlinks, and there are also indications of effects on telecommunication, e.g. mobile phone networks, on Earth. The interference and absorption depend on wavelength, and the effects can be minimised by proper choice of wavelength bands, which has been done for critical services like the GPS satellite system. Also the long range, long wavelength radio transmission on Earth, which utilises the reflection from the lower boundary of the ionosphere, suffers from disturbances and even blackouts due to these events, since the absorption of the propagating signal is increased. The very highly energetic photons from a flare also penetrate deeply in material exposed to it in space. The intensity of such gamma-ray photons from solar flares is, however, so low that the effects are mainly detected as extra background in e.g. X-ray and gamma-ray sensors on satellites.

The energetic particles from solar events and magnetic storms heat the ionosphere similarly to the UV and X-ray photons. During magnetic storms that follow CMEs there are also strong electric currents flowing through the upper atmosphere, which dissipate energy and cause very efficient heating of the ionosphere. Both of the above phenomena cause heating and expansion of the ionosphere, and other consequences that follow it, but with the respective time delay following the propagation of a CME instead of the flare photons. Another specific effect is observed at polar areas, where a solar energetic particle event (SEPE) may lead to increased ionisation of the lower ionosphere due to highly energetic electrically charged particles that enter the magnetosphere near the poles. The increased ionisation can lead to severe HF and VHF signal absorption, which is called a polar cap absorption (PCA) event. During a PCA event, which may last from days to weeks, the lower boundary of the ionosphere is depressed to lower altitudes and polar HF radio propagation may be completely blocked.

Large and rapidly varying electric currents flowing in the ionosphere during space weather storms, in particular at high latitudes, are propagated downwards, and finally manifest themselves as disturbances in the geomagnetic field on the ground. The event is called a *geomagnetic storm*. By Faraday's law of induction, a variable magnetic field causes an electric current. This is how also geomagnetic storms and disturbances cause *geomagnetically induced currents* (*GIC*) in conductors such as electric power transmission grids, pipelines, telecommunication cables and railway systems. In power systems, GIC may cause saturation of transformers with various possible consequences, the worst of which are a blackout of the whole system and a permanent damage of transformers. The basic reason is that the GICs are slowly varying direct currents (DC), which may be so large that they damage the power line components which are designed for AC currents. The effects of GIC are most probable in long power lines at high latitudes. Such an event caused the total blackout in the national service of electrical power in Canada in 1989. Oil and gas pipelines may experience corrosion due to galvanic effects in pipeline joints, which is activated if the grounding is poor. Ultimately, this leads to permanent damages and leakages in the pipelines. Communication systems and railways may suffer from voltage overflow [Pirjola et al. 2003]. The first observations of a GIC were made about 150 years ago in telegraph systems, long before there was any knowledge on what would later be defined as space weather.

Due to the effective shielding provided by the Earth's atmosphere and the magnetosphere, space radiation from solar events (flares and CMEs) does not reach the surface of the Earth. However, there are a few examples of phenomena, which are related to this. First of them is actually not a phenomenon but atmospheric constituent, the ozone in the Earth's atmosphere. The ozone layer is affected by particle radiation from solar events, and the thickness can be temporarily depressed by it in the polar areas. This is not, however, the reason for the systematic long term decrease of ozone in the Earth's atmosphere, and even complete disappearance of ozone in certain areas (ozone holes). The reason is the increase of chemical compounds called CFCs (i.e., Freons or Halons) in the atmosphere by e.g. the leaking old cooler devices (refridgerators). If the decrease of shielding from solar ultraviolet radiation by the ozone layer continues, we will very probably have a serious global problem in surprisingly near future. This problem is temporarily amplified during solar events (especially flares), when the input of the radiation is increased and the shielding decreased.

The effects of exposure to enhanced UV radiation are destructive to live tissue and increase significantly the risk of e.g. skin cancer and problems with the eyesight. In the ionosphere there are also irregular variations due to the dynamics of the underlying atmosphere that are similar to effects of solar origin, resulting in a superimposed combination effect. Thus the field includes more players than expected on the basis of a simple approach, and space weather chain cannot be fully modelled with unidirectional flow of processes.

There are also a few types of cosmic events, such as a potential impact of a comet or an asteroid, and a supernova explosion in the neighbourhood (e.g. within ~100 pc) of the Sun, or a gamma-ray burst (GRB) in the Milky Way, which could cause a very serious hazard, and even a total extinction of life on the Earth by mechanisms that involve dramatically enhanced space radiation and/or climatic changes. They are, however, issues separate from current definition of space weather. Nevertheless, they definitely belong to a serious wider scope discussion on the future of human civilization and its relation to the surrounding cosmic environment, which could be called "the Earth-Universe connection".

Space Weather Instruments

Space radiation, magnetic field, and the solar wind can be observed directly only in space outside the Earth atmosphere, although some of the related phenomena like the aurora, properties of the ionosphere, and the geomagnetic field can be measured to some extent from ground. Space-based observational data was fairly scarce and temporally restricted in 1950's and beginning of 1960's when only sounding rockets with scientific equipment penetrated the ionosphere for an hour or less at a time. More systematic and extensive space-based observations of space radiation and magnetic field began in the turn of 1970's with extended space missions and satellites. NASA's Apollo program and Skylab space station played an important role in this progress. Since then, an expanding wealth of data has accumulated with an increasing number of versatile multi-instrument satellites and space probes at Earth orbit and beyond.

Instruments for space weather parameters have to be highly tolerant against space radiation, since they must be operable during the heaviest magnetic storms, and repeat this for years without significant deterioration or degradation of performance. They also must be able to measure accurately such radiation, which normally causes hazards, malfunctioning, and even permanent damages to detectors. The requirements are thus hard, and careful design with proper material and component selection are in key role for such detectors.

There are roughly speaking three categories of instruments for space-based space weather observations. First category includes instruments, which have been designed for scientific research. These instruments are well calibrated, are highly sensitive, with spectral resolution, and often conduct measurements of more than one parameter of interest. While these instruments provide adequate information for scientific research and quantitative modelling of space weather, and suit also well for simple triggering purposes, they also are complex and heavy, may consume a lot of electric power, and produce high amounts of telemetry. Examples of instruments falling in this category are used on e.g. SOHO and ACE.

Another other category includes simple instruments, which have been designed for triggering of alerts for carefully selected and very limited set of parameters, are very poorly calibrated for signals below triggering level, but they also may provide rough numerical parameter values. This type of instrument aims at measuring a parameter, which is easily derivable to the strength of the final effect of interest, thus avoiding all but a very simple signal analysis. These instruments suit poorly for scientific research, but they can be, on the other hand, lightweight and robust, small in size, consume only low amount of power, and produce a very limited amount of telemetry. Due to their simplicity, the triggering monitors can also be designed to operate only locally to advice the spacecraft to take quick measures, should an alert flag arise. Examples of this type are simple proton flux monitors, which are used on numerous satellites, and so-called "tissue equivalent proportional counters" (TEPC), which are used on high altitude airplanes and manned spaceflight.

A third category of instruments falls between the above two categories. They may be nearly as versatile as the high quality science instruments, but they are not so sensitive, have fever energy channels, have not so good spatial resolution, and may not be sufficiently well calibrated for scientific use in the full range of parameters provided. These instruments suit well for space weather services and triggering, and, to a limited extent, also for scientific research. As the requirements for performance and accuracy is not extremely high, the third category instruments may apply compromises in design and avoid complexity, which leads to lower power consumption and mass, and also saves telemetry. Majority of e.g. the space environment monitoring instruments on GOES satellites are in this category.

Observational Facilities

There is a large number of satellites and satellite systems, which collect systematic and extensive data for specific use in space weather research and services. They are at Earth orbit, or launched to the Lagrange 1 (L1) point. The L1 point is 1.5 million km from the Earth towards the Sun, thus a perfect site for purely scientific investigations as well as space weather forecasting and service purposes. Since the Earth's magnetosphere has a complicated spatial structure, and it is also highly variable in time, measurements of particle fluxes and magnetic field with a population of satellites at high as well as low orbits is necessary to obtain the full picture. Some of the satellites have a very complicated and highly evolving orbit to enable reaching the space far away in the Earth's magnetotail or the solar wind and radiation environment at large distances and various directions from the Earth. The list of satellites described here is not exhaustive, but rather represents the facilities within the biggest collaborative efforts and organisations for scientific data and services related with space weather. There is national activity and satellite projects also in e.g. Japan, India, China, and Canada, and Sweden, which are not included in the detailed description below.

ACE (Advanced Composition Explorer) is NASA's satellite launched in 1997 to L1 point for versatile investigation of the solar wind. It has propellant to maintain its orbit until 2019, thus being potentially a very long time facility for space weather research and services. ACE provides data for the US national Space Environment Center (SEC), which is administered by the US National Oceanic and Atmospheric Administration (NOAA). The other satellite systems within the SEC framework are POES and GOES. SEC is presently the most advanced and extensive public service in the world for space weather information and related data. The scientific tasks for ACE are in four main areas, the elemental and isotopic composition of matter, the origin and subsequent evolution of chemical elements, formation of the solar corona and acceleration of the solar wind, and particle acceleration and transport in nature. ACE has three instruments measuring various energetic particle fluxes, and a sensor for magnetic field, which give continuous in advance knowledge of upcoming space weather effects at Earth. **EPAM** (Electron, Proton and Alpha Monitor) is a system of five telescopes, which measure the flux and direction of electrons above 30 keV, the flux and direction of ions greater than 50 keV, and the elemental composition of the ions. EPAM is the flight spare model of the HI-SCALE instrument from the ULYSSES mission. **MAG** (Magnetometer instrument) measures the local interplanetary magnetic field (IMF) direction and magnitude and provides continuous monitoring information on the large scale structure and fluctuation characteristics of the IMF at very high time resolution. **SIS** (Solar Isotope Spectrometer) performs isotopically resolved measurements of the elements from lithium to zinc over the energy range 10 - 100 MeV/nucleon. **SWEPAM** (Solar Wind Electron, Proton and Alpha Monitor) is a solar wind instrument and performs high quality measurements of electron and ion fluxes in the low-energy solar wind range (electrons 1 to 1240 eV and ions 0.26 to 35 keV). SWEPAM supplies data for the NOAA Real Time Solar Wind 24-hour monitoring of Space Weather along with three other instruments.

POES (Polar Orbiting Environmental Satellite, formerly TIROS) system orbits the Earth at 850 km orbit, with inclination 98 degrees. The instruments measure the charged particle fluxes near the Earth, and image the aurora activity on the way across the polar regions of the Earth. There is also a very wide field monitor for charged particles coming from the Sun. Currently operating satellites in the POES program are NOAA-16 and NOAA-17. The satellites carry an instrument package called Space Environment Monitor (SEM). The package includes detectors for low energy electrons and positively charged ions in the range 50 eV - 20 keV, which measure the auroral activity, medium energy detectors in the range 30 keV – 1 MeV for electrons, and 30 keV – 6.9 MeV for protons, and high energy proton detectors for the range 16 MeV – 275 MeV. The high-energy proton detectors are designed to detect and monitor energetic solar particles that cause severe ionosphere disturbances, which are called "blackouts", during solar particle events.

GOES (Geostationary Orbiting Environmental Satellite) system flies at higher, geostationary orbit. The main task of GOES satellite system is to provide data on the surface of the Earth. The satellites also carry a set of instruments suitable for use in space weather monitoring and scientific use in studying the space particle environment, although they are not as sensitive as those onboard the dedicated facilities like ACE. GOES satellites (currently operating GOES 10-12) measure proton, electron, and X-ray fluxes in a few energy bands, and they also have a magnetometer (Hp component measurement). From 2001 there has also been an X-ray imager (Solar X-ray Imager, SXI) for real-time soft X-ray imaging of the whole Sun. Both GOES 10 and GOES 12 are located on the geographic equator, but due to the tilt of Earth's dipole magnetic field, GOES 10 (at 135 degrees west longitude) is at lower geomagnetic latitude than GOES 12 (at 75 degrees west longitude). The GOES system provides a wide range of parametric data on weather and atmospheric phenomena in addition to space weather. Because the satellites stay above a fixed position on the Earth, they provide a good way to monitor the beginning of severe weather conditions such as tornados, flash floods, hail storms, and hurricanes, and other phenomena on the Earth's surface.

SOHO (Solar and Heliospheric Observatory) is a NASA/ESA satellite, which was launched to L1 point in 1995, with a nominal lifetime of only two years. The magnificent success of the mission has been a strong driver for extensions, and present plan is to continue until 2007. If all goes well, the mission will thus be able to celebrate covering the extent of a complete 11-year solar cycle. The main science tasks for SOHO are related to investigating the solar interior through observations of oscillations in both velocity and irradiance domain, imaging and spectroscopy of the inner and outer corona of the Sun, and measurements of the solar wind. SOHO carries a very extensive set of instruments, some of which are no more operating. The instruments onboard are **CDS** (Coronal Diagnostic Spectrometer), **CELIAS** (Charge, Element, and Isotope Analysis System), **COSTEP** (Comprehensive Suprathermal and Energetic Particle Analyzer), **EIT** (Extreme ultraviolet Imaging Telescope), **ERNE** (Energetic and Relativistic Nuclei and Electron experiment), **GOLF** (Global Oscillations at Low Frequencies), **LASCO** (Large Angle and Spectrometric Coronagraph), **MDI/SOI** (Michelson Doppler Imager/Solar Oscillations Investigation), **SUMER** (Solar Ultraviolet Measurements of Emitted Radiation), **SWAN** (Solar Wind Anisotropies), **UVCS** (Ultraviolet Coronagraph Spectrometer), and **VIRGO** (Variability of Solar Irradiance and Gravity Oscillations), Unlike ACE, SOHO does not have a magnetometer. SOHO is the central facility within the framework of the international Solar Terrestrial Science Program (STSP). The other satellites within the program are Cluster, Geotail, Wind, and Polar.

Cluster is a system of four satellites, which flies in a long, highly elliptic orbit with perigee about 20000 km, and apogee over 100000 km. The first launch of Cluster satellites failed and the satellites were destroyed, but the project was recovered, and the four Cluster II satellites were launched successfully in 2000. Present plan is to continue operations until the end of 2009. All four Cluster satellites carry identical sets of 11 science instruments for investigating the space weather inside and outside the Earth’s magnetosphere. Cluster satellites investigate the small-scale structure of the Earth's plasma environment along the orbit of the satellites, and long term data enable building the first detailed three dimensional map of the Earth’s magnetosphere.

Geotail is a satellite with long and highly evolved orbit for investigation of the Earth’s magnetotail, which was launched in 1992 and is still in operation. The scientific purpose of Geotail is to achieve better understanding of fundamental processes in the Earth’s magnetosphere via in-situ high resolution measurements of the solar wind and hot plasma in the magnetotail.

Wind satellite was launched in 1994, with a goal to measure near Earth, solar induced phenomena including solar wind, particles that have been ejected by the Sun and magnetic fields. The satellite carries 9 science instruments. Like Geotail, also WIND has a long and highly evolving orbit, with largest distance from Earth more than 400 Earth radii (about 2.5 million km).

Polar satellite was launched in 1996 in a highly eccentric polar orbit, with apogee at 9 and perigee at 1.8 Earth radii. The nominal mission for Polar was two years, but it has been extended, and the satellite is still operating. Polar carries 3 scientific instruments for imaging the aurora in various wavelengths especially when the satellite is above the north pole, and 9 instruments for in-situ measurements of particles, and the magnetic and electric fields around the whole orbit.

Interball is a Russian magnetospheric satellite project, which includes two satellites, a Tail Probe launched in 1995, and an Auroral Probe launched in 1996. The satellites orbit the Earth at 63 degrees inclination. The Tail Probe has a very eccentric orbit with apogeum at nearly 200000 km height, while the apogeum of the Auroral Probe is at a distance one tenth of this. The tasks for Interball project are to study the mechanisms responsible for the transport of solar-wind energy to the Earth's magnetosphere, and the accumulation and dissipation of this energy. Special emphasis in the research is on dynamical processes. Interball is an international collaborative project, with close connection to the international STSP program.

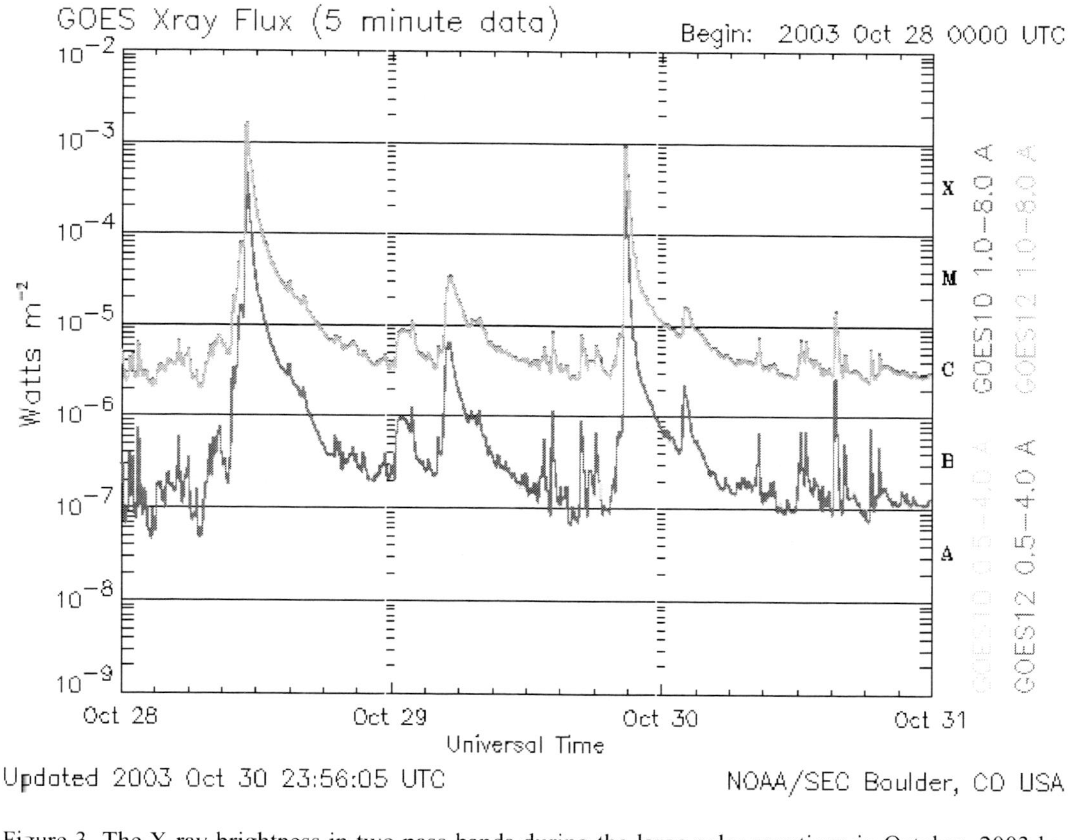

Figure 3. The X-ray brightness in two pass-bands during the large solar eruptions in October, 2003 by GOES (http://www.ssec.wisc.edu/datacenter/archive.html).

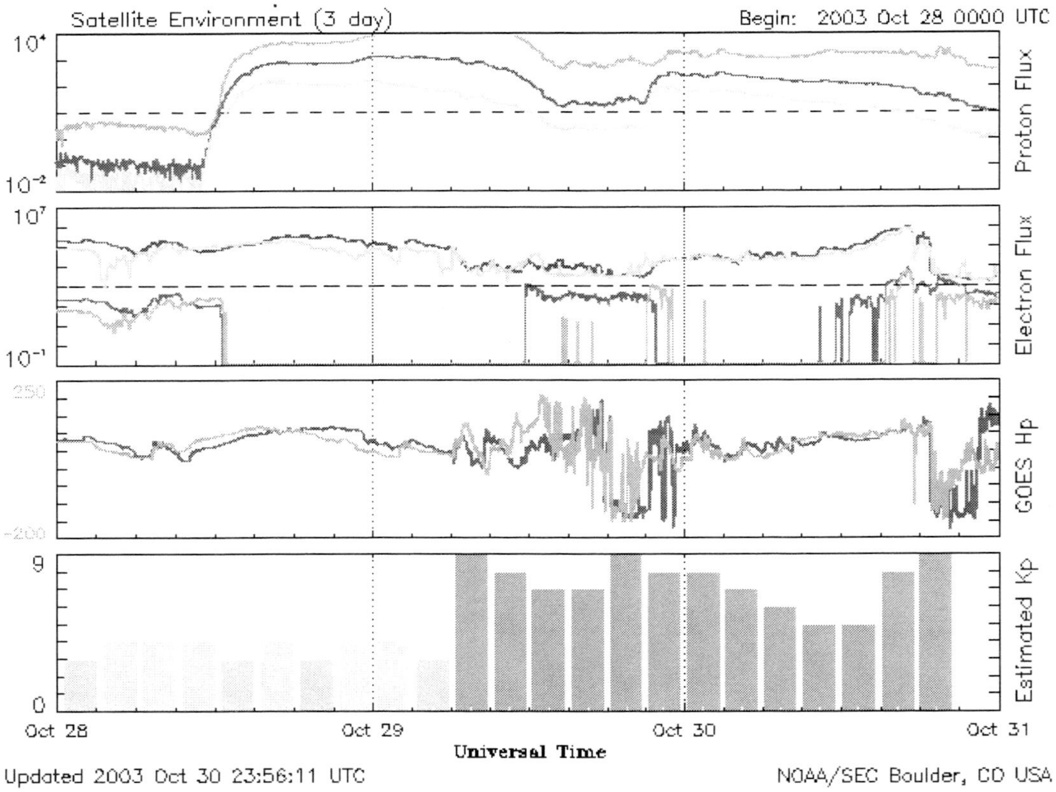

Figure 4. Proton and electron fluxes, and magnetic field during the large solar eruptions in October, 2003 by GOES (http://www.ssec.wisc.edu/datacenter/archive.html).

Space Weather Forecasting and Services

Space weather services provide parametric information from which it is possible to derive effects and forecasts, and direct information on effects that can be used right away for e.g. alerts. A prerequisite for such service is timely access to information for facilitating predictions and forecasts of subsequent evolution and propagation of e.g. a magnetic storm. Thus a space weather service must have access to e.g. solar wind and magnetic field at some distance away from the Earth towards the Sun. Satellites located at L1 point are very suitable for this purpose, as the particles and the shock front from a solar eruption arrive at L1 location about one hour before they enter the Earth's magnetosphere.

As the structure of the magnetic field and also the changes that occur in the radiation distribution and the field are complicated and currently impossible to model with sufficient reliability and accuracy using only the in-situ data from L1 satellites, there are satellites that give additional local information on various Earth orbits. This element of in-situ data does not give information on space weather in advance, and is not effective for forecasts. It is vital, however, for our knowledge and understanding of the details of space weather phenomena near the Earth.

Observations of the Sun (XUV, X-rays and coronagraphs) can be used to certain extent onboard any satellite to see the changes of solar activity, which give rise to the disturbances at

Earth. As the propagation path of a somewhat collimated and spatially limited CME or particle burst from a solar flare cannot be predicted with certainty, it is not possible to make reliable space weather forecasts by merely using the observations of the Sun. Sometimes big eruptions flow away from the Earth, and thus a Sun-based alert would be false. It is very important to include Sun observations in the space weather service, since direct observations of the Sun are a prerequisite for a complete picture of the space weather chain. It is also possible to estimate the severity of a possibly upcoming space weather storm from the X-ray intensity of a flare, and a rough probability of a space weather storm by the location of the eruption on the solar disk. The biggest advantage of using the solar X-ray and UV observations is the time gap between the solar event, seen in 8 minutes from its occurrence by these observation, to the beginning of the space weather storm at Earth, beginning in earnest typically a day later. This provides a possibility for forecasts with time spans comparable to the "ordinary" weather forecasts.

The most advanced and versatile public space weather service in the world is currently NOAA's (National Oceanic and Atmospheric Administration) Space Environment Center (SEC). It is currently a clear leader and in its own class, providing complete on-line space weather information including solar, interplanetary, geomagnetic and near-Earth data. There is also an alert service for significant space weather events, a space weather forecast service, and archives for past time data. The SEC service utilizes ACE, POES, and GOES as basic data resources.

An advanced European correspondence of the US space weather service, called "Space Weather European Network" (SWENET), is under development by ESA. It provides already limited parameter monitoring information obtained mainly by SOHO. Further development is made in many European countries, with the Space Weather Working Team (SWWT) collecting progress information from each participant.

Several other initiatives for a multinational collaboration in space weather services have also been made. The widest of these is the *International Space Environment Service* (ISES) initiative. It aims at collecting together all services in the world, and has currently support from 11 space weather service providers. The Regional Warning Centres are in China (Beijing), USA (Boulder), Russia (Moscow), India (New Delhi), Canada (Ottawa), Czech Republic (Prague), Japan (Tokyo), Australia (Sydney), Sweden (Lund), Belgium (Brussels) and Poland (Warsaw). A procedure has been scheduled for efficient data exchange between the centres, with the NOAA Space Environment Centre (SEC) as the coordinating centre and "World Warning Agency".

The national space weather services (i.e. regional centres) within ISES collaboration currently provide their services independently. An example is the Regional Warning Centre of Sweden maintained by the Lund section of the Swedish Space Research Institute (IRF–Lund). Another example of national space weather services relying mostly on external data by e.g. NOAA/SEC, but appended by local data from ionosondes and radiotelescopes is the space weather service by the Australian Space Weather Agency.

Conclusion

A good motivation for increased investments in space weather research, as well as forecasting and public space weather services is carried by the fact that sensitivity and susceptibility to failures and malfunctioning in electrical systems both on Earth and in space increase with the complexity of these systems. More and more sophisticated electronic systems are being used in various individual applications, for large-scale support of critical maintenance systems of the society, and also for economical and military purposes. The explosive increase of advanced electronics applications marketed for common people, and the explosive increase of our dependence on information technology combined with ubiquitous telecommunication activity may be even more important in this respect.

With commercial space flights, human activity outside the Earth's atmosphere is also becoming an achievable goal for an increasing number of people. It is vital for this business to have access to reliable space weather forecasts. Increasing levels of space radiation are, in practice, equally dangerous to living organisms as strong radioactivity. People transported to large heights above the Earth, even at suborbital (< 200 km) level, are in great danger should a space storm occur during a flight experiment or "holiday in heavens". Even during present time long distance flights at (11-12 km altitude) it is possible to receive a dangerous level of space radiation.

The population of operational unmanned satellites is very large. By applying space weather data to modify their operation it is generally possible to increase their lifetime and reliability, and also predict and determine potential time periods with technical problems and possible corruption of satellite information. Therefore all individual satellites have a good motivation to use space weather information. There are also critical satellite systems, which may have specific, application-related reasons for use of space weather monitoring services or local on-board monitoring of space weather. Good examples are the GPS satellite system and the future similar system in Europe (Galileo), which are critical space-borne services for the whole world, and can suffer from corruption of navigation information on Earth during space weather storms. Space weather effects are accounted for in the operation of GPS system, and it should be of common interest to both USA and Europe to apply the user experience of GPS to the design of Galileo taking account for space weather effects. Hopefully this is already in plans and actions have been taken in this, although issues related with competition and politics may cause additional work.

At present virtually every scientific satellite, and also many other satellites have an instrument for monitoring of charged particle radiation. This is to ensure integrity of data and telemetry, and for precautions due to possible damages and malfunctioning of electronic devices caused by enhanced charged particle (mainly proton and electron) radiation. Provided that a space radiation monitor is properly calibrated, the data, albeit covering only a narrow range of parameters, may also be useful as supplementary data for solar-terrestrial research, and for improving the spatial and temporal coverage of space weather monitoring information. Particle monitor data from satellites not dedicated for this purpose is, however, currently not usually collected and saved for use in research or space weather services, although it may contain useful information for e.g. improving the coverage of radiation information in the Earth's environment and also the model of the magnetosphere. If not already in plans, measures should be taken for coordination of optimised use of this data in the future. This could start by e.g. first creating an international committee to conduct a survey of what is potentially available, and then by negotiating agreements with the resource holders for an access of their data.

There are still major unsolved scientific questions within the path of space weather effects from the Sun to near Earth space and finally to the surface of the Earth. In the Sun, the energy transport and release mechanisms are only qualitatively understood, and proper quantitative models for the energy transport to the corona and the release mechanisms during flares and CMEs should be developed. Improved observational data to support theoretical work is mandatory for this. Observations with significantly improved spatial resolution in carefully selected wavelength bands connected with moderate spectroscopic resolution would be essential for this. Going further from the Sun, the spatial structure and dynamics of the solar wind throughout the inner solar system should be modelled. Significant improvement of observational data is needed also here, putting emphasis on filling the gaps in spatial coverage of solar wind data and measurements of the interplanetary magnetic field. In the neighbourhood of the Earth and its magnetosphere, sophisticated models and a lot of data are already available. For detailed modelling, however, the spatial coverage and the time continuity are not sufficient. In here, more efficient and coordinated exploitation of standard space weather monitors distributed in satellites at various orbits may provide an inexpensive solution, as many satellites will have a radiation monitor in any case.

A related activity already going on at ESA is aiming at creating proper standards and harmonisation of the development of such technology and monitor applications. One issue is to find and specify a sufficiently flexible design to allow easy implementation for different platforms. Defining standards also for the measurement data would help in more efficient use of the data for science and modelling, as well as organised collection of the data from the telemetry of each satellite. These aspects carry along with it a problem, however, as commercial and military satellite operators do not have any obligation to support scientific research or public services by collecting data for them. It is difficult to assess offhand whether this becomes a major obstacle for realizing the idea. Sufficient spatial coverage of data would probably be available even if only a subset of future satellites will support this and provide their data.

To fully understand the Solar-Terrestrial connection, theoretical models supported with proper data on the solar wind and also the magnetic field throughout this spatial volume need to be developed. In both USA (NASA) and Europe (ESA) there are similar strategic means to improve the modelling, and by that way improve the reliability of space weather forecasts in the future. They contain both direct means and supportive background activities. The ways forward include, 1) improving of data collection and modelling, 2) implementation and use of radiation monitors at all possible satellites, 3) improving of the radiation tolerance of electronics by developing processes for their production and test facilities to qualify them, 4) developing of long lasting batteries and slowly degrading solar cells for use in space, 5) establishing of advanced test facilities for on-ground combined effects testing of equipment and materials for space, 6) flying missions beyond low-Earth orbit, and 7) improving the radiation environment models via organised efforts, preferably by open networking which should include USA, EU countries, Russia, Japan, and other countries.

The "traditional" effects like corrosion of gas pipelines, and damages for certain types of electrical power station systems during solar storms have already for some time given motivation for research and forecasting of space weather, and our space activities have provided significant additional value to it. Now it is clear that human civilization and its activities will be influenced in a rapidly growing rate by space weather effects, when a space weather storm is triggered. A clear implication of this is that there is an increasing need for accurate and systematic space weather data and forecasts all over the world, and global scale planning and actions are needed to meet the challenges ahead. While activities towards this goal have already begun with collaboration in alert services (ISES), there are areas where more open policy is needed, like collaboration in modelling efforts and services, and in technology development and transfer of related knowledge.

Acknowledgements

This work is partly based on a survey and related unpublished report by the author made for Finnish space industry for a pre-study of potential markets of space weather monitoring devices. I am deeply grateful to Dr Risto Pirjola and other collaborators at the Finnish Meteorological Institute for discussions on space weather, and support in the form of unpublished as well as published material on the subject. Private communication with Alan Owens from ESA and Seppo Nenonen from Oxford Instruments Analytical Ltd has helped in getting insight in technology-related issues. The internet search machine Google is acknowledged for making it easy to find updated information on various subjects published on WWW. In addition to the listed references, the general background information resources used are public reports and presentations on the WWW pages of ESA (European Space Agency), NASA (National Aeronautical and Space Administration), AGU (American Geophysical Union), and UCLA (University of California, Los Angeles).

References

[1] Fedder, J.A.; Lyon, J. G.; Slinker, S. P.; Mobarry, C. M. *J. Geophys. Res.* 1995, 100, 3613–3622

[2] Gosling, J. T. *J. Geophys. Res.* 1993, Vol. 98, No.18, 937

[3] Gosling, J.T. *Eos* 1993, Vol. 74, No. 52, 611-612

[4] Hale, G. E. *Astrophys. J.* 1931, 73, 379

[5] Huovelin J.; Saar S.; Tuominen I. *Astrophys. J.* 1988, 329, 882-893

[6] Ip, W.-H.; Kopp, A. *J. Geophys. Res.* 2002, Vol 107, No. A11, 1348

[7] Janhunen, P., in *Proceedings of the ESA 1996 Symposium on Environmental Modelling for Space–Based Applications*, SP-392, ESA, ESTEC, The Netherlands, 1996, pp 233–239

[8] Kahler, S. W. *Ann. Rev. Astron. Astrophys.* 1992, 30, 113

[9] Kappenman, J.G.; Radasky, W.A.; Gilbert, J.L.; Erinmez, I.A. *IEEE Trans. Plasma Sci.* 2000, 28, 2114–2121

[10] Klimas, A.J.;Vassiliadis, D.; Baker, D.N.; Valdivia, J.A. *Phys. Chem. Earth* 1999, 24, 37–44

[11] Liou, K.; Newell, P.T.; Meng, C.I.; Wu, C.-C.; Lepping, R.P. *J. Geophys. Res.* 2004, 109, A06306, doi: 10.1029/2004JA010400

[12] Luhmann, J. G., in *Introduction to Space Physics*; Kivelson, M.G., Russell, C.T.; ed.; Cambridge University Press, Cambridge, UK, 1995, pp 203-226

[13] Lundstedt, H., in Proc . *ESA Workshop on Space Weather*, WPP-155, ESA, ESTEC, The Netherlands, 1998, 197–205

[14] O'Neal, D.; Saar, S.H.; Neff, J.E. *Astrophys. J.* 1996, 463, 766

[15] Palmroth, M.; Pulkkinen, T.I.; Janhunen, P.; Wu, C.-C. *J. Geophys. Res.* 2003, 108(A1), 1048, doi:101029/2002JA009446

[16] Parker, E.N. *Astrophys. J.* 1955, 121, 491

[17] Parker, E.N., *Astrophys. J.* 1958, 128, 664

[18] Parker, E.N., *Astrophys. J.* 1960, 132, 175

[19] Parker, E.N. *Astrophys. J.* 1988, 330, 474

[20] Pirjola, R.; Lappalainen, H.; Kauristie, K. in *Proc.54th International Astronautical Congress*, Bremen, Germany, 2003, IAC-03-IAA.6.3.01, 9 pp.

[21] Powell, K.G.; Roe, P.L.; Linde, T.J.; Gombosi, T.I.; DeZeeuw, D.L. J. Comp. Phys. 1999, 154, 284–309

[22] Raeder, J., Bercgem, J.; Ashour–Abdalla, M.. *J. Geophys. Res.* 1998, 103, 14787–14797

[23] Reames, D. V. *Lect. Notes Phys.* 1992, 399, 180-185

[24] Russell, C.T.; Vaisberg, O., in Venus, Hunton, D.M.; Colin, L.; Donahue, T.M., Moroz, V.I., ed.; pp. 873-940,University of Arizona Press, Tucson, AZ, 1983, pp 873-940

[25] Vassiliadis, D.; Klimas, A.J.; Valdivia, J.A.; Baker, D.N. *Adv. Space Res.* 2000, 26, 197–207

[26] Østgaard, N.; Tanskanen, E., in *Geospace: The Storm–Substorm Relationship, AGU Geophysical Monograph* **142**, doi: 10.1029/142GM14, 2003

INDEX

A

B

C

D

E

F

G

H

I

J

K

L

M

N

O

P

T

U

V

W

Y

Z